Herbal Treatment of Major Depression

Scientific Basis and Practical Use

Clinical Pharmacognosy Series

Series Editors
Navindra P. Seeram and Luigi Antonio Morrone

Botanical medicines are rapidly increasing in global recognition, with significant public health and economic implications. For instance, in developing countries, a vast majority of the indigenous populations use medicinal plants as a major form of healthcare. Also, in industrialized nations, including Europe and North America, consumers are increasingly using herbs and botanical dietary supplements as part of integrative health and complementary and alternative therapies. Moreover, the paradigm shifts occurring in modern medicine, from mono-drug to multi-drug and poly-pharmaceutical therapies, has led to renewed interest in botanical medicines and botanical drugs.

Natural Products and Cardiovascular Health
Catherina C. Caballero George

Aromatherapy
Basic Mechanisms and Evidence-Based Clinical Use
Giacinto Bagetta, Marco Cosentino, and Tsukasa Sakurada

Herbal Medicines
Development and Validation of Plant-Derived Medicines for Human Health
Giacinto Bagetta, Marco Cosentino, Marie Tiziana Corasaniti, and Shinobu Sakurada

Natural Products Interactions on Genomes
Siva G. Somasundaram

Principles and Practice of Botanicals as an Integrative Therapy
Anne Hume and Katherine Kelly Orr

Herbal Treatment of Major Depression: Scientific Basis and Practical Use
Scott D. Mendelson

Herbal Treatment of Major Depression

Scientific Basis and Practical Use

Scott D. Mendelson, MD, PhD

CRC Press
Taylor & Francis Group
Boca Raton London New York

CRC Press is an imprint of the
Taylor & Francis Group, an **informa** business

CRC Press
Taylor & Francis Group
6000 Broken Sound Parkway NW, Suite 300
Boca Raton, FL 33487-2742

First issued in paperback 2021

ISBN-13: 978-0-367-37532-4 (hbk)
ISBN-13: 978-1-03-208715-3 (pbk)
ISBN-13: 978-0-429-35551-6 (eBook)

Library of Congress Cataloging-in-Publication Data

Names: Mendelson, Scott D, author.
Title: Herbal treatment of major depression : scientific basis and
practical use / Scott D. Mendelson, M.D., Ph.D.
Description: Boca Raton, FL : CRC Press, [2020] I Series: Clinical
pharmacognosy series I Includes bibliographical references and index.
Identifiers: LCCN 2019031641 I ISBN 9780367375324 (hardback) I ISBN
9780429355516 (ebook)
Subjects: LCSH: Depression, Mental--Alternative treatment. I
Herbs--Therapeutic use.
Classification: LCC RC537 .M464 2020 I DDC 616.85/27061--dc23
LC record available at https://lccn.loc.gov/2019031641

Visit the Taylor & Francis Web site at
http://www.taylorandfrancis.com

and the CRC Press Web site at
http://www.crcpress.com

Contents

Author xxv

Chapter 1   Major Depression: A Brief History of Western Medical Treatment 1

               References 6

Chapter 2   How Antidepressants Work, but Often Do Not 7

               References 10

Chapter 3   Clues Revealed by Ketamine 13

               References 16

Chapter 4   New Understanding of the Nature and Causes of Major
               Depression 19

               4.1   Oxidative and Nitrosative Damage 19
               4.2   Inflammation 21
               4.3   Chronic Stress 23
               4.4   Insulin Resistance 24
               4.5   Metabolic Syndrome 25
               4.6   Summary 26
               References 26

Chapter 5   Phytochemicals: Some Basics 31

               5.1   Carbohydrates 31
               5.2   Lipids 32
               5.3   Terpenes 32
               5.4   Phenolics 33
                        5.4.1   Flavonoids 34
                        5.4.2   Non-Flavonoid Phenolics 34
               5.5   Alkaloids 35
               5.6   Summary 36
               References 36

Chapter 6   Models and Paradigms for Assessment of Antidepressant Effects 39

               6.1   Antioxidant Effects 39
               6.2   Anti-Inflammatory Effects 40
               6.3   Antidiabetic/Anti-Metabolic Syndrome Effects 42
               6.4   Preclinical Antidepressant-Like Effects 42

          6.4.1   Forced Swim Test...42
          6.4.2   Tail Suspension Test..44
          6.4.3   Sucrose Consumption Test 44
          6.4.4   Test Conditions...45
     References 46

Chapter 7   Herbs with Antidepressant Effects..49
     7.1   *Allium sativum* (Garlic) 49
          7.1.1   Antioxidant..50
          7.1.2   Anti-Inflammatory 50
          7.1.3   Antidiabetic/Anti-Metabolic Syndrome 51
          7.1.4   Preclinical Antidepressant-Like Effects 51
          7.1.5   Human Antidepressant Effects....................................52
          7.1.6   Dosage 52
          7.1.7   Toxicity 53
          7.1.8   Safety in Pregnancy..53
          7.1.9   Drug Interactions..53
     7.2   *Angelica sinensis* 55
          7.2.1   Antioxidant..55
          7.2.2   Anti-Inflammatory 56
          7.2.3   Antidiabetic/Anti-Metabolic Syndrome 56
          7.2.4   Preclinical Antidepressant-Like Effects 56
          7.2.5   Human Antidepressant Effects....................................57
          7.2.6   Dosage 57
          7.2.7   Toxicity...57
          7.2.8   Safety in Pregnancy..57
          7.2.9   Drug Interactions..57
     References 57
     7.3   *Apium graveolens* (Celery) 59
          7.3.1   Antioxidant..59
          7.3.2   Anti-Inflammatory 59
          7.3.3   Antidiabetic/Anti-Metabolic Syndrome 60
          7.3.4   Preclinical Antidepressant-Like Effects 60
          7.3.5   Human Antidepressant Effects....................................61
          7.3.6   Dosage 61
          7.3.7   Toxicity...61
          7.3.8   Safety in Pregnancy..61
          7.3.9   Drug Interaction 61
     7.4   *Astragalus membranaceus* 63
          7.4.1   Antioxidant..63
          7.4.2   Anti-Inflammatory 64
          7.4.3   Antidiabetic/Anti-Metabolic Syndrome 64
          7.4.4   Preclinical Antidepressant-Like Effects 64
          7.4.5   Human Antidepressant Effects....................................65
          7.4.6   Dosage...65

	7.4.7	Toxicity	65
	7.4.8	Safety in Pregnancy	65
	7.4.9	Drug Interactions	66
References			66
7.5	*Atractylodes macrocephala*		67
	7.5.1	Antioxidant	67
	7.5.2	Anti-Inflammatory	68
	7.5.3	Antidiabetic/Anti-Metabolic Syndrome	68
	7.5.4	Preclinical Antidepressant-Like Effects	68
	7.5.5	Human Antidepressant Effects	68
	7.5.6	Dosage	69
	7.5.7	Toxicity	69
	7.5.8	Safety in Pregnancy	69
	7.5.9	Drug Interactions	69
References			69
7.6	*Avena sativa* (Common Oat)		70
	7.6.1	Antioxidant	70
	7.6.2	Anti-Inflammatory	70
	7.6.3	Antidiabetic/Anti-Metabolic Syndrome	71
	7.6.4	Preclinical Antidepressant-Like Effects	71
	7.6.5	Human Antidepressant Effects	71
	7.6.6	Dosage	71
	7.6.7	Toxicity	72
	7.6.8	Safety in Pregnancy	72
	7.6.9	Drug Interactions	72
References			72
7.7	*Bacopa monnieri*		73
	7.7.1	Antioxidant	73
	7.7.2	Anti-Inflammatory	74
	7.7.3	Antidiabetic/Anti-Metabolic Syndrome	74
	7.7.4	Preclinical Antidepressant-Like Effects	74
	7.7.5	Human Antidepressant Effects	75
	7.7.6	Dosage	75
	7.7.7	Toxicity	75
	7.7.8	Safety in Pregnancy	75
	7.7.9	Drug Interactions	75
References			76
7.8	*Borage officinalis* (European Borage)		77
	7.8.1	Antioxidant	77
	7.8.2	Anti-Inflammatory	77
	7.8.3	Antidiabetic/Anti-Metabolic Syndrome	78
	7.8.4	Preclinical Antidepressant-Like Effects	78
	7.8.5	Human Antidepressant Effects	78
	7.8.6	Dosage	78
	7.8.7	Toxicity	78
	7.8.8	Safety in Pregnancy	79

        7.8.9   Drug Interactions 79
References 79
7.9   *Bupleurum chinense* 80
        7.9.1   Antioxidant 80
        7.9.2   Anti-Inflammatory 80
        7.9.3   Antidiabetic/Anti-Metabolic Syndrome 81
        7.9.4   Preclinical Antidepressant-Like Effects 81
        7.9.5   Human Antidepressant Effects..................................... 81
        7.9.6   Dosage 82
        7.9.7   Toxicity 82
        7.9.8   Safety in Pregnancy 82
        7.9.9   Drug Interactions.. 82
References 82
7.10   *Camellia sinensis* (Tea) 83
        7.10.1 Antioxidant 84
        7.10.2 Anti-Inflammatory 84
        7.10.3 Antidiabetic/Anti-Metabolic Syndrome 84
        7.10.4 Preclinical Antidepressant-Like Effects 84
        7.10.5 Human Antidepressant Effects 85
        7.10.6 Dosage 85
        7.10.7 Toxicity 85
        7.10.8 Safety in Pregnancy.. 86
        7.10.9 Drug Interactions 86
References 86
7.11   *Cannabis* 87
        7.11.1 Antioxidant... 88
        7.11.2 Anti-Inflammatory 88
        7.11.3 Antidiabetic/Anti-Metabolic Syndrome 88
        7.11.4 Preclinical Antidepressant-Like Effects 89
        7.11.5 Human Antidepressant Effects 90
        7.11.6 Dosage 91
        7.11.7 Toxicity 91
        7.11.8 Safety in Pregnancy.. 92
        7.11.9 Drug Interactions.. 92
References 92
7.12   *Cecropia* 95
        7.12.1 Antioxidant... 95
        7.12.2 Anti-Inflammatory 95
        7.12.3 Antidiabetic/Anti-Metabolic Syndrome 95
        7.12.4 Preclinical Antidepressant-Like Effects 96
        7.12.5 Human Antidepressant Effects..................................... 96
        7.12.6 Dosage 97
        7.12.7 Toxicity 97
        7.12.8 Safety in Pregnancy.. 97
        7.12.9 Drug Interactions 97
References 97

7.13 *Centella asiatica* (Gotu Kola)..98
    7.13.1 Antioxidant 99
    7.13.2 Anti-Inflammatory 99
    7.13.3 Antidiabetic/Anti-Metabolic Syndrome....................99
    7.13.4 Preclinical Antidepressant-Like Effects 100
    7.13.5 Human Antidepressant Effects..........................100
    7.13.6 Dosage 100
    7.13.7 Toxicity..101
    7.13.8 Safety in Pregnancy...................................101
    7.13.9 Drug Interactions.....................................101
  References 101
7.14 *Chrysactinia mexicana* 102
    7.14.1 Antioxidant 103
    7.14.2 Anti-Inflammatory 103
    7.14.3 Antidiabetic/Anti-Metabolic Syndrome..............103
    7.14.4 Preclinical Antidepressant-Like Effects 103
    7.14.5 Human Antidepressant Effects......................103
    7.14.6 Dosage 104
    7.14.7 Toxicity..104
    7.14.8 Safety in Pregnancy................................104
    7.14.9 Drug Interactions..................................104
  References 104
7.15 *Cimicifuga racemosa* (Black Cohosh).......................105
    7.15.1 Antioxidant 105
    7.15.2 Anti-Inflammatory 106
    7.15.3 Antidiabetic/Anti-Metabolic Syndrome............106
    7.15.4 Preclinical Antidepressant-Like Effects 107
    7.15.5 Human Antidepressant Effects....................107
    7.15.6 Dosage 108
    7.15.7 Toxicity..108
    7.15.8 Safety in Pregnancy..............................108
    7.15.9 Drug Interactions................................108
  References 108
7.16 *Cinnamomum zeylanicum* (Cinnamon)........................110
    7.16.1 Antioxidant......................................110
    7.16.2 Anti-Inflammatory 111
    7.16.3 Antidiabetic/Anti-Metabolic Syndrome 111
    7.16.4 Preclinical Antidepressant-Like Effects 112
    7.16.5 Human Antidepressant Effects....................112
    7.16.6 Dosage...113
    7.16.7 Toxicity...113
    7.16.8 Safety in Pregnancy 113
    7.16.9 Drug Interactions................................113
  References 113
7.17 *Coffea arabica* (Coffee)................................115
    7.17.1 Antioxidant 115

7.17.2   Anti-Inflammatory 116
7.17.3   Antidiabetic/Anti-Metabolic syndrome 116
7.17.4   Preclinical Antidepressant-Like Effects 116
7.17.5   Human Antidepressant Effects................................ 117
7.17.6   Dosage 117
7.17.7   Toxicity.. 118
7.17.8   Safety in Pregnancy 118
7.17.9   Drug Interactions... 118
References 118
7.18   *Coriandrum sativum* (Coriander)...................................... 120
7.18.1   Antioxidant 120
7.18.2   Anti-Inflammatory 121
7.18.3   Antidiabetic/Anti-Metabolic Syndrome.................. 121
7.18.4   Preclinical Antidepressant-Like Effects 121
7.18.5   Human Antidepressant Effects................................ 122
7.18.6   Dosage 122
7.18.7   Toxicity.. 122
7.18.8   Safety in Pregnancy.. 122
7.18.9   Drug Interactions... 122
References 122
7.19   *Corydalis yanhusuo* 123
7.19.1   Antioxidant 124
7.19.2   Anti-Inflammatory 124
7.19.3   Antidiabetic/Anti-Metabolic Syndrome.................. 124
7.19.4   Preclinical Antidepressant-Like Effects 125
7.19.5   Human Antidepressant Effects................................ 125
7.19.6   Dosage 125
7.19.7   Toxicity.. 125
7.19.8   Safety in Pregnancy.. 125
7.19.9   Drug Interactions 126
References 126
7.20   *Crocus sativa* (Saffron) 127
7.20.1   Antioxidant.. 127
7.20.2   Anti-Inflammatory 127
7.20.3   Antidiabetic/Anti-Metabolic Syndrome.................. 128
7.20.4   Preclinical Antidepressant-Like Effects 128
7.20.5   Human Antidepressant Effects................................ 129
7.20.6   Dosage 129
7.20.7   Toxicity.. 129
7.20.8   Safety in Pregnancy.. 130
7.20.9   Drug Interactions 130
References 130
7.21   *Curcuma longa* (Turmeric).. 132
7.21.1   Antioxidant 132
7.21.2   Anti-Inflammatory 132
7.21.3   Antidiabetic/Anti-Metabolic Syndrome.................. 133

7.21.4 Preclinical Antidepressant-Like Effects 133
7.21.5 Human Antidepressant Effects.............................. 133
7.21.6 Dosage 133
7.21.7 Toxicity... 133
7.21.8 Safety in Pregnancy.. 134
7.21.9 Drug Interactions.. 134
References 134
7.22 *Cyperus rotundus* 135
7.22.1 Antioxidant 135
7.22.2 Anti-Inflammatory 136
7.22.3 Antidiabetic/Anti-Metabolic Syndrome 136
7.22.4 Preclinical Antidepressant-Like Effects 136
7.22.5 Human Antidepressant Studies 137
7.22.6 Dosage 137
7.22.7 Toxicity... 137
7.22.8 Safety in Pregnancy.. 137
7.22.9 Drug Interactions.. 137
References 137
7.23 *Echium amoenum*.. 139
7.23.1 Antioxidant... 139
7.23.2 Anti-Inflammatory 139
7.23.3 Antidiabetic/Anti-Metabolic Syndrome............. 139
7.23.4 Preclinical Antidepressant-Like Effects 139
7.23.5 Human Antidepressant Effects............................ 140
7.23.6 Dosage.. 140
7.23.7 Toxicity... 140
7.23.8 Safety in Pregnancy.. 140
7.23.9 Drug Interactions.. 140
References 141
7.24 *Eleutherococcus senticoccus* (Siberian Ginseng) 141
7.24.1 Antioxidant... 142
7.24.2 Anti-Inflammatory 142
7.24.3 Antidiabetic/Anti-Metabolic Syndrome............. 142
7.24.4 Preclinical Antidepressant-Like Effects 143
7.24.5 Human Antidepressant Effects 143
7.24.6 Dosage 143
7.24.7 Toxicity... 144
7.24.8 Safety in Pregnancy 144
7.24.9 Drug Interactions.. 144
References 144
7.25 *Epimedium brevicornum* (Horny Goat Weed) 145
7.25.1 Antioxidant... 145
7.25.2 Anti-Inflammatory 146
7.25.3 Antidiabetic/Anti-Metabolic Syndrome............. 146
7.25.4 Preclinical Antidepressant-Like Effects 147
7.25.5 Human Antidepressant Effects............................ 147

         7.25.6 Dosage 148
         7.25.7 Toxicity.. 148
         7.25.8 Safety in Pregnancy.. 148
         7.25.9 Drug Interactions.. 148
    References 148
    7.26 *Foeniculum vulgare* (Fennel) 149
         7.26.1 Antioxidant.. 150
         7.26.2 Anti-Inflammatory 150
         7.26.3 Antidiabetic/Anti-Metabolic Syndrome.................... 150
         7.26.4 Preclinical Antidepressant-Like Effects 151
         7.26.5 Human Antidepressant Effects............................... 151
         7.26.6 Dosage 151
         7.26.7 Toxicity.. 151
         7.26.8 Safety in Pregnancy.. 152
         7.26.9 Drug Interactions.. 152
    References 152
    7.27 *Ginkgo biloba* 153
         7.27.1 Antioxidant 154
         7.27.2 Anti-Inflammatory 154
         7.27.3 Antidiabetic/Anti-Metabolic Syndrome.................... 154
         7.27.4 Preclinical Antidepressant-Like Effects 154
         7.27.5 Human Antidepressant Effects............................... 155
         7.27.6 Dosage 156
         7.27.7 Toxicity.. 156
         7.27.8 Safety in Pregnancy.. 156
         7.27.9 Drug Interactions.. 156
    References 156
    7.28 *Glycyrrhiza* (Licorice)... 158
         7.28.1 Antioxidant 158
         7.28.2 Anti-Inflammatory 158
         7.28.3 Antidiabetic/Anti-Metabolic Syndrome 159
         7.28.4 Preclinical Antidepressant-Like Effects 159
         7.28.5 Human Antidepressant Effects............................... 159
         7.28.6 Dosage 159
         7.28.7 Toxicity.. 159
         7.28.8 Safety in Pregnancy.. 160
         7.28.9 Drug Interactions 160
    References 160
    7.29 *Hedyosmum brasiliense*.. 161
         7.29.1 Antioxidant/Anti-Inflammatory 162
         7.29.2 Antidiabetic/Anti-Metabolic Syndrome 162
         7.29.3 Preclinical Antidepressant-Like Effects 162
         7.29.4 Human Antidepressant Effects............................... 162
         7.29.5 Dosage 162
         7.29.6 Toxicity 163
         7.29.7 Safety in Pregnancy... 163

| | 7.29.8 | Drug Interaction | 163 |

7.30 *Hemerocallis citrina* (Daylily) 163
    7.30.1 Antioxidant 164
    7.30.2 Anti-Inflammatory 164
    7.30.3 Antidiabetic/Anti-Metabolic Syndrome 164
    7.30.4 Preclinical Antidepressant-Like Effects 164
    7.30.5 Human Antidepressant Effects................................ 165
    7.30.6 Dosage 165
    7.30.7 Toxicity 165
    7.30.8 Safety in Pregnancy.. 165
    7.30.9 Drug Interactions.. 166
References 166
7.31 *Hericium erinaceus* (Lion's Mane)... 167
    7.31.1 Antioxidant.. 167
    7.31.2 Anti-Inflammatory 167
    7.31.3 Antidiabetic/Anti-Metabolic Syndrome 168
    7.31.4 Preclinical Antidepressant-Like Effects 168
    7.31.5 Human Antidepressant Effects 169
    7.31.6 Dosage 169
    7.31.7 Toxicity.. 169
    7.31.8 Safety in Pregnancy.. 170
    7.31.9 Drug Interactions 170
References 170
7.32 *Hibiscus rosa-sinensis* (Hibiscus) 171
    7.32.1 Antioxidant 171
    7.32.2 Anti-Inflammatory 172
    7.32.3 Antidiabetic/Anti-Metabolic Syndrome 172
    7.32.4 Preclinical Antidepressant-Like Effects 172
    7.32.5 Human Antidepressant Effects 172
    7.32.6 Dosage.. 173
    7.32.7 Toxicity.. 173
    7.32.8 Safety in Pregnancy.. 173
    7.32.9 Drug Interactions.. 173
References 173
7.33 *Humulus lupulus* (Hops)... 174
    7.33.1 Antioxidant.. 174
    7.33.2 Anti-Inflammatory 175
    7.33.3 Antidiabetic/Anti-Metabolic Syndrome 175
    7.33.4 Preclinical Antidepressant-Like Effects 175
    7.33.5 Human Antidepressant Effects................................ 176
    7.33.6 Dosage 176
    7.33.7 Toxicity.. 176
    7.33.8 Safety in Pregnancy.. 176
    7.33.9 Drug Interactions.. 176
References 177
7.34 *Huperzia serrata*.. 178

        7.34.1   Antioxidant/Anti-Inflammatory Effects 178
        7.34.2   Antidiabetic/Anti-Metabolic Syndrome................... 178
        7.34.3   Preclinical Antidepressant-Like Effects 179
        7.34.4   Human Antidepressant Effects 179
        7.34.5   Dosage 180
        7.34.6   Toxicity... 180
        7.34.7   Safety in Pregnancy.. 180
        7.34.8   Drug Interactions.. 180
    References 180
    7.35   *Hypericum perforatum* (St John's wort) 181
        7.35.1   Antioxidant 182
        7.35.2   Anti-Inflammatory 182
        7.35.3   Antidiabetic/Anti-Metabolic Syndrome 182
        7.35.4   Preclinical Antidepressant-Like Effects 183
        7.35.5   Human Antidepressant Effects 184
        7.35.6   Dosage 185
        7.35.7   Toxicity... 185
        7.35.8   Safety in Pregnancy.. 185
        7.35.9   Drug Interactions.. 185
    References 185
    7.36   *Ilex paraguariensis* (Yerba Mate)...................................... 187
        7.36.1   Antioxidant 187
        7.36.2   Anti-Inflammatory 188
        7.36.3   Antidiabetic/Anti-Metabolic Syndrome................... 188
        7.36.4   Preclinical Antidepressant-Like Effects 188
        7.36.5   Human Antidepressant Effects................................. 189
        7.36.6   Dosage 189
        7.36.7   Toxicity... 189
        7.36.8   Safety in Pregnancy.. 189
        7.36.9   Drug Interactions.. 189
    References 189
    7.37   *Lavandula* (Lavender) 190
        7.37.1   Antioxidant 191
        7.37.2   Anti-Inflammatory 191
        7.37.3   Antidiabetic/Anti-Metabolic Syndrome 191
        7.37.4   Preclinical Antidepressant-Like Effects 192
        7.37.5   Human Antidepressant Effects 192
        7.37.6   Dosage 193
        7.37.7   Toxicity... 193
        7.37.8   Safety in Pregnancy.. 193
        7.37.9   Drug Interactions 193
    References 193
    7.38   *Ligusticum chuanxiong* 195
        7.38.1   Antioxidant... 195
        7.38.2   Anti-Inflammatory 196
        7.38.3   Antidiabetic/Anti-Metabolic Syndrome................... 196

7.38.4   Preclinical Antidepressant-Like Effects 196
7.38.5   Human Antidepressant Effects................................ 196
7.38.6   Dosage 196
7.38.7   Toxicity.. 197
7.38.8   Safety in Pregnancy.. 197
7.38.9   Drug Interactions 197
References 197
7.39   *Magnolia officinalis* 198
7.39.1   Antioxidant 198
7.39.2   Anti-Inflammatory 198
7.39.3   Antidiabetic/Anti-Metabolic Syndrome 199
7.39.4   Preclinical Antidepressant-Like Effects 199
7.39.5   Human Antidepressant Effects 200
7.39.6   Dosage 200
7.39.7   Toxicity.. 200
7.39.8   Safety in Pregnancy.. 201
7.39.9   Drug Interactions.. 201
References 201
7.40   *Matricaria recutita* (Chamomile)................................. 203
7.40.1   Antioxidant... 203
7.40.2   Anti-Inflammatory 203
7.40.3   Antidiabetic/Anti-Metabolic Syndrome 204
7.40.4   Preclinical Antidepressant-Like Effects 204
7.40.5   Human Antidepressant Effects............................. 205
7.40.6   Dosage 205
7.40.7   Toxicity.. 206
7.40.8   Safety in Pregnancy.. 206
7.40.9   Drug Interaction 206
References 206
7.41   *Melissa officinalis* (Lemon Balm)................................. 208
7.41.1   Antioxidant... 208
7.41.2   Anti-Inflammatory 208
7.41.3   Antidiabetic/Anti-Metabolic Syndrome 209
7.41.4   Preclinical Antidepressant-Like Effects 209
7.41.5   Human Antidepressant Effects............................. 209
7.41.6   Dosage 210
7.41.7   Toxicity.. 210
7.41.8   Safety in Pregnancy.. 210
7.41.9   Drug Interactions.. 210
References 210
7.42   *Mimosa pudica* 211
7.42.1   Antioxidant... 212
7.42.2   Anti-Inflammatory 212
7.42.3   Antidiabetic/Anti-Metabolic Syndrome 212
7.42.4   Preclinical Antidepressant-Like Effects 212
7.42.5   Human Antidepressant Effects............................. 213

          7.42.6 Dosage 213
          7.42.7 Toxicity 213
          7.42.8 Safety in Pregnancy 213
          7.42.9 Drug Interactions 214
      References 214
      7.43 *Ocimum basilicum* (Sweet Basil) 215
          7.43.1 Antioxidant 215
          7.43.2 Anti-Inflammatory 215
          7.43.3 Antidiabetic/Anti-Metabolic Syndrome 215
          7.43.4 Preclinical Antidepressant-Like Effects 216
          7.43.5 Human Antidepressant Effects 217
          7.43.6 Dosage 217
          7.43.7 Toxicity 217
          7.43.8 Safety in Pregnancy 217
          7.43.9 Drug Interactions 217
      References 217
      7.44 *Origanum vulgare* (Oregano) 219
          7.44.1 Antioxidant 219
          7.44.2 Anti-Inflammatory 219
          7.44.3 Antidiabetic/Anti-Metabolic Syndrome 220
          7.44.4 Preclinical Antidepressant-Like Effects 220
          7.44.5 Human Antidepressant Effects 220
          7.44.6 Dosage 221
          7.44.7 Toxicity 221
          7.44.8 Safety in Pregnancy 221
          7.44.9 Drug Interactions 221
      References 221
      7.45 *Paeonia lactiflora* (Peony) 222
          7.45.1 Antioxidant 223
          7.45.2 Anti-Inflammatory 223
          7.45.3 Antidiabetic/Anti-Metabolic Syndrome 223
          7.45.4 Preclinical Antidepressant-Like Effects 223
          7.45.5 Human Antidepressant Effects 224
          7.45.6 Dosage 224
          7.45.7 Toxicity 224
          7.45.8 Safety in Pregnancy 225
          7.45.9 Drug Interactions 225
      References 225
      7.46 *Panax ginseng* (Ginseng) 226
          7.46.1 Antioxidant 226
          7.46.2 Anti-Inflammatory 227
          7.46.3 Antidiabetic/Anti-Metabolic Syndrome 227
          7.46.4 Preclinical Antidepressant-Like Effects 227
          7.46.5 Human Antidepressant Effects 228
          7.46.6 Dosage 229
          7.46.7 Toxicity 229

7.46.8   Safety in Pregnancy..229
7.46.9   Drug Interactions..229
References 229
7.47   *Passifloraceae incarnata* (Passionflower)............................231
7.47.1   Antioxidant..231
7.47.2   Anti-Inflammatory 231
7.47.3   Antidiabetic/Anti-Inflammatory 231
7.47.4   Preclinical Antidepressant-Like Effects 232
7.47.5   Human Antidepressant Effects.................................232
7.47.6   Dosage..232
7.47.7   Toxicity..232
7.47.8   Safety in Pregnancy..233
7.47.9   Drug Interactions..233
References 233
7.48   *Piper methysticum* (Kava) 234
7.48.1   Antioxidant..235
7.48.2   Anti-Inflammatory 235
7.48.3   Antidiabetic/Anti-Metabolic Syndrome 235
7.48.4   Preclinical Antidepressant-Like Effects 235
7.48.5   Human Antidepressant Effects.................................236
7.48.6   Safety/Toxicity..236
References 237
7.49   *Piper nigrum* (Black Pepper)..238
7.49.1   Antioxidant..238
7.49.2   Anti-Inflammatory 239
7.49.3   Antidiabetic/Anti-Metabolic Syndrome 239
7.49.4   Preclinical Antidepressant-Like Effects 239
7.49.5   Human Antidepressant Effects.................................240
7.49.6   Dosage..240
7.49.7   Toxicity..240
7.49.8   Safety in Pregnancy..240
7.49.9   Drug Interactions..240
References 241
7.50   *Polygala tenuifolia* 242
7.50.1   Antioxidant 242
7.50.2   Anti-Inflammatory 243
7.50.3   Antidiabetic/Anti-Metabolic Syndrome 243
7.50.4   Preclinical Antidepressant-Like Effects 243
7.50.5   Human Antidepressant Effects 245
7.50.6   Dosage..245
7.50.7   Toxicity..245
7.50.8   Safety in Pregnancy..245
7.50.9   Drug Interactions..245
References 246
7.51   *Poria cocos* 247
7.51.1   Antioxidant..248

          7.51.2 Anti-Inflammatory 248
          7.51.3 Antidiabetic/Anti-Metabolic Syndrome...................248
          7.51.4 Preclinical Antidepressant-Like Effects 248
          7.51.5 Human Antidepressant Effects................................248
          7.51.6 Dosage 249
          7.51.7 Toxicity...249
          7.51.8 Safety in Pregnancy...249
          7.51.9 Drug Interactions...249
     References 249
     7.52 *Psoralea corylifolia* 250
          7.52.1 Antioxidant...250
          7.52.2 Anti-Inflammatory 251
          7.52.3 Antidiabetic/Anti-Metabolic Syndrome...................251
          7.52.4 Preclinical Antidepressant-Like Effects 252
          7.52.5 Human Antidepressant Effects................................253
          7.52.6 Dosage 253
          7.52.7 Toxicity...253
          7.52.8 Safety in Pregnancy...253
          7.52.9 Drug Interactions...253
     References 254
     7.53 *Rhodiola rosea* 255
          7.53.1 Antioxidant...256
          7.53.2 Anti-Inflammatory 256
          7.53.3 Antidiabetic/Anti-Metabolic Syndrome...................256
          7.53.4 Preclinical Antidepressant-Like Effects 257
          7.53.5 Human Antidepressant Effects................................257
          7.53.6 Dosage 258
          7.53.7 Toxicity...258
          7.53.8 Safety in Pregnancy...258
          7.53.9 Drug Interactions...258
     References 258
     7.54 *Rosmarinus officinalis* (Rosemary).....................................260
          7.54.1 Antioxidant 260
          7.54.2 Anti-Inflammatory 260
          7.54.3 Antidiabetic/Anti-Metabolic Syndrome 261
          7.54.4 Preclinical Antidepressant-Like Effects 261
          7.54.5 Human Antidepressant Effects................................262
          7.54.6 Dosage 262
          7.54.7 Toxicity...262
          7.54.8 Safety in Pregnancy...262
          7.54.9 Drug Interactions...262
     References 263
     7.55 *Salvia divinorum* 264
          7.55.1 Antioxidant 265
          7.55.2 Anti-Inflammatory 265
          7.55.3 Antidiabetic/Anti-Metabolic Syndrome...................265

7.55.4   Preclinical Antidepressant-Like Effects 265
7.55.5   Human Antidepressant Effects............................266
7.55.6   Dosage 267
7.55.7   Toxicity..267
7.55.8   Safety in Pregnancy.....................................267
7.55.9   Drug Interactions.......................................267
References 268
7.56   *Sceletium tortuosum*.....................................269
7.56.1   Antioxidant 269
7.56.2   Anti-Inflammatory 269
7.56.3   Antidiabetic/Antimetabolic Syndrome....................270
7.56.4   Preclinical Antidepressant-Like Effects 270
7.56.5   Human Antidepressant Effects............................270
7.56.6   Dosage..271
7.56.7   Toxicity..271
7.56.8   Safety in Pregnancy.....................................271
7.56.9   Drug Interactions.......................................271
References 272
7.57   *Schisandra chinensis*....................................273
7.57.1   Antioxidant 273
7.57.2   Anti-Inflammatory 273
7.57.3   Antidiabetic/Anti-Metabolic Syndrome...................273
7.57.4   Preclinical Antidepressant-Like Effects 274
7.57.5   Human Antidepressant Effects............................274
7.57.6   Dosage..274
7.57.7   Toxicity..274
7.57.8   Safety in Pregnancy.....................................274
7.57.9   Drug Interactions.......................................275
References 275
7.58   *Scutellaria lateriflora* (Skullcap).....................276
7.58.1   Antioxidant 276
7.58.2   Anti-Inflammatory 276
7.58.3   Antidiabetic/Anti-Metabolic Syndrome...................277
7.58.4   Preclinical Antidepressant-Like Effects 277
7.58.5   Human Antidepressant Effects 277
7.58.6   Dosage 278
7.58.7   Toxicity..278
7.58.8   Safety in Pregnancy.....................................278
7.58.9   Drug Interactions.......................................278
References 279
7.59   *Silybum marianum* (Milk Thistle)........................280
7.59.1   Antioxidant 280
7.59.2   Anti-Inflammatory 280
7.59.3   Antidiabetic/Anti-Metabolic Syndrome 281
7.59.4   Preclinical Antidepressant-Like Effects 281
7.59.5   Human Antidepressant Effects 282

7.59.6   Dosage 282
7.59.7   Toxicity 283
7.59.8   Safety in Pregnancy...283
7.59.9   Drug Interactions...283
References 283
7.60   *Theobroma cacao* (Chocolate) 284
7.60.1   Antioxidant 285
7.60.2   Anti-Inflammatory 285
7.60.3   Antidiabetic/Anti-Metabolic Syndrome 285
7.60.4   Preclinical Antidepressant-Like Effects 286
7.60.5   Human Antidepressant Effects 286
7.60.6   Dosage...287
7.60.7   Toxicity..287
7.60.8   Safety in Pregnancy...287
7.60.9   Drug Interaction 287
References 288
7.61   *Tilia* (Linden)...289
7.61.1   Antioxidant..289
7.61.2   Anti-Inflammatory 289
7.61.3   Antidiabetic/Anti-Metabolic Syndrome 290
7.61.4   Preclinical Antidepressant-Like Effects 290
7.61.5   Human Antidepressant Effects...............................290
7.61.6   Dosage 290
7.61.7   Toxicity..291
7.61.8   Safety in Pregnancy...291
7.61.9   Drug Interactions...291
References 291
7.62   *Trigonella foenum-graecum* (Fenugreek)......................292
7.62.1   Antioxidant..293
7.62.2   Anti-Inflammatory 293
7.62.3   Antidiabetic/Anti-Metabolic Syndrome 294
7.62.4   Preclinical Antidepressant-Like Effects 294
7.62.5   Human Antidepressant Effects 295
7.62.6   Dosage 295
7.62.7   Toxicity..295
7.62.8   Safety in Pregnancy...296
7.62.9   Drug Interactions 296
References 296
7.63   *Valeriana officinalis* (Valerian)......................................298
7.63.1   Antioxidant..298
7.63.2   Anti-Inflammatory 299
7.63.3   Antidiabetic/Anti-Metabolic Syndrome 299
7.63.4   Preclinical Antidepressant-Like Effects 299
7.63.5   Human Antidepressant Effects 300
7.63.6   Dosage 300
7.63.7   Toxicity..301

    7.63.8 Safety in Pregnancy...301
    7.63.9 Drug Interactions...301
  References 301
  7.64 *Verbena officinalis* (Vervain) 302
    7.64.1 Antioxidant...303
    7.64.2 Anti-Inflammatory 303
    7.64.3 Antidiabetic/Anti-Metabolic Syndrome..................304
    7.64.4 Preclinical Antidepressant-Like Effects 304
    7.64.5 Human Antidepressant Effects.................................305
    7.64.6 Dosage...305
    7.64.7 Toxicity...305
    7.64.8 Safety in Pregnancy...305
    7.64.9 Drug Interactions...305
  References 305
  7.65 *Vitex agnus-castus* (Chaste Tree) 307
    7.65.1 Antioxidant 307
    7.65.2 Anti-Inflammatory 307
    7.65.3 Antidiabetic/Anti-Metabolic Syndrome..................307
    7.65.4 Preclinical Antidepressant-Like Effects 308
    7.65.5 Human Antidepressant Effects 308
    7.65.6 Dosage...309
    7.65.7 Toxicity...309
    7.65.8 Safety in Pregnancy 309
    7.65.9 Drug Interactions...309
  References 309
  7.66 *Withania somnifera* (Ashwagandha) 311
    7.66.1 Antioxidant 311
    7.66.2 Anti-Inflammatory 311
    7.66.3 Antidiabetic/Anti-Metabolic Syndrome 312
    7.66.4 Preclinical Antidepressant-Like Effects 312
    7.66.5 Human Antidepressant Effects 313
    7.66.6 Dosage 313
    7.66.7 Toxicity...313
    7.66.8 Safety in Pregnancy...313
    7.66.9 Drug Interactions...313
  References 313

Chapter 8 The Antidepressant Effects of *Yueue* and the Herbs of
Traditional Chinese Medicine...317

  8.1 Fundamental Considerations 317
  8.2 *Yueju*...318
  8.3 *Xiao Yao San* 319
  8.4 *Chai Hu Shu Gan* 320
  8.5 *Gan Mai Da Zao* 320
  8.6 *Gui Pi* 321

8.7   *Shi Wei Wen Dan Tang* 321
8.8   *Ban Xia Hou Pu*... 322
8.9   *Chai Hu Jia Long Gu Mu Li* 322
8.10   *Tiao Qi*.. 322
8.11   *Yi Pi* 323
8.12   *Tang Shen Kang*.. 323
8.13   *Kai Xin San*... 323
8.14   *Shu Gan Jie Yu* 323
8.15   *Si Ni San* 324
8.16   *Wu Ling*... 324
8.17   Other TCM Herbs Used in the Treatment of MDD 325
8.18   A Medical "Theory of Everything" 326
References 327

Chapter 9   Flavonoids with Preclinical Antidepressant-Like Effects............... 331

9.1   Amentoflavone 331
9.2   Apigenin 332
9.3   Astilbin 332
9.4   Baicalein and Baicalin... 332
9.5   Chrysin 332
9.6   7,8,Dihydroxyflavone... 333
9.7   Fisetin 334
9.8   Heptomethoxyflavone 334
9.9   Hesperidin and Hesperitin 334
9.10   Hyperoside 335
9.11   Icariin 335
9.12   Isosakurentin-5-O-rutinoside 336
9.13   Kaempferol 336
9.14   Liquiritin and Isoliquirtin.. 336
9.15   Luteolin 336
9.16   Miquelianin 336
9.17   Myricetin 337
9.18   Naringenin and Naringin.. 337
9.19   Nobiletin 337
9.20   Orientin.. 338
9.21   Quercetin 338
9.22   Vitexin 339
9.23   Wogonin and Wogonoside 339
9.24   Synthetic Flavonoids 340
9.25   Mechanisms of Flavonoid Antidepressant Action................. 340
References 343

Chapter 10   Preclinical Antidepressant-Like Effects of Terpenes,
                 Polyphenolics, and Other Non-Flavonoid Phytochemicals.............. 351

10.1   Amyrins.. 352
10.2   Bacopasides 353

10.3   Berberine 353
10.4   3-n-Butylphthalide 354
10.5   Caffeic Acid 354
10.6   β-Carotene 355
10.7   Carvacrol 355
10.8   β-Caryophyllene 356
10.9   Chlorogenic Acid 356
10.10   Crocin 356
10.11   Curcumin 356
10.12   3,6′-Disinapoyl Sucrose 357
10.13   Ellagic Acid 357
10.14   Eugenol 358
10.15   Ferulic Acid 358
10.16   Gallic Acid 359
10.17   Gastrodin 360
10.18   Genipin 360
10.19   Ginsenoside Rg1 361
10.20   Glycyrrhizin 361
10.21   4-Hydroxyisoleucine 362
10.22   Hyperfoliatin 362
10.23   Linalool 362
10.24   Macranthol 363
10.25   Methyl Jasmonate 363
10.26   Mitragynine 363
10.27   Oleanolic Acid 364
10.28   Orcinol 364
10.29   Paeoniflorin 365
10.30   Paeonol 365
10.31   Palmatine 366
10.32   Plumbagin 366
10.33   Podoandin 367
10.34   Punarnavine 367
10.35   Resveratrol 367
10.36   Riparin 368
10.37   Rosmarinic Acid 369
10.38   Safranal 369
10.39   Salidroside 369
10.40   Sarsasapogenin 370
10.41   Scopoletin 371
10.42   Sulphoraphane 371
10.43   Tetrandrine 371
10.44   L-theanine 372
10.45   Uliginosin B 372
10.46   Ursolic acid 372
10.47   Vanillin 373
References 373

Chapter 11 Choosing Herbal Treatments...387

11.1 Efficacy of Herbal Treatments of MDD................................388
11.2 Herbs for Which There Is Less than Compelling
       Evidence of Efficacy...390
11.3 Combinations of Herbs..391
11.4 Addressing Comorbidities...392
       11.4.1 Anxiety and Insomnia..392
       11.4.2 Obsessive-Compulsive Disorder 394
       11.4.3 Premenstrual and Perimenopausal Symptoms.........395
       11.4.4 Dementia 395
       11.4.5 Diabetes and Metabolic Syndrome 396
       11.4.6 Fatigue, Lack of Resiliency, and General Malaise.....397
11.5 Augmentation of Standard Antidepressant Treatment
       with Herbs 398
11.6 Safety..399
References 401

Index..409

Author

Scott D. Mendelson, MD, PhD, earned his PhD in Biopsychology at the University of British Columbia, in Vancouver, British Columbia. He then worked as a post-doctoral fellow at the Rockefeller University in the Laboratory of Neuroendocrinology. During his doctoral and post-doctoral work, he published papers on the subjects of serotonergic and hormonal regulation of sexual behavior and the effects of stress on serotonin receptor subtypes in the brain. He subsequently attended medical school at the University of Illinois. After graduating in 1996 he did his residency in psychiatry at the University of Virginia. His first book, *Metabolic Syndrome and Psychiatric Illness: Interactions, Pathophysiology, Assessment and Treatment*, was published in 2007. His second book, *Beyond Alzheimer's: How to Avoid the Modern Epidemic of Dementia*, was published in 2009. He has worked for 20 years as an inpatient and outpatient psychiatrist, and is currently practicing in Roseburg, Oregon. In his practice he uses both herbal treatments and standard psychiatric medications.

1 Major Depression: A Brief History of Western Medical Treatment

Everyone feels down now and then. At times, the stresses, losses, and disappointments of life can make such feelings almost unavoidable. However, when painful feelings of sadness, lack of enjoyment, guilt, fatigue, and hopelessness persist for more than two weeks and cause impairment of function at work and in relationships, then the criteria are met for the diagnosis of Major Depression (Major Depressive Disorder, MDD).[1] The severity of MDD may vary from person to person, with symptoms classified as mild, moderate, or severe. In mild cases, symptoms cause only minor impairment of function. In the most severe cases, suffering is intense and disabling. It may even be accompanied by psychotic symptoms. MDD may also vary in how often a person suffers the illness. It may occur only once during a stressful time in a person's life, or it may appear recurrently. In some individuals, MDD recurs in predictable cycles, such as in Seasonal Affective Disorder, in which symptoms of depression emerge during the fall and winter months. In others, depression may return in seemingly unpredictable fashion.

MDD is a common illness, and in any one year it affects nearly 7% of the US population. One in five people, or 20% of the American public can expect to suffer MDD at some time in their lives. Women are roughly twice as likely to be diagnosed with MDD as men, though it is unclear if women actually suffer the illness more often or are simply more willing to seek diagnosis and treatment. MDD is the leading cause of disability in the United States among people 15 to 44 years of age. Suicide, which in most cases occurs in the context of MDD, is the second leading cause of death among people in that age group. Aside from personal cost, MDD is also expensive for society. The economic cost of the illness to the US economy is estimated to be over $200 billion a year.[2] Despite many advances in psychiatry and neuroscience, rates of depression in the United States have remained about the same. Sadly, evidence shows the rates of suicide have been increasing.

Medical science has long tried to understand the nature and causes of MDD. Though we tend to think of it as a disease rising from the pressures of modern life, the ancients also suffered from and puzzled over this illness. There is perhaps no better description of MDD than one from a nearly 4,000-year-old cuneiform text from Babylonia.

If he has had misfortune suffering losses and knows not how it came upon him; if he has frequent nervous breakdowns from giving orders with none complying, calling with none answering, and striving to achieve his desires while having to look after his household. If he shakes with fear and his limbs have become weak; if he cannot sleep by day or night and sees disturbing dreams; if he is weak from not eating; if he forgets the word he is trying to say; then the anger of god and goddess is upon him.[3]

The Babylonians, Egyptians, and many other ancient cultures thought the illness was due to having offended the gods. Accordingly, appeasing the gods was the standard treatment. However, some cultures saw physical components of depression. The ancient Greek physician, Hippocrates, recognized MDD, which he referred to as melancholia. In the fourth century BCE, he described symptoms of "aversion to food, despondency, sleeplessness, irritability and restlessness." The word melancholia arose from the ancient notion of the four humors, or life fluids, that circulated in the body. Melancholia referred to "black bile," and it was thought that overabundance of black bile in the spleen caused the symptoms of depression. As treatment, Hippocrates advised bloodletting, bathing, exercise, mild diet, and herbs. Among the herbs he prescribed for melancholia were black hellebore, borage, bugloss, marigold, and epithyme.[4] However, when the patient did not respond to such measures, Hippocrates, too, entertained the notion of demonic possession.

The Greeks and Romans came to see melancholia as arising from perturbations of spiritual, physical, and psychological factors, and they developed treatments to address these various components. Though shackled by superstitions and lacking the scientific knowledge to understand the physiological complexities of the illness, they attempted to treat the patient in his entirety, not unlike modern "holistic" approaches. The Epicurean and Stoic schools arose in Greek culture a century or so after Hippocrates. They believed that peace and tranquility of mind – the avoidance of despondency and anxiety – could be achieved by developing proper perspectives on life and reasonable management of one's affairs. Thus, for the educated Greeks, engagement in a form of what we might now call cognitive behavioral psychotherapy prevented and treated depressed mood.

Several hundred years after Hippocrates, the Roman author Aulus Cornelius Celsus wrote a text on ailments of men including various forms of mental illness. He carried forward many Hippocratic notions of physical and mental health. He saw the need to calm the minds of the anxious and fearful, and to enliven the minds of the withdrawn and sorrowful. For mental illnesses, he advocated darkness, calm, and music therapy. "The sorrowful thoughts of others must be dispelled, for which purpose concerts of music, and cymbals and noise are useful."[5] Yet, with more recalcitrant patients, treatment went on to include shackling, starvation, and beating. During this time, herbal treatments continued to be explored and refined. Dioscorides (40–90 AD), a near-contemporary of Celsus, published the text, *On Medicinal Substances*, in which he discussed the use of nearly a thousand different medicinal plants and thus expanded and consolidated knowledge of the use of herbs to treat human illness. Galen, the great philosopher, anatomist, and physician to emperors of Rome, also amassed a knowledge of herbal treatments. Thus, in the first century, the Greeks and Romans were well-versed in the uses of herbs such as poppy,

St John's wort, hops, valerian, saffron, balm, basil, marjoram, and others that affect the nervous system. Indeed, Dioscorides and Galen are both alternatively known as the Father of Pharmacy.

In Europe during the Middle Ages, the pendulum swung back to religious explanations of mental illness in general. Demons, devils, and falling short of the glory of God were seen as the major causes of madness and melancholia. Exorcisms, beatings, near-drowning and other harsh treatments were applied as penitence and to drive the demons from the bodies of the victims. The Arabs kept alive the views and practices of the Greeks, including the sophisticated herbalism of Galen, and here and there in Europe were Christian oases of kindness and care. Still, the clouds of fear and superstition persisted over Europe, and the brutal practices they spawned persisted well into the Renaissance.

During the Renaissance, the more holistic philosophies of the Greeks were revived. New scientific methods and discoveries changed our ways of seeing the world. An age of humanism began. Man – no longer a mere servant of God – became a focus of study. The discoveries of Galileo whittled away at the power of religion and superstition. The anatomical discoveries of Vesalius demystified the human body. Francis Bacon and the scientific method were born. The work of Ambroise Paré, the first advocate of experimental medicine, provided new means to resolve questions of efficacy of treatment. Knowledge of human anatomy and function grew. By the middle of the 1600s, Dr Thomas Willis, the English anatomist, had explored in detail the structure of the brain. However, diseases of the mind and brain required exploration on microscopic, and even finer, neurochemical levels. The underlying defects in brain function that gave rise to MDD and other mental illnesses thus continued to defy understanding.

Throughout this time, herbs remained a means to relieve human suffering. In 1652, Nicholas Culpeper wrote his famous treatise on herbal medicines, *The English Physician*. Although steeped in pseudoscientific notions of the astrological significance of plants, his descriptions of the uses of hundreds of European herbs became a mainstay of treatment among healers and laypeople alike. He described 28 different herbs to treat melancholia. His book remains in print to this day.

By the time of Dr Benjamin Rush – signer of the Declaration of Independence and widely recognized Father of American Psychiatry – the treatment of psychiatric illness had not greatly progressed. Rush recommended a variety of herbs, teas, and spirits to calm the spirit. His favorite medication was laudanum, a preparation of opium. In reference to a depressed patient, he noted "Ten drops of laudanum, taken occasionally, saved him from being devoured by melancholy." His method of treatment included bloodletting, purgatives, emetics, hot baths, cold baths, strict bland diets, music, solitude, exercise, pain, blistering, and other ancient practices. Indeed, the recommendations were similar regardless of the type and manifestation of the psychiatric illness.[6]

The modern era of medicine can be said to have begun in the mid-1800s. At that time, the great physiologist, Rudolf Virchov, laid the foundations of our modern understanding of the phenomenon of inflammation. Pasteur and Koch developed the germ theory of disease, and Joseph Lister recognized and promoted the notion of antiseptics. In the 1880s, Ivan Pavlov began his study of the conditioned reflex that later earned him the 1904 Nobel Prize. Ramon y Cajal drew and cataloged the microscopic details of the brain's neurons that led to his Nobel Prize in 1906. In the 1890s, Sir Charles Sherrington began his Nobel-winning research into the mechanisms of action of the nervous system. It was during the late 1800s that German neurologists developed the field that later gave birth to psychiatry as a field unto itself. The German neuropsychiatrist, Wilhelm Griesinger, set the stage for modern psychiatry in his 1848 statement that "mental diseases are brain diseases."

One of the most important events in the history of psychiatry was the development of psychoanalytic theory by Sigmund Freud. His famous book, *The Interpretation of Dreams,* was published in 1889. It established the basic tenets of psychoanalytic theory, which were that the mind contains both conscious and subconscious material. He postulated that childhood traumas thwarted natural drives, which led to conflicts between conscious and subconscious processes. As part of this process, individuals build resistance to making these conflicts accessible to awareness. Thus, special techniques were required to free, recognize, experience, and work through those traumas. Freud further proposed that systematic but unfettered release of freely associated thoughts allows unconscious material to emerge, and that skilled interpretation of material at the right stages allows the patient to regain awareness and control.

Nonetheless, Freud was the heir of the great German and French schools of neurology, and he himself believed that the ultimate causes of psychiatric illnesses, such as MDD, were physiological abnormalities of brain function. In 1895, in his book, *Project for a Scientific Psychology*, he outlined his attempts to understand how consciousness might evolve out of neuronal activity.[7] He also experimented with cocaine as a potential miracle drug in the treatment of psychiatric conditions, particularly "nervous exhaustion." He published his first paper on the subject, Über Coca[8] in 1884. His abandonment of medication as a part of psychiatric treatment was thought to have been due to disillusionment after the death of a friend on whom he used cocaine to treat morphine addiction.

At the time of Freud's ascendency, the armamentarium of medications to treat mental illness remained very limited. The Danish physicians, Carl and Fritz Lange, had successfully used lithium salts to treat mania and depression as early as the 1880s. However, they based this use on the false notion that mania was a manifestation of "gout of the brain." The drug fell out of favor and did not return to use in psychiatry until Dr John Cade's landmark publication in 1949.[9]

Emil Kraepelin is best known for his careful and detailed observations of patients with various mental illnesses, and his categorization of types of those conditions. It was he who first distinguished schizophrenia from manic depressive illness. In his 1917 book, *Lectures on Clinical Psychiatry*, Kraepelin described the treatment of severe melancholia:

Patients who show such tendencies require the closest watching, day and night. They are kept in bed and given plenty of food, though this is often very difficult on account of their resistance. Care is also taken to regulate their digestion, and, as far as possible, to secure them sufficient sleep by means of baths and medicines. Paraldehyde is generally to be recommended, or, under some circumstances, alcohol, or occasional doses of trigonal (a weak, and somewhat toxic sedative first synthesized in 1888 and used into the 1940s). Opium is employed to combat the apprehensiveness, in gradually increasing doses, which are then by degrees reduced.

Whereas Kraepelin's ability to appreciate the differences among the various illnesses and to describe their natural history was groundbreaking, his methods of treatment had not progressed much beyond those used by Benjamin Rush from a century before.

For the first half of the twentieth century, Dr Adolf Meyer led opinion in American psychiatry. Unlike the German school of thought led by Kraeplin, Meyer saw mental illness more as product of a dysfunctional personality than a dysfunctional brain. Although head of psychiatry at the pre-eminent Johns Hopkins University from 1908 to 1941, his views were often seen as more "anti-psychiatric" than psychiatric. He was not a champion of psychiatric medication and did not serve to advance that aspect of treatment.

Sophisticated psychiatric medications were not possible until the chemistry of the nervous system was understood. Acetylcholine was the first neurotransmitter to be discovered, by Sir Henry Dale in 1915, and was subsequently identified as a neurotransmitter by the pharmacologist Otto Loewi. Serotonin was discovered in 1935 by the Italian pharmacologist Vittorio Erspamer, but it wasn't until 1953 that Betty Twarog and Irvine Page found serotonin in the mammalian central nervous system. Dopamine was first discovered in the human brain in 1957. It was identified as a neurotransmitter in 1958 by Arvid Carlsson and Nils-Ake Hillarp. Norepinephrine, one of the "sympathomimetic amines," was discovered in 1898. However, it was not until 1961 that Julius Axelrod fully explained how epinephrine acts as a neurotransmitter in the brain. Those accomplishments led to the understanding of the chemical nature of brain activity. It made it possible to both understand the mechanisms of action of the antidepressant drugs discovered in the 1940s and 1950s and to rationally design drugs for stronger and safer effects. Most of those scientists were rewarded with Nobel Prizes.

It was in 1929, in the midst of this new age of neuropharmacological exploration, that Gordon Alles patented his new discovery, amphetamine.[10] The use of amphetamine for the treatment of depression was first reported in 1936 by Guttman[11] and Meyerson.[12] It helped for some, but side effects sometimes included anxiety and agitation. Initial well-being was sometimes followed by worse depression, tolerance, and addiction. Methylphenidate, pipradol, and deanol were other stimulant drugs used in the 1950s for monotherapy of MDD prior to the predominant use of the soon to follow MAO (monoamine oxidase) inhibitors and tricyclics antidepressants.[13] The discovery and use of the "modern" antidepressants, MAO inhibitors and tricyclics in the 1950s, and the later discovered selective serotonin reuptake inhibitors (SSRIs), will be discussed at length in the following chapter.

The current view of MDD is that it is a complicated illness having biological, psychological, and social components. This defines the bio-psycho-social model of MDD[14] in which the illness is seen as rising out of combinations of those factors. Abnormal physiological processes in the brain and body, disease, poor nutrition, and hormonal imbalances can exacerbate or even cause the condition. From the psychological standpoint, unreasonable expectations from life and other people, rigidity in values, and unjustified opinions of one's self can predispose one to conflicts, struggles, and despondency that lead to the illness. Social issues, such as persistent financial problems, stressful work conditions, poor access to education and health resources, and institutionalized injustices contribute to the development of MDD.

There is no basis to suggest that a person with MDD must be treated with medications. There is a variety of effective treatments – medication, psychotherapy, exercise, meditation – and each form of therapy can be effective for different people and for different severities of illness. Nonetheless, the focus of modern psychiatry is the biological treatment of MDD, primarily with medications. Unfortunately, antidepressants are slow in taking effect. They can also cause intolerable or even dangerous side effects. Most concerning is that antidepressants are not as effective as some doctors and pharmaceutical companies would have us believe.

REFERENCES

1. American Psychiatric Association. *Diagnostic and Statistical Manual of Mental Disorders*. 5th ed, Text Revision. Washington, DC: American Psychiatric Association; 2013.
2. Greenburg PE, Fournier AA, Sisitsky T, et al. The economic burden of adults with Major Depressive Disorder in the United States 2005 and 2010. *J Clin Psychiatry*. 2015;76(2):155–162.
3. Reynolds EH, Kinnier-Wilson JV. Depression and anxiety in Babylon. *J R Soc Med*. 2013;106(12):478–481.
4. Rees L. Treatment of depression by drugs and other means. *Nature*. 1960;186:114–120.
5. Farrington C. Music and madness: from Kontakte to The Cure. *The Lancet*. 2015;2(5):388–390.
6. Rush B. *Medical Inquiries and Observations, Upon the Diseases of the Mind*. Philadelphia, PA: Grigg and Elliot; 1935. p. 100.
7. Schore AN. A century after Freud's project: is a rapprochement between psychoanalysis and neurobiology at hand? *J Am Psychoanal Assoc*. 1997;45:807–840.
8. Markel H. Über coca: Sigmund Freud, Carl Koller, and Cocaine. *JAMA*. 2011;305(13):1360–1361.
9. Mitchell PB, Hadzi-Pavlovic D. Lithium treatment for bipolar disorder. *Bull World Health Organ*. 2000;78(4):515–517.
10. Rasmussen N. Making the first anti-depressant: amphetamine in American medicine, 1929–1950. *J Hist Med Allied Sci*. 2006;61(3):288–323.
11. Guttman E. The effect of benzedrine on depressive states. *J Ment Sci*. 1936;82:618–625.
12. Meyerson A. Effect of benzedrine sulfate on mood and fatigue in normal and neurotic persons. *Arch Neurol Psychiatry*. 1936;36:816–822.
13. Rees L. Treatment of depression by drugs and other means. *Nature*. 1960;186:114–120.
14. Engel GL. The clinical application of the biopsychosocial model. *Am J Psychiatry*. 1980;137(5):535–544.

2 How Antidepressants Work, but Often Do Not

In 1951, Drs Irving Selikoff and Edward Robitzek began experimenting with a new antibiotic drug, isoniazid, to treat tuberculosis.[1] Along with improvements in the patients' tuberculosis, the doctors serendipitously noted improvements in the moods of many of the patients being given the medication. Even stronger mood-elevating effects were seen with a sister tuberculosis medication, iproniazid. It was later determined that these drugs acted in the brain as inhibitors of the monoamine oxidase enzymes. Thus, it seemed possible that these new drugs produced their antidepressant effects by increasing brain levels of the monoamines serotonin, norepinephrine, and dopamine.

With the apparent success of iproniazid as an antidepressant, it was marketed in 1958 as the drug Marsilid. However, it was withdrawn in 1961 after it was found to cause severe liver damage in some individuals. The search began for other monoamine oxidase inhibitors (MAOIs) without toxicity to the liver. Unfortunately, while newer, more liver-friendly MAOIs were being discovered, another, sometimes fatal, danger was soon recognized. In 1962, a man taking an MAOI died after eating a large sandwich of well-aged cheese. It was then realized that the MAOIs inhibited enzymes not only in the brain, but in the gut as well. This allowed buildup in the blood of a toxic substance, tyramine, that could cause a sudden and fatal rise in blood pressure.

The MAOI antidepressants were almost abandoned because of the so-called "cheese effect."[2] However, it was found that by eliminating foods that contained or generated tyramine, such as cheese, red wine, and preserved or fermented meats, the drugs could be used relatively safely. Indeed, many psychiatrists still see the MAOIs as being among the most effective antidepressant medications. One of the more recent developments in psychiatry is a patch form of the MAOI selegiline that greatly reduces concern for dietary restrictions. Nonetheless, older psychiatrists often note with dismay that many younger psychiatrists have never prescribed MAOIs, even for their more difficult patients.

The first major antipsychotic medication, chlorpromazine, was discovered in 1950. It was being developed as a potential anesthetic agent for surgery when it was found by chance to calm the turmoil of schizophrenia. It was marketed in 1953 as the drug Largactil, and later in the US as Thorazine. Scientists simultaneously began to evaluate similar molecules in hope of developing newer and better antipsychotic medications. Imipramine was discovered in this search in 1951. It has a three ring, or "tricyclic," chemical structure very similar to chlorpromazine, and was first evaluated as a medication for schizophrenia. Its effects on schizophrenia were disappointing. However, it was recognized that it improved the moods of patients. In 1958, it was marketed as the antidepressant Tofranil.[3] Several years later, the tricyclic antidepressant (TCA) amitriptyline hit the market as Elavil. Since that time, there have been many TCAs and related tetracyclic antidepressants produced and sold around the world.

The TCAs are effective treatments of depression for many people. However, like the MAOIs, there are significant dangers inherent in their use, and for some there are intolerable side effects. TCAs are dangerous in overdose, and a month's supply of medication is often more than enough to poison a patient. Even therapeutic doses can cause problems with heart rhythm. Psychiatrist have to be vigilant in obtaining EKGs to monitor effects on heart function, and to make certain that patients at risk for suicide have limited supplies of the medication. There are no dietary restrictions with the tricyclics as is the case with MAOIs. However, the side effects of dry mouth, grogginess, and constipation often drive patients to stop treatment.

The discovery of the MAOIs and TCAs not only provided some of the first medicinal treatments for depression, but also seemed to provide insight into the nature and cause of the illness. In the case of the MAOIs, it was assumed that the increases in concentrations of monoamine neurotransmitters in the brain were responsible for their antidepressant effects. In turn, it stood to reason that if increasing the monoamines relieved depression, then depression itself must be caused by depletion of the brain's monoamines.

The TCAs were also thought to work by increasing the availability of monoamines in the brain. However, they did so by a different mechanism than the MAOIs. The TCAs have been thought to act primarily by blocking monoamine reuptake sites in the brain. After neurons release their neurotransmitters, they can take them back up through special portals to either be recycled or destroyed. The TCAs act primarily as blockers of those reuptake sites for serotonin and norepinephrine. Thus, it was concluded that they produced their antidepressant effects by increasing the levels of those neurotransmitters. This suspected mechanism of action lent further credence to the hypothesis that depression was caused by depletion of the monoamine transmitters, especially serotonin and norepinephrine.

Because of the dangers of the MAOIs and TCAs, scientists pursued medications that could treat MDD more safely. The first alternative to those older medications was fluoxetine. As early as 1972, fluoxetine was known to be a potent and selective blocker of serotonin reuptake sites in the brain. Because it did not have effects on MAO enzymes or on heart rhythm, hopes were high that it would act as a safe and effective antidepressant. Initial clinical trials were positive, and the drug became available as Prozac in the United States in 1987. Since that time, a variety of selective serotonin reuptake inhibitors, or SSRIs, have become available.

Many psychiatrists believe that the older MAOIs and TCAs may be slightly more effective medications for severe depression. Nonetheless, fluoxetine and subsequent SSRIs have gained predominance in frequency of use. A newer class of antidepressants, serotonin and norepinephrine reuptake inhibitors, or SNRIs, are now another commonly used type of antidepressant medication. As their name suggests, they increase the activity of both serotonin and norepinephrine in the brain, but without the dangers of overdose inherent in the older TCAs. The relatively pure noradrenergic reuptake inhibitor, reboxetine, is a poor antidepressant.[4]

It wasn't the discovery of low serotonin and norepinephrine levels in the brain that drove the search for antidepressant drugs to correct the low levels of those neurotransmitters. Rather, the antidepressants were discovered by accident, and the recognition of their ability to enhance monoaminergic activity led to the assumption that depression was caused by low levels of monoamines. The fact that some patients were helped by serotonin-selective medications suggested that it was depletion of serotonin in particular that led to symptoms of depression. However, there has never been strong, independent evidence that MDD is caused by low concentrations of serotonin.

First, there is no clear evidence that people suffering MDD actually have low levels of serotonin in their brains. Measurement of serotonergic activity inside the brains of living subjects has been a technically difficult task. Most of the early studies have examined levels of the main metabolite of serotonin, 5-hydroxyindoleacetic acid (5-HIAA), in the cerebrospinal fluid. Not all such studies have found the expected decreased levels of 5-HIAA. Moreover, when decreases are found, the degree of depletion has not correlated with severity of depression. Curiously, the cerebrospinal fluid of depressed patients treated with antidepressants has tended to contain decreased levels of 5-HIAA, suggesting that these medications reduced serotonergic activity.[5] For example, in a 2008 study, levels of the serotonin metabolite 5-HIAA were measured in blood coming from the brain in the jugular vein. It was found that those levels were higher in depressed patients, and that levels decreased after treatment with SSRIs.[6]

The time course of the effects of standard antidepressants has also been difficult to explain in the context of low serotonin theories of MDD. Antidepressants increase levels of serotonin fairly quickly. In rhesus monkeys, administration of the SSRI sertraline increases concentrations of serotonin in cerebrospinal fluid by 290% in only three hours.[7] However, when antidepressant effects are achieved in human patients, they may take weeks to materialize.[8] One would think that if a simple lack of serotonin were responsible for depression, then the replenishing of serotonin would quickly remedy the situation. Conversely, dietary depletion of serum levels of tryptophan, the precursor molecule of serotonin, does not cause symptoms of depression in individuals who have not previously suffered MDD, though it can cause relapse in those who have suffered the illness.[9]

Perhaps the most damning fact about the hypothesis that depletion of serotonin causes MDD is that antidepressant medications that increase serotonergic activity do not work well for all depressed patients. In a large study published in the *British Journal of General Practice* in 1998,[10] 60% of patients being treated with standard antidepressants continued to meet criteria for diagnosis of MDD after one year of treatment. In a more recent review of the effectiveness of antidepressants in primary care patients, the study found most patients, 56% to 60%, improved on those medications. However, 42% to 47% of patients also improved when merely treated with placebo. The suspicion that these medications are more effective in the hands of experienced psychiatrists was quashed by the STAR*D studies published in the *American Journal of Psychiatry*.[11] It was noted that only 36% of patients showed remission of symptoms after five weeks of treatment with the first antidepressant. Eventually, 67% of patients experienced remission after four different antidepressants were sequentially tried. Unfortunately, those that did not show a strong early response to antidepressant treatment relapsed at rates of up to 70% within a year despite remaining on the medication. All in all, it is safe to conclude that about a third of patients prescribed antidepressants do very well after a month or so, another third is helped somewhat, and a final unfortunate third is not helped at all. Those grim findings must also be viewed in light of the fact that many individuals experience distressing side effects of these medications, including sexual dysfunction, nausea, dry mouth, insomnia, weight gain, fatigue, and others.

That depletion of serotonin in the brain is not the underlying cause of MDD is not to say that serotonergic activity is irrelevant in the etiology and treatment of the illness. Whereas there is no compelling evidence that serotonin itself is depleted in depressed humans and animals, there is evidence that the activities of certain subtypes of serotonin receptors are abnormal in depression. For example, the density of 5-HT1A receptors is decreased in the frontal cortex and hippocampus of individuals suffering MDD.[12] Despite some inconsistency in the literature, there is evidence that 5-HT2A receptor density decreases in the brains of depressed patients.[13] Receptors often downregulate due to overactivity, and many atypical antipsychotics used to augment the effects of antidepressants, e.g., olanzapine,[14] quetiapine ,[15] and aripiprazole ,[16] act as potent 5-HT2A receptor antagonists. There is also growing evidence that activity at 5-HT3[17] and 5-HT7[18] receptors may exacerbate MDD. The MAOI, TCA, SSRI, and SNRI antidepressants have helped people, and it would be imprudent and wrong to dismiss them. Indeed, as will be discussed, many herbs with antidepressant effects have been shown to enhance monoaminergic activity, and this may contribute to their therapeutic effects. Nonetheless, it is becoming clear that some antidepressants, including some herbs, can relieve depression by means that have no direct effects on serotonergic, noradrenergic, or dopaminergic activity. Such changes in monoaminergic activity appear to be neither necessary nor sufficient for antidepressant effects.

REFERENCES

1. López-Muñoz F, Alamo C. Monoaminergic neurotransmission: the history of the discovery of antidepressants from 1950s until today. *Curr Pharm Des.* 2009;15:1563–1586.
2. Simpson GM, White K. Tyramine studies and the safety of MAOI drugs. *J Clin Psychiatry.* 1984;45(7):59–61.
3. Boschmans SA, Perkin MF, Terblanche SE. Antidepressant drugs: imipramine, mianserin and trazodone. *Comp Biochem Physiol C: Comp Pharmacol Toxicol.* 1987;86(2):225–232.
4. Cipriani A, Furukawa TA, Salanti G, et al. Comparative efficacy and acceptability of 12 new-generation antidepressants: a multiple-treatments meta-analysis. *The Lancet.* 2009;373(9665):746–758.
5. Mann IJ. Role of the serotonergic system in the pathogenesis of major depression and suicidal behavior. *Neuropsychopharmacology.* 1999;21:99S–105S.
6. Barton DA, Esler MD, Dawood T, et al. Elevated brain serotonin turnover in patients with depression: effect of genotype and therapy. *Arch Gen Psychiatry.* 2008;65(1):38–46.
7. Anderson GM, Barr CS, Lindell S, et al. Time course of the effects of the serotonin-selective reuptake inhibitor sertraline on central and peripheral serotonin neurochemistry in the rhesus monkey. *Psychopharmacology.* 2005;178(2–3):339–346.

8. van Calker D, Dykierek P, Deimel CM, et al. Time course of response to antidepressants: predictive value of early improvement and effect of additional psychotherapy. *J Affect Disord.* 2009;114(1–3):243–253.
9. Delgado PL, Price LH, Miller HL, et al. Serotonin and the neurobiology of depression: effects of tryptophan depletion in drug-free depressed patients. *Arch Gen Psychiatry.* 1994;51(11):865–874.
10. Goldberg D, Privett M, Ustun B, et al. The effects of detection and treatment on the outcome of major depression in primary care: a naturalistic study in 15 cities. *Br J Gen Pract.* 1998;48(437):1840–1844.
11. Rush AJ, Trivedi MH, Wisniewski SR, et al. Acute and longer-term outcomes in depressed outpatients requiring one or several treatment steps: a STAR*D report. *Am J Psychiatry.* 2006;163:1905–1917.
12. López-Figueroa AL, Norton CS, López-Figueroa MO, et al. Serotonin 5-HT1A, 5-HT1B, and 5-HT2A receptor mRNA expression in subjects with major depression, bipolar disorder, and schizophrenia. *Biol Psychiatry.* 2004;55(3):225–233.
13. Yatham LN, Liddle PF, Shiah IS, et al. Brain serotonin2 receptors in major depression: a positron emission tomography study. *Arch Gen Psychiatry.* 2000;57(9):850–858.
14. Zhang W, Bymaster FP. The *in vivo* effects of olanzapine and other antipsychotic agents on receptor occupancy and antagonism of dopamine D1, D2, D3, 5HT2A and muscarinic receptors. *Psychopharmacology.* 1999;141(3):267–278.
15. Jones HM, Travis MJ, Mulligan R, et al. *In vivo* 5-HT2A receptor blockade by quetiapine: an R91150 single photon emission tomography study. *Psychopharmacology.* 2001;157(1):60–66.
16. Stark AD, Jordan S, Allers KA, et al. Interaction of the novel antipsychotic aripiprazole with 5HT1A and 5-HT2A receptors: functional receptor-binding and *in vivo* electrophysiological studies. *Psychopharmacology.* 2007;190(3):373–382.
17. Rajkumar R, Mahesh R. Review: the auspicious role of the 5-HT3 receptor in depression: a probable neuronal target? *J Psychopharmacol.* 2010;24(4):455–469.
18. Guscott M, Bristow LJ, Hadingham K, et al. Genetic knockout and pharmacological blockade studies of the 5-HT7 receptor suggest therapeutic potential in depression. *Neuropharmacology.* 2005;48(4):492–502.

3 Clues Revealed by Ketamine

Ketamine has a unique ability to quickly reverse severe depression in individuals who have failed standard treatments. This is likely one of the most important developments in psychiatry in the past 50 years. Curiously, ketamine is not new, nor it was ever intended for use in psychiatry. It was first synthesized in 1962, and since 1970 it has been used in the United States as a surgical anesthetic. It has commonly been used in Emergency Departments for procedures on small children. This is owing to its ability to produce "conscious sedation," a trance-like state in which the central nervous system is cut off from sensations of pain, sight, and sound, but spontaneous respirations and protective airway reflexes are maintained.[1] It is considered to be safer than general anesthesia.

One of the drawbacks in using ketamine for anesthesia has been the frequent reports of adverse psychiatric side effects.[2] Dissociation and depersonalization are often reported by people administered the drug. These people feel unreal and separate from themselves. So-called "out of body experiences" may occur. These effects can be extremely frightening.

It is not surprising that ketamine can produce bizarre mental experiences, as the drug is similar in its action to phencyclidine, or PCP. PCP has no recognized medical use but is used by some individuals to get high. Indeed, the hallucinogenic effects of PCP have been seen as similar to the signs and symptoms of schizophrenia. Like PCP, ketamine has been used "recreationally" and has been known as a "club drug" with the name "Special K."[3] Descriptions of the ketamine experience gleaned from online conversations among recreational users include:

"It's the one of most insanely bizarre and alien experiences."

"Basically, the outside world either ceases to exist, or just seems flat out distant or unreal. Everything you know about yourself and reality fades away in a very difficult to explain process."

"The room gets really big ... like it's the size of a football field, and the furniture in the middle of the room is very tiny. Emotionally and mentally there isn't much too it, unless you 'hole,' then you think you are God and the universe has ended."

The term, "hole" or "fall through the K-hole" refers to the point of ego dissolution that some experience with high doses of the drug. Considering that ketamine is an old anesthetic known to produce frightening, even psychotic, symptoms in some people, it has seemed an unlikely candidate for a psychiatric breakthrough. However, at doses lower than those generally used for anesthesia – or out of intent to "fall through the K-hole" – the drug can produce remarkable effects in individuals suffering severe, intractable depression.

Researchers had been aware that ketamine and similar drugs were effective in animal models of depression. However, the first formal study of the effects of ketamine in patients with MDD was described in 2000.[4] Results showed that over half of patients treated with ketamine enjoyed a 50% or more reduction in symptoms within 72 hours. When standard antidepressants are effective, symptoms of depression are generally seen to diminish only after several weeks, with the maximal benefits often not seen for a month or more. With refinement of methods over the last 18 years, ketamine has been found to produce moderate-to-large effects in about 70% of depressed patients that have not responded to standard treatment. About 30% of patients are completely relieved of symptoms within 24 hours.

Ketamine has no direct effects on serotonin, and thus its action cannot be attributed to correction of what was once considered to be the cause of MDD, that is, a deficit of serotonin in the brain. Ketamine acts primarily as an antagonist at the N-methyl-D-aspartate (NMDA) receptor.[5] However, glutamate is the most abundant neurotransmitter in the human brain, and the NMDA receptor is only one of at least a dozen subtypes of such receptors.[6] It appears that blockade of the NMDA receptor, possibly situated on GABAergic interneurons, disinhibits glutamate transmission with the net result of activation of another glutamate receptor, the AMPA receptor.[7] Thus, the antidepressant effect of ketamine is the result of a series of biochemical events that occur downstream from its action at the NMDA receptor and, indirectly, at the AMPA receptor.

The simplified version of neuronal activity is that neurotransmitters activate receptors on the surface membranes of neurons in the brain and turn those cells on or off like light switches. In fact, the activations of surface receptors by neurotransmitters are only the starting places of long and complicated cascades of intracellular activity. Receptor activation triggers enzymes. One enzyme affects another, then another, with the results being altered by what other enzymes may have interacted inside the neuron microseconds before. The end results of those various biochemical cascades are what drive and direct neurons to express genes; synthesize proteins; repair themselves; utilize energy; make neurotransmitters and receptors; and maintain the critical dendritic processes through which they communicate with each other.

A recent paper described evidence for three main actions of ketamine in the intracellular machinery.[8] Ketamine activates the mammalian target of rapamycin (mTOR), inhibits glycogen synthase kinase 3beta (GSK-3β), and disinhibits eukaryotic elongation factor 2 (eEF2) which is essential for the synthesis of proteins by cells. A fourth essential component of the antidepressant effect of ketamine is the production and release of brain-derived neurotrophic factor (BDNF), which plays a major role in maintaining the health and structural integrity of neurons in the brain.[9] The exact course of events involving these four substances that mediate the antidepressant effect of ketamine is not entirely clear. However, it is suspected that multiple intracellular pathways may be activated, with convergent and synergistic effects taking place. For example, whereas activation of mTOR by ketamine can lead to increased synthesis of BDNF, BDNF itself is stimulated by ketamine and is an activator of mTOR. Moreover, whereas mTOR stimulates protein synthesis, ostensibly through downstream disinhibition of eEF2, blockade of NMDA activity by ketamine may directly disinhibit the eEF2 necessary for protein synthesis.[10]

The intracellular mediators of ketamine, mTOR, GSK-3β, and eEF2, are evolutionarily ancient and deeply embedded in cell biology. mTOR and GSK-3β, in particular, are major biochemical hubs and play critical roles in many intracellular signaling pathways. The stimulation of mTOR is essential for the antidepressant effect of ketamine and has been shown to increase synaptic signaling proteins and the number and function of new spine synapses in the prefrontal cortex of rats.[11] However, mTOR also plays a role in many cellular functions. It is critical in maintaining a balance between cell growth and apoptosis, which is the programmed process of cell death in the brain and other organs. It regulates neuronal responses to insulin and changes in availability of nutrients such as glucose, fat, and amino acids. It regulates protein synthesis and plays an important role in learning and memory. mTOR is activated by insulin, glutamate, and a variety of cellular growth factors, including BDNF. However, mTOR also plays a role in the inflammatory response,[12] in the defense against oxidative damage,[13] and in managing the glucocorticoid stress response.[14]

Kinases, such as GSK-3β, are a class of enzymes that act by donating a phosphate group to other enzymes and proteins inside the cell and thereby activating or inactivating them. The name GSK-3β seems to have been poorly chosen. Along with phosphorylating and thus inhibiting glycogen synthase – the action from which its name is derived – GSK-3β also phosphorylates dozens of other proteins and enzymes inside the cell.[15] Some of the targets of GSK-3β turn genes on or off, whereas others modify various cell-signaling cascades occurring within the cells. Certainly, one of the most important roles of GSK-3β is being part of the cellular response to insulin. Insulin inhibits the activity of GSK-3β and thereby turns on the cellular machinery to store energy by producing glycogen. Although neurons do not produce glycogen, GSK-3β remains an important component of the so-called "insulin response system" of neurons. Current evidence suggests that resistance to insulin in the brain contributes to mood disorders.[16] Although deeply involved in the insulin response, GSK-3β also helps activate the inflammatory response,[17] is stimulated by oxidative stress,[18] and is activated by glucocorticoid.[19]

Although ketamine appears unique in its ability to quickly reverse symptoms of depression, it is not unique in affecting mTOR, GSK-3β, eEF2, and BDNF. Although the most common antidepressants – the TCAs, SSRIs, SNRIs, and MAOIs – have been thought to act by increasing the monoamine neurotransmitters, particularly serotonin, many of those medications have also recently been found to act upon some of the same intracellular systems as ketamine. There are reports that some antidepressants – escitalopram, paroxetine, and tranylcypromine – stimulate mTOR, while others – fluoxetine, sertraline, and imipramine – do not.[20] Like ketamine, the antidepressants citalopram, fluoxetine, and venlafaxine inhibit GSK-3β.[21] Curiously, both fluoxetine and imipramine inhibit GSK-3β in mouse brains within only a few hours after treatment, which makes one wonder why those drugs don't produce the rapid antidepressant effects seen with ketamine.[22] Like ketamine, antidepressants of various types, such as tranylcypromine, sertraline, desipramine, and mianserin, increase

BDNF in rat dentate gyrus and frontal cortex.[23] Antidepressants also increase levels of BDNF in the brains of human subjects.[24]

When an antidepressant doesn't work, or if it is only partially effective, it is common practice to add a second medication as an augmentation strategy. Among the most common medications that are added to antidepressants are lithium, the so-called "atypical antipsychotics," thyroid hormone, and lamotrigine. None of these augmentations cause increases in levels of serotonin or other monoamine neurotransmitters. However, all produce effects in neuronal intracellular chemistry similar to those of ketamine.

Lithium was used as a treatment of both depression and mania in the 1880s. It was often used in the form of a bromide, with both lithium and bromine contributing to the therapeutic effects. After falling out of favor, it came back into use in the 1940s after John Cade discovered its powerful antimanic effects. However, its mechanism of action remained unknown until modern techniques of molecular biology unraveled its neurochemical mysteries. It has been learned that, like ketamine, lithium inhibits GSK-3β.[25] It also increases activity of mTOR,[26] stimulates eEF2,[27] and stimulates production and release of BDNF.[28]

The atypical antipsychotics, such as aripiprazole, olanzapine, quetiapine, risperidone, and others, produce their antidepressant effects by blocking both dopamine type 2 and serotonin type 2 receptors. Together, those effects combine to activate the Akt signaling system within neurons and, in turn, activate mTOR and deactivate GSK-3β.[29] Thus, the augmentation with atypical antipsychotics can hasten and enhance the effects of antidepressants by producing effects similar to those of ketamine.

The active form of the thyroid hormone, triiodothyronine, is often successful as an augmentation when antidepressants are ineffective.[30] Among its effects on neurons in the brain are decreases in GSK-3β, increases in BDNF,[31] and activation of mTOR.[32] Lamotrigine can also be effective when added to an antidepressant in patients with treatment-resistant MDD.[33] Lamotrigine increases BDNF, decreases activity of GSK-3β,[34] and increases activity of mTOR.[35] Tianeptine, an antidepressant that enhances serotonin reuptake, also induces many of the same neuronal changes as the SSRIs. Both tianeptine and fluoxetine inhibit activation of GSK-3β in cortical neurons of prenatally stressed mice.[36] Both fluoxetine[37] and tianeptine[38] increase the activity of mTOR. Finally, both fluoxetine and tianeptine reverse stress and inflammation-induced decreases in neuronal BDNF.[39]

It appears that the feature common among drugs that improve depression is not enhancement of serotonin or the other monoamines, but rather the ability to activate the mTOR pathway, inhibit GSK-3β, and enhance production of BDNF and other proteins that maintain the integrity of the neurons. The question remains as to what conditions upset the balance of the activities of those intracellular systems that then results in MDD.

REFERENCES

1. Craven R. Ketamine. *Anesthesia.* 2007;62(1):48–53.
2. Morgan CJA, Mofeez A, Brandner B, et al. Acute effects of ketamine on memory systems and psychotic symptoms in healthy volunteers. *Neuropsychopharmacology.* 2004;29:208–218.

3. Muetzelfeldt L, Kamboj SK, Rees H, et al. Journey through the K-hole: phenomeno-logical aspects of ketamine use. *Drug Alcohol Depend.* 2008;95(3):219–229.
4. Berman RM, Cappiello A, Anand A, et al. Antidepressant effects of ketamine in depressed patients. *Biol Psychiatry.* 2000;47(4):351–354.
5. Hirota K, Lambert DG. Ketamine: its mechanism(s) of action and unusual clinical uses. *Br J Anaesth.* 1966;77(4):441–444.
6. Willard SS, Koochekpour S. Glutamate, glutamate receptors, and downstream signal-ing pathways. *Int J Biol Sci.* 2013;9(9):948–959.
7. Moghaddam B, Adams B, Verma A, et al. Activation of glutamatergic neurotransmis-sion by ketamine: a novel step in the pathway from NMDA receptor blockade to dopa-minergic and cognitive disruptions associated with the prefrontal cortex. *J Neurosci.* 1997;17(8):2921–2927.
8. Duman RS, Aghajanian GK, Sanacora G, et al. Synaptic plasticity and depression: new insights from stress and rapid-acting antidepressants. *Nat Med.* 2016;22(3):238–249.
9. Lepack AE, Fuchikami M, Dwyer JM, et al. BDNF release is required for the behav-ioral actions of ketamine. *Int J Neuropsychopharmacol.* 2015;18(1):1–6.
10. Zanos P, Thompson SM, Duman RS, et al. Convergent mechanisms underlying rapid antidepressant action. *CNS Drugs.* 2018;32(3):197–227.
11. Li N, Lee B, Liu RJ, et al. mTOR-dependent synapse formation underlies the rapid antidepressant effects of NMDA antagonists. *Science.* 2010;329(5994):959–964.
12. Maya-Monteiro C, Bozza P. Leptin and mTOR: partners in metabolism and inflamma-tion. *Cell Cycle.* 2008;7(12):1713–1717.
13. Filomeni G, De Zio D, Cecconi F. Oxidative stress and autophagy: the clash between damage and metabolic needs. *Cell Death Differ.* 2015;22:377–388.
14. Pazini FL, Cunha MP, Rosa JM, et al. Creatine, similar to ketamine, counteracts depressive-like behavior induced by corticosterone via PI3K/Akt/mTOR pathway. *Mol Neurobiol.* 2016;53(10):6818–6834.
15. Cohen P, Frame S. The renaissance of GSK3. *Nat Rev Mol Cell Biol.* 2001;2:769–776.
16. Krabbe KS, Nielsen AR, Krogh-Madsen R, et al. Brain-derived neurotrophic factor (BDNF) and type 2 diabetes. *Diabetologia.* 2007;50(2):431–438.
17. Jope RS, Yuskaitis CJ, Beurel E. Glycogen synthase kinase-3 (GSK3): inflammation, diseases, and therapeutics. *Neurochem Res.* 2007;32(4–5):577–595.
18. King TD, Jope RS. Inhibition of glycogen synthase kinase-3 protects cells from intrin-sic but not extrinsic oxidative stress. *Neuroreport.* 2005;16(6):597–601.
19. Jope RS, Cheng Y, Lowell JA, et al. Stressed and inflamed, can GSK3 be blamed? *Trends Biochem Sci.* 2017;42(3):180–192.
20. Park SW, Lee JG, Seo MK, et al. Differential effects of antidepressant drugs on mTOR signalling in rat hippocampal neurons. *Int J Neuropsychopharmacol.* 2014;17(11):1831–1846.
21. Okamoto H, Voleti B, Banasr M, et al. Wnt2 expression and signaling is increased by different classes of antidepressant treatments. *Biol Psychiatry.* 2010;68(6):521–527.
22. Li X, Zhu W, Roh MS, et al. *In vivo* regulation of glycogen synthase kinase-3β (GSK-3β) by serotonergic activity in mouse brain. *Neuropsychopharmacology.* 2004;29(8):1426–1431.
23. Nibuya M, Morinobu S, Duman RS, et al. Regulation of BDNF and trkB mRNA in rat brain by chronic electroconvulsive seizure and antidepressant drug treatments. *J Neurosci.* 1995;75(11):7539–7547.
24. Chen B, Dowlatshahi D, MacQueen GM, et al. Increased hippocampal BDNF immu-noreactivity in subjects treated with antidepressant medication. *Biol Psychiatry.* 2001;50(4):260–265.
25. Lenox RH, Wang L. Molecular basis of lithium action: integration of lithium-respon-sive signaling and gene expression networks. *Mol Psychiatry.* 2003;8:135–144.

26. Chiu C-T, Scheuing L, Liu G, et al. The mood stabilizer lithium potentiates the antide-pressant-like effects and ameliorates oxidative stress induced by acute ketamine in a mouse model of stress. *Int J Neuropsychopharmacol.* 2015;18(6):18. doi:10.1093/ijnp/pyu102.

27. Karyo R, Eskira Y, Pinhasov A, et al. Identification of eukaryotic elongation factor-2 as a novel cellular target of lithium 2 and glycogen synthase kinase-3. *Mol Cell Neurosci.* 2010;45(4):449–455.

28. Fukumoto T, Morinobu S, Okamoto Y, et al. Chronic lithium treatment increases the expression of brain-derived neurotrophic factor in the rat brain. *Psychopharmacology.* 2001;158(1):100–106.

29. Li X, Rosborough KM, Friedman AB, et al. Regulation of mouse brain glycogen syn-thase kinase-3 by atypical antipsychotics. *Int J Neuropsychopharmacol.* 2007;10:7–19.

30. Aronson R, Offman HJ, Joffe RT, et al. Triiodothyronine augmentation in the treatment of refractory depression: a meta-analysis. *Arch Gen Psychiatry.* 1996;53(9):842–848.

31. Prieto-Almeida F, Panveloski-Costa AC, Crunfli F, et al. Thyroid hormone improves insulin signaling and reduces the activation of neurodegenerative pathway in the hip-pocampus of diabetic adult male rats. *Life Sci.* 2018;192(1):253–258.

32. Varela L, Martínez-Sánchez N, Gallego R, et al. Hypothalamic mTOR pathway mediates thyroid hormone-induced hyperphagia in hyperthyroidism. *J Pathol.* 2012;227(2):209–222.

33. Rocha FL, Hara C. Lamotrigine augmentation in unipolar depression. *Int Clin Psychopharmacol.* 2003;18(2):97–99.

34. Abelairaa HM, Réus GZ, Ribeiro KF, et al. Effects of acute and chronic treatment elicited by lamotrigine on behavior, energy metabolism, neurotrophins and signaling cascades in rats. *Neurochem Int.* 2011;59(8):1163–1174.

35. Ignácio ZM, Réus GZ, Arent CO, et al. New perspectives on the involvement of mTOR in depression as well as in the action of antidepressant drugs. *Br J Clin Pharmacol.* 2016;82(5):1280–1290.

36. Szymańska M, Suska A, Budziszewska B, et al. Prenatal stress decreases glyco-gen synthase kinase-3 phosphorylation in the rat frontal cortex. *Pharmacol Rep.* 2009;61(4):612–620.

37. Chen Z, Wang Z, Pan W, et al. Fluoxetine regulates hippocampal synaptic plasticity in CUMS depression rats. *Chin J Pathophysiol.* 2016;32(9):1642–1647.

38. Seo MK, McIntyre RS, Cho HY, et al. Effects of tianeptine on mTOR signaling in rat hippocampal neurons. *Int J Neuropsychopharmacol.* 2016;19(Suppl 1):39.

39. Nowacka-Chmielewska MM, Kasprowska D, Paul-Samojedny M, et al. The effects of desipramine, fluoxetine, or tianeptine on changes in bulbar BDNF levels induced by chronic social instability stress and inflammation. *Pharmacol Rep.* 2017;69(3):520–525.

4 New Understanding of the Nature and Causes of Major Depression

It had long been thought that MDD was a manifestation of low levels of serotonin and, to some degree, norepinephrine and dopamine in the brain. However, antidepressant treatments directed toward correcting such deficits in monoaminergic activity have been only marginally effective. Moreover, it has recently been shown that rapid improvement in even treatment-resistant MDD can occur after administration of ketamine, a treatment that has little to do with increasing monoaminergic activity. Rather, ketamine acts on the glutamatergic systems in the brain, and the necessary components of its therapeutic effect are activation of mTOR, inhibition of GSK-3β, disinhibition of eEF2, and increases in the activity of BDNF. Review of the mechanisms of action of various other antidepressants and augmentation medications show that these effects are common among them. Thus, we must ask, "What are the underlying causes of MDD, and how are those causes reflected in our new understanding of the changes in neuronal chemistry that must occur for treatment to be successful?"

Among the conditions that can cause both depressed mood and aberrations in the intracellular chemistry of neurons are oxidative and nitrosative damage, inflammation, prolonged stress, and insulin resistance.[1] These conditions often interact in the brain and periphery. These conditions are also often seen together in the overarching condition referred to as Metabolic Syndrome. This condition has some genetic basis, but more commonly arises out of poor diet, stress, and unwise lifestyle choices.[2]

4.1 OXIDATIVE AND NITROSATIVE DAMAGE

Increases in oxidative and nitrosative stress are seen in MDD. Sufferers of MDD tend to have depleted stores of antioxidants, including zinc, coenzyme Q10, vitamin E, and glutathione. Such individuals also tend to show biomarkers indicative of oxidative and nitrosative damage. They exhibit signs of increased lipid peroxidation, oxidized and nitrosylated proteins, and oxidative damage to DNA and mitochondria. Serum levels of 8-Hydroxy-2′-deoxyguanosine and F2-isoprostanes, measures of oxidative DNA and lipid damage respectively, have been found to be increased.[3] Antibodies directed against redox modified nitrosylated proteins and oxidative specific epitopes are also seen in serum.[4]

Some of the increases in oxidative and nitrosative stress in MDD may result from genetic predisposition. For example, associations have been made between depression and certain polymorphisms in genes affecting the enzymatic activities of manganese superoxide dismutase and catalase.[5] Some, but not all, researchers have seen decreases in the activity of antioxidant enzymes, such as glutathione peroxidase, during episodes of depression.[6] Severe chronic or repeated psychological stress increase blood levels of oxidation biomarkers, and these changes correspond to increases in salivary cortisol.[7] Metabolic Syndrome is also associated with oxidative stress, due in part to disturbances in energy metabolism.[8] Inflammatory conditions cause oxidative and nitrosative damage and, in turn, debris from cells damaged by oxidative stress stimulates TLR4 receptors that initiate the inflammatory cascade.[9]

The increase in oxidative stress in MDD is due to an increase in production of reactive oxidative species (ROS) as well as deficits in the neutralization of ROS. Recently, attention is being paid to the potential role of mitochondrial dysfunction in MDD. Mitochondria are the source of ATP, the molecular currency of energy in cells that is produced by the process of oxidative phosphorylation. Mitochondrial oxidative phosphorylation involves respiratory complexes that donate electrons to oxygen, and in that process necessarily produce superoxide anions and peroxides. In animal models of depression, deficits that predispose to loss of control of oxidative phosphorylation can be seen in the protein complexes of the electron transport chain of brain.[10] In humans with MDD, increased levels of cortisol stimulate activity of mitochondria, which increases production of ROS.[11] Glucocorticoids also inhibit the Nrf2-dependent antioxidant pathways that would otherwise help control oxidative damage, thus compounding the problem.[12] Accumulation of ROS can damage mitochondria and lead to fragmentation of those organelles.[13] The NLRP inflammasome is also activated by mitochondrial dysfunction and ROS, both of which are caused by oxidative stress.[14] Deficits in Nrf2 contribute to depression due to oxidative damage stimulating inflammatory processes. However, those effects are dampened by the anti-inflammatory COX-2 inhibitor, rofecoxib. Nrf2 knock-out mice are susceptible to depression-like syndromes, whereas substances such as the phytochemical sulforaphane that induce Nrf2 have antidepressant-like effects in rodents.[15]

Some antioxidants have antidepressant effects. Among other effects, N-acetylcysteine increases levels of the natural antioxidant glutathione in the brain. It has been found to have beneficial effects in both bipolar and unipolar depression.[16] α-Tocopherol, also known as vitamin E, shows an antidepressant-like effect in mice by reducing immobility in the forced swim test. It is a powerful antioxidant and increases activities of glutathione peroxidase and glutathione reductase activity in the hippocampus and prefrontal cortex. It also increases levels of glutathione.[17] However, despite the existence of low serum levels of vitamin E in depressed patients, there is no compelling evidence of antidepressant effects of the substance in human patients. However, a study in children found that the addition of vitamin C, a strong antioxidant, significantly enhanced the antidepressant effect of fluoxetine.[18]

Most antidepressant medications themselves possess significant antioxidant effects *in vitro* and *in vivo*. Among those antidepressants are amitriptyline, bupropion, citalopram, desipramine, duloxetine, fluoxetine, fluvoxamine, imipramine, maprotiline, milnacipran, mirtazapine, moclobemide, nefazodone, nortriptyline, paroxetine, reboxetine, sertraline, tianeptine, trazodone, and venlafaxine.[19] Thus, virtually every class of antidepressant exhibits this effect. As examples, fluoxetine attenuated oxidative damage in rat pheochromocytoma cells that were exposed to H_2O_2, and also spared the antioxidant enzyme superoxide dismutase.[20] It also suppressed interferonγinduced production of nitric oxide in cultured microglia.[21] Chronic or acute treatment with imipramine increased superoxide dismutase and catalase activities in the prefrontal cortex and hippocampus and decreased lipid and protein oxidative damage in those areas.[22] Chronic treatment with venlafaxine decreased MDA and total nitrite levels in the hippocampus, while maintaining levels of glutathione and the antioxidant enzyme, glutathione S-transferase. It also reduced hippocampal levels of 8-OHdG, a marker of oxidative DNA damage.[23]

In the current context, it is important to note that oxidative damage decreases levels of BDNF, whereas antioxidants of various types restore BDNF in a manner similar to that of ketamine. For example, it has been found that diets rich in antioxidants increase BDNF in temporal cortex of aging dogs.[24] The carotenoid pigment, astaxanthin, a potent antioxidant, restored the activities of glutathione peroxidase and superoxide dismutase, and increased glutathione in the brains of aged rats. It decreased concentrations of the oxidation biomarkers malondialdehyde, protein carbonylation, and 8-hydroxy-2- deoxyguanosine. It also increased levels of BDNF in the hippocampus and other brain areas of those animals.[25] The antioxidant vitamin E, or α-tocopherol, increased levels of BDNF in brains of rats subjected to toxic levels of phenytoin.[26]

4.2 INFLAMMATION

Inflammation is a complicated series of chemical and cellular reactions the body performs to defend itself against a variety of insults. It is a non-specific form of body defense that likely predates the more sophisticated and precise actions of the immune system, such as the production of antibodies against specific proteins and bacteria. When initial inflammatory steps, such as release of histamine and eicosanoids, fail to control irritation and infection, the body changes course and switches to a different set of substances and defensive cells. Primary among such cells are the macrophages. In brain tissue, the microglia play the primary role in generating the inflammatory response. These cells migrate to damaged tissue and release a plethora of inflammatory substances.[27] Both neuronal and non-neuronal cells in the brain express receptors for these mediators of inflammation.[28] Elevations of the inflammatory cytokines, TNF-α, IL-1, IL-6, IL-4, IL-2, IL-8, IL-10, and IFN-γ are seen in the blood and brain tissues of patients with MDD.[29]

Resident microglia are the primary source of neuroinflammatory effects in the brain. However, peripheral inflammation also affects the brain. Peripheral cytokines activate primary afferent nerves, such as the vagus nerve, that in turn affects the brain. Macrophage-like cells residing in the circumventricular organs and the choroid plexus – the porous boundary between the brain and periphery – respond to inflammatory stimuli and release cytokines into the brain. Inflammatory cytokines can also be actively transported across the blood–brain barrier into brain tissue. Finally, activation of perivascular macrophages of brain venules can release eicosanoids into brain tissue. In any case, it is well established that peripheral administration of potent inflammatory agents, such as lipopolysaccharide, can later induce the expression of IL-1β and other pro-inflammatory cytokine mRNAs and proteins in the brain.[30]

Suspicions that inflammation might be involved in MDD arose from seeing similarities between the illness and what is referred to as "sickness behavior." The syndrome of sickness behavior includes malaise, fatigue, lack of motivation, decrease in appetite, withdrawal from the environment, and depressed mood. Ostensibly, these symptoms of illness serve to force us to rest and keep us away from dangers we may be incapable of surviving because of loss of strength and vitality. Sickness behavior is due to the effects of cytokines. One of the first cytokines to be identified was alpha-interferon. Alpha-interferon stimulates the immune system to fight off viral infection, and it was one of the first successful treatments of the viral infection hepatitis C. However, the most common and predictable side effects of interferon treatment were depressed mood, malaise, fatigue, irritability, insomnia, poor memory, and difficulty concentrating.[31] Indeed, psychiatrists were often enlisted into treatment teams to help manage the sometimes severe psychiatric side effects of interferon. Individuals already suffering MDD, or those with histories of the condition, were sometimes eliminated as candidates for interferon therapy.

Along with high serum levels of inflammatory cytokines, sufferers of MDD have also been found to have high serum levels of C-reactive protein, a protein released by the liver during episodes of stress and illness.[32] The serum levels of these substances return toward normal with effective treatment with antidepressants. Acute psychological stress has been found to increase serum levels of the inflammatory cytokines IL-1 and IL-6, and to some extent CRP in human subjects.[33] Autopsies have found that depressed patients that commit suicide, an event likely to mark the apex of psychological stress, have increased densities of activated microglial cells in their frontal cortex.[34]

Inflammation affects intracellular processes in neurons in ways opposite to that of antidepressant doses of ketamine. For example, both TNF-α[35] and IFN-γ[36] stimulate GSK-3β. In turn, GSK-3β activity is necessary for maximal production of IL-6, IL-1β, and TNF-α, in monocytes and peripheral blood mononuclear cells.[37] GSK-3β also enhances release of IL-1β, TNF-α, and IL-10 from cortical glial cells.[38] In mice, stress activates GSK-3β in the hippocampus. However, this does not occur in mice in which TLR4, the receptor that initiates much of the inflammatory response, has been inactivated.[39] Thus a stress-induced inflammatory response enhances GSK-3β, whereas ketamine and most antidepressants inhibit it.

TNF-α reduces the activity of BDNF in the brain. This occurs when the primary source of inflammation is either in the brain or the periphery.[40] The inflammatory cytokine IL-1β also reduces levels of BDNF in the hippocampi of mice. This effect of IL-1β is caused not only by the usual experimental technique of injecting lipopolysaccharide into the animal, but also by social isolation.[41] This suggests that inflammation is a more general process than simply fighting infection and injury. Indeed, in the mammalian brain, inflammation may have evolved to also play a role in buffering against social and emotional injury as well.[42]

Effects of inflammation on mTOR are complex. One arm of mTOR activity plays an important role in neuroinflammation. Activation of the TLR4 can activate mTOR that in turn can activate NF-κB, a major player in the inflammatory process. Thus, in some contexts, mTOR is pro-inflammatory, which seems at odds with the notion of inflammation being a factor in the etiology of depression.[43] The complexity of the intracellular signaling pathways is evident in the fact that while activation of mTOR is an essential component of the ketamine-induced antidepressant effect, rapamycin, an antagonist of mTOR, can itself exhibit antidepressant-like effects in rodents.[44]

There is evidence that anti-inflammatory agents, such as NSAIDs, may exert antidepressant effects. For example, the addition of acetylsalicylic acid to an existing SSRI helped a large percentage of nonresponding patients achieve response and even remission.[45] Addition of the cyclooxygenase-2 inhibitor celecoxib to reboxetine also enhanced the otherwise weak antidepressant effects of that medication.[46] Such antidepressant effects may be confounded with effects such as pain relief, which would have little bearing on inflammation being part of the pathology of MDD. Nonetheless, mild antidepressant effects have been seen with anti-inflammatory cytokine inhibitors infliximab, adalimumab, ustekinumab, and etanercept, as well as the anti-inflammatory antibiotic minocycline, that are not thought to have any specific pain-relieving qualities.[47] In any case, it has been observed that cases in which there is clear evidence of involvement of inflammation in depression, such as the depression that may occur during treatment with interferon, may provide the most prudent context for consideration of antidepressant effects of anti-inflammatory agents.[48]

Although standard antidepressants are thought to work primarily by enhancing monoaminergic activity, many have been found to exert anti-inflammatory effects. For example, the antidepressants fluoxetine, sertraline, paroxetine, fluvoxamine, and citalopram all attenuated the production of TNFα and nitric oxide by microglia stimulated by lipopolysaccharide.[49] Imipramine, venlafaxine, and fluoxetine all suppressed the ratio of inflammatory IFN-γ to anti-inflammatory IL-10 in blood samples to which the inflammation-inducing agents phytohemagglutinin and lipopolysaccharide were added.[50] Bupropion has also been found to reduce synthesis of TNF-α, IL-1β, and IFN-γ in a mouse model of lipopolysaccharide stimulation.[51] The antidepressants amitriptyline, imipramine, clomipramine, trazodone, and fluoxetine all attenuated carrageenan-induced paw edema in rats, a standard assay of anti-inflammatory action. Curiously, sertraline enhanced edema in that test.[52] Levels of TLR4, the receptors that mediate the inflammatory effect of lipopolysaccharide, have been found to be increased in peripheral blood mononuclear cells and postmortem brains of depressed and suicidal patients. However, levels of the entire family of inflammation-mediating toll-like receptors were normalized on peripheral blood mononuclear cells of depressed patients treated with antidepressants.[53]

4.3 CHRONIC STRESS

When the body is met with unusual demands from the environment, it generates the "stress response" to maintain balance and resiliency. Stress is unavoidable in this life, and stress, *per se*, is neither good nor bad. However, unrelenting stress eventually causes damage to the body and brain. Prolonged exposure to high levels of the primary stress hormone cortisol can lead to obesity, high blood glucose, and impairment of the immune system. Persistent high levels of cortisol also cause changes in the brain that can lead to depression. Up to 75% of sufferers of MDD exhibit some degree of hypercortisolemia.[54] In Cushing's Disease, a disorder of the pituitary in which high levels of cortisol are found in the blood, patients can exhibit severe, even psychotic, forms of depression. RU486, a drug that blocks cortisol receptors in the brain, is useful for the treatment of severe psychotic depression.[55]

There are several ways by which cortisol may cause depression. As was noted above, persistent stress and high levels of glucocorticoids lead to activation of mitochondria and subsequent oxidative stress. Glucocorticoids also interact with the intracellular systems that control activity of mTOR, eEF2, and BDNF. Prolonged treatment with corticosterone, the rodent analogue of cortisol, causes a depression-like syndrome in mice. Ketamine quickly reverses this, but drugs that block activation of mTOR, prevent this remedial effect of ketamine on corticosterone-induced depression.[56] Other studies have similarly shown that glucocorticoids inhibit mTOR activity, and that managing chronic stress, such as by exercise, restores this activity.[57] In the brains of those corticosterone-treated mice, the BDNF factor was also reduced. Neurogenesis, the BDNF-dependent regrowth of neurons in the brain, is thought to play a role in the therapeutic effects of antidepressants. Neurogenesis is disrupted by high levels of cortisol.[58] Chronic exposure to high levels of corticosterone also reduces the activity of the protein producing factor, eEF2. Corticosterone-induced decreases in growth factors and protein production eventually lead to atrophy of neurons and loss of connectivity.[59] Together, those findings suggest that the depression-like effects of prolonged exposure to cortisol are due in part to inactivation of mTOR, diminished activity of eEF2, decreases in BDNF production, and overall loss of neuronal plasticity. These effects are opposite to those of ketamine and many other antidepressant treatments.

4.4   INSULIN RESISTANCE

It has long been known that a disproportionately large percentage of individuals with diabetes suffer MDD. The prevalence of MDD among diabetics, regardless of it being Type I or Type II, is roughly three times that seen in the general population. Some of this may simply be due to the difficulties of suffering a stressful chronic illness. However, there is a strong relationship between insulin resistance, *per se*, and MDD.[60] Insulin resistance is four times more likely to occur in individuals with MDD than in those not suffering this illness.[61] Certain illnesses, such as Polycystic Ovary Syndrome (PCOS), are characterized as often including both MDD and insulin-resistance.[62]

Neurons do not require insulin to take up glucose. Indeed, the two main glucose transporters in the brain, GLUT1 and GLUT3, have classically been seen as not being responsive to insulin. In turn, it has been assumed that insulin has no important role in brain activity. However, insulin is actively transported into the brain, and insulin receptors are found in a variety of important areas of the brain, particularly the frontal cortex.[63] There are also areas of the brain, such as the hippocampus, where insulin-sensitive GLUT4 reside.[64] Recent studies have shown that insulin can increase the uptake and metabolism of glucose in the brain.[65] Thus, one might expect that insulin resistance would dampen glucose metabolism in the brain and contribute to mood disorder. However, effects of insulin on the cell signaling pathways within the neurons and glial cells of the brain may be more important than effects on glucose metabolism. Indeed, insulin treatment increases the expression of BDNF and its transducer, tropomyosin receptor kinase B (TrkB) receptors in the hippocampi of young rats.[66] Another study showed that in the brains of rats made insulin-resistant by high fat diets, insulin no longer fully activated the Akt and mTOR, nor inhibited GSK-3β pathways in the cerebral cortex, which led to decreases in dendritic spine density.[67]

A variety of medications that reduce insulin-resistance have been found to improve treatment-resistant MDD. A 2002 case report revealed that a young woman who for years suffered both PCOS and treatment-resistant MDD was relieved of the symptoms of both conditions with metformin and spironolactone.[68] Subsequently, the diabetes medications rosiglitazone[69] and pioglitazone[70] improved mood in patients suffering treatment-resistant MDD. Both rosiglitazone and pioglitazone stimulate the intracellular protein PPARγ, which, among other things, stimulates mTOR and inhibits GSK-3β.[71]

A low serum level of L-acetylcarnitine is a biomarker of both MDD and insulin resistance. It has been shown that treatment with L-acetylcarnitine produces a rapid antidepressant effect in patients with treatment-resistant depression. It acts in part to reduce insulin resistance, increase BDNF, and modulate glutamatergic activity in the brain resulting in changes somewhat similar to those seen with ketamine.[72] Indeed, like ketamine, L-acetylcarnitine treatment can buffer the overactivity of NMDA receptors.[73]

4.5 METABOLIC SYNDROME

Oxidative damage, inflammation, stress, and insulin resistance can occur independently of each other. Nonetheless, the occurrence of one tends to induce and exacerbate the others. All four states occur in an overarching condition known as Metabolic Syndrome. The defining signs of Metabolic Syndrome include hypertension, hypertriglyceridemia, low HDL, and high fasting glucose. However, of importance to the current context, Metabolic Syndrome also includes insulin resistance and increases in visceral fat.

Exposure of brain proteins and lipids to the high levels of glucose seen in Metabolic Syndrome and states of insulin resistance leads to the creation of advanced glycation end products. In microglia, activation of receptors for advanced glycation end products stimulates neuroinflammatory effects through translocation of NfKb with subsequent enhanced synthesis of the inflammatory cytokines TNF-α, IL-6, and PGE2.[74] Insulin resistance also leads to oxidative stress in the body that stimulates peripheral inflammatory effects. Inflammation then causes further oxidative damage. Together, inflammation and oxidative damage exacerbate insulin resistance, thus helping to perpetuate an upward spiral of pathophysiology.

Visceral adipocytes manage food intake, energy, and metabolic state through the production and release of various cytokines. Among these cytokines are adiponectin and leptin. Adiponectin enhances sensitivity to insulin, reduces inflammation,[75] and has antidepressant effects.[76] Serum levels of adiponectin are low in viscerally obese subjects with Metabolic Syndrome. Leptin also enhances sensitivity to insulin and has antidepressant effects. Leptin levels are often high in Metabolic Syndrome. However, as with insulin, sufferers of Metabolic Syndrome become resistant to leptin.[77] Visceral adipocytes also release the inflammatory cytokines TNFα, IL-1, IL-6, and others. Adipocytes too full of fat do not oxygenate well, experience oxidative stress, and serve sources of inflammation. Migration of activated macrophages into inflamed adipose tissue adds further to the inflammatory milieu.[78]

4.6 SUMMARY

Ketamine has been found to be remarkably effective as an antidepressant, even in patients that had been resistant to standard, monoaminergic treatments. Preclinical studies have revealed that the necessary components of ketamine's therapeutic effect are activation of mTOR, inhibition of GSK-3β, disinhibition of eEF2, and increases in the activity of BDNF. The question has remained as to what pathological conditions leading to MDD produce abnormalities in those neurochemical systems such that they require the therapeutic changes provided by medications such as ketamine.

There is compelling evidence that oxidative and nitrosative damage, inflammation, prolonged stress, and insulin resistance can interact to produce abnormalities in the deep, intracellular chemistry of the brain. These conditions are seen in large percentages of patients with MDD, and are shown to cause the deficits in the activities of mTOR, GSK-3β, eEF2, and BDNF that ketamine corrects. It is also found that many standard antidepressants exert effects similar to ketamine on those particular systems as well as acting to reverse oxidative and nitrosative damage, inflammation, stress, and insulin resistance. At the same time, medications that act primarily as antioxidant, antiinflammatory, or antidiabetic agents have also been shown exhibit antidepressant effects.

Herbs have been used for thousands of years to treat the various manifestations of melancholia, despondency, and depressed mood. Many herbs contain phytochemicals with significant antioxidant, anti-inflammatory, anti-stress, and antidiabetic effects that would be expected to result in antidepressant effects. In the following pages, we shall explore the effects and likely mechanisms of action of herbs that have been found to be effective in animal models of depression and, in many cases, in human sufferers of MDD.

REFERENCES

1. Duman RS, Aghajanian GK, Sanacora G, et al. Synaptic plasticity and depression: new insights from stress and rapid-acting antidepressants. *Nat Med.* 2016;22(3):238–249.
2. Eckel RH, Grundy SM, Zimmet PZ. The metabolic syndrome. *Lancet.* 2005;365(9468):1415–1428.
3. Blackad CN, Bot M, Scheffer PG, et al. Is depression associated with increased oxidative stress? A systematic review and meta-analysis. *Psychoneuroendocrinology.* 2015;51:164–175.
4. Moylanah S, Berk M, Dean OM, et al. Oxidative & nitrosative stress in depression: why so much stress? *Neurosci Biobehav Rev.* 2014;45:46–62.

5. Maes M, Galecki P, Chang YS, et al. A review on the oxidative and nitrosative stress (O&NS) pathways in major depression and their possible contribution to the (neuro) degenerative processes in that illness. *Prog Neuropsychopharmacol Biol Psychiatry.* 2011;35(3):676–692.

6. Pandya CD, Howell KR, Pillai A. Antioxidants as potential therapeutics for neuropsychiatric disorders. *Prog Neuropsychopharmacol Biol Psychiatry.* 2013;46:214–223.

7. Aschbacher K, O'Donovan A, Wolkowitz OM, et al. Good stress, bad stress and oxidative stress: insights from anticipatory cortisol reactivity. *Psychoneuroendocrinology.* 2013;38(9):1698–1708.

8. Roberts CK, Sindhub KK. Oxidative stress and metabolic syndrome. *Life Sci.* 2009;84(21–22):705–712.

9. Berk M, Kapczinski F, Andreazza AC, et al. Pathways underlying neuroprogression in bipolar disorder: focus on inflammation, oxidative stress and neurotrophic factors. *Neurosci Biobehav Rev.* 2011;35:804–817.

10. Rezina GT, Cardoso MR, Gonçalves CL, et al. Inhibition of mitochondrial respiratory chain in brain of rats subjected to an experimental model of depression. *Neurochem Int.* 2008;53(6–8):395–400.

11. McIntosh LJ, Sapolsky RM. Glucocorticoids increase the accumulation of reactive oxygen species and enhance adriamycin-induced toxicity in neuronal culture. *Exp Neurol.* 1996;141(2):201–206.

12. Kratschmar DV, Calabrese D, Walsh J, et al. Suppression of the Nrf2-dependent antioxidant response by glucocorticoids and 11β-HSD1-mediated glucocorticoid activation in hepatic cells. *PLoS One.* 2012;7:e36774.

13. Tobe EH. Mitochondrial dysfunction, oxidative stress, and major depressive disorder. *Neuropsychiatr Dis Treat.* 2013;9:567–573.

14. Zhou R, Yazdi AS, Menu P, et al. A role for mitochondria in NLRP3 inflammasome activation. *Nature.* 2011;469:221–225.

15. Martín-de-Saavedra MD, Budni J, Cunha MP, et al. Nrf2 participates in depressive disorders through an anti-inflammatory mechanism. *Psychoneuroendocrinology.* 2013;38(10):2010–2022.

16. Berk M, Malhi GS, Gray LJ, et al. The promise of N-acetylcysteine in neuropsychiatry. *Trends Pharmacol Sci.* 2013;34(3):167–177.

17. Lobato KR, Cardoso CC, Binfaré RW, et al. α-Tocopherol administration produces an antidepressant-like effect in predictive animal models of depression. *Behav Brain Res.* 2010;209(2):249–259.

18. Amr M, El-Mogy A, Shams T, et al. Efficacy of vitamin C as an adjunct to fluoxetine therapy in pediatric major depressive disorder: a randomized, double-blind, placebo-controlled pilot study. *Nutr J.* 2013;12:31–39.

19. Behr GA, Moreira JCF, Frey BN. Preclinical and clinical evidence of antioxidant effects of antidepressant agents: implications for the pathophysiology of major depressive disorder. *Oxid Med Cell Longev.* 2012;ID 609421.

20. Kolla N, Wei Z, Richardson JS, et al. Amitriptyline and fluoxetine protect PC12 cells from cell death induced by hydrogen peroxide. *J Psychiatry Neurosci.* 2005;30(3):196–201.

21. Hashioka S, Klegeris A, Monji A, et al. Antidepressants inhibit interferon-γ-induced microglial production of IL-6 and nitric oxide. *Exp Neurol.* 2007;206(1):33–42.

22. Reus GZ, Stringari RB, de Souza B, et al. Harmine and imipramine promote antioxidant activities in prefrontal cortex and hippocampus. *Oxid Med Cell Longev.* 2010;3(5):325–331.

23. Abdel-Wahab BA, Salama RH. Venlafaxine protects against stress-induced oxidative DNA damage in hippocampus during antidepressant testing in mice. *Pharmacol Biochem Behav.* 2011;100(1):59–65.

24. Fahnestock M, Marchese M, Head E, et al. BDNF increases with behavioral enrichment and an antioxidant diet in the aged dog. *Neurobiol Aging.* 2012;33(3):546–554.
25. Wu W, Wang X, Xiang Q, et al. Astaxanthin alleviates brain aging in rats by attenuating oxidative stress and increasing BDNF levels. *Food Funct.* 2013;5:158–166.
26. Nagib MM, Tadros MG, Rahmo RM, et al. Ameliorative effects of α-tocopherol and/or coenzyme Q10 on phenytoin-induced cognitive impairment in rats: role of VEGF and BDNF-TrkB-CREB pathway. *Neurotox Res.* 2019;35(2):451–462.
27. Rubio-Perez JM, Morillas-Ruiz JM. A review: inflammatory process in Alzheimer's disease, role of cytokines. *Sci World J.* 2012;ID 756357.
28. Dantzer R. Expression and Action of Cytokines in the Brain: Mechanisms and Pathophysiological Implications. In: Ader R, editor. *Psychoneuroimmunology.* Amsterdam: Elsevier; 2007. p. 271–280.
29. Dowlati Y, Herrmann N, Swardfager W, et al. A meta-analysis of cytokines in major depression. *Biol Psychiatry.* 2010;67(5):446–457.
30. Dantzer R, O'Connor JC, Freund GG, et al. From inflammation to sickness and depression: when the immune system subjugates the brain. *Nat Rev Neurosci.* 2008;9(1):46–56.
31. Raison CL, Capuron L, Miller AH. Cytokines sing the blues: inflammation and the pathogenesis of depression. *Trends Immunol.* 2006;27(1):24–31.
32. Howren MB, Lamkin DM, Suls J. Associations of depression with C-reactive protein, IL-1, and IL-6: a meta-analysis. *Psychosom Med.* 2009;71(2):171–186.
33. Steptoe A, Hamer M, Chida Y. The effects of acute psychological stress on circulating inflammatory factors in humans: a review and meta-analysis. *Brain Behav Immun.* 2007;21(7):901–912.
34. Steiner J, Bielau H, Brisch R, et al. Immunological aspects in the neurobiology of suicide: elevated microglial density in schizophrenia and depression is associated with suicide. *J Psychiatr Res.* 2008;42(2):151–157.
35. Park SH, Park-Min KH, Chen J, et al. TNF induces endotoxin tolerance mediated by GSK3 in macrophages. *Nat Immunol.* 2011;12(7):607–615.
36. Hu X, Paik PK, Chen J, et al. IFN-γ suppresses IL-10 production and synergizes with TLR2 by regulating GSK3 and CREB/AP-1 proteins. *Immunity.* 2006;24(5):563–574.
37. Martin M, Rehani K, Jope RS, et al. Toll-like receptor-mediated cytokine production is differentially regulated by glycogen synthase kinase 3. *Nat Immunol.* 2005;6:777–784.
38. Green HF, Nolan YM. GSK-3 mediates the release of IL-1β, TNF-α and IL-10 from cortical glia. *Neurochem Int.* 2012;61(5):666–671.
39. Cheng Y, Pardo M, de Souza Armini R, et al. Stress-induced neuroinflammation is mediated by GSK3-dependent TLR4 signaling that promotes susceptibility to depression-like behavior. *Brain Behav Immun.* 2016;53:207–222.
40. Li C, Li M, Yu H, et al. Neuropeptide VGF C-terminal peptide TLQP-62 alleviates lipopolysaccharide-induced memory deficits and anxiety-like and depression-like behaviors in mice: the role of BDNF/TrkB signaling. *ACS Chem Neurosci.* 2017;8(9):2005–2018.
41. Barrientos RM, Sprunger DB, Campeau S, et al. Brain-derived neurotrophic factor mRNA downregulation produced by social isolation is blocked by intrahippocampal interleukin-1 receptor antagonist. *Neuroscience.* 2003;121:847–853.
42. Iwata M, Ota KT, Duman RS. The inflammasome: pathways linking psychological stress, depression, and systemic illnesses. *Brain Behav Immun.* 2013;31:105–114.
43. Temiz-Resitoglu M, Kucukkavruk SP, Guden DS, et al. Activation of mTOR/IκB-α/NF-κB pathway contributes to LPS-induced hypotension and inflammation in rats. *Eur J Pharmacol.* 2017;802:7–19.
44. Cleary C, Linde JA, Hiscock KM, et al. Antidepressive-like effects of rapamycin in animal models: implications for mTOR inhibition as a new target for treatment of affective disorders. *Brain Res Bull.* 2008;76:469–473.

45. Mendlewicz J, Kriwin P, Oswald P, et al. Shortened onset of action of antidepressants in major depression using acetylsalicylic acid augmentation: a pilot open-label study. *Int Clin Psychopharmacol.* 2006;21(4):227–231.
46. Muller N, Schwarz MJ, Dehning S, et al. The cyclooxygenase-2 inhibitor celecoxib has therapeutic effects in major depression: results of a double-blind, randomized, placebo controlled, add-on pilot study to reboxetine. *Mol Psychiatry.* 2006;11:680–684.
47. Köhler O, Benros ME, Nordentoft M, et al. Effect of anti-inflammatory treatment on depression, depressive symptoms, and adverse effects: a systematic review and meta-analysis of randomized clinical trials. *JAMA Psychiatry.* 2014;71(12):1381–1391.
48. Raison CL, Miller AH. Anti-inflammatory agents as antidepressants: truth or dare. *Psychiatr Ann.* 2015;45(5):255–261.
49. Tynan RJ, Weidenhofer J, Hinwood M, et al. A comparative examination of the anti-inflammatory effects of SSRI and SNRI antidepressants on LPS stimulated microglia. *Brain Behav Immun.* 2012;26(3):469–479.
50. Marta K, Lin AH, Kenis G, et al. Anti-inflammatory effects of antidepressants through suppression of the interferon-γ/interleukin-10 production ratio. *J Clin Psychopharmacol.* 2001;21(2):199–206.
51. Brustolim D, Ribeiro-dos-Santos R, Kast RE, et al. A new chapter opens in anti-inflammatory treatments: the antidepressant bupropion lowers production of tumor necrosis factor-alpha and interferon-gamma in mice. *Int Immunopharmacol.* 2005;6(6):903–907.
52. Abdel-Salam OM, Nofal SM, El-Shenawy SM, et al. Evaluation of the anti-inflammatory and anti-nociceptive effects of different antidepressants in the rat. *Pharmacol Res.* 2003;48(2):157–165.
53. Hung Y-Y, Huang KW, Kang HY, et al. Antidepressants normalize elevated Toll-like receptor profile in major depressive disorder. *Psychopharmacology.* 2016;233(9):1707–1714.
54. Nemeroff CB. Clinical significance of psychoneuroendocrinology in psychiatry: focus on the thyroid and adrenal. *J Clin Psychiatry.* 1989;50:13–20.
55. Belanoff JK, Kalehzan M, Sund B, et al. Cortisol activity and cognitive changes in psychotic major depression. *Am J Psychiatry.* 2001;158:1612–1616.
56. Pazini FL, Cunha MP, Rosa JM, et al. Creatine, similar to ketamine, counteracts depressive-like behavior induced by corticosterone via PI3K/Akt/mTOR pathway. *Mol Neurobiol.* 2016;53(10):6818–6834.
57. Watson K, Baar K. mTOR and the health benefits of exercise. *Semin Cell Dev Biol.* 2014;36:130–139.
58. Alfonso J, Pollevick GD, Van Der Hart MG, et al. Identification of genes regulated by chronic psychosocial stress and antidepressant treatment in the hippocampus. *Eur J Neurosci.* 2004;19:659–666.
59. Magariños AM, McEwen BS. Stress-induced atrophy of apical dendrites of hippocampal CA3c neurons: involvement of glucocorticoid secretion and excitatory amino acid receptors. *Neuroscience.* 1995;69(1):89–98.
60. Everson-Rose SA, Meyer PM, Powell LH, et al. Depressive symptoms, insulin resistance, and risk of diabetes in women at midlife. *Diabetes Care.* 2004;27:2856–2862.
61. Petrlova B, Rosolova H, Hess Z, et al. Depressive disorders and the metabolic syndrome of insulin resistance. *Semin Vasc Med.* 2004;4:161–165.
62. Rasgon NL, Rao RC, Hwang S, et al. Depression in women with polycystic ovary syndrome: clinical and biochemical correlates. *J Affect Dis.* 2003;74(3):299–304.
63. Craft S, Watson GS. Insulin and neurodegenerative disease: shared and specific mechanisms. *Lancet Neurol.* 2004;3:169–178.
64. Pearson-Leary J, Jahagirdar V, Sage J, et al. Insulin modulates hippocampally-mediated spatial working memory via glucose transporter-4. *Behav Brain Res.* 2018;338(15):32–39.

65. Bingham EM, Hopkins D, Smith D, et al. The role of insulin in human brain glucose metabolism: an 18fluoro-deoxyglucose positron emission tomography study. *Diabetes.* 2002;51(12):3384–3390.

66. Haas CB, Kalinine E, Zimmer ER, et al. Brain insulin administration triggers distinct cognitive and neurotrophic responses in young and aged rats. *Mol Neurobiol.* 2016;53(9):5807–5817.

67. Arnold SE, Lucki I, Brookshire BR, et al. High fat diet produces brain insulin resistance, synaptodendritic abnormalities and altered behavior in mice. *Neurobiol Dis.* 2014;67:79–87.

68. Rasgon NL, Carter MS, Elman S, et al. Common treatment of polycystic ovarian syndrome and major depressive disorder: case report and review. *Curr Drug Targets Immune Endocr Metabol Disord.* 2002;2(1):97–102.

69. Rasgon NL, Kenna HA, Williams KE, et al. Rosiglitazone add-on in treatment of depressed patients with insulin resistance: a pilot study. *Sci World J.* 2010;10:321–328.

70. Kemp DE, Ismail-Beigi F, Ganocy SJ, et al. Use of insulin sensitizers for the treatment of major depressive disorder: a pilot study of pioglitazone for major depression accompanied by abdominal obesity. *J Affect Disord.* 2012;136(3):1164–1173.

71. Landreth G. Therapeutic use of agonists of the nuclear receptor PPARγ in Alzheimer's disease. *Curr Alzheimer Res.* 2007;4(2):159–164.

72. Watson K, Nasca C, Aasly L, et al. Insulin resistance, an unmasked culprit in depressive disorders: promises for interventions. *Neuropharmacology.* 2018;136(Part B):327–334.

73. Forloni G, Angeretti N, Smiroldo S. Neuroprotective activity of acetyl-L-carnitine: studies *in vitro. J Neurosci Res.* 1994;37(1):92–96.

74. Wang T, Fu F, Han B, et al. Danshensu ameliorates the cognitive decline in streptozotocin-induced diabetic mice by attenuating advanced glycation end product-mediated neuroinflammation. *J Neuroimmunol.* 2012;245(1–2):79–86.

75. Tilg H, Moschen AR. Adipocytokines: mediators linking adipose tissue, inflammation and immunity. *Nat Rev Immunol.* 2006;6:772–783.

76. Liu J, Guo M, Zhang D, et al. Adiponectin is critical in determining susceptibility to depressive behaviors and has antidepressant-like activity. *Proc Natl Acad Sci USA.* 2012;109(30):12248–12253.

77. Lu X-Y. The leptin hypothesis of depression: a potential link between mood disorders and obesity? *Current Opin Pharmacol.* 2007;7(6):648–652.

78. Mendelson SD. *Metabolic Syndrome and Psychiatric Illness: Interactions, Pathophysiology, Assessment and Treatment.* London, UK: Academic Press; 2008.

5 Phytochemicals: Some Basics

The experts at the Royal Botanic Garden at Kew estimate there to be about 390,000 different species of plants in the world. They have further estimated that of those plant species, about 18,000 have significant medicinal value. Some physicians dismiss the values of herbal medicine. However, for centuries, healers – both folk and formal – have used plants now known to contain morphine, quinine, pseudoephedrine, cocaine, colchicine, digoxin, atropine, galantamine, cannabidiol, salicylate, reserpine, sennosides, and other currently common medications that have been proven to relieve various forms of human suffering. The modern pharmaceutical industry has also looked to the plant world as a source for new medications. The Center for Biological Diversity in Tuscon, Arizona has stated that of the top 150 prescribed medications in the United States, at least 118 were originally derived from plant sources. Some proponents of herbal medicine speak of herbs with undue reverence, as if being "natural" lends them special, even magical properties. However, the medicinal substances in plants are not miracles – they are molecules. These molecules, like those of synthesized pharmaceuticals, have medicinal properties related to their chemical structures. The molecules in plants, collectively known as phytochemicals, are classified into the general categories of carbohydrates; lipids, including fatty acids, terpenes, and steroids; phenolics; and alkaloids. This chapter is not an in-depth review of phytochemicals. Rather, it may serve as a primer for health care providers not fluent in the nature of these substances.

5.1 CARBOHYDRATES

Carbohydrates are found in all plants and share the basic chemical formula of $C_n(H_2O)_n$. They can be found as simple sugars or as the complex interconnected sugar molecules of polysaccharides, starches, and cellulose. They serve as structural elements of plant cells and tissues, as well as primary sources of energy for metabolism in both plants and animals. Simple sugars may bind to non-sugar phytochemicals, thus forming glycosides. Some glycosides have important medicinal properties, with the cardiac glycosides, such as digoxin, being perhaps the best known and most important of this class. Inside the plant, the bound sugars may compartmentalize and control the activity of the glycosidic phytochemical, often inactivating it, with release of the active component by enzymatic, hydrolytic splitting. In herbal medicines, glycosides have different pharmacokinetics and bioavailabilities than the aglycone molecules. For example, most flavonoids, discussed below, exist naturally as glycosides. The glycosidic forms can have different pharmacological properties than the aglycone forms. Flavonoid glycosides often produce higher plasma levels and have longer halflives than those of aglycones.

Carbohydrates themselves have not generally been seen as possessing significant pharmacological effects. There have been reports of various carbohydrates affecting glucose metabolism, though often through simple mechanisms, such as inhibiting gastrointestinal α-amylase and α-glucosidase activities to slow glucose absorption and improve its metabolism. There are exceptions. The oriental medicinal herb, *fuzi*, or *Aconitum carmichaeli*, has for centuries been used to treat chronic wounds, poor circulation, spasms, and mood disorders.[1] It had been assumed that active principles of the plant were aconitine and other constituent alkaloids. In the West, *fuzi* is known as monkshood or wolf's bane, which throughout the ages even expert herbalists have hesitated to use due to the high toxicity of its alkaloids. Four of its polysaccharides were found to reduce serum glucose in both normal and alloxan-induced diabetic mice. These effects were seen after intraperitoneal injection that bypassed the gastrointestinal system, thus eliminating trivial mechanisms of action.[2] Another polysaccharide isolated from *fuzi* has recently been reported to have antidepressant-like effects in mice. It not only showed antidepressant-like effects in behavioral assays, but it also restored levels of BDNF and stimulated neurogenesis in the hippocampus.[3]

5.2 LIPIDS

Lipids include the fatty acids, terpenes, and steroids. The fatty acids and related triglycerides are generally more important as nutrients than as substances pharmacologically active in modest concentration. Nonetheless, many edible plant and seed oils have important effects on mental health and mood state, due to their containing physiologically essential fatty acids. Low serum concentration of polyunsaturated fatty acids is a risk factor for MDD. This condition can be improved by supplementation with the food oils of walnuts, avocado, canola, peanut, safflower, and sunflower that are rich in those fatty acids.[4] The oils of herbs borage, blackcurrant seed, evening primrose, and flaxseed are also used to supplement intake of essential fatty acids.[5]

Modest but significant improvements of MDD are seen when antidepressants are supplemented with the omega-3 polyunsaturated fatty acids, eicosapentaenoic acid, or docosahexaenoic acid.[6] Humans can synthesize eicosapentaenoic and ocosahexaenoic acids from a-linolenic acid, but with poor efficiency.[7] Thus, many if not most people can benefit from supplementation, especially if symptomatic. The best sources of those two omega-3 fatty acids are ocean fish and fish oils. However, certain forms of marine algae are the original source of eicosapentaenoic acid and docosahexaenoic acid that work their way up the marine food chain. Thus, it is possible to obtain those fatty acids from marine plant sources.[8] Fatty acids will not be discussed further in this text.

5.3 TERPENES

Terpenes are common in plants. They are hydrocarbons synthesized from various combinations of the 5-carbon molecular building block isoprene. They are often aromatic, and may serve plants by deterring herbivores or attracting pollinators. They are responsible for the pleasant fragrances of pine and citrus fruits, the floral aroma of lavender, and the pungent odors of marijuana. Terpenes can be simple in structure, such as the hemiterpene isovaleric acid, a four-carbon chain with a methyl group attached. They can also be large and complex, multi-ringed, steroid-like molecules. Terpenes, as a class, are highly lipophilic and penetrate the brain where they can manifest psychiatrically significant effects.

Roughly 90% of the terpenes in the essential oils of plants are monoterpenes, which are single pairs of bound isoprene molecules. Many have been found to have analgesic, anti-inflammatory, anesthetic, and antioxidant effects. Among the better-known pharmacologically active monoterpenes are linalool, from lavender; myrcene, from ylang-ylang; citronellal, from kaffir lime leaves; carvacrol, from oregano and thyme; limonene, from citrus fruits; and menthol, from plants of the mint family.[9]

Animals also produce terpenes, and many, such as cholesterol and other steroids, are essential to life. This may explain how various steroid-like tri- and tetraterpenes of plant origin can affect human health, as they may mimic or block the effects of substances natural to the human body. Many of the so-called "adaptogen" herbs, such as ashwagandha and ginseng, contain triterpene sterols, saponins, and steroids, that buffer activity in the hypothalamic-pituitary-adrenal axis.[10] Glycyrrhetic acid, a steroid triterpene from licorice, binds to mineralocorticoid and glucocorticoid receptors. It can act as an anti-inflammatory agent, but may also cause severe hypertension due to salt retention.[11] Diosgenin, a triterpene phytosteroid extracted from wild yam, was once used as an inexpensive starting point in the manufacturing of contraceptive steroids.[12]

Other well-known pharmacologically active terpenes are: beta-carotene, the common terpene that is the precursor to vitamin A; valerenic acid, a component of valerian root extract thought to be partially responsible for its sedative effects; gossypol, the extract of cotton seed that has been investigated as a male contraceptive pill; and paclitaxel (perhaps better known as Taxol), the large and complex anticancer terpene from the yew tree. The medically important heart drug digoxin and the ginsenosides from the revered ginseng plant are modified terpenes, i.e., terpene glycosides. The hallucinogenic drug, salvinorin A, extracted from the leaves of *Salvia divinorum*, is a bicyclic diterpenoid.

Because of similarities among plants in fundamental enzymes and metabolic pathways, it is quite common for the same terpenes to exist across a wide variety of plants. For example, the monoterpene linalool is found in lavender, cinnamon, marijuana, birch trees, mint, rosewood, laurel, citrus, and perhaps as many as 200 other plants.[13] In many cases, a symptom may be helped by a number of different plants because of a shared active molecule – a terpene or other type of phytochemical. For the same reason, it may be possible to use one herb rather than two or three, because of redundancy of the active principles.

5.4 PHENOLICS

Phenolics are molecules that range from the simple phenol – a benzene ring with a hydroxyl group – to complicated multiple-ringed structures known as polyphenols. The multiple carbon double bonds in phenolic rings lend color to plants, including leaves, fruits, and flowers. The deep colors of the phenols attract pollinators, protect the plant's tissues from the sun's radiation and, in some cases, help fight off bacterial infection. They may also act as chemical messengers, physiological regulators, and cell cycle modulators.

5.4.1 FLAVONOIDS

There are more than 8,000 naturally occurring phenolic compounds in plants, and roughly half of them are of the important class of phenolics known as flavonoids. Flavonoids, a major subtype of polyphenol, readily cross from the bloodstream into brain tissue where they might act to reduce the symptoms of MDD.[14] The common underlying structure of flavonoids lends itself to nearly endless permutations by replacement of functional groups on the core 15-carbon skeleton of two phenyl rings and a heterocyclic ring. Indeed, they form the largest group of phytonutrients. The Linus Pauling Institute at Oregon State University has noted over 5,000 different variations of flavonoid molecules. The flavonoids are themselves categorized into six different subclasses, including: flavonols, such as catechin, epicatechin, and epigallocatechin gallate, found in green tea, red wine, and chocolate; deeply blue and purple anthocyanidins, such as cyanidin, delphinidin, and malvidin, from red wine and berries; flavanols, such as kaempferol and quercetin, found in cruciferous vegetables; flavones, such as apigenin and luteolin found in chamomile and celery; flavanones, including hesperidin and naringenin from citrus; and isoflavones, such as daidzein and genistein, from soy. While the categorization is complicated, the differences among the groups are simply in substitutions of hydrogens with hydroxyl versus ketone groups, and where on the skeleton these substitutions take place.

Flavonoids are major contributors to the medicinal effects of herbs. These molecules tend to exhibit strong antioxidant effects. A number of them have been shown to have anti-inflammatory and antidiabetic effects, as well as affecting important cell signaling pathways in neurons that can be important in the treatment of mood disorders.[15] A variety of flavonoids, such as the common quercetin, activate the PI3K and MAPK pathways that in turn result in inactivation of GSK-3β and activations of mTOR and BDNF.[16] Some, such as quercetin and apigenin, have potent nanomolar and low micromolar abilities to inhibit monoamine oxidase.[17] In preclinical studies, many isolated and purified flavonoids have been found to exhibit antidepressant effects.

5.4.2 NON-FLAVONOID PHENOLICS

Some well-known non-flavonoid phenols include: capsaicin, the topical pain-relieving agent extracted from hot peppers; salicylic acid, the pain- and fever-relieving substance found in willow that led to the synthesis of the stronger acetylsalicylic acid; and sennosides, the laxative polyphenols in the senna plant. There is also a variety of so-called phytoestrogens whose complex polyphenolic structures mimic both the structures and effects of estrogens produced in animals.

It has long been known that inclusion of deep-green vegetables in the diet is protective against the development of cancers. The polyphenols in these vegetables protect not only by acting as antioxidants, but also by affecting cell signaling processes that martial the body's natural abilities to fight off kill aberrant, rapidly multiplying

cells.[18] The National Cancer Institute has recommended including cruciferous vegetables, such as cabbage, broccoli, Brussel sprouts, kale, and others, for their abilities
to ward off cancer. In addition to polyphenols, these plants contain the nonflavonoid, sulfur-containing phenolic compounds isothiocyanates that have been found
to inhibit cancer growth by other mechanisms. Some isothiocyanates, for example,
sulforaphane,[19] appear to ward off cancer by inhibiting mTOR and the protein synthesis it can stimulate. To some extent, this is at odds with the antidepressant effect
of ketamine and other antidepressants in stimulating mTOR. However, sulforaphane
has been found to produce antidepressant-like effects in mice.[20] Indeed, the various
cell signaling processes are complex and interwoven, and there is evidence that some
of the end products of mTOR stimulation, such as BDNF, are enhanced by isothiocyanates. For example, one study has shown that the isothiocyanates of onion and
garlic enhance BDNF and neural plasticity by dampening abnormal inflammatory
processes in the brain.[21]

5.5 ALKALOIDS

Alkaloids are nitrogen containing molecules produced by about a quarter of plants.
Many alkaloids are quite toxic to animals. The alkaloid, nicotine, milligram for
milligram, may be as toxic as cyanide.[22] Thus, it is likely that alkaloids serve to
protect the plants from animal predators. The nitrogen groups in alkaloids make
them similar to nitrogen-containing neurotransmitters and other molecules active
in the brain. Thus, it is not surprising that many well-known hallucinogenic and
otherwise mind-altering plant substances are alkaloids. Examples are nicotine from
tobacco, morphine from the poppy, psilocybin from mushrooms, lysergic acid from
ergot mold, mescaline from cactus, and ephedrine from *Ma Huang*. Galantamine,
a cholinesterase inhibitor that is an FDAapproved medication to treat Alzheimer's
disease, was originally obtained from bulbs of *Galanthus caucasicus*, the Caucasian
snowdrop flower.[23]

The anticholinergic alkaloids, such as atropine, hyoscyamine, and scopolamine,
can cause hallucinations at high dose. They can also be fatal at relatively low doses.
Every summer, there are reports of boys dying while trying to get "high" with
Datura, a delirium-producing plant and source of all three of the tropane alkaloids,
atropine, hyoscyamine, and scopolamine. These alkaloids act by blocking muscarinic receptors in the brain. However, it has long been known that the anticholinergic
effects of many of the older tricyclic antidepressants contribute to the efficacy of
those medications. Moreover, it has been found that scopolamine can produce a rapid
antidepressant effect in some otherwise treatment-resistant sufferers of MDD, and it
does so in part by activating mTOR.[24]

Many of the hallucinogenic alkaloids, including psilocybin, mescaline, and
dimethyltryptamine from *ayahuasca* or the *Banisteriopsis caapi* vine, derive their
unique effects on consciousness largely by stimulating the serotonin type 2A receptor (5-HT2A) in the brain. Although often used for recreational purposes, there has
recently been a revival of interest in using modest doses of these drugs in shaman-
istic or even standard psychotherapeutic settings for conditions such as obsessive-
compulsive disorder or treatment-resistant MDD.[25] However, in the current context,

the actions of these 5-HT2A agonists are unusual. The receptors do not mediate inhibition of GSK-3β and stimulation of mTOR and eEF2, as we are coming to expect of antidepressant medications. Indeed, they do precisely the opposite.[26] While this may be an over-simplified distinction, the psychotherapeutic effects of the 5-HT2A agonist hallucinogens are likely less due to ongoing changes in intracellular signaling than to temporary changes in the brain circuitry that underlies consciousness, including shifts in awareness and the so-called default mode network.[27]

5.6 SUMMARY

In the scientific literature, there are many reports of terpenes, alkaloids, and phenolics interacting with the same systems that mediate the antidepressant effects of ketamine. That is, these phytochemicals affect the activities of mTOR, GSK-3β, eEF2, and BDNF. Many have also been shown to affect the pathologies that underlie depression, i.e., oxidative damage, inflammation, stress, and insulin resistance. Some have been found to exhibit antidepressant-like effects when administered in pure form, or are constituents of plants that have been found to exhibit antidepressant effects. In the following pages, we will explore the antidepressant effects of herbs. There will also be discussion of the use of herbs in Traditional Chinese Medicine. Although many specific herbal treatments have been used for thousands of years in that medical tradition, recent research shows that some such treatments exhibit rapid action similar to that of ketamine and may act by similar mechanisms.

REFERENCES

1. Zhang H. The clinical and theory study of rouse yang-qi method on depression. *J Shangdong Univ Tradit Chin Med.* 2006;30:140–143.
2. Hikino H, Murayama M, Sugiyama K, et al. Isolation and hypoglycemic activity of aconitans A, B, C and D, glycans of *Aconitum carmichaeli* roots. *Planta Med.* 1985;51:160–161.
3. Yan H-C, Qu HD, Sun LR, et al. Fuzi polysaccharide-1 produces antidepressant-like effects in mice. *Int J Neuropsychopharmacol.* 2010;13(5):623–633.
4. De Vriese SR, Christophe AB, Maes M. Lowered serum n-3 polyunsaturated fatty acid (PUFA) levels predict the occurrence of postpartum depression: further evidence that lowered n-PUFAs are related to major depression. *Life Sci.* 2003;73(25):3181–3187.
5. Hoffmann D. *Medical Herbalism: The Science and Practice of Herbal Medicine.* Rochester, VT: Healing Arts Press; 2003.
6. Mocking RJT, Harmsen I, Assies J, et al. Meta-analysis and meta-regression of omega-3 polyunsaturated fatty acid supplementation for major depressive disorder. *Transl Psychiatry.* 2016;6:e756.
7. Gerster H. Can adults convert a-linolenic acid (18:3n-3) to eicosapentaenoic acid (20:5n-3) and docosahexaenoic acid (22:6n-3)? *Int J Vitam Nutr Res.* 1998;68:159–173.
8. Gladyshev MI, Sushchik NN, Makhutova ON, et al. Production of EPA and DHA in aquatic ecosystems and their transfer to the land. *Prostaglandins Other Lipid Mediat.* 2013;107:117–126.
9. Guimarães AG, Quintans JSS, Quintans-Júnior LJ. Monoterpenes with analgesic activity—a systematic review. *Phytother Res.* 2013;27:1–15.
10. Wagner H, Nörr H, Winterhoff H. Plant adaptogens. *Phytomedicine.* 1994;1(1):63–76.

11. Ulmann A, Menard J, Corvol P. Binding of glycyrrhetinic acid to kidney mineralocorticoid and glucocorticoid receptors. *Endocrinology.* 1975;97(1):46–51.
12. Djerassi C. Steroid research at Syntex: "the pill" and cortisone. *Steroids.* 1992;57(12):631–641.
13. Peana AT, Moretti MDL. Linalool in Essential Plant Oils: Pharmacological Effects. In: Watson RR, Preedy VR, editors. *Botanical Medicine in Clinical Practice.* Wallingford, UK: CABI; 2008. p. 716–724.
14. Youdim KA, Dobbie MS, Kuhnle G, et al. Interaction between flavonoids and the blood-brain barrier: *in vitro* studies. *J Neurochem.* 2003;85:180–192.
15. Spencer JPE. The interactions of flavonoids within neuronal signaling systems. *Genes Nutr.* 2007;2:257–273.
16. Rao JS, Ertley RN, Lee HJ, et al. n-3 Polyunsaturated fatty acid deprivation in rats decreases frontal cortex BDNF via a p38 MAPK-dependent mechanism. *Mol Psychiatry.* 2007;12:36–46.
17. Chimenti F, Fioravanti R, Bolasco A, et al. A new series of flavones, thioflavones, and flavanones as selective monoamine oxidase-B inhibitors. *Bioorgan Med Chem.* 2010;18:1273–1279.
18. Kampa M, Nifli AP, Notas G, et al. Polyphenols and cancer cell growth. *Rev Physiol Biochem Pharmacol.* 2007;159:79–113.
19. Wiczk A, Hofman D, Konopa G, et al. Sulforaphane, a cruciferous vegetable-derived isothiocyanate, inhibits protein synthesis in human prostate cancer cells. *Biochim Biophys Acta: Mol Cell Res.* 2012;1823(8):1295–1305.
20. Wua S, Gao Q, Zhao P, et al. Sulforaphane produces antidepressant- and anxiolytic-like effects in adult mice. *Behav Brain Res.* 2016;301:55–62.
21. Juturu V, Caglayan B, Kilic E, et al. Allyl isothiocyanate enhances brain neuronal plasticity proteins via inhibition of inflammation proteins: 2928 Board# 211. *Med Sci Sports Exerc.* 2018;50(5S):724.
22. Mayer B. How much nicotine kills a human? Tracing back the generally accepted lethal dose to dubious self-experiments in the nineteenth century. *Arch Toxicol.* 2014;88(1):5–7.
23. Atanasovaa M, Stavrakov G, Philipova I, et al. Galantamine derivatives with indole moiety: docking, design, synthesis and acetylcholinesterase inhibitory activity. *Bioorgan Med Chem.* 2015;23(17):5382–5389.
24. Voleti B, Navarria A, Liu RJ, et al. Scopolamine rapidly increases mammalian target of rapamycin complex 1 signaling, synaptogenesis, and antidepressant behavioral responses. *Biol Psychiatry.* 2013;74(10):742–749.
25. Wilcox JA. Psilocybin and obsessive compulsive disorder. *J Psychoactive Drugs.* 2014;46(5):393–395.
26. Beaulieu J-M. A role for Akt and glycogen synthase kinase-3 as integrators of dopamine and serotonin neurotransmission in mental health. *J Psychiatry Neurosci.* 2012;37(1):1–16.
27. Carhart-Harris RL, Erritzoe D, Williams T, et al. Neural correlates of the psychedelic state as determined by fMRI studies with psilocybin. *Proc Natl Acad Sci USA.* 2012;109(6):2138–2143.

6 Models and Paradigms for Assessment of Antidepressant Effects

The current chapter will prepare the reader for the following discussions of antidepressant effects of specific herbs by presenting the format in which these herbs are discussed and by analyzing the models and paradigms from which conclusions are drawn. First discussed for each herb are the general effects that lend themselves to antidepressant effects, as per the most recent theories about the etiology and pathophysiology of MDD. The three primary mechanisms discussed are antioxidant, anti-inflammatory, and antidiabetic/anti-Metabolic Syndrome effects. The standard experimental techniques used to evaluate those effects will be reviewed here. Those discussions are followed by analyses of studies specifically demonstrating antidepressant effects. In animal studies, the paradigms by which these effects are demonstrated are the tail suspension and forced swim tests, and the sucrose preference test. In this chapter, those experimental methods will be reviewed. Herbs have also been shown to prevent or relieve depression-like syndromes caused by persistent stress, such as restraint or unpredictable mild shock; chronic exposure to stress hormones; systemic inflammation, such as that induced by injection of lipopolysaccharide; or by diabetic/Metabolic Syndrome-like states caused by high calorie diets or treatment with diabetes-inducing substances such as streptozocin. These testing paradigms reflect common human pathologies that lead to the development of MDD, and their theoretical bases will presently be discussed.

6.1 ANTIOXIDANT EFFECTS

Free radicals are molecules with an unpaired electron that can readily participate in either donation or acceptance of electrons to become stabilized. Free radicals can thereby alter the chemical nature of lipids, proteins, carbohydrates, and DNA, resulting in stress, inflammation, and loss of function. Free radicals are the natural result of cell metabolism, but can be increased in concentration by various pathological processes, including inflammation and diabetes. The body has substances and enzymatic processes to protect it from the damaging effects of free radicals. However, an important source of protection is dietary intake of antioxidants. An antioxidant is a molecule stable enough to donate an electron to a free radical – that is, to itself become oxidized – and thus neutralize the free radical and reduce its capacity to do cellular damage. Many dietary antioxidants are polyphenolic compounds from plants. Many medicinal herbs are rich in such compounds.[1]

A variety of *in vitro* techniques have been used to assess the antioxidant activities of herbal extracts and their active components. In many cases, the goals of antioxidant studies have been to grade the abilities of spices and essential oils in herbs to preserve food and protect it from oxidation and rancidity. Nonetheless, such information is valuable in assessing the abilities of herbs to protect living systems from oxidative and nitrosative damage. One of the most commonly used tests to evaluate herbal extracts is the DPPH (diphenyl-1-picrythydrazyl) test. It is a straightforward colorimetric test that measures the product of the antioxidant reaction with an organic radical. Another test is simply the measurement of total phenolic content, as it corresponds well with antioxidant effect.[2] The TBARS (thiobarbituric reactive substances) assay is the most common test for measuring ability to protect lipids from peroxidation. TBARS are formed as byproducts of lipid peroxidation and can be generated using thiobarbituric acid as a reagent. It is often used *ex vivo* after exposure of animals to oxidative stress. Liver, brain, or other tissue can be homogenized and the presence of TBARS compared in treated and untreated animals.[3] Common oxidative stresses are carbon tetrachloride (CCl_4) stress of the liver and ischemic stress of a cerebral hemisphere induced by tying off a carotid artery. Often, the TBARS test is performed along with assays of the activities of antioxidant enzymes, such as superoxide dismutase, catalase, and glutathione peroxidase, and for levels of the endogenous antioxidant glutathione. High levels of those substances remaining in tissue entail the antioxidant effects of the herb's phenolic content having spared the tissue from oxidative damage. In the context of antidepressant effects, antioxidant effects seen in the brain show that the substance is centrally active. However, protection from oxidation in the liver not only establishes antioxidant capacity, but also shows the ability to protect from oxidative effects of non-alcoholic steatohepatitis and diabetes. In addition, reduction of lipid peroxidation in visceral adipose tissue would suggest ability to reduce inflammatory effects that arise from poor oxygen diffusion in enlarged adipose cells found in Metabolic Syndrome.[4] Interestingly, in human subjects, serum levels of TBARS and BDNF have been found to be inversely correleated.[5]

6.2   ANTI-INFLAMMATORY EFFECTS

Inflammation plays a major role in the etiology of MDD. It increases oxidative stress and increases insulin resistance. Inflammation also impedes processes that are necessary components of the antidepressant mechanisms of treatments such as ketamine. For example, in both humans and animals, increased glucocorticoid levels and inflammation are associated with reduced BDNF and reduced hippocampal volume.[6]

There are several methods by which the anti-inflammatory effects of herbs are evaluated. One of the most common techniques is to gauge the ability of substances to attenuate lipopolysaccharide-induced inflammation. The lipopolysaccharide can be administered to intact animals, or placed into cultured peripheral immune cells, microglia, or neuron-like pheochromocytoma cell lines. Lipopolysaccharide is a major molecular component of the outer membrane of gram-negative bacteria, and is the primary activator of the evolutionarily ancient innate immunity system. Lipopolysaccharide activates the toll-like receptor type4 (TLR4) that then initiates the intracellular inflammatory cascade. The translocation of NF-kβ to the nucleus enhances synthesis of the inflammatory cytokines TNF-α and IL-1β, as well as prostaglandins, leukotrienes, and Reduced Oxygen Species. In T-lymphocytes, NK cells and many other cell types, the release of IL-12 or IL-18 induces release of IFN-γ. The IFN-γ hypersensitizes mononuclear phagocytes to lipopolysaccharide, thus amplifying the inflammatory process.[7] In some cases, the mechanism by which herbs interfere with the lipopolysaccharide-induced inflammatory cascade are explored, such as antagonism of TLR4, or prevention of translocation of NF-kβ. In most cases, the anti-inflammatory effects are established by assessing the degree to which the levels of the inflammatory cytokines are contained.

The technique of causing oxidative damage from artificially induced cerebral ischemia also produces inflammatory effects mediated by TLR4. Although lipopolysaccharide is a primary activator of TLR4, the receptor is also activated by products of cell damage from ischemia, as well as entities such as oxidized LDL cholesterol complexes that would be increased in Metabolic Syndrome.[8] TLR4 receptors in the brain are found primarily on microglia, the brain's primary resident immune cells. However, these receptors are expressed, to a lesser extent, by neurons, astroglia, oligodendroglia, and cerebral vascular endothelium. Thus, conditions that stimulate these receptors can have wideranging effects in the brain.[9] Recent research is suggesting a stronger than expected role for TLR4 in the etiology of human depression. Of note are recent reports showing significantly higher expression of TLR3, 4, 5, and 7 mRNA in patients with MDD compared to healthy controls.[10] Along with their prevention of oxidative damage in the brain, a number of studies have evaluated the ability of herbs to attenuate the inflammatory effects of cerebral ischemia. Activation of the TLR4 pathway also causes deactivation of Akt and activation of GSK-3β, and in some studies, the ability of herbs to ameliorate the impact of that chain of events has also been evaluated.

Another common method used to assess anti-inflammatory effects is the carrageenan-induced edema procedure. In this method, carrageenan is injected into the hind paw of a rodent and the ability of an agent to prevent inflammation is assessed.[11] Classically, the inflammatory response to carrageenan is a local process that is mediated initially by serotonin, histamine, and kinins, but primarily by prostaglandins and leukotrienes. Thus, much of the anti-inflammatory effect of herb treatment in this method is due to inhibition of COX-2 and 5-lipoxygenase or 5-LOX. Infiltration by neutrophils with release of the neutrophil-derived free radicals, nitric oxide, and pro-inflammatory cytokines, such as TNF-α and IL-1β, can also be involved in the later phases of carrageenan-induced acute inflammation.[12] However, it is known that carrageenan, a complex polysaccharide, also activates the TLR4, and thus stimulates NF-kβ and the consequent cascade of inflammatory cytokines similar to the effect of lipopolysaccharide.[13] Effectiveness in carrageenan-induced paw edema does not entail a general anti-inflammatory effect. For example, the flavonoid quercetin is quite effective in preventing carrageenan-induced edema, but relatively ineffective against xylene-induced ear edema. The latter is a neurogenic inflammatory process initiated by release of Substance P.[14] Croton oil-induced ear edema is primarily due to activation of phospholipase A2, which supplies substrate for the synthesis of leukotrienes and inflammatory prostaglandins.[15]

6.3   ANTIDIABETIC/ANTI-METABOLIC SYNDROME EFFECTS

Another area of evidence discussed for each herb are its effects on glucose metabolism and its abilities to reverse the effects of experimentally induced diabetes and Metabolic Syndrome. Metabolic Syndrome is known to play a role in the development of MDD due to inflammation; oxidative effects of disturbed lipid and glucose metabolism; and direct effects on pathways controlling synthesis of BDNF in brain tissue. For example, hyperglycemia prevents reparative increases in BDNF in the hippocampus and cingulate cortex of rats subjected to transient forebrain ischemia.[16] Plasma levels of BDNF are decreased in humans with type 2 diabetes, and these levels of BDNF vary inversely with fasting levels of plasma glucose.[17]

Insulin resistance is the hallmark of Metabolic Syndrome. Indeed, the discoverer of Metabolic Syndrome, Dr Gerald Reaven, has forgone the phrase Metabolic Syndrome in favor of Insulin Resistance Syndrome. The phenomenon of insulin resistance is complex. The various underlying causes include oxidative stress,[18] inflammation,[19] persistent hyperinsulinemia[20] hyperlipidemia,[21] hypercortisolemia[22] and hypoadiponectinemia.[23] Whereas impaired glucose metabolism decreases expression of BDNF, there is also evidence that decreases in BDNF may in turn increase insulin resistance in human subjects.[17] Each of the above abnormalities has been shown to be improved by one or more herbs. However, most studies that have addressed the effects of herbs on diabetes have only looked indirectly at insulin resistance. Thus, it is frequently shown that herbs lower serum glucose levels and generally improve energy metabolism in streptozotocin- or alloxan-induced diabetic rodents.

6.4   PRECLINICAL ANTIDEPRESSANT-LIKE EFFECTS

The antidepressant-like effects of the herbs in animals, mostly rodents, were demonstrated by one or more standard techniques, including the forced swim test, the tail suspension test, and the sucrose consumption test.

6.4.1   FORCED SWIM TEST

One of the primary methods of screening for antidepressant activity is the forced swim test. This method was first described by Porsolt et al. in 1977.[24] The original method was described thus:

> …mice were dropped into the cylinder (height 25 cm, diameter 10 cm, 6 cm of water at 21–23°C) and left for 6 min. Because little immobility is observed during first 2 min, only that occurring during the last 4 min was counted. The duration of immobility occurring in each minute was scored. A mouse was judged to be immobile when it ceased struggling and remained floating motionless in the water making only movements necessary to keep its head above water.

In interpreting the meaning of the results gained from the method, Porsolt wrote,

> A depressed state can be induced in mice by forcing them to swim in a narrow cylinder from which they cannot escape. After a brief period of vigorous activity, the

mice adopt a characteristic immobile posture which is readily identifiable. Immobility was reduced by tricyclic antidepressants, monoamine oxidase inhibitors and atypical antidepressants, as well as by electroconvulsive shock. Psychostimulants also reduced immobility but in contrast to antidepressants caused marked motor stimulation. Immobility was not affected by minor or major tranquilizers. These findings, closely parallel to those we have previously reported in rats, suggest that the procedure is selectively sensitive to antidepressant treatments.

He further noted that the mouse procedure was very similar in results to using rats, but was simply, "more rapid and less costly than that with rats and is thus more suitable for the primary screening of antidepressant drugs."

The reliability of the forced swim test, that is, the likelihood of getting the same results in repeated trials, appears to be high.[25] However, the question of validity in drawing conclusions about antidepressant activity from the test is more complicated. We cannot conclude that "antidepressants" reverse something in rodents in the swim test that is the same as, or even similar to, MDD. First, it is generally normal, "nondepressed" rats that are used in the forced swim tests. Ostensibly, the futility of their efforts to escape the situation produces what Porsolt referred to as behavioral despair in rodents that resembles aspects of MDD. However, in some cases, animals that have already been subjected to conditions that model human MDD, such as chronic unpredictable stress, chronic treatment with corticosteroids, olfactory bulbectomy, or social defeat, are used in the forced swim tests. Such animals later being subjected to forced swimming may come closer to mimicking the exacerbation of symptoms of humans in the state of depression.

Also problematic is that the administration of test substances in the forced swim protocol is unlike the clinical use of antidepressants. The most commonly followed protocol in the testing of substances for antidepressant effects is to give a single acute administration of the test substance just prior to the forced swim. This is at odds with clinical use in which administrations over weeks of time is what is required to affect antidepressant action in human patients. Still, questions of predictive validity arising from this difference may be moot, as standard antidepressants that are effective in humans do reduce immobility when administered acutely just prior to the behavioral test. Also, given the rapid action of ketamine, it is entirely possible that some antidepressant effect of any test substance can be appreciated in a single, acute administration.

Given the complexity of the human mind, it seems safe to say that rodents do not experience the condition we humans call MDD. However, in view of the facts that Traditional Chinese Medicine does not recognize MDD as a diagnostic entity, and that traditional herbalism does not recognize "antidepressant" as a mode of herbal action, there is reasonable doubt that MDD exists as a specific illness rather than a mere construct of mainstream Western Medicine. This would tend to question the so-called, "construct validity" of the swim test. On the other hand, in testing a substance that has not yet been categorized in terms of human response, it is possible that a substance represents a category of drugs that substantially reduces immobility in rodents but has no antidepressant effects whatsoever in humans. Indeed, many substances subjected to the swim test are those already suspected of having antidepressant effects in humans. Thus, the so-called "predictive validity" is also somewhat in question until enough randomly chosen herbs are tested to provide some indication of the value of the test. Perhaps, the best position to take is simply an operational one. That is, that since most, if not all, antidepressant medications also reduce immobility in the swim test in rodents, then an herb that reduces immobility is one that at least deserves testing as an antidepressant in humans. For a more in-depth discussion of validity of the forced swim test, see the paper by Petit-Demouliere et al.[26]

6.4.2 TAIL SUSPENSION TEST

In their initial paper on the tail suspension test of antidepressant effect, Steru et al[27] noted the essence of the test to be, "a mouse is suspended by the tail from a lever, the movements of the animal being recorded. The total duration of the test can be divided into periods of agitation and immobility." It was then noted that, "Antidepressant drugs decrease the duration of immobility, as do psychostimulants and atropine. If coupled with measurement of locomotor activity in different conditions, the test can separate the locomotor stimulant doses from antidepressant doses."

In an exhaustive review of the tail suspension test, Cryan et al.[28] noted some problems, such as genetic variations in behaviors between strains of mice in the test, and even variation within animals of the same strain. They warned against inadequate numbers of animals per test, ostensibly to avoid false negatives due to statistical variance. The group also pointed out apparent inconsistencies, such as reduction of immobility by nortriptyline and desipramine, but not by amitriptyline in their hands. Those results did differ from those of Steru et al., who first proposed the tail suspension as a test of antidepressant action. Nonetheless, the results of extensive testing of standard antidepressants showed reduction of immobility by these agents, with no such effect from lithium, methylphenidate, benzodiazepines, or neuroleptics. There was a false positive with amphetamine, which reduced immobility, but without the typical effects on "power and energy" of movement, all of which can be identified by computerized versions of the testing apparatus. That observation is in keeping with that reported by Steru et al. in their seminal paper. The overall conclusion of Cryan et al. was, "The TST has great utility as a model to assess antidepressant-related behaviour, however its use is best exploited when used in battery style fashion with other depression models such as FST, learned helplessness, anhedonia models and olfactory bulbectomy." Nonetheless, it is again a logical fallacy to assume that since antidepressants almost uniformly reduce immobility in the tail suspension test, a substance that reduces such immobility is then an effective antidepressant.

6.4.3 SUCROSE CONSUMPTION TEST

One of the effects of chronic unpredictable stress, a rodent model of MDD, is disruption of appetite and decrease in food consumption. In particular, both sucrose and saccharine consumption have been observed to diminish up to 50%.[29] In a 1982 study, it was shown that this decrease in the appetite for sucrose could be reversed by the tricyclic antidepressant imipramine.[30] The use of the effect of unpredictable stress on sucrose intake as a test for antidepressant-like effects was further developed by Willner et al.[31] A wide variety of standard antidepressants have been found to restore appetite for sucrose in chronically stressed rodents.[32]

Of some significance is the fact that the time course of the development of the diminished taste for sucrose, as well as the gradual reversal of the deficit through chronic antidepressant treatment, more clearly resembled MDD in humans and the time course of patient response to antidepressant treatment. This, in turn, leads to somewhat stronger validity as a test of antidepressant action. Because of the recent findings of rapid antidepressant effects seen with ketamine in human subjects, this similarity to more standard antidepressant treatments in the time course of reversal of the hedonic deficit seems less significant.

In a 1997 review[33] Willner noted that,

The model has good predictive validity (behavioural changes are reversed by chronic treatment with a wide variety of antidepressants), face validity (almost all demonstrable symptoms of depression have been demonstrated), and construct validity (CMS causes a generalized decrease in responsiveness to rewards, comparable to anhedonia, the core symptom of the melancholic subtype of major depressive disorder).

While stating that the method appeared at least as valid as any other animal model of depression, his only reservation was the time and effort necessary to adequately perform the technique. Despite the difficulties inherent in the method, more than 20% of published studies utilizing the chronic mild stress model of depression were concerned with evaluation of potential novel antidepressants. Interestingly, between 2010 and 2015, there was a 30% increase in such studies, largely reflecting the growth in studies involving traditional Chinese medicines.[34]

6.4.4 TEST CONDITIONS

The original methods of evaluating potential antidepressant effects in the forced swim and tail suspension tests utilized healthy animals with no interventions other than the acute administration of a test substance just prior to the behavioral testing. However, in some cases, animals underwent treatments that are themselves inducers of animal models of MDD. One such procedure is olfactory bulbectomy.[35] More often, the model of depression is a week or more of chronic stress prior to the testing procedure. Methods of inducing chronic stress include intermittent mild shock; periods of physical restraint; or chronic administration of corticosteroids that to some degree mimic the effects of chronic stress. In the standard forced swim and tail suspension tests, such extra steps are not necessary to establish the potential antidepressant effects of a substance. However, in some cases, the test substance itself was given in chronic basis. Thus, any attenuation of depression-like effects caused by procedures such as chronic mild stress or restraint might allow one to consider the possibility that the chronic herbal treatment helped prevent the development of depression-like behaviors. Moreover, in many cases, the brains of animals given specific herbs for evaluation of antidepressant effects in behavioral assays underwent further neurochemical evaluations. Effects of those herbs were seen on neuronal metabolism, maintenance, and growth, including impacts on BDNF, neurogenesis, and the intracellular signaling systems, e.g., PI3K, Akt, mTor, GSK-3β, etc., that control that activity. In many cases, not only depression-like behaviors, but also the pathological neurochemical sequelae of stress-inducing procedures were reversed by herbal treatment.

REFERENCES

1. Lobo V, Patil A, Phatak A, et al. Free radicals, antioxidants and functional foods: impact on human health. *Pharmacogn Rev.* 2010;4(8):118–126.
2. Moharram HA, Youssef MM. Methods for determining the antioxidant activity: a review. *Alex J Food Sci Technol.* 2014;11(1):31–42.
3. Gutteridge JMC. Lipid peroxidation and antioxidants as biomarkers of tissue damage. *Clin Chem.* 1995;41:1819–1828.
4. Mendelson SD. *Metabolic Syndrome and Psychiatric Illness: Interactions, Pathophysiology, Assessment and Treatment.* London, UK: Academic Press; 2008.
5. Kapczinski F, Frey BN, Andreazza AC, et al. Increased oxidative stress as a mechanism for decreased BDNF levels in acute manic episodes. *Braz J Psychiatry.* 2008;30(3):243–245.
6. Mondelli V, Cattaneo A, Murri MB, et al. Stress and inflammation reduce BDNF expression in firstepisode psychosis: a pathway to smaller hippocampal volume. *J Clin Psychiatry.* 2011;72(12):1677–1684.
7. Alexander C, Rietschel ET. Bacterial lipopolysaccharides and innate immunity. *J Endotoxin Res.* 2001;7(3):167–202.
8. Hua F, Ma J, Ha T, et al. Activation of toll-like receptor 4 signaling contributes to hippocampal neuronal death following global cerebral ischemia/reperfusion. *J Neuroimmunol.* 2007;190(1–2):101–111.
9. García Bueno B, Caso JR, Madrigal JLM, et al. Innate immune receptor Toll-like receptor 4 signaling in neuropsychiatric diseases. *Neurosci Biobehav Rev.* 2016;64:134–147.
10. Hung YY, Kang HY, Huang KW, et al. Association between toll-like receptors expression and major depressive disorder. *Psychiatry Res.* 2014;220(1–2):283–286.
11. Winter C, Risley E, Nuss G. Carrageenan-induced edema in hind paw of rats as an assay for antiinflammatory drug. *Proc Soc Exp Biol Med.* 1962;111:207–210.
12. Mansouri MT, Hemmati AA, Naghizadeh B, et al. A study of the mechanisms underlying the anti-inflammatory effect of ellagic acid in carrageenan-induced paw edema in rats. *Indian J Pharmacol.* 2015;47(3):292–298.
13. Bhattacharyya S, Borthakur A, Anbazhagan AN, et al. Specific effects of BCL10 serine mutations on phosphorylations in canonical and noncanonical pathways of NF-kB activation following carrageenan. *Am J Physiol Gastrointest Liver Physiol.* 2011;301(3):G475–G486.
14. Rotelli AE, Guardia T, Juárez AO, et al. Comparative study of flavonoids in experimental models of inflammation. *Pharmacol Res.* 2003;48(6):601–606.
15. De Melo GO, Malvar Ddo C, Vanderlinde FA, et al. Antinociceptive and antiinflammatory kaempferol glycosides from *Sedum dendroideum. J Ethnopharmacol.* 2009;124(2):228–232.
16. Uchino H, Lindvall O, Siesjö BK, et al. Hyperglycemia and hypercapnia suppress BDNF gene expression in vulnerable regions after transient forebrain ischemia in the rat. *J Cereb Blood Flow Metab.* 1997;17(12):1303–1308.
17. Krabbe KS, Nielsen AR, Krogh-Madsen R, et al. Brain-derived neurotrophic factor (BDNF) and type 2 diabetes. *Diabetologia.* 2007;50(2):431–438.
18. Evans JL, Maddux BA, Goldfine ID. The molecular basis for oxidative stress-induced insulin resistance. *Antioxid Redox Signal.* 2005;7(7–8):1040–1052.

19. Shoelson SE, Lee J, Goldfine AB. Inflammation and insulin resistance. *J Clin Investig.* 2006;116(7):1793–1801.
20. Shanik MH, Xu Y, Škrha J, et al. Insulin resistance and hyperinsulinemia: is hyperinsulinemia the cart or the horse? *Diabetes Care.* 2008;31(Suppl 2):S262–S268.
21. Stranahan AM, Norman ED, Lee K, et al. Diet-induced insulin resistance impairs hippocampal synaptic plasticity and cognition in middle-aged rats. *Hippocampus.* 2008;18(11):1085–1088.
22. Rizza RA, Mandarino LJ, Gerich JE. Cortisol-induced insulin resistance in man: impaired suppression of glucose production and stimulation of glucose utilization due to a postreceptor detect of insulin action. *J Clin Endocrinol Metab.* 1982;54(1):131–138.
23. Weyer C, Funahashi T, Tanaka S, et al. Hypoadiponectinemia in obesity and Type 2 diabetes: close association with insulin resistance and hyperinsulinemia. *J Clin Endocrinol Metab.* 2001;86(5):1930–1935.
24. Porsolt RD, Bertin A, Jalfre M. Behavioral despair in mice: a primary screening test for antidepressants. *Arch Int Pharmacodyn Ther.* 1977;229(2):327–336.
25. Borsini F, Meli A. Is the forced swimming test a suitable model for revealing antidepressant activity? *Psychopharmacology.* 1988;94(2):147–160.
26. Petit-Demouliere B, Chenu F, Bourin M. Forced swimming test in mice: a review of antidepressant activity. *Psychopharmacology.* 2005;177(3):245–255.
27. Steru L, Chermat R, Thierry B, et al. The tail suspension test: a new method for screening antidepressants in mice. *Psychopharmacology.* 1985;85(3):367–370.
28. Cryan JF, Mombereau C, Vassout A. The tail suspension test as a model for assessing antidepressant activity: review of pharmacological and genetic studies in mice. *Neurosci Biobehav Rev.* 2005;29(4–5):571–625.
29. Katz RJ, Roth KA, Mefford IA, et al. The chronically stressed rat: a novel animal model of endogenomorphic depression. *Proc. 3rd Int. Cong. Biol. Psychiat.* Stockholm, Sweden; 1981.
30. Katz RJ. Animal model of depression: pharmacological sensitivity of an hedonic deficit. *Pharmacol Biochem Behav.* 1982;16(6):965–968.
31. Willner P, Towell A, Sampson D, et al. Reduction of sucrose preference by chronic unpredictable mild stress, and its restoration by a tricyclic antidepressant. *Psychopharmacology.* 1987;93(3):358–364.
32. Vollmayr B, Henn FA. Stress models of depression. *Clin Neurosci Res.* 2003;3(4–5):245–251.
33. Willner P. Validity, reliability and utility of the chronic mild stress model of depression: a 10-year review and evaluation. *Psychopharmacology.* 1997;134(4):319–329.
34. Willner P. The chronic mild stress (CMS) model of depression: history, evaluation and usage. *Neurobiol Stress.* 2017;6:78–93.
35. Kelly JP, Wrynn AS, Leonard BE. The olfactory bulbectomized rat as a model of depression: an update. *Pharmacol Ther.* 1997;74(3):299–316.

7 Herbs with Antidepressant Effects

In this chapter, the antidepressant effects of specific herbs are discussed. The herbs were included under two primary criteria. The first criterion was evidence of traditional use of the herb to treat depression or mood disorders. This was established through review of the herbalism literature, and with internet search engines using combinations of key words including antidepressant, depression, herb, alternative medicine, folk medicine, phytochemical, flavonoid, and related terms. The second essential criterion was at least two well-performed preclinical studies demonstrating antidepressant-like effects through standard methods. For some of the herbs, it has been possible also to present results of human studies on MDD or depressed mood.

A primary goal of this book is to provide information for the practical use of herbs in the treatment of MDD. Therefore, at the end of the discussion of each herb, there is provided data on dosage, safety, and adverse effects. In most cases, there is information about use in pregnancy as well as potentially dangerous interactions with commonly prescribed medications. When available, animal LD50 data are also presented and extrapolated to human doses. For ease of reference, the sources cited in these discussions are listed separately for each herb.

7.1 *ALLIUM SATIVUM* (GARLIC)

Allium sativum (garlic) is a species native to Central Asia and northeastern Iran, but it has been spread worldwide largely due to its appeal as a unique flavoring for food. Along with its use as a spice, it has for thousands of years been used to treat various human ailments. The ancient medical writings of Egypt, Greece, Rome, China, and India each describe medical applications for *Allium sativum*. It was recommended by Hippocrates for maintaining strength and vitality, and it was fed to athletes of ancient Greece prior to competition. It was commonly used in the ancient world to rid patients of parasites, treat arthritis, and to aid digestion.[1] The close relatives of *Allium sativum*, onion, shallot, leek, chive, and Chinese onion, share many of its culinary and medical properties.

Allium sativum contains a wide variety of phytochemicals, including flavonoids, phenolics, saponins, and glycosylated steroidal compounds. The common flavonoids apigenin, isorhamnetin, kaempferol, luteolin, quercetin, and myricetin are found in *Allium sativum*, as are the less common steroid-like sativosides, erubosides, tropeosides, ascalonicosides, and alliospirosides. It is uniquely rich in organosulfur compounds. Most notable among those compounds are allicin, alliin, isoalliin, ajoene, diallyl polysulfides, vinyldithiins, and S-allylcysteine.[2]

Aging *Allium sativum* alters the phytochemical content, but without loss of the primary medicinal components. The primary changes that occur during aging are: (1) complete hydrolysis of the γ-glutamylcysteines to SAC and S-1-propenylcysteine; (2) increase in cystine due to protein hydrolysis and increase in S-allylmercaptocysteine, probably due to the reaction of allicin with protein derived cysteine; (3) initial loss of alliin to thiosulfinates formation (allicin); (4) complete loss of thiosulfinates after three months due to the fact that they are converted into volatile allyl sulfides, which evaporate almost completely.[3]

7.1.1 ANTIOXIDANT

The antioxidant activity of raw *Allium sativum* extract has been found to be quite high by the DPPH radical scavenging method.[4] The ethanolic extract of *Allium sativum* also powerfully inhibits nitrosation *in vitro*.[5]

In vivo, chronic treatment with extract of *Allium sativum* showed potent antioxidant and hepatoprotective effects in rabbits that had been exposed to CCl_4. Administration of SCG significantly reduced serum levels of liver enzymes, indicating protection of liver tissue. Levels of catalase and superoxide dismutase were higher in liver homogenates of animals treated with *Allium sativum*, whereas levels of the lipid peroxidation marker malondialdehyde were lower.[6]

Antioxidant effects of *Allium sativum* were also observed in rats subjected to reversible cerebral ischemia. Treatment with *Allium sativum* prior to the experimental ischemia maintained brain levels of glutathione peroxidase, glutathione reductase, glutathione S-transferase, superoxide dismutase, and catalase. Levels of malondialdehyde were reduced in *Allium sativum* treated animals, whereas levels of glutathione were maintained. The neurobehavioral activities (grip strength, spontaneous motor activity, and motor coordination) were also spared in rats that received *Allium sativum*, thus showing functional significance of treatment.[7] Pretreatment with S-Allyl-L-cysteine, purified from *Allium sativum* extract, similarly spared hippocampal tissue of rats from the oxidative damage of ischemia and reperfusion. S-Allyl-L-cysteine maintained ATP content and spared activity of mitochondrial respiratory complexes in rats that were otherwise severely impaired by cerebral ischemia. Also noted in rats that received S-Allyl-L-cysteine were significant decreases in mitochondrial lipid peroxidation, protein carbonyl, and concentration of H_2O_2.[8]

Both aged *Allium sativum* and its main constituent, S-Allylcysteine, are potent antioxidants. Evidence shows that a major component of this effect is the ability to activate Nrf2 – a master regulator of the cellular redox state. Activation of Nrf2 results in the induction of many cytoprotective, antioxidant enzymes, including NAD(P)H quinone oxidoreductase, glutamate-cysteine ligase, heme oxygenase-1, glutathione S-transferase, and others.[9] As has been previously noted, activation of Nrf2 is associated with antidepressant-like effects.[10]

7.1.2 ANTI-INFLAMMATORY

Ex vivo, extract of *Allium sativum* attenuates the inflammatory effect of lipopolysaccharide in human blood. Pretreatment with *Allium sativum* reduced lipopolysaccharide-induced activation of NF-kβ, as well as production of inflammatory cytokines IL-1 and TNF-α. The expression of the anti-inflammatory cytokine IL-10 was unchanged.[11] *Allium sativum* also blocked the activation of TLR4 by lipopolysaccharide in mouse macrophages, resulting in the inhibition of NF-kβ activation, and attenuation of the expression of cyclooxygenase 2 and inducible nitric oxide synthase.[12]

The organosulfur compound thiacremonone purified from *Allium sativum* has a particularly potent anti-inflammatory effect. Pretreatment with the compound suppressed the 12-O-tetradecanoylphorbol-13-acetate-induced ear edema, whereas administration of thiacremonone directly into the hind paw of rats suppressed carrageenan-induced edema and arthritic responses. It also prevented expression of iNOS and COX-2, as well the nuclear actions of NF-κB. In vitro, thiacremonone inhibited lipopolysaccharide-induced activation of nitric oxide and nuclear translocation of NF-kβ in mouse macrophages.[13]

Of significance to the current discussion, raw, but not steamed, *Allium sativum* exercises anti-inflammatory effects in microglia, the brain's resident immune cells. Pretreatment with *Allium sativum* extract, or one of several organosulfur compounds purified from *Allium sativum*, attenuated lipopolysaccharide-induced stimulation of nitric oxide, IL-1β, TNF-α, and monocyte chemoattractant protein-1 in cultured mouse microglia cells.[14]

7.1.3 ANTIDIABETIC/ANTI-METABOLIC SYNDROME

Daily administration of raw *Allium sativum* paste to rats made diabetic and insulin resistant by high fructose diets improved insulin sensitivity while attenuating signs of metabolic syndrome and oxidative stress. Serum glucose, insulin, triglyceride, and uric acid levels were all reduced by treatment. *Allium sativum* also normalized the increased levels of lipid peroxidation and decreased levels of glutathione in hepatic tissues of the fructose-treated rats.[15]

Studies of *Allium sativum* in the treatment of Alzheimer's disease have revealed further relationships between this herb and insulin. Loss of insulin sensitivity is thought to play a role in the etiology of Alzheimer's disease. For example, insulin helps prevent phosphorylation of tau protein that precipitates the so-called "tangles" of tau in the brains of sufferers of Alzheimer's disease. Insulin reduces this phosphorylation largely by inhibiting GSK-3β.[16] There is evidence that aged extract of *Allium sativum*, and one of its active ingredients, S-allyl-L-cysteine, also inhibit GSK-3β in brain tissue. S-allyl-L-cysteine is also in fresh *Allium sativum*, but the concentration is higher in the aged product. In a mouse model of Alzheimer's disease, dietary administration of S-allyl-L-cysteine for four months decreased amyloid-β load, IL-1β reactive, plaque-associated microglia, Tau2 reactivity, and levels of GSK-3β protein.[17] PPARγ, a target of many diabetes medications,[18] decreases levels of pro-inflammatory cytochemokine activity in microglia and macrophages.[19] Diallyl disulfide, found in aged *Allium sativum*, increases the expression of PPARγ in cell culture and hence can act as an anti-inflammatory agent.[20] It should be recalled that activators of PPARγ have been found to have antidepressant effects in human subjects.[21]

7.1.4 PRECLINICAL ANTIDEPRESSANT-LIKE EFFECTS

Oral administration of an aqueous extract of *Allium sativum* for 20 successive days significantly decreased immobility time of rats in both the tail suspension and forced swim tests. The efficacy of the extract was found to be comparable to that of imipramine.[22] The chronic administration of *Allium sativum* extract also decreased immobility time in streptozotocin-induced diabetic rats. Further study showed that the treatment also increased activities of superoxide dismutase and glutathione peroxidase in brain tissue, and decreased levels of the oxidation marker malondialdehyde.[23] Because of its phylogenetic and phytochemical similarities to *Allium sativum*, onion has been found to exhibit antidepressant-like effects in rats. Daily administration of quercetin-rich onion powder decreased immobility in the forced swim test.[24] A similar antidepressant-like effect of onion is seen in mice.[25]

Chronic administration of ethanolic extract of *Allium sativum* also decreased immobility in mice in the tail suspension and forced swim tests. At the highest dose, those effects of *Allium sativum* were no different than those of chronic fluoxetine or imipramine. Efforts to determine the mechanism of action revealed that *Allium sativum* extract significantly decreased brain MAO-A and MAO-B levels, as compared to the control group.[26]

In a particularly in-depth study, allicin, the major organosulfur compound in *Allium sativum*, exhibited antidepressant-like effects in mice that had suffered chronic social defeat. Intraperitoneal administration of allicin prior to each instance of social defeat later decreased immobility times in the force swimming test and normalized activity in the social interaction test. It also restored appetite for sucrose. Evaluations of the brains of the mice revealed mechanisms by which the allicin likely produced the antidepressant-like effects. Antioxidant effects were evident, in that allicin decreased the production of reactive oxygen species, malondialdehyde, and protein carbonyl in brain tissue. It also upregulated the activities of superoxide dismutase and Nrf2/HO-1 pathway. Anti-inflammatory effects were also observed. Allicin attenuated the hyperactivity of the NLRP3 inflammasome in the socially defeated mice, with subsequent decreases of caspase-1, IL-1β, and other inflammatory cytokines in the hippocampus.[27]

7.1.5   HUMAN ANTIDEPRESSANT EFFECTS

There are no published studies of the use of *Allium sativum* or its primary constituent phytochemicals in the treatment of MDD in human subjects.

7.1.6   DOSAGE

The German E Commission text describes *Allium sativum* in quite benign terms. It lists no contraindications, and notes the only compelling side effects to be bad breath and occasional but rare gastrointestinal upset. I note that aged *Allium sativum* is alleged to be odorless. The book gives the common daily dosage as 4 g fresh *Allium sativum*. The authors do not give recommendations for use for mood or cognitive effects.[28]

For dosage more germane for the central nervous system, I note a study of healthy women in whom 400mg of dried powdered garlic twice daily had significant effects on memory and attention without adverse effects.[29]

7.1.7 Toxicity

In rabbits, the LD50 of dried aqueous extract of *Allium sativum* has been found to be 3 g/kg.[30] When extrapolated to a 150-pound human adult, that LD50 would be about 200 g.

7.1.8 Safety in Pregnancy

The *Botanical Safety Handbook* notes references of caution about the use of *Allium sativum* during pregnancy in the ancient Ayurvedic literature. However, they note a general lack of indications for concern for its use, as well as a modern study showing safety in the third trimester. Overall, they suggest that moderate consumption is generally regarded as safe.[31]

7.1.9 Drug Interactions

The *Botanical Safety Handbook* notes that high doses of *Allium sativum* in humans can affect INR values in patients on coumadin. It may also interfere with the blood levels of certain antivirals and antibiotics, including ritonavir, saquinavir, and rifampin.[32]

REFERENCES

1. Rivlin RS. Historical perspective on the use of garlic. *J Nutr.* 2001;131(3):951S–954S.
2. Lanzotti V. The analysis of onion and garlic. *J Chromatog.* 2006;1112:3–22.
3. Lawson LD. Garlic: a review of its medicinal effects and indicated active compounds. In: Lawson LD, Bauer R, editors. *Phytomedicines of Europe: Chemistry and Biological Activity*, ACS Symposium Series 691. Washington, DC: American Chemical Society; 1998. p. 179–209.
4. Rahman MM, Fazlic V, Saad NW, et al. Antioxidant properties of raw garlic (*Allium sativum*) extract. *Int Food Res J (Malaysia).* 2012;19(2):589–591.
5. Im KJ, Park DK, Rhee MS, et al. Inhibitory effects of garlic extracts on the nitrosation. *Appl Biol Chem.* 2000;43(2):110–115.
6. Naji KM, Al-Shaibani ES, Alhadi FA, et al. Hepatoprotective and antioxidant effects of single clove garlic against CCl_4-induced hepatic damage in rabbits. *BMC Complement Altern Med.* 2017;17:411.
7. Saleem S, Ahmad M, Ahmad AS, et al. Behavioral and histologic neuroprotection of aqueous garlic extract after reversible focal cerebral ischemia. *J Med Food.* 2006; 9(4):537–544.
8. Fahim Atif F, Yousuf S, Agrawal SK. S-Allyl L-cysteine diminishes cerebral ischemia-induced mitochondrial dysfunctions in hippocampus. *Brain Res.* 2009;1265:128–137.
9. Colin-Gonz AL, Santana RA, Silva-Islas CA, et al. The antioxidant mechanisms underlying the aged garlic extract- and S-allylcysteine-induced protection. *Oxid Med Cell Longev.* 2012: Article ID 907162.
10. Maes M, Fišar Z, Medina M, et al. New drug targets in depression: inflammatory, cell-mediated immune, oxidative and nitrosative stress, mitochondrial, antioxidant, and neuroprogressive pathways. And new drug candidates—Nrf2 activators and GSK-3 inhibitors. *Inflammopharmacology.* 2012;20(3):127–150.
11. Keiss HP, Dirsch VM, Hartung T, et al. Garlic (*Allium sativum* L.) modulates cytokine expression in lipopolysaccharide-activated human blood thereby inhibiting NF-kβ activity. *J Nutr.* 2003;133:2171–2175.

12. Youn HS, Lim HJ, Lee HJ, et al. Garlic (*Allium sativum*) extract inhibits lipopoly-saccharide-induced toll-like receptor 4 dimerization. *Biosci Biotechnol Biochem.* 2008;72(2):368–375.
13. Ban JO. Oh JH, Kim TM, et al. Anti-inflammatory and arthritic effects of thiacremo-none, a novel sulfurcompound isolated from garlic via inhibition of NF-κB. *Arthritis Res Ther.* 2009;11:R145.
14. Ho SC, Su, MS. Evaluating the anti-neuroinflammatory capacity of raw and steamed garlic as well as five organosulfur compounds. *Molecules* 2014. 19(11), 17697–17714.
15. Padiya R, Khatua TN, Bagul PK, et al. Garlic improves insulin sensitivity and associ-ated metabolic syndromes in fructose fed rats. *Nutr Metab.* 2011;8:53.
16. Li ZG, Zhang W, Sima AA. Alzheimer-like changes in rat models of spontaneous dia-betes. *Diabetes.* 2007;56(7):1817–1824.
17. Chauhan NB. Effect of aged garlic extract on APP processing and tau phosphorylation in Alzheimer's transgenic model Tg2576. *J Ethnopharmacol.* 2006;108(3):385–394.
18. Kota BP, Huang TH, Roufogalis BD. An overview on biological mechanisms of PPARs. *Pharmacol Res.* 2005;51(2):85–94.
19. Pisanu A, Lecca D, Mulas G, et al. Dynamic changes in pro- and anti-inflammatory cytokines in microglia after PPAR-γ agonist neuroprotective treatment in the MPTPp mouse model of progressive Parkinson's disease. *Neurobiol Dis.* 2014;71:280–291.
20. Lee JH, Kim KA, Kwon KB, et al. Diallyl disulfide accelerates adipogenesis in 3T3-L1 cells. *Int J Mol Med.* 2007;20(1):59–64.
21. Kemp DF, Ismail-Beigi F, Ganocy SJ, et al. Use of insulin sensitizers for the treatment of major depressive disorder: a pilot study of pioglitazone for major depression accom-panied by abdominal obesity. *J Affect Disord.* 2012;136(3):1164–1173.
22. Singh A, Singh A. Antidepressant activity of aqueous extract of *Allium sativum* Linn. in albino rats. *Int J Sci Res.* 2018;7(4):36–38.
23. Rahmani G, Farajdokht F, Mohaddes G, et al. Garlic (*Allium sativum*) improves anxi-ety- and depressive-related behaviors and brain oxidative stress in diabetic rats. *Arch Physiol Biochem.* 2018;1–6.
24. Sakakibara H, Yoshino S, Kawai Y, et al. Antidepressant-like effect of onion (*Allium cepa* L.) powder in a rat behavioral model of depression. *Biosci Biotech Biochem.* 2008;72(1):94–100.
25. Samad N, Saleem, A. Administration of *Allium cepa* L. bulb attenuates stress-pro-duced anxiety and depression and improves memory in male mice. *Met Brain Dis.* 2018;33(1):271–281.
26. Dhingra D, Kumar V. Evidences for the involvement of monoaminergic and GABAergic systems in antidepressant-like activity of garlic extract in mice. *Indian J Pharmacol.* 2008;40(4):175–179.
27. Gao W, Wang W, Liu G, et al. Allicin attenuated chronic social defeat stress induced depressive-like behaviors through suppression of NLRP3 inflammasome. *Met Brain Dis.* 2018;1–11.
28. Blumenthal M, Busse WR, Goldberg A, et al. *The Complete German Commission E Monograph: Therapeutic Guide to Herbal Medicines.* Austin, TX: American Botanical Council; 1998. p.134.
29. Tasnim S, Haque PS, Bari M, et al. *Allium sativum* L. improves visual memory and attention in healthy human volunteers. *Evid-Based Compl Alt.* 2015: Article ID 103416.
30. Mikail HG. Phytochemical screening, elemental analysis and acute toxicity of aque-ous extract of *Allium sativum* L. bulbs in experimental rabbits. *J Med Plant Res.* 2010;4(4):322–326.
31. Gardner Z, McGuffin M, editors. *Botanical Safety Handbook*, 3rd edition. Boca Raton, FL: CRC Press; 2013. p. 40.

32. Gardner Z, McGuffin M, editors. *Botanical Safety Handbook*, 3rd edition. Boca Raton, FL: CRC Press; 2013. p. 41.

7.2 ANGELICA SINENSIS

Angelica sinensis, also known as *dong quai*, grows in cool high-altitude mountains in China, Japan, and Korea. The dried root of this plant from the carrot and celery family has been used for nearly 2,000 years in Traditional Chinese Medicine (TCM). The earliest record of its use was in the *Divine Husbandman's Classic of the Materia Medica (Shen Nong Ben Cao Jing)*, published during the period of the Han Dynasty (AD 25–225). It is most widely used for various women's health problems, e.g., menstrual disorders, amenorrhea, dysmenorrhea, premenstrual syndrome, and menopause, but has been used for inflammatory conditions, anemia, fatigue, and hypertension. It is also noted by Yeung et al. as being one of the most commonly used herbs in the various combinations prescribed for treatment of MDD in TCM.[1]

The herb contains a variety of phytosterols, polysaccharides, and flavonoids. However, ferulic acid, ligustilide and butylidenephthalide are likely the main biologically active components.[2] Of note is the reputation of *Angelica sinensis* of acting as a phytoestrogen. Indeed, this has been the basis for its use in various gynecological conditions. However, modern evaluation of this ancient herbal remedy has found little basis for this. For example, in a study of postmenopausal women, treatment with *Angelica sinensis* was no different than placebo in effects on endometrial thickness, vaginal maturation index, number of vasomotor flushes, or the Kupperman Index, a widely used rating scale of menopausal symptoms.[3]

7.2.1 ANTIOXIDANT

Unpurified aqueous extracts of *Angelica sinensis* exhibited antioxidant activities in a concentration-dependent manner in several standard assays. It significantly inhibited $FeCl_2$-ascorbic acid-induced lipid peroxidation *in vitro* in rat liver homogenates, with up to half the efficacy of α-tocopherol. It showed similar antioxidant effects in the Cytochrome c assay of superoxide anion scavenging activity, and the xanthine oxidase test of anti-superoxide activity.[4] Ligustilide, a major bioactive component of *Angelica sinensis*, also exhibited antioxidant effects *in vitro* by protecting cultured PC12 neuronal cells from H_2O_2-induced cytotoxicity.[5]

In intact rats subjected to induced cerebral ischemia, intraperitoneal administration of extract of *Angelica sinensis* reduced the size of cerebral infarction, improved neurological deficit scores, and increased blood flow and superoxide dismutase activity.[6] Ligustilide extracted from the herb was similarly effective in sparing brain tissue from oxidative damage after cerebral ischemia and reperfusion.[7] Ferulic acid, another major component of *Angelica sinensis*, also exhibits potent anti-oxidative effects, as well as reducing expression of nitric oxide synthase expression after focal cerebral ischemia, thus reducing nitrosative damage.[8]

7.2.2 ANTI-INFLAMMATORY

In vitro, purified extracts of *Angelica sinensis* stimulated activity of NrF2, which mediates the expression of cellular anti-oxidative stress genes. It also suppressed lipopolysaccharide (LPS)-induced IL-1β and TNF-α expression in cultured macrophages.[9] In the carrageenan-induced paw edema model of inflammation, treatment with the volatile oils of *Angelica sinensis* attenuated the inflammatory response, likely due to dampening of prostaglandin E2 synthesis.[10]

Specific phytochemical components of *Angelica sinensis* have also been found to have anti-inflammatory properties that could contribute to antidepressant effects. Ferulic acid inhibits release of macrophage inflammatory protein-2 (MIP-2) from mouse macrophages. Z-ligustilide inhibits activity of the inflammatory cytokine TNF-α and the intracellular activator of inflammation, NF-κB.[11]

7.2.3 ANTIDIABETIC/ANTI-METABOLIC SYNDROME

Angelica sinensis exhibits significant antidiabetic and insulin-sensitizing effects. In both prediabetic and streptozotocin (STZ)-induced diabetic mice, chronic treatment with *Angelica sinensis* reduced fasting blood glucose levels and abnormal fasting serum insulin concentrations. Insulin-sensitivity was increased as shown by the homeostasis model assessment of insulin resistance. Part of this anti-diabetic effect was likely due to decreases in serum levels of the inflammatory cytokines IL-6 and TNF-α that are known to contribute to insulin-resistance.[12] However, extracts of the herb have also been shown to stimulate the PI3K/Akt/mTOR pathway in mouse muscle cells.[13] This pathway forms a major part of the intracellular insulin response system. In the brain, activation of this system contributes to the antidepressant effects of ketamine and other medications.

7.2.4 PRECLINICAL ANTIDEPRESSANT-LIKE EFFECTS

Whereas *Angelica sinensis* is a component of several TCM herbal combinations used in the treatment of MDD, recent studies have shown that the herb may have its own antidepressant properties. In rats subjected to chronic unpredictable stress, alcohol extracts of *Angelica sinensis* restored appetite for sucrose and reduced immobility in the forced swim test. These effects were similar to those of the fluoxetine active control group. Both *Angelica sinensis* and fluoxetine partially reversed the stress-induced reductions in CREB and ERK proteins with resulting treatment-induced increases in hippocampal BDNF.[14] Interestingly, extract of *Angelica sinensis* was found to stimulate PI3K and inhibit GSK-3β in cultured cortical neurons.[15]

Studies have also been performed using ferulic acid, a primary active component of *Angelica sinensis*. Ferulic acid (4-hydroxy-3-methoxycinnamic acid) is a phenolic compound present in a variety of food and medicinal plants, including chocolate, coffee, red cabbage, grapefruit, eggplant, spinach, and whole grains. Particularly rich sources are popcorn and bamboo shoots.[16] In Japan ferulic acid has been approved as an antioxidant additive and food preservative.[17] There is also evidence that ferulic acid, like ketamine, may act as an NMDA receptor antagonist.[18]

Acute treatment with ferulic acid decreased immobility in mice in the tail suspension test. Intracerebroventricular administrations of inhibitors of various cell signaling pathways involved in neuroplasticity, neurogenesis, and cell survival blocked that antidepressant-like effect. These pathways included protein kinase A, Ca2 +/calmodulin-dependent protein kinase II, protein kinase C, MAPK/ERK, and PI3K.[19] It should be noted that the reversals of the antidepressant effects were defined by increases of immobility, which could have been non-specific.

In another study, acute treatment with ferulic acid decreased immobility in the tail suspension and forced swim tests in mice, and these effects were thought to be due to increases in serotonergic and noradrenergic activities. Serotonin and noradrenaline levels of treated animals were increased in the hippocampus and frontal cortex.[20]

7.2.5   HUMAN ANTIDEPRESSANT EFFECTS

There are no published studies of *Angelica sinensis* or its main phytochemical components in the treatment of MDD in human subjects.

7.2.6   DOSAGE

The German E Commission makes no mention of the use of *Angelica sinensis* for the treatment of mood disorders. It lists its use for loss of appetite and gastrointestinal upset. The daily dosage it recommends for such is 4.5 g powder, 1.5 to 3 g fluid extract, 1.5 g of tincture, or 10 to 20 drops of essential oil.[21]

7.2.7   TOXICITY

The toxicity of *Angelica sinensis* appears to be quite low. The LD50 value of the volatile oil of *Angelica sinensis* in mice was found to be 1.76 mL (about 35 drops)/kg.[22] This is about 100 times the recommended dose.

7.2.8   SAFETY IN PREGNANCY

Easley and Horne[23] advise against using the herb during pregnancy. However, two references from Chinese medicine give no indications of danger in its use by pregnant women.[24,25] Thus, it seems reasonable to suggest that benefits must outweigh any concerns for risk.

7.2.9   DRUG INTERACTIONS

The *Botanical Safety Handbook* reports a possible interaction between *Angelica sinensis* and warfarin. Thus, if the herb is considered necessary, INRs must be monitored accordingly.[26]

REFERENCES

1. Yeung WF, Chung KF, Ng KY, Yu YM, Ziea ET, Ng BF. A systematic review on the efficacy, safety and types of Chinese herbal medicine for depression. *J Psychiat Res* 2014;57:165–175.

2. Lao SC, Li SP, Kan KKW, et al. Identification and quantification of 13 components in *Angelica sinensis* (Danggui) by gas chromatography-mass spectrometry coupled with pressurized liquid extraction. *Anal Chim Acta* 2004;526(2):131–137.

3. Hirata JD, Swiersz LM, Zell B, Small R, Ettinger B. Does dong quai have estrogenic effects in postmenopausal women? A double-blind, placebo-controlled trial. *Fertil Steril* 1997;68(6):981–986.

4. Wu SJ, Ng LT, Lin CC. Antioxidant activities of some common ingredients of traditional Chinese medicine, *Angelica sinensis*, *Lycium barbarum* and *Poria cocos*. *Phytother Res* 2004;18(12):1008–1012.

5. Yu Y, Du JR, Wang CY, Qian ZM. Protection against hydrogen peroxide-induced injury by Z-ligustilide in PC12 cells. *Exp Brain Res* 2008;184(3):307–312.

6. Zhao XQ, Ji XM, Shi WJ. Angelica against cerebral ischemia/reperfusion injury. *Zhongguo Meitan Gongye Yixue Zazhi* 2009;12(11):1773–1774.

7. Peng HY, Du JR, Zhang GY, et al. Neuroprotective effect of Z-ligustilide against permanent focal ischemic damage in rats. *Biol Pharm Bull* 2007;30(2):309–312.

8. Koh PO. Ferulic acid modulates nitric oxide synthase expression in focal cerebral ischemia. *Lab Anim Res* 2012;28(4):273–278.

9. Saw CLL, Wu Q, Su ZY, et al. Effects of natural phytochemicals in *Angelica sinensis* (Danggui) on Nrf2-mediated gene expression of phase II drug metabolizing enzymes and anti-inflammation. *Biopharm Drug Dispos* 2013;34(6):303–311.

10. Zhang WQ, Hua Y, Zhang M, et al. Metabonomic analysis of the anti-inflammatory effects of volatile oils of *Angelica sinensis* on rat model of acute inflammation. *Biomed Chromatogr* 2015;29(6):902–910.

11. Chao WW, Lin B-F. Bioactivities of major constituents isolated from *Angelica sinensis* (Danggui). *Chin Med* 2011;6(1):29.

12. Wang K, Cao P, Shui W, Yang Q, Tang Z, Zhang Y. *Angelica sinensis* polysaccharide regulates glucose and lipid metabolism disorder in prediabetic and streptozotocin-induced diabetic mice through the elevation of glycogen levels and reduction of inflammatory factors. *Food Funct* 2015;6(3):902–909.

13. Yeh TS, Hsu CC, Yang SC, Hsu MC, Liu JF. *Angelica sinensis* promotes myotube hypertrophy through the PI3K/Akt/mTOR pathway. *BMC Compl ALT Med* 2014;14(1):144–153.

14. Shen J, Zhang J, Deng M, Liu Y, Hu Y, Zhang L. The antidepressant effect of *Angelica sinensis* extracts on chronic unpredictable mild stress-induced depression is mediated via the upregulation of the BDNF signaling pathway in rats. *Evid-Based Compl Alt* 2016;1–8:Article ID 7434692.

15. Zhang Z, Zhao R, Qi J, Wen S, Tang Y, Wang D. Inhibition of glycogen synthase kinase-3β by *Angelica sinensis* extract decreases β-amyloid-induced neurotoxicity and tau phosphorylation in cultured cortical neurons. *J Neurosci Res* 2011;89(3):437–447.

16. Kumar N, Pruthi V. Potential applications of ferulic acid from natural sources. *Biotechnol Rep (Amst)* 2014;4:86–93.

17. Graf E. Antioxidant potential of ferulic acid. *Free Radic Biol Med* 1992;13(4):435–448.

18. Yu L, Zhang Y, Liao M, et al. Neurogenesis-enhancing effect of sodium ferulate and its role in repair following stress-induced neuronal damage. *WJNS* 2011;1:9–18.

19. Zeni ALB, Zomkowski ADE, Maraschin M, Rodrigues AL, Tasca CI. Involvement of PKA, CaMKII, PKC, MAPK/ERK and PI3K in the acute antidepressant-like effect of ferulic acid in the tail suspension test. *Pharmacol Biochem Behav* 2012;103(2):181–186.

20. Chen J, Lin D, Zhang C, et al. Antidepressant-like effects of ferulic acid: Involvement of serotonergic and norepinergic systems. *Met Brain Dis* 2015;30(1):129–136.

21. Blumenthal M, Busse WR, Goldberg A, et al. *The Complete German Commission E Monographs*. Austin, TX: American Botanical Council; 1998.

22. Li J, Hua Y, Ji P, et al. Effects of volatile oils of *Angelica sinensis* on an acute inflammation rat model. *Pharml Biol* 2016;54(9):1881–1890.
23. Easley T, Horne S. *The Mod Herb Dispensary*. Berkeley, CA: North Atlantic Books. p. 224; 2016.
24. Bensky D, Barolet R. *Chin Herb Med Formulas Strateg*. Seattle, WA: Eastland Press; 1990.
25. Chen JK, Chen TT. *Chinese Medical Herbology and Pharmacology*. City of Industry, CA: Art of Medicine Press; 2004.
26. Gardner Z, McGuffin M, editors. *Botanical Safety Handbook*. 3rd ed. Boca Raton, FL: CRC Press. p. 264; 2013.

7.3   *APIUM GRAVEOLENS* (CELERY)

Apium graveolens (celery) is a marshland plant in the family *Apiaceae*, which includes angelica, anise, caraway, carrot, coriander, cumin, dill, fennel, hemlock, and parsley. Like many plants in this family, *Apium graveolens* has for thousands of years been used as food, spice, and medicine. The leaves, stalks, and seeds of the plant have traditionally been used to treat rheumatism, gout, urinary tract inflammation, agitation, anorexia, exhaustion, and depression.[1] *Apium graveolens* contains a wide variety of phytochemicals including limonene, selenine, celerin, bergapten, apiumoside, apiumetin, apigravrin, osthenol, isopimpinellin, isoimperatorin, celereoside, graveobiosides, apigenin, isoquercitrin, pinene, camphene, cymene, α-thuyene, β-phellendrene, γ-terpinene, sabinene, terpinolene, myristicic, myristic, santalol, eudesmol, sedanenolide, and 3-n-butylphthalide.[2] It is likely sedanenolide and 3-n-butylphthalide that are responsible for the characteristic flavor and aroma of *Apium graveolens*.

7.3.1   ANTIOXIDANT

Apium graveolens contains potent antioxidants, and methanolic extracts have been shown to have strong radical scavenging capabilities in the standard *in vitro* DPPH assay. In the β-carotene-linoleate antioxidant assay, the extract was nearly as effective as the commercial antioxidant butylated hydroxyl toluene.[3] When administered to streptozotocin-induced diabetic rats, extract of seed was found to restore activities of the antioxidant enzymes superoxide dismutase, glutathione-transferase, and glutathione-reductase in liver. It also reduced levels of malondialdehyde but increased glutathione concentrations.[4] Juice from the roots and leaves of *Apium graveolens* similarly spared the activity of antioxidant liver enzymes in rats that had received toxic doses of the antibiotic doxorubicin.[5]

7.3.2   ANTI-INFLAMMATORY

Anti-inflammatory effects of *Apium graveolens* extracts are also well documented. Intraperitoneal administration of a hydroalcoholic extract of *Apium graveolens* seed dampened inflammation in the standard carrageenan-induced paw edema in rats. The highest dose of the extract, 100 mg/kg, had an anti-inflammatory effect similar to that of acetylsalicylic acid.[6] Extract of *Apium graveolens* seed also attenuated the inflammatory response of murine macrophages exposed to oxidized low-density lipoprotein. The extract decreased the secretion of the inflammatory cytokines TNF-α and IL-6 by these cells, and this was found to be due in part to suppression of the NF-κB pathway.[7]

7.3.3 ANTIDIABETIC/ANTI-METABOLIC SYNDROME

Chronic administration of hexane extract of *Apium graveolens* seed reduced serum glucose levels and restored insulin levels toward normal in streptozotocin-induced diabetic rats.[8] *Apium graveolens* extract also has antidiabetic effects in human patients. In elderly, pre-diabetic subjects, *Apium graveolens* reduced both pre- and post-prandial plasma glucose levels without affecting levels of plasma insulin, the most reasonable conclusion being that *Apium graveolens* improved insulin sensitivity.[9] Further evidence of salutary effects of *Apium graveolens* on diabetes and Metabolic Syndrome are seen in a study showing that chronic supplementation with extract of *Apium graveolens* accelerates weight loss and increases serum levels of adiponectin. Adiponectin has been found to enhance sensitivity to insulin and reduce inflammation,[10] as well as producing antidepressant effects.[11]

7.3.4 PRECLINICAL ANTIDEPRESSANT-LIKE EFFECTS

Several studies have confirmed antidepressant-like effects of extract of *Apium graveolens* seed. Acute administration of methanolic extract of *Apium graveolens* seed decreased immobility of rats in both the forced swim and tail suspension tests.[12] This treatment has also reduced immobility in the forced swim test and tail suspension tests in mice.[13,14] Chronic daily oral consumption of *Apium graveolens* extract also decreased immobility time in both the forced swimming and tail suspension tests in mice. Further evaluation found that activities of the antioxidant enzymes glutathione peroxidase and superoxide dismutase were increased with concomitant decreases in lipid peroxidation in brain tissues. On the other hand, activities of MAO-A and acetylcholinesterase were inhibited.[15] Similar findings came from a study of plant species used in Danish folk medicine for treatment of depression and anxiety. Aqueous and ethanolic extracts of these plants, including *Apium graveolens*, were tested for affinity to the serotonin transporter and for inhibition of MAO-A, as both have been targets for antidepressive treatment. The ethanolic extract of leaves of *Apium graveolens* had no effect on the 5-HT transporter. However, it was found to inhibit MAO-A with a surprisingly potent IC50 of 5 μg/ml.[16]

Recent studies have focused on 3-n-butylphthalide, a phytochemical of which *Apium graveolens* is uniquely well supplied. Studies in China have found 3-n-butylphthalide to be a potent antioxidant and anti-inflammatory agent. It also exerts powerful neuroprotective effects and improves outcomes after strokes, which has led it to be approved for clinical use as an anti-ischemic agent.[17]

3-n-Butylphthalide has also been found to have an antidepressant-like effect in rats. Specifically, it attenuated the depression-like behavior of rats that had been subjected to two weeks of injections of the inflammatory agent lipopolysaccharide. This antidepressant-like effect of 3-n-Butylphthalide was associated with antioxidant, anti-inflammatory, and anti-apoptotic effects in the hippocampi of these animals. 3-n-Butylphthalide reduced the inflammatory response with inhibited expression of pro-inflammatory cytokines, including IL-1β and IL-6, and downregulation of the NF-κB signal pathway. Concurrent with the anti-inflammation action, 3-n-Butylphthalide reduced LPS-induced oxidative reactions in the hippocampus and enhanced Nrf2-targeted signals, as evidenced by increased transcription of antioxidant enzymes and decreased malondialdehyde production.[18] Moreover, in APP/PS1 double transgenic mice, an animal model of Alzheimer's disease, 3-n-butylphthalide improved cognitive function and concomitantly activated the PI3K/AKT pathway, increased BDNF levels, and stimulated TrkB. All of these effects are strongly associated with antidepressant effects.[19]

7.3.5 HUMAN ANTIDEPRESSANT EFFECTS

There are no studies of the use of *Apium graveolens* in the treatment of MDD in human subjects. 3-n-Butylphthalide, found in *Apium graveolens*, has shown remarkable effects in the treatment of stroke, and has neurochemical effects that would also be expected to improve MDD. There have been no tests of 3-n-Butylphthalide in the treatment of MDD. Doses of 200 mg 3-n-Butylphthalide three times daily have been safely and effectively used in patients with vascular cognitive impairment without dementia, stroke, or MDD.[20]

7.3.6 DOSAGE

Due to lack of documentation of efficacy, the German E Commission has not approved *Apium graveolens* as a treatment of any human illness. It also notes the possibility of allergic reaction, the so-called, "celery-carrot-mugwort syndrome," and thus does not recommend its use.[21] Easley and Horne[22] recommend 500 to 1500 mg three times a day, or a 1:5 tincture in 10 drops to 2 ml three times a day.

7.3.7 TOXICITY

In a study in rats, the LD50 of a dried ethanolic extract of *Apium graveolens* seed was found to be 4.25 grams per kilogram, which in turn was equivalent to about 90 grams of whole seed per kilogram.[23]

7.3.8 SAFETY IN PREGNANCY

Easley and Horne advise against the medicinal use of *Apium graveolens* during pregnancy. The *Botanical Safety Handbook* does not warn against its use in pregnancy.[24] They cite no human data (nor do Easley and Horne), but do note studies showing even high doses to be benign in pregnant rodents.

7.3.9 DRUG INTERACTION

The *Botanical Safety Handbook* notes no known interactions with drugs. One study did show potential for such interaction. Mice pretreated with celery juice exhibited a

prolonged response to pentobarbital. Prolongation of the analgesic effects of amino-
pyrine and paracetamol were also observed. In those animals, exposure to the celery
juice caused a significant decrease of cytochrome P450 in the liver homogenates.[25]

REFERENCES

1. Ebadi M. *Pharmacodynamic Basis of Herbal Medicine*. Boca Raton, FL: CRC press; 2007. p. 89.
2. Al-Snafi AE. The pharmacology of *Apium graveolens*.—A review. *IJPRS*. 2014;3(1):671–677.
3. Jung WS, Chung IM, Kim SH, Praveen N. In vitro antioxidant activity, total phenolics and flavonoids from celery (*Apium graveolens*) leaves. *J Med Plants Res*. 2011;5(32):7022–7030.
4. Al-Sa'aidi JAA, Alrodhan MNA, Ismael AK. Antioxidant activity of n-butanol extract of celery (*Apium graveolens*) seed in streptozotocin-induced diabetic male rats. *Res Pharm Biotech*. 2012;4(2):24–29.
5. Kolarovic J, Popovic M, Zlinská J, Trivic S, Vojnovic M. Antioxidant activities of celery and parsley juices in rats treated with doxorubicin. *Molecules*. 2010;15(9): 6193–6204.
6. Arzi A, Hemmati AA, Karampour NS, Nazari Z, Baniahmad B. Anti-inflammatory effects of celery seed hydroalcoholic extract on carrageenan-induced paw edema in rats. *Res J Pharm Biol Chem Sci*. 2014;5(6):24–29.
7. Si Y, Guo S, Fang Y, et al. Celery seed extract blocks peroxide injury in macrophages via Notch1/NF-κB pathway. *Am J Chin Med*. 2015;43(3):443–455.
8. Tashakori-Sabzevar F, Ramezani M, Hosseinzadeh H, et al. Protective and hypoglycemic effects of celery seed on streptozotocin-induced diabetic rats: Experimental and histopathological evaluation. *Acta Diabetol*. 2016;53(4):609–619.
9. Yusni Y, Zufry H, Meutia F, Sucipto K. The effects of celery leaf (*Apium graveolens* L.) treatment on blood glucose and insulin levels in elderly pre-diabetics. *Saudi Med J*. 2018;39(2):154–160.
10. Tilg H, Moschen AR. Adipocytokines: Mediators linking adipose tissue, inflammation and immunity. *Nat Rev Immunol*. 2006;6(10):772–783.
11. Liu J, Guo M, Zhang D, et al. Adiponectin is critical in determining susceptibility to depressive behaviors and has antidepressant-like activity. *PNAS*. 2012;109(30):12248–12253.
12. Srinivasa B, Desu R, Sivaramakrishna K. Antidepressant activity of methanolic extract of Apium graveolens seeds. *IJRPC*. 2012;2(4):1124–1127.
13. Sutrisna E, Azizah T, Wuryaningrum A, Sari MP. The potency of *Lactuca sativa* Linn. And *Apium graveolens* L. from Indonesia as tranquilizer. *Int J Ayur Pharma Res*. 2015;3(4):6–11.
14. Farshid BP, Abbasi MS, Efthekhari SK, Morteza H, Zafar NS. Antidepressant-like potential of ethanolic extract of Apium graveolens L. in the forced swim and tail suspension test in mice. *National Congress on Medicinal Plants*. 2013;2.
15. Boonruamkaew P, Sukketsiri W, Panichayupakaranant P, et al. *Apium graveolens* extract influences mood and cognition in healthy mice. *J Nat Med*. 2017;71(3): 492–505.
16. Jäger AK, Gauguin B, Andersen J, Adsersen A, Gudiksen L. Screening of plants used in Danish folk medicine to treat depression and anxiety for affinity to the serotonin transporter and inhibition of MAO-A. *J Ethnopharmacol*. 2013;145(3):822–825.
17. Abdoulaye IA, Guo YJ. A review of recent advances in neuroprotective potential of 3-N-butylphthalide and its derivatives. *BioMed Res Int*. 2016;5012341.

18. Yang M, Dang R, Xu P, et al. Dl-3-n-Butylphthalide improves lipopolysaccharide-induced depressive-like behavior in rats: Involvement of Nrf2 and NF-κB pathways. *Psychopharmacology.* 2018;235(9):2573–2585.
19. Xiang J, Pan J, Chen F, et al. L-3-n-butylphthalide improves cognitive impairment of APP/PS1 mice by BDNF/TrkB/PI3K/AKT pathway. *Int J Clin Exp Med.* 2014;7(7):1706–1713.
20. Jia J-P, Wei C, Liang J, et al. The effects of DL-3-n-butylphthalide in patients with vascular cognitive impairment without dementia caused by subcortical ischemic small vessel disease: A multicentre, randomized, double-blind, placebo-controlled trial. *Alzheimers Dement.* 2016;12(2):89–99.
21. Blumenthal M, Busse WR, Goldberg A, et al. *The Complete German Commission E Monographs.* Austin, TX: American Botanical Council; 1998. p. 320.
22. Easley T, Horne S. *The Modern Herbal Dispensary.* Berkeley, CA: North Atlantic Books; 2016;p. 208.
23. Aburjai T, Mansi K, Abushoffa A, Disi A. Hypolipidemic effects of seed extract of celery (*Apium graveolens*) in rats. *Pharmacogn Mag.* 2009;5(20):301–305.
24. Gardner Z, McGuffin M, editors. *Botanical Safety Handbook.* 3rd ed. Boca Raton, FL: CRC Press; 2013;p. 72.
25. Jakovljevic V, Raskovic A, Popovic M, Sabo J. The effect of celery and parsley juices on pharmacodynamic activity of drugs involving cytochrome P450 in their metabolism. *Eur J Drug Metab Pharmacokinet.* 2002;27(3):153–156.

7.4 *ASTRAGALUS MEMBRANACEUS*

Astragalus membranaceus is the species of Astragalus used in traditional Chinese medicine. It is alternatively known as *A. propinquus* or *A. mongholicus*. As is typical in the Chinese tradition, *Astragalus membranaceus* has generally been used as part of combinations of herbs for treatment of various illness. It has most commonly found use in treatments of what are seen as in TCM as spleen and lung deficiencies contributing to diarrhea, fatigue, poor appetite, sweating, colds, shortness of breath, and various wasting disorders. However, given the role thought to be played by the spleen meridian in symptoms of fatigue and despondency, *Astragalus membranaceus* has also been part of what might be considered antidepressant treatments, such as Gui Pi Wan, or Restore the Spleen Pills, and Bu Zhong Yi Qi Wan, or Pill to Strengthen the Center and Boost Qi.[1]

Astragalus membranaceus contains many types of phytochemicals including: formononetin, astraisoflavan, astrapterocarpan, 2'-3'-dihydroxy-7,4'-dimethooxyisoflavone, isoliquiritigenin, D-ß-asparagine, calycosin, cycloastragenol, astragalosides, kumatakenin, ß-sitosterol, soyasaponin, and astragalin.[2] Extracts of *Astragalus membranaceus* exhibit a variety of pharmacological effects of significance for the treatment of MDD.

7.4.1 ANTIOXIDANT

Extracts of *Astragalus membranaceus* also show potent antioxidant effects. The aqueous extract from *Astragalus membranaceus* inhibited lipid peroxidation and production of malondialdehyde in rat heart mitochondria.[3] In rat liver, saponins from *Astragalus membranaceus* protected cells against the toxic oxidative effects of carbon tetrachloride and acetaminophen, as evidenced by decreases in malondialdehyde and increases in protective reduced glutathione.[4] Peripherally administered *Astragalus membranaceus* also offers neuroprotective effects against the oxidative and inflammatory effects of hypoxic damage in rodent brain. Pretreatment of mice with *Astragalus membranaceus* prior to arterial occlusion and reperfusion decreased concentrations of the oxidation product malondialdehyde, and increased activity of superoxide dismutase in brain homogenates.[5] Pretreatment with the herb similarly protected brain tissue of rats subjected to arterial occlusion and reperfusion. Levels of the inflammatory cytokines IL-1β, IL-6, and TNF-α were reduced in both serum and cortical homogenates of treated animals.[6] Astragalosides purified from *Astragalus membranaceus* extract also increased levels of BDNF and RNA for its target receptor, the TrkB receptor, in brains of rats that had suffered arterial occlusion and reperfusion.[7]

7.4.2 ANTI-INFLAMMATORY

Astragalus membranaceus exhibits anti-inflammatory effects. Extracts reduce the production of TNF-α and IL-1β in macrophages exposed to lipopolysaccharide.[8] The extracts also block release of those two inflammatory cytokines from macrophages stimulated by Advanced Glycation End Products. This was mediated by blocking phosphorylation of p38 MAPK protein and reducing activation of NF-κB.[9] Polysaccharides extracted from the herb also attenuate the lipopolysaccharide-induced inflammatory response of cultured microglia. The polysaccharides blocked lipopolysaccharide-stimulated release of nitric oxide and PGE2, as well as iNOS and cyclooxygenase-2 gene expression. They also attenuated generation of the pro-inflammatory cytokines IL-1β and TNF-α. The translocation of NF-κB, which is an important component of the inflammatory cascade was also dampened.[10]

7.4.3 ANTIDIABETIC/ANTI-METABOLIC SYNDROME

Astragalus membranaceus improves cognitive function in rats previously made diabetic by intraperitoneal injection of streptozotocin, thus demonstrating an ability to ameliorate central effects of diabetes and insulin resistance. Treatment decreased serum levels of both glucose and fasting insulin, showing improvement of insulin sensitivity. Performance in the Morris water maze improved in treated rats, indicating improvement of memory in comparison to untreated diabetic rats. In the hippocampi of treated animals, levels of the oxidation marker malondialdehyde were reduced, whereas activity of the antioxidant enzyme superoxide dismutase was increased. Thus, *Astragalus membranaceus* ameliorated the ongoing oxidative damage of diabetes.[11] Extracts of the herb also increase cortical levels of NGF in diabetic rats.[12]

7.4.4 PRECLINICAL ANTIDEPRESSANT-LIKE EFFECTS

Administration of *Astragalus membranaceus* to rats and mice has been shown to ameliorate depression-like behaviors induced by various mechanisms. The depression-like behavior in rats caused by administration of lipopolysaccharide was reduced by pretreatment with *Astragalus membranaceus*. Treated rats showed reduced immobility in the forced swim test and increased preference for sucrose. Treatment also inhibited the lipopolysaccharide-induced activation of the inflammatory NF-κB and MAPK signaling pathways in the hippocampus and hypothalamus. Levels of TNF-α, IL-1β, and IL-6 were consequently reduced in those areas.[13] In rats subjected to chronic unpredictable mild stress, concomitant treatment with *Astragalus membranaceus* extract maintained sucrose preference. That effect of the herb was similar to that seen with fluoxetine.[14]

Astragaloside IV (AS-IV) is an active component purified from *Astragalus membranaceus* that has shown anti-inflammatory, anti-oxidative, and anti-apoptotic effects. When chronically administered to mice undergoing repeated restraint stress, the typical depression-like effects of that procedure were dampened. Preference for sucrose was maintained, and immobility in the tail suspension and forced swim tests were reduced. Furthermore, astragaloside IV significantly reduced the levels of TNF-α and IL-1β in the hippocampus of the mice, which likely was the result of inhibition of NF-κB and deactivation of the NLRP3 inflammasome in hippocampal tissue. In the same animals, astragaloside IV increased hippocampal PPARγ and GSK-3β inhibition. Virtually the same behavioral and anti-neuroinflammatory effects of astragaloside IV were observed in the mice when the restraint stress procedure was replaced by induction of the depression-like effects of administration of lipopolysaccharide.[15]

7.4.5 HUMAN ANTIDEPRESSANT EFFECTS

A Chinese study showed *Astragalus membranaceus* to be an effective treatment for primary insomnia in geriatric patients.[16] However, there are no published reports of the effects of *Astragalus membranaceus* alone or any of its constituent phytochemicals on MDD or other mood disorders in human subjects.

7.4.6 DOSAGE

Hoffman suggests 4 to 8 ml of tincture three times a day. He also notes that in Traditional Chinese Medicine, the root is boiled in soup and removed before serving.[17] Easley and Horne suggest the tincture is not as effective as other forms, and recommend powdered root, 1 to 3 g three times daily.[18]

7.4.7 TOXICITY

There are many toxic species in the genus *Astragalus*.[19] However, the traditional medical plant, *Astragalus membranaceus*, is low in toxicity. The LD50 of *Astragalus membranaceus* is approximately 40 g/kg when administered by intraperitoneal injection. Overall, it is safe, and doses as high as 100 g/kg of the raw herb have been given to rats by lavage with no adverse effects.[20]

7.4.8 SAFETY IN PREGNANCY

The *Botanical Safety Handbook* notes a lack of information on the safety of *Astragalus membranaceus* in pregnancy.[21]

7.4.9 Drug Interactions

Astragalus membranaceus has been found to affect various immunosuppressive treatments. For example, it can enhance the inhibitory effect of corticosteroids on apoptosis.[22] It has also been found to enhance the effects of alpha-interferon in human subjects.[23] On the other hand, it has also been reported to dampen the immunosuppressive effects of cyclophosphamide.[24]

REFERENCES

1. Schnyer EN, Flaws B. *Curing Depression Naturally with Chinese Medicine.* Boulder, CO: Blue Poppy Press; 1998.
2. Sinclair S. Chinese herbs: A clinical review of *Astragalus, Ligusticum*, and *Schizandrae. Altern Med Rev.* 1998;3(5):338–344.
3. Hong CY, Lo YC, Tan FC, Wei YH, Chen CF. *Astragalus membranaceus* and *Polygonum multiflorum* protect rat heart mitochondria against lipid peroxidation. *Am J Chin Med.* 1994;22(1):63–70.
4. Zang YD, Shen JP, Zhu SH, Huang DK, Ding Y, Zhang XL. Effects of Astragalus (ASI, SK) on experimental liver injury. *Acta Pharm Sin.* 1992;27:401–406.
5. Ting H. Protective effects of extract of astragalus on injuries of global cerebral ischemia and reperfusion in rats and anoxia in mice. *Chin Pharm Bull.* 2004-05.
6. Jing L, Liang M, Rongrong H. Effects of extract of astragalus on IL-1β, TNF-α and IL-6 expressions after global cerebral ischemia and reperfusion. *Acta Universitis Med Nahui.* 2005-06.
7. Yin YY, Weiping LI, Weizu LI, Gong H, Zhu F, Guocui WU. Effects of astragalosides on the expression of BDNF, TrkB and p75NTR mRNA against focal cerebral ischemia-reperfusion injury. *Chin Pharm Bull.* 2009;25:672–676.
8. Ko JK, Chik CW. The protective action of radix *Astragalus membranaceus* against hapten-induced colitis through modulation of cytokines. *Cytokine.* 2009;47(2): 85–90.
9. Qin Q, Niu J, Gu Y, Wang Z, Xu W, Qiao Z. *Astragalus membranaceus* Inhibits Inflammation via phospho-P38 mitogen-activated protein kinase (MAPK) and nuclear factor (NF)-κB pathways in advanced glycation end product-stimulated macrophages. *Int J Mol Sci.* 2012;13(7):8379–8387.
10. Luo T, Qin J, Liu M, et al. Astragalus polysaccharide attenuates lipopolysaccharide-induced inflammatory responses in microglial cells: Regulation of protein kinase B and nuclear factor-κB signaling. *Inflamm Res.* 2015;64(3–4):205–212.
11. Dun CP, Liu J, Qiu F, et al. Effects of Astragalus polysaccharides on memory impairment in a diabetic rat model. *Neuropsychiatr Dis Treat.* 2016;12:1617–1621.
12. Liu HF, Cui R, Shen M, et al. Effect of astragalus polysaccharides on expression of nerve growth factor gene of cerebrum in Type 2 diabetic rats. *Food Nutr China.* 2011–08.
13. Li C, Wang Y, Qu J, et al. Astragalus polysaccharide inhibits lipopolysaccharide-induced depressive-like behaviors and inflammatory response through regulating NF-κB and MAPK signaling pathways in rats. *Int J Clin Exp Med.* 2018;11(3):2361–2370.
14. Ai Q, Zhang H, Tian G, Yang Y, Li D, Gao M. Effect of astragalus—Implant on actions of mice with chronic unpredictable mild stress depression. *J Dalian Med Univ.* 2010;32(4):405–407.
15. Song MT, Ruan J, Zhang RY, Deng J, Ma ZQ, Ma SP. Astragaloside IV ameliorates neuroinflammation-induced depressive-like behaviors in mice via the PPARγ/NF-κB/ NLRP3 inflammasome axis. *Acta Pharmacol Sin.* 2018;39(10):1559–1570.

16. Wang J-C, Song X, Yang S, et al. The mechanism of action of astragalus injection to treat chronicity primary insomnia. *Chin Gen Pract.* 2005-02.
17. Hoffmann D. *Medical Herbalism: The Science and Practice of Herbal Medicine.* Rochester, VT: Healing Arts Press; 2003.
18. Easley T, Horne S. *The Modern Herbal Dispensary.* Berkeley, CA: North Atlantic Books; 2016. p. 279
19. Rios JL, Waterman PG. A review of the pharmacology and toxicology of astragalus. *Phytother Res.* 1997;11(6):411–418.
20. Bensky D, Gamble A. *Chinese Herbal Medicine: Materia Medica.* Revised edition. Seattle, WA: Eastland Press; 1993.
21. Gardner Z, McGuffin M, editors. *Botanical Safety Handbook.* 3rd ed. Boca Raton, FL: CRC Press; 2013. p. 274
22. Cai XY, Xu YL, Lin XJ. Effects of radix Astragali injection on apoptosis of lymphocytes and immune function in patients with systemic lupus erythematosus. *Chin J Integ Trad West Med.* 2006;26(5):443–445.
23. Qian ZW, Mao SJ, Cai XC, et al. Viral etiology of chronic cervicitis and its therapeutic response to a recombinant interferon. *Chin Med J.* 1990;103(8):647–651.
24. Chu DT, Wong WL, Mavligit GM. Immunotherapy with Chinese medicinal herbs. II. Reversal of cyclophosphamide-induced immune suppression by administration of fractionated Astragalus membranaceus *in vivo. J Clin Lab Immunol.* 1988;25(3):125–129.

7.5 ATRACTYLODES MACROCEPHALA

Atractylodes macrocephala, called *Baizhu* in China, is a plant of the sunflower family native to Asia. It has long been used as a tonic agent in various ethno-medical systems, especially in China. It is noted by Yeung et al. to be among the most commonly used ingredients in TCM herbal treatments of depression.[1] More than 79 chemical compounds have been isolated from *Atractylodes macrocephala*, including sesquiterpenoids, triterpenoids, polyacetylenes, coumarins, phenylpropanoids, flavonoids and flavonoid glycosides, steroids, benzoquinones, and polysaccharides. The rhizome also contains a number of unique terpenoid lactones – referred to as atractylenolides. Extracts of the herb have been found to exhibit anti-tumor, anti-inflammatory, hypoglycemic, anti-aging, anti-oxidative, neuroprotective activity, and immunomodulatory effects, as well as improving gastrointestinal function and gonadal hormone regulation.

7.5.1 ANTIOXIDANT

A number of studies have demonstrated the antioxidative effects of *Atractylodes macrocephala in vitro*, in tissue culture, and in intact animals. *Atractylodes macrocephala* exerts potent antioxidant effects in DPPH and TBARS assays. Its antioxidant and radical-scavenging capacities are largely due to its constituent flavonoids and phenolic acids.[2] Extracts of *Atractylodes macrocephala* also protected cultured liver cells from the oxidative damage of CCl_4.

In rats that had been subjected to ligation-induced cerebral ischemia, polysaccharides extracted from *Atractylodes macrocephala* reduced oxidative damage in cerebral tissue. Levels of malondialdehyde were decreased, whereas superoxide dismutase activity was increased in affected brain regions.[3]

7.5.2 ANTI-INFLAMMATORY

Atractylenolides I and III purified from *Atractylodes macrocephala* are as potent as prednisone in relieving xylene-induced ear edema in mice.[4] Stimulation of the TLR4 activates NF-κB, which in turn promotes the expression of multiple genes in the inflammatory cascade. Atractylenolide I prevents the inflammatory response of white blood cells by blocking the TLR4.[5] TLR4 density is increased in the post-mortem brains of depressed and suicidal individuals.[6] Interestingly, at least two SSRIs, fluoxetine and citalopram, exhibit potent anti-inflammatory effects through inhibition of the TLR4.[7]

7.5.3 ANTIDIABETIC/ANTI-METABOLIC SYNDROME

Atractylodes macrocephala reduces serum glucose, and the herb is currently being used by TCM practitioners to treat diabetes type II. The Atractylenolides I, II, and II, and other terpenes in the herb bind to insulin receptors, the PPAR receptor, and the dipeptidyl peptidase-IV enzyme, all of which are deeply involved in the insulin response.[8]

7.5.4 PRECLINICAL ANTIDEPRESSANT-LIKE EFFECTS

Several studies have shown efficacy of atractylenolides in animal models of depression. Atractylenolide I significantly reduced depression like behaviors induced by chronic unpredictable stress in mice. It increased sucrose preference and shortened immobility time in both the forced swimming and the tail suspension test.[9] Neurochemically, atractylenolide I reduced stress induced decreases in the concentrations of serotonin and norepinephrine in the hippocampi of those mice. It also inhibited the activation of the inflammasome and reduced the resulting increased concentration of the pro inflammatory cytokine interleukin (IL) 1β in the hippocampus. Not only stress, but also metabolic dysregulation of cells and mitochondrial dysfunction activates the inflammasome.[10]

Atractylenolide III has also been found to exhibit antidepressant and cognitive enhancing effects in bulbectomized mice, an animal model of MDD. Oral administration of atractylenolide III for 18 days decreased immobility of the bulbectomized mice in the tail suspension test. This treatment also improved performance in the step-through passive avoidance, Y-Maze, and Novel Object Recognition tasks. Chemical analyses further showed that atractylenolide III attenuated decreases in Ca2+/calmodulin-dependent protein kinase II (CaMKII) autophosphorylation, phosphorylation of Ca2+/calmodulin-dependent protein kinase IV (CaMKIV), and cyclic AMP response element binding protein (CREB), all of which enhance activity of BDNF.[11] The atractylenolides have significant positive allosteric effects on the GABA A receptor that are independent of benzodiazepines. Thus, they could contribute to anxiolytic and soporific effects of the herb that can be important in the treatment of MDD.[12]

7.5.5 HUMAN ANTIDEPRESSANT EFFECTS

In a study designed to chromatographically isolate and identify the active components of the effective Traditional Chinese antidepressant herbal treatment, *xiaoyaosan*, atractylenolide I and atractylenolide II were identified as two of the four most likely candidates.[13] Thus, it is likely that much of the antidepressant effect of *xiaoyaosan* is due to the inclusion of *Atractylodes macrocephala*. Nonetheless, there are no published clinical studies of the use of the herb for the treatment of MDD in human subjects.

7.5.6 DOSAGE

Easley and Horne recommend 500 to 1000 mg of powdered root or 4 oz of weak infusion three times daily.[14]

7.5.7 TOXICITY

Atractylodes macrocepala appears to have low toxicity. The acute oral LD50 in rats was estimated to be over 4,000 mg/kg.[15] This would amount to approximately 280 g in a 75 kg human adult.

7.5.8 SAFETY IN PREGNANCY

In Traditional Chinese Medicine, *Atractylodes macrocepala* has been used during pregnancy to "calm the fetus."[16]

7.5.9 DRUG INTERACTIONS

The *Botanical Safety Handbook* notes no known interactions with drugs.[17]

REFERENCES

1. Yeung W-F, Chung KF, Ng KY, Yu YM, Ziea ET, Ng BF. A systematic review on the efficacy, safety and types of Chinese herbal medicine for depression. *J Psychiat Res.* 2014;57:165–175.
2. Li X, Lin J, Han W, et al. Antioxidant ability and mechanism of rhizoma *Atractylodes macrocephala*. *Molecules.* 2012;17(11):13457–13472.
3. Wang G-W, Feng Y, Qiu X-M, Liu Y-L. Effects of polysaccharide of *Atractylodes macrocephala* Koidz on brain edema after traumatic brain injury in rats. *Food Sci.* 2008-12.
4. Dong H, He L, Huang M, Dong Y. Ant-iinflammatory components isolated from *Astractylodes macrocephala*. *Nat Prod Res.* 2008;22(16):1418–1427.
5. Li C, He L. Establishment of the model of white blood cell membrane chromatography and screening of antagonizing TLR4 receptor component from *Atractylodes macrocephala* Koidz. *Sci China C Life Sci.* 2006;49(2):182–189.
6. Hung YY, Huang KW, Kang HY, et al. Antidepressants normalize elevated toll-like receptor profile in major depressive disorder. *Psychopharmacology.* 2016;233(9):1707–1714.
7. Sacre S, Medghalchi M, Gregory B, Brennan F, Williams R. Fluoxetine and citalopram exhibit potent anti-inflammatory activity in human and murine models of rheumatoid arthritis and inhibit toll-like receptors. *Arthritis Rheum.* 2010;62(3):683–693.
8. Chen C, Zhou S, Meng Q. A molecular docking study of rhizoma Atractylodis and rhizoma *Atractylodis Macrocephalae* herbal pair with respect to type 2 diabetes mellitus. *J Trad Chin Med Sci.* 2018;5(2):185–198.

9. Gao H, Zhu X, Xi Y, Li Q, Shen Z, Yang Y. Anti-depressant-like effect of atractyleno-lide I in a mouse model of depression induced by chronic unpredictable mild stress. *Exp Ther Med*. 2018;15(2):1574–1579.

10. Xie Q, Wei M, Zhang B, et al. MicroRNA-33 regulates the NLRP3 inflammasome signaling pathway in macrophages. *Mol Med Rep*. 2018;17(2):3318–3327.

11. Izumi H, Sasaki Y, Yabuki Y, et al. Memory improvement by Yokukansankachimpihange and atractylenolide III in the olfactory bulbectomized mice. *Adv Alzheimers Dis*. 2016;05(2):35–45.

12. Singhuber J, Baburin I, Kählig H, Urban E, Kopp B, Hering S. GABA-A receptor modulators from Chinese herbal medicines traditionally applied against insomnia and anxiety. *Phytomedicine*. 2012;19(3–4):334–340.

13. Zhou Y, Ren Y, Ma Z, et al. Identification and quantification of the major volatile con-stituents in antidepressant active fraction of xiaoyaosan by gas chromatography-mass spectrometry. *J Ethnopharmacol*. 2012;141(1):187–192.

14. Easley T, Horne S. *The Modern Herbal Dispensary*. Berkeley, CA: North Atlantic Books; 2016. p. 180.

15. Choi H-K, Roh HS, Jeong JY, Ha H. Acute oral toxicity of *Atractylodes macrocepala* KOIDZ. *Korean J Plant Res*. 2014;27(1):11–21.

16. Bensky D, Clavey S, Stoger E. *Chinese Herbal Medicine: Materia Medica*. 3rd ed. Seattle, WA: Eastland Press; 2004.

17. Gardner Z, McGuffin M, editors. *Botanical Safety Handbook*. 3rd edn. Boca Raton, FL: CRC Press; 2013. p. 111.

7.6 *AVENA SATIVA* (COMMON OAT)

Avena sativa, or the common oat, is a member of the grass family. It is native to Europe, Asia, and northwest Africa, and has been cultivated for thousands of years as a food grain. It is the same oat as in oatmeal, though for medicinal purposes, it is harvested earlier, when the seeds are green, in the so-called "milky stage". It is perhaps too common and simple to be seen as a serious herbal treatment. However, it contains a variety of flavonoids, flavonolignans, triterpenoid saponins, sterols, and tocols, as well as the somewhat unique indole alkaloids, the avenanthramides. Flavonoids isolated from *Avena sativa* include apigenin, luteolin, caffeic, p-cou-maric, and ferulic acids.[1] Traditionally, *Avena sativa* has been used as a stimulant, antispasmodic, antitumor, diuretic, and neurotonic.[2]

7.6.1 ANTIOXIDANT

Avena sativa has significant antioxidant effects, owing largely to the effects of the avenanthramides it contains. These substances were particularly effective *in vitro* in scavenging singlet oxygen and superoxide radicals.[3] These antioxidant effects of *Avena sativa* are exhibited in living cells. For example, extracts of the herb were shown to both have high DPPH radical-scavenging activity and to help prevent steatosis in cul-tured human liver cells, which is believed to be due in part to oxidative stress.[4]

7.6.2 ANTI-INFLAMMATORY

Methanolic extracts of *Avena sativa* prevent conversion of arachidonic acid into inflammatory metabolites through inhibition of cyclooxygenase and lipoxygenase enzymes.[5] The inflammatory products of those enzymes have been linked to MDD.[6] The avenanthramides in *Avena sativa* are particularly potent anti-inflammatory polyphenols. In concentrations as low as one part per billion, they facilitated degradation of NF-kβ activity in human keratinocytes. They also prevent the stimulation of NF-kβ by TNF-α, with subsequent reduction in levels of inflammatory IL-8.[7]

7.6.3 ANTIDIABETIC/ANTI-METABOLIC SYNDROME

When eaten as food, *Avena sativa* improves glycemic status in individuals suffering diabetes type II. It leads to a decline in glucose, insulin, and glucagon responses to carbohydrates loads. To some extent, this is due to the mechanical effects of the soluble fiber, b-glucan, it contains. The fiber slows digestion and absorption of carbohydrate.[8]

However, tests in mice made diabetic through exposure to streptozotocin show that chronic consumption of β-glucan in *Avena sativa* can significantly stimulate insulin secretion, improve insulin resistance, and glucose tolerance ability, and promote synthesis of liver and muscle glycogen.[9] Moreover, certain flavonoids in *Avena sativa*, such as apigenin and luteolin, have antidiabetic effects. Both flavonoids have anti-oxidant and anti-inflammatory effects that decrease insulin resistance, whereas apigenin also acts as an insulin secretagogoue.[10]

7.6.4 PRECLINICAL ANTIDEPRESSANT-LIKE EFFECTS

In a study of rats, the effects of extract of wild green *Avena sativa* on immobility in the forced swim were inconsistent. In one component of the study, the herb decreased immobility, but that result was not seen in the subsequent component. Nonetheless, treatment did lead to diminution of stress responses, enhancement of shock avoidance learning, and increased pro-social behaviors among the rats.[11]

Antidepressant-like effects of *Avena sativa* were more clearly seen in mice. A methanolic extract of the herb significantly decreased immobility in the forced swim and tail suspension tests.[12] Another group found the same result with a methanolic extract,[13] whereas yet another group saw reduced immobility in those tests utilizing an ethanolic extract of the herb[14].

7.6.5 HUMAN ANTIDEPRESSANT EFFECTS

Consumption of extract of *Avena sativa* does affect the human brain. For example, two doses of extract significantly improved the performance of elderly subjects in the Stroop Color Word Test – a test of cognitive function.[15] Moreover, in a survey of expert herbalists, all practicing members of The American Herbalists Guild, the majority saw *Avena sativa* to be useful in the treatment of MDD.[16] Nonetheless, there are no human studies of *Avena sativa* as a treatment for MDD. The German Commission E has concluded that the herb cannot be considered an effective treatment for any psychiatric illness.[17]

7.6.6 DOSAGE

Pursell recommends one to two droppers of tincture one to three times a day. Alternatively, one may make an infusion using one to two teaspoons of oatstraw per cup, then drinking one to four cups a day.[18]

7.6.7 TOXICITY

In one study, authors reported some difficulty in determining the LD50. None of the doses of dried methanolic extract of *Avena sativa* that they administered to mice, in doses up to 2000 mg/kg, had any obvious toxic effects.[19] That extrapolated LD50 dose (which was not actually lethal to mice), is roughly 150 g for a 75 kg human adult. That is far beyond the recommended dose.

7.6.8 SAFETY IN PREGNANCY

The *Botanical Safety Handbook* notes no evidence to confirm either safety or adverse effects of *Avena sativa* during pregnancy.[20]

7.6.9 DRUG INTERACTIONS

The *Botanical Safety Handbook* notes no known interactions with drugs.

REFERENCES

1. Gheorghe P, Gottfried W, Marie-Louise BG, Georgette D, Jean C. Isolation and characterization of flavonoids from *Avena sativa* L. *ZeitschriftfürPflanzenphysiologie.* 1977;85(2):103–115.
2. Singh R, De S, Belkheir A. *Avena sativa* (Oat), a potential neutraceutical and therapeutic agent: An overview. *Crit Rev Food Sci Nutr.* 2013;53(2):126–144.
3. Yang J, Ou B, Wise ML, Chu Y. In vitro total antioxidant capacity and anti-inflammatory activity of three common oat-derived avenanthramides. *Food Chem.* 2014;160:338–345.
4. Cai S, Huang C, Ji B, et al. In vitro antioxidant activity and inhibitory effect, on oleic acid-induced hepatic steatosis, of fractions and subfractions from oat (*Avena sativa* L.) ethanol extract. *Food Chem.* 2011;124(3):900–905.
5. Ahmed S, Gul S, Gul H, Bangash MH. Anti-inflammatory and antiplatelet activities of *Avena sativa* are mediated through the inhibition of cyclooxygenase and lipoxygenase enzymes. *Int J Endorsing Health Sci.* 2013;1(2):62–65.
6. Oxenkrug GF. Metabolic syndrome, age-associated neuroendocrine disorders, and dysregulation of tryptophan-kynurenine metabolism. *Ann N Y Acad Sci.* 2010;1199:1–14.
7. Sur R, Nigam A, Grote D, Liebel F, Southall MD. Avenanthramides, polyphenols from oats, exhibit anti-inflammatory and anti-itch activity. *Arch Dermatol Res.* 2008;300(10):569–574.
8. Tapola N, Karvonen H, Niskanen L, Mikola M, Sarkkinen E. Glycemic responses of oat bran products in type 2 diabetic patients. *Nutr Metab Cardiovas.* 2005;15(4):255–261.
9. Stzdong JL. The effects of oat β-glucan on insulin resistance in type 2 diabetic mice caused by STZ. *J Zhengzhou University Light Industry.* 2011-02.
10. Vinayagam R, Xu B. Antidiabetic properties of dietary flavonoids: A cellular mechanism review. *Nutr Metab.* 2015;12:60.
11. Schellekens C, Perrinjaquet-Moccetti T, Wullschleger C, Heyne A. An extract from wild green oat improves rat behavior. *Phytother Res.* 2009;23(10):1371–1377.
12. Uppala PK, Latha SM, Reddy SR, Chakravarthi G. Evaluation of anti-depressant activity of methanolic seed extract of *Avena sativa* L. In mice. *Res J Pharmacol Pharmacodyn.* 2013;5(4):212–217.
13. Niloufar M, Mehrdad F, Abdolmajid A. The most effective fraction of Avena sativa as antidepressant in mouse model of depression. National Congress on Medicinal Plants; 2015;4: Poster #1692.

14. Hamid A, Abbasi MS, Efthekhari SK, et al. Antidepressant-like effect of ethanolic extract of Avena sativa L. in the forced swim test and tail suspension test in male mice. 2nd National Congress on Medicinal Plants. 2013;2.
15. Berry NM, Bryan J, Buckley JD, Murphy KJ, Howe PRC, Robinson MJ. Acute effects of an *Avena sativa* herb extract on responses to the Stroop Color Word Test. *J Altern Complem Med*. 2011;17(7):635–637.
16. Einerson LS. *A Delphi Study on Herbs Used to Address Depression and Anxiety According to Master Herbalists*. Lubbock, TX: Texas Tech University; 2017.
17. Blumenthal M, Busse WR, Goldberg A, et al. *The Complete German Commission E Monographs*. Austin, TX: American Botanical Council; 1998.
18. Pursell JJ. *The Herbal Apothecary*. Portland, OR: Timber Press; 2016. p. 139
19. Usha Rani K, Ramaiah M, Nagaphani K, Preethi V, Srinadh M. Screening for antidepressant-like effect of methanolic seed extract of *Avena sativa* using animal models. *Phcog J*. 2014;6(3):86–92.
20. Gardner Z, McGuffin M, editors. *Botanical Safety Handbook*. 3rd ed. Boca Raton, FL: CRC Press; 2013. p. 115.

7.7   BACOPA MONNIERI

Bacopa monnieri, also known as brahmi or water hyssop, is a creeping perennial commonly found as a weed in rice fields throughout India and east Asia. It has had a particularly important role in Indian Ayurvedic medicine, and was initially described in Sanskrit texts from the sixth century AD. It was classed as a *medhya rasayana* – an herb taken to sharpen intellect and attenuate mental deficits.[1] Along with its alleged brain rejuvenating effects, *Bacopa monnieri* has also been used for treatment of inflammation, fatigue, pain, diabetes, asthma, and infection.

Bacopa monnieri contains a unique group of complex triterpenoid saponins called bacosides, as well as the triterpenoids jujubogenins and pseudojujubogenins and their glycosides. The latter triterpenoids are also found in the herb *Ziziphus jujube* that is frequently used in Chinese medicine, as previously discussed. *Bacopa monnieri* also contains the steroid-like betulinic acid, stigmastanol, β-sitosterol and stigmasterol, and the alkaloids brahmine, nicotinine, and herpestine.[2]

7.7.1   ANTIOXIDANT

Bacopa monnieri has significant capacity to reduce the damage of reactive oxygen species. It was shown to have free radical-scavenging capacity in the DPPH assay, as well as to effectively scavenge for superoxide anion. In the same study, methanolic extract of *Bacopa monnieri* was found to attenuate hydrogen peroxide-induced cytotoxicity and DNA damage in cultured human fibroblasts.[3]

Bacopa monnieri exerts antioxidant effects *in vivo*. Fourteen days of pretreatment of rats with methanolic extract of the herb protected their liver and kidney tissues from the oxidative damage of CCl_4. Serum levels of ALT, AST, and creatinine were reduced toward normal, and histopathological changes were substantially reduced.[4]

Of interest to the current discussion, *Bacopa monnieri* also has potent antioxidant effects in brain tissue of living animals. Administration of the hydroethanolic extract to rats over 21 days caused increases in the activities of superoxide dismutase, catalase, and glutathione peroxidase in the frontal cortex, striatum, and hippocampus.[5]

7.7.2 ANTI-INFLAMMATORY

Bacopa monnieri has shown significant systemic anti-inflammatory effects. The methanol extract and aqueous fractions showed a significant reduction of swelling in the carrageenan-induced edema test in rat's hind paws.[6] In the rat collagen-induced arthritis model of rheumatoid arthritis, *Bacopa monnieri* significantly inhibited foot-pad swelling and arthritic symptoms. There were decreases in neutrophil infiltration and myeloperoxidase activity, as well as improvement in joint architecture. *Bacopa monnieri* inhibited cyclooxygenase and lipoxygenase activities in the arthritic rats.[7] Extract of the herb also inhibits the production of pro-inflammatory cytokines TNF-α and IL-6 in monocytes stimulated with LPS *in vitro*.[8] *Bacopa monnieri* also exhibits anti-inflammatory effects in microglia that populate the brain. The tea, infusion, and alkaloid extract of *Bacopa monnieri*, as well as isolated Bacoside A significantly inhibited the release of pro-inflammatory cytokines TNF-α and IL-6 from activated microglial cells *in vitro*.[9]

7.7.3 ANTIDIABETIC/ANTI-METABOLIC SYNDROME

Bacopa monnieri has antidiabetic effects that counter insulin resistance and other aspects of the metabolic syndrome that contribute to MDD. In the alloxan-induced diabetes model in rats, the triterpene bacosine, extracted from the herb, decreases blood glucose levels. Bacosine increased glycogen content in the liver of diabetic rats and peripheral glucose utilization in the diaphragm of diabetic rats *in vitro*, comparable with the action of insulin.[10]

A number of studies have purported to show adaptogenic, stress-resisting effects of extracts of *Bacopa monnieri*. In one such study, the effects of *Bacopa monnieri* were evaluated in both acute and chronic unpredictable stress. Subjecting animals to a single course of immobilization for 150 min, or different stressors for seven days, resulted in significant elevation in plasma corticosterone levels that was countered by treatment with the herb. In the chronic stress regimen, levels of NA, DA, and 5-HT were depleted in cortex and hippocampus. Treatment with *Bacopa monnieri* attenuated the chronic unpredictable stress induced changes NA, DA, and 5-HT levels in cortex and levels of NA and 5-HT in the hippocampus.[11]

7.7.4 PRECLINICAL ANTIDEPRESSANT-LIKE EFFECTS

In a study of rats, it was found that daily administration of *Bacopa monnieri* extract reversed the behavioral effects of 28 days of chronic unpredictable stress. It restored sucrose consumption and increased locomotor behavior in the open field. Neurochemically, bacopa increased to normal the levels of BDNF, akt, and CREB in the hippocampus of stressed animals.[12] Extracted bacoside triterpenes alone exhibited antidepressant effects in mouse models of depression.[13]

7.7.5 HUMAN ANTIDEPRESSANT EFFECTS

Despite the traditional and time-honored use of *Bacopa monnieri* in Ayurvedic Medicine for the treatment of depression and cognitive decline, there are few, if any, compelling human studies in the scientific literature. In one such marginal study, 20 healthy adults between the ages of 60 and 75 were administered Bacopa extract containing 11% of Bacopa glycosides or a placebo once daily for 12 weeks. The extract produced significant improvement compared with placebo in short-term memory, processing speed, attention, and depression, effects that were also observed at 16 weeks, four weeks after dosing stopped. No adverse events were reported in either group. Unfortunately, the means of measuring degree of depression was not described in this poster presentation.[14]

In a randomized, double-blind, placebo-controlled clinical trial, 300 mg/day of standardized *Bacopa monnieri* extract improved cognitive function in elderly adults as measured by standard cognitive performance tests. CESD-10 Depression scores, per the Center for Epidemiologic Studies Depression scale, and anxiety scores, per the State-Trait Anxiety Inventory were said to decrease over time in the treatment group, whereas those scores increased in the placebo group. The treatment was said to be well tolerated with few adverse events, primarily stomach upset.[15] In a double-blind, placebo-controlled cross-over study of healthy adults, an extract of *Bacopa monnieri* improved cognitive performance. There were also said to be "positive mood effects" and a reduction in cortisol levels.[16]

7.7.6 DOSAGE

Easley and Horne recommend dosage of 1–3 ml of dried leaf tincture up to three times daily, or 400 to 500mh capsules of standardized extract 2 times a day.[17] They also warn against use in hyperthyroidism.

7.7.7 TOXICITY

The LD50 of dried ethanolic extract of *Bacopa monnieri* in mice has been reported to be 520 mg/kg.[18] This would extrapolate to roughly 40 g for a human adult weighing 75 kg.

7.7.8 SAFETY IN PREGNANCY

The *Botanical Safety Handbook* notes no evidence to confirm either safety or adverse effects of *Bacopa monnieri* during pregnancy.[19]

7.7.9 DRUG INTERACTIONS

A study shows that *Bacopa monnieri* extract can inhibit the activities of the cytochrome P450 enzymes CYP1A2, CYP3A4, CYP2C9, and CYP2C19. Thus, use of the herb could increase blood levels of drugs metabolized by those enzymes.[20]

REFERENCES

1. Aguiar S, Borowski T. Neuropharmacological review of the nootropic herb Bacopa monnieri. *Rejuvenation Res.* 2013;16(4):313–326.
2. Al-Snafi AE. The pharmacology of Bacopa monniera. A review. *IJPSR.* 2013;4(12):154–159.
3. Russo A, Izzo AA, Borrelli F, Renis M, Vanella A. Free radical scavenging capacity and protective effect of Bacopa monniera L. on DNA damage. *Phytother Res.* 2003;17(8):870–875.
4. Shahid M, Subhan F. Protective effect of Bacopa monniera methanol extract against carbon tetrachloride induced hepatotoxicity and nephrotoxicity. *PhOL.* 2014;2:18–28.
5. Bhattacharya SK, Bhattacharya A, Kumar A, Ghosal S. Antioxidant activity of Bacopa monniera in rat frontal cortex, striatum and hippocampus. *Phytother Res.* 2000;14(3):174–179.
6. Mathur A, Verma SK, Purohit R, Singh SK, Mathur D. Pharmacological investigation of *Bacopa monnieri* on the basis of antioxidant, antimicrobial and anti-inflammatory properties. *J Chem Pharm Res.* 2010;2(6):191–198.
7. Viji V, Kavitha SK, Helen A. Bacopa monniera (L.) wettst inhibits type II collagen-induced arthritis in rats. *Phytother Res.* 2010;24(9):1377–1383.
8. Viji V, Helen A. Inhibition of pro-inflammatory mediators: Role of Bacopa monniera (L.) Wettst. *Inflammopharmacology.* 2011;19(5):283–291.
9. Nemetchek MD, Stierle AA, Stierle DB, Lurie DI. The Ayurvedic plant *Bacopa monnieri* inhibits inflammatory pathways in the brain. *J Ethnopharmacol.* 2017;197:92–100.
10. Ghosh T, Maity TK, Singh J. Antihyperglycemic activity of bacosine, a triterpene from Bacopa monnieri, in alloxan-induced diabetic rats. *Planta Med.* 2011;77(8):804–808.
11. Sheikh N, Ahmad A, Siripurapu KB, Kuchibhotla VK, Singh S, Palit G. Effect of Bacopa monniera on stress induced changes in plasma corticosterone and brain monoamines in rats. *J Ethnopharmacol.* 2007;111(3):671–676.
12. Hazra S, Kumar S, Saha GK, Mondal AC. Reversion of BDNF, Akt and CREB in hippocampus of chronic unpredictable stress induced rats: Effects of phytochemical, Bacopa monnieri. *Psychiatry Investig.* 2017;14(1):74–80.
13. Zhou Y, Shen YH, Zhang C, Su J, Liu RH, Zhang WD. Triterpene saponins from *Bacopa monnieri* and their antidepressant effects in two mice models. *J Nat Prod.* 2007;70(4):652–655.
14. Hingorani L, Patel S, Ebersole B. Sustained cognitive effects and safety of HPLC-standardized Bacopa monnieri extract: A randomized, placebo controlled clinical trial. *Planta Med.* 2012;78:PH22.
15. Calabrese C, Gregory WL, Leo M, Kraemer D, Bone K, Oken B. Effects of a standardized *Bacopa monnieri* extract on cognitive performance, anxiety, and depression in the elderly: A randomized, double-blind, placebo-controlled trial. *J Altern Complement Med.* 2008;14(6):707–713.
16. Benson S, Downey LA, Stough C, Wetherell M, Zangara A, Scholey A. An acute, double-blind, placebo-controlled cross-over study of 320 mg and 640 mg doses of *Bacopa monnieri* (CDRI 08) on multitasking stress reactivity and mood. *Phytother Res.* 2014;28(4):551–559.
17. Easley T, Horne S. *The Modern Herbal Dispensary.* Berkeley, CA: North Atlantic Books; 2016. p. 180.
18. Dar A, Channa S. Relaxant effect of ethanol extract of Bacopa monniera on trachea, pulmonary artery and aorta from rabbit and guinea-pig. *Phytother Res.* 1997;11(4):323–325.
19. Gardner Z, McGuffin M, editors. *Botanical Safety Handbook.* 3rd ed. Boca Raton, FL: CRC Press; 2013. p. 123.

20. Ramasamy S, Kiew LV, Chung LY. Inhibition of human cytochrome P450 enzymes by *Bacopa monnieri* standardized extract and constituents. *Molecules.* 2014;19(2):2588–2601.

7.8 *BORAGE OFFICINALIS* (EUROPEAN BORAGE)

Borago officinalis or borage is an annual herb native to Europe, North Africa, and Asia Minor. *Borago officinalis* is also known as European Borage and may be easily confused with *Echium amoenum*, which is sometimes referred to as Iranian Borage. Both plants are from the *Boraginaceae* family, but they are from different genera. *Echium amoenum* is from the *Echium* genus, grows in the Alborz hillsides, has not yet been domesticated, and cannot be cultivated. *Borago officinalis* grows in Isfahan, Azerbaijan, and other cities of Iran. It was imported from Spain to northern Africa, cultivated there, and then incorporated into European traditional medicine. The medicinal properties of these two plants are slightly different.

The use of *Borago officinalis* is ancient, and it is one the herbs used by Hippocrates for the treatment of melancholia.[1] The leaves of borage contain a variety of compounds including pyrrolizidine alkaloids, licosamin, intermedin, sopinin, sopindian, yezan, colin, δ-bornesitol, rosmarinic acid, and cianozhens. The leaves of borage in seeding stage also contain modest amounts of gamma-linolenic acid and stearidonic acid. Along with primrose oil, borage seed oil is a major source of gamma-linolenic acid.[2]

The plant is reputed to have antispasmodic, antihypertensive, antipyretic, aphrodisiac, demulcent, and diuretic effects, and is used to treat asthma, bronchitis, cramps, diarrhea, palpitations, and kidney ailments. Decoction of the plant is used as a nerve and cardiac tonic.[3]

7.8.1 ANTIOXIDANT

In rats administered CCl_4, prior treatment with extract of *Borago officinalis* spared liver tissue from oxidative and inflammatory damage. The extract reduced lipid peroxidation and restored levels of glutathione in liver tissue. Tissue levels of TNF-α and NFκB were also reduced, as were serum levels of hepatic transaminase enzymes.[4]

7.8.2 ANTI-INFLAMMATORY

In an *in vitro* study, addition of extract of *Borago officinalis* to the culture medium dampened the response of mouse macrophages to lipopolysaccharide and /INF-γ. The extracts inhibited induced NO production in a dose-dependent manner.[5]

Borage oil has long been used as a folk remedy for arthritis and other inflammatory conditions. Certainly, some of this effect is due to the herb being a uniquely rich source of gamma linolenic acid. This fatty acid serves as substrate for synthesis of anti-inflammatory prostaglandin E that in turn dampens synthesis of TNF-α and other cytokines.[6] In this role, *Borago officinalis* would act not as a pharmacological mechanism, but rather as a nutritional supplement.

7.8.3   ANTIDIABETIC/ANTI-METABOLIC SYNDROME

There are no reports in the literature of effects of *Borago officinalis* on diabetes, insulin sensitivity or Metabolic Syndrome.

7.8.4   PRECLINICAL ANTIDEPRESSANT-LIKE EFFECTS

Rosmarinic acid, which is found in leaves of *Borago officinalis* at a concentration of about 15 mg/g, has been found to have a variety of pharmacological effects.[7] Animal studies suggest it may produce antidepressant effects. Pretreatment with rosmarinic acid (RA) reduced the length of immobility in the tail suspension test in mice, downregulated Mkp-1, and increased BDNF in cortical tissue to levels of those of unstressed mice. Treatment also reduced serum corticosterone and maintained dopamine levels in the levels in the limbic system.[8]

It has been reported that pre-treatment with the ethanol extract of *Borago officinalis* prevented the depletion of serotonin, norepinephrine, and dopamine in the brains of mice subjected to either isolation stress or chronic unpredictable mild stress.[9]

7.8.5   HUMAN ANTIDEPRESSANT EFFECTS

Borago officinalis was one of the herbs recommended by Hippocrates for the treatment of melancholy. It also has a long history of use in the Islamic medical tradition, where it is seen as a *mufarrehat*, an exhilarant herb, to be used in the treatment of MDD.[10] Unfortunately, there are no scientific human trials of *Borago officinalis* in the treatment of MDD. There is a trial of *Borago officinalis* extract in women for the treatment of Premenstrual Syndrome.[11] One of the outcomes in that study was an improvement in emotional wellbeing. However, as noted above, the extract of the seeds was used because of its high content of γ-linolenic acid. It is not clear what phytochemicals unique to the herb may have been contained in the seed oil or if they contributed to the beneficial effects.

7.8.6   DOSAGE

Easley and Horne note that *Borago officinalis* "lifts sadness and depression." They add that the herb contains pyrrolizidine alkaloids and advise caution, but do not further warn of potential adverse liver effects. They advise against its use in pregnancy. For general use, they recommend 300 mg capsules of borage oil taken three or four times a day, or a 1:5 tincture of dried leaf and flowers, six to twelve drops three times a day.[12]

The German E Commission considers the claims of medicinal properties of *Borago officinalis* to be baseless, and does not approve its use. They also note the herb contains pyrrolizidine alkaloids with potential hepatotoxic effects.[13]

7.8.7   TOXICITY

No LD50 data appears in the literature. However, due to the toxic effect the leaves and flowers can have on the liver, many herbalists have suggested that nothing other than the pressed oil of Borage seed be used for treatment of human ailments. The oil obtained from the ripe seeds of *Borago officinalis* by cold pressing is devoid of the toxic pyrrolizidine alkaloids.[14] Seeds of *Borago officinalis* are rich in phenolic acids, such as ferulic, chlorogenic, gallic, and caffeic acids, which likely remain in crude oil pressing.[15]

7.8.8   SAFETY IN PREGNANCY

The *Botanical Safety Handbook* gives no definitive information on the safety of using borage during pregnancy. However, they offer an opinion against its use due to its content of pyrrolizidine alkaloids.[16]

7.8.9   DRUG INTERACTIONS

The *Botanical Safety Handbook* notes no known interactions with drugs.

REFERENCES

1. Rees L. Treatment of depression by drugs and other means. *Nature.* 1960;186:114–120.
2. Asadi-Samani M, Bahmani M, Rafieian-Kopaei M. The chemical composition, botanical characteristic and biological activities of *Borago officinalis*: A review. *Asian Pac J Trop Med.* 2014;7S1(Suppl 1):S22–S28.
3. Gilani AH, Bashir S, Khan AU. Pharmacological basis for the use of *Borago officinalis* in gastrointestinal, respiratory and cardiovascular disorders. *J Ethnopharmacol.* 2007;114(3):393–399.
4. Hamed ANE, Wahid A. Hepatoprotective activity of *Borago officinalis* extract against CCl_4-induced hepatotoxicity in rats. *J Nat Prod.* 2015;8:113–122.
5. Karimi E, Oskoueian E, Karimi A, Noura R, Ebrahimi M. *Borago officinalis* L. flower: A comprehensive study on bioactive compounds and its health-promoting properties. *J Food Meas Charact.* 2017:1–13.
6. Kast RE. Borage oil reduction of rheumatoid arthritis activity may be mediated by increased cAMP that suppresses tumor necrosis factor-alpha. *Int Immunopharmacol.* 2001;1(12):2197–2199.
7. Bandoniene D, Murkovic M, Venskutonis PR. Determination of rosmarinic acid in sage and borage leaves by high-performance liquid chromatography with different detection methods. *J Chromatogr Sci.* 2005;43(7):372–376.
8. Kondo S, El Omri A, Han J, Isoda H. Antidepressant-like effects of rosmarinic acid through mitogen-activated protein kinase phosphatase-1 and brain-derived neurotrophic factor modulation. *J Funct Food.* 2015;14:758–766.
9. Gang H-L, He Z, Liu X, Ma Y. Effects of the ethanol extractive of *Borago officinalis* on neurotransmitter in the brain tissue of mouse model of chronic depression. *J Jiangsu Univ.* 2012-02.
10. Anwar N, Ahmed NZ, Shahida T, Kabiruddin K, Aslam H. The role of Mufarrehat (exhilarants) in the management of depression: An evidence based approach. *J Psychiatry.* 2017;20(5):420.
11. Gama CRB, Lasmar R, Gama GF, et al. Premenstrual syndrome: Clinical assessment of treatment outcomes following *Borago officinalis* extract therapy. *RBM.* 2014;71:211–217.
12. Easley T, Horne S. *The Modern Herbal Dispensary.* Berkeley, CA: North Atlantic Books; 2016. p. 194.

13. Blumenthal M, Busse WR, Goldberg A, et al. The Complete German Commission E Monographs. Austin, TX: American Botanical Council; 1998. p. 316.
14. Pieszak M, Mikolajczak PL, Manikowska K. Borage (Borago officinalis L.)—a valuable medicinal plant used in herbal medicine. *Herba Pol.* 2012;58(4):95–103.
15. Zadernowski R, Naczk M, Nowak-Polakowska H. Phenolic acids of borage (*Borago officinalis* L.) and evening primrose (*Oenothera biennis* L.). *JAOCS.* 2002;79(4):335–338.
16. Gardner Z, McGuffin M, editors. *Botanical Safety Handbook.* 3rd ed. Boca Raton, FL: CRC Press; 2013. p. 161.

7.9 *BUPLEURUM CHINENSE*

Bupleurum chinense, known in the West as thorowax, is a genus of plants in the carrot family. These plants have been used for centuries in Asia for their medicinal properties. They contain a variety of essential oils, phenolics, triterpene saponins, alkaloids, and polyacetylenes. Curiously, *Bupleurum* species are officially listed in the Chinese and Japanese Pharmacopoeias, but none in this genus were selected by the German Commission E, the British Pharmacopoeia 2009, or the British National Formulary.[1] *Bupleurum chinense* is noted by Yeung to be one of the most commonly used herbs in combinations prescribed for treatment of MDD in Traditional Chinese Medicine.[2]

Some phytochemicals, such as the large, steroid-like saikosaponin triterpenes and bupleurosides, are apparently unique to the *Bupleurum* genus. The plants also contain a number of phytochemicals common among medicinal plants, such as the flavonol quercetin and the related glycoside rutin.

7.9.1 ANTIOXIDANT

Bupleurum chinense has a number of biological activities. The water-soluble polysaccharide fraction from the roots of *Bupleurum chinense* has significant antioxidant effects. In the rat model of hepatic injury caused by d-galactosamine, oral administration of the polysaccharides provided substantial protection to the liver. This was largely due to elevations in the hepatic activities of the antioxidant enzymes, glutathione reductase, γ-glutamylcysteine synthetase, glutathione S-transferase, and superoxide dismutase. Glutathione levels were increased and markers of lipid peroxidation, i.e., thiobarbituric acid reactive substances, were decreased. The antioxidant effect secondarily reduced inflammatory effects, as evidenced by reduced serum levels of TNF-α.[3] Extracts of the closely related herb, *Bupleurum kaoi* Liu, also offered hepatoprotective antioxidant effects in rats suffering CCl_4-induced liver damage.[4]

7.9.2 ANTI-INFLAMMATORY

The herb has substantial anti-inflammatory effects, and has been shown to reduce concentrations of the inflammatory cytokines IL-6, tumor necrosis factor, and interferon (IFN)-g in mice.[5]

Anti-inflammatory effects of *Bupleurum chinense* are also seen in the report that polysaccharides isolated from the herb protect against lipopolysaccharide-induced acute lung injury in mice. It attenuated lung injury, reduced exudate, and improved lung morphology. It also reversed the lipopolysaccharide-induced increases levels of myeloperoxidase, TNF-α, and serum nitric oxide.[6]

Saikosaponin A, a major triterpenoid saponin isolated from *Bupleurum chinense*, also blocked the inflammatory effects of lipopolysaccharide in cultured macrophages. It markedly inhibited the lipopolysaccharide-induced expression of COX-2, inducible nitric-oxide synthase, TNF-α, IL-1β, and IL-6, as well as preventing the translocation of NF-κB that contributes to the inflammatory response. It also upregulated the expression of the anti-inflammatory cytokine IL-10.[7]

7.9.3 ANTIDIABETIC/ANTI-METABOLIC SYNDROME

Bupleurum chinense improves glucose tolerance in streptozotocin-induced diabetic mice. However, this may have been secondary to its antioxidant and anti-inflammatory effects.[8] One traditional Chinese method of preparing *Bupleurum chinense* has been to bake minced root in vinegar. This is reported to strengthen the effect, though it is not clear if this alters certain phytochemicals or simply allows for better bioavailability. In any case, chronic feeding of vinegar-baked *Bupleurum chinense* to rats made obese by high fat diets greatly diminished signs of Metabolic Syndrome. Body weight gain, visceral fat-pad weights, plasma lipid levels, liver triglycerides, and cholesterol content were all significantly reduced. This was found to be due to increases in fatty acid oxidation resulting from *Bupleurum chinense* upregulating PPARα and downregulating sterol regulatory element binding proteins expression in the livers of those animals.[9]

7.9.4 PRECLINICAL ANTIDEPRESSANT-LIKE EFFECTS

In a study of rats, chronic administration of extract of *Bupleurum chinense* reduced immobility time in the forced swim test. Moreover, treating cultured SH-SY5Y cells (derived from human bone marrow neuroblasts) with *Bupleurum chinense* extract increased CREB and BDNF levels. It activated phosphatidylinositol 3-kinase (PI3K) and Akt and inhibited GSK-3β.[10] *Bupleurum chinense* also restores BDNF levels in the brains of rats subjected to long periods of unpredictable stress.[11]

Several closely related *Bupleurum* species have also been so studied. Methanolic extracts of *Bupleurum falcatum*, which contains similar bupleurosides, saikosaponins, and polyphenolic phytochemicals, had antidepressant-like effects in mice. Treated mice exhibited less immobility in the forced swim test.[12] Daily administration of *Bupleurum falcatum* to rats subjected to chronic restraint stress similarly reduced immobility in those animals during the forced swim test.[13]

Total saikosaponins extracted from closely related *Bupleurum yinchowense* also significantly reduced the immobility time in the forced swim and tail suspension tests in mice after seven-day treatment. It also restored preference for sucrose.[14]

7.9.5 HUMAN ANTIDEPRESSANT EFFECTS

There are no studies of *Bupleurum* species in the treatment of MDD in humans.

7.9.6 Dosage

Easley and Horne recommend standard decoction 2 to 4 ounces two to three times daily, or tincture of dried bark (1:5; 50% alcohol) 1 to 2 ml up to three times a day. They offer no warnings for its use.[15]

7.9.7 Toxicity

The LD50 of the volatile oil of *Bupleurum chinense* in mice is reported to be about 3.0 ml/kg.[16]

7.9.8 Safety in Pregnancy

The *Botanical Safety Handbook* notes there to be no definitive guidance from the literature concerning the safety of *Bupleurum chinense* during pregnancy.[17]

7.9.9 Drug Interactions

The *Botanical Safety Handbook* notes no known interactions with drugs. One study demonstrated that various extracts of vinegar baked root of *Bupleurum chinense*, a common potentizing method of preparing the herb in TCM, had negligible effects on activities of CYP1A2 or CYP3A4 enzymes. However, potential for interference of metabolism of CYP2C9 substrates was noted.[18]

REFERENCES

1. Ashoura ML, Winka M. Genus *Bupleurum*: A review of its phytochemistry, pharmacology and modes of action. *JPP*. 2011;63:305–321.
2. Yeung WF, Chung KF, Ng KY, Yu YM, Ziea ET, Ng BF. A systematic review on the efficacy, safety and types of Chinese herbal medicine for depression. *J Psychiat Res*. 2014;57:165–175.
3. Zhao W, Li JJ, Yue SQ, Zhang LY, Dou KF. Antioxidant activity and hepatoprotective effect of a polysaccharide from Bei Chaihu (*Bupleurum chinense* DC). *Carbohyd Polym*. 2012;89(2):448–452.
4. Wang BJ, Liu CT, Tseng CY, Wu CP, Yu ZR. Hepatoprotective and antioxidant effects of *Bupleurum kaoi Liu* (Chao et Chuang) extract and its fractions fractionated using supercritical CO2 on CCl$_4$-induced liver damage. *Food Chem Toxicol*. 2004;42(4):609–617.
5. Wong VKW, Zhou H, Cheung SSF, Li T, Liu L. Mechanistic study of saikosaponin-d (ssd) on suppression of murine T lymphocyte activation. *J Cell Biochem*. 2009;107(2):303–315.
6. Xie JY, Di HY, Li H, Cheng XQ, Zhang YY, Chen DF. *Bupleurum chinense* DC polysaccharides attenuates lipopolysaccharide-induced acute lung injury in mice. *Phytomedicine*. 2012;19(2):130–137.
7. Zhu J, Luo C, Wang P, He Q, Zhou J, Peng H. Saikosaponin A mediates the inflammatory response by inhibiting the MAPK and NF-κB pathways in LPS-stimulated RAW 264.7 cells. *Exp Ther Med*. 2013;5(5):1345–1350.
8. Pan L, Weng H, Li H, et al. Therapeutic effects of Bupleurum polysaccharides in streptozotocin induced diabetic mice. *PLOS ONE*. 2015;10(7):e0133212.

9. Tzeng T-F, Lu HJ, Liou SS, Chang CJ, Liu IM. Vinegar-baked Radix Bupleuri regulates lipid disorders via a pathway dependent on peroxisome-proliferator-activated receptor-α in high-fat-diet-induced obese rats. *Evid-Based Compl Alt.* 2012; Article ID 827278.

10. Seo MK, Song JC, Lee SJ, et al. Antidepressant-like effects of *Bupleuri Radix* extract. *Eur J Integr Med.* 2012;4(4):e392–e399.

11. Mao QQ, Huang Z, Zhong XM, et al. Effects of SYJN, a Chinese herbal formula, on chronic unpredictable stress-induced changes in behavior and brain BDNF in rats. *J Ethnopharmacol.* 2010;128(2):336–341.

12. Kwon S, Lee B, Kim M, Lee H, Park HJ, Hahm DH. Antidepressant-like effect of the methanolic extract from *Bupleurum falcatum* in the tail suspension test. *Prog Neuropsychopharmacol Biol Psychiatry.* 2010;34(2):265–270.

13. Lee BB, Yun HY, Shim IS, Lee HJ, Hahm DH. *Bupleurum falcatum* prevents depression and anxiety-like behaviors in rats exposed to repeated restraint stress. *J Microbiol Biotechnol.* 2012;22(3):422–430.

14. Sun X, Shi Z, Li T, et al. Antidepressant-like effects of total saikosaponins of *Bupleurum yinchowense* in mice. *J Med Plant.* 2012;6(26):4308–4316.

15. Easley T, Horne S. *The Modern Herbal Dispensary.* Berkeley, CA: North Atlantic Books; 2016; p. 198.

16. Sun R, Wang L, Yang Q, Huang W, Lv LL. Acute toxicity of volatile oil from *Bupleurum chinense* in rats and mice. *Chin J Exp Trad Med Formulae.* 2010–2011.

17. Gardner Z, McGuffin M, editors. *Botanical Safety Handbook.* 3rd ed. Boca Raton, FL: CRC Press; 2013; p. 148.

18. Yu T, Chen X, Wang Y, Zhao R, Mao S. Modulatory effects of extracts of vinegar-baked Radix Bupleuri and saikosaponins on the activity of cytochrome P450 enzymes in vitro. *Xenobiotica.* 2014;44(10):861–867.

7.10 *CAMELLIA SINENSIS* (TEA)

Tea is the most popular brewed beverage in the world. It is brewed from the leaves of the *Camellia sinensis* plant that is native to East Asia, the Indian Subcontinent, and Southeast Asia. Due to its popularity, the plant is now cultivated in most tropical and subtropical regions of the world. There are hundreds, if not thousands, of cultivars, or propagated strains, of *Camellia sinensis*, each with slightly unique characteristics. There are also differences in how tea is processed into finished product. Green teas are steamed to inactivate oxidizing enzymes, and thus they contain the highest amounts of certain beneficial polyphenols. Other types of teas, for example, oolong, red, and black, are fermented to varying degrees, thus somewhat altering the phytochemical content. As has been found with coffee, epidemiological studies have shown that the daily drinking of tea reduces the risk of developing MDD. This benefit appears to be regardless of the type of tea, such as green versus black tea.[1,2]

Tea contains significant amounts of caffeine. As is noted in the coffee section of this book, caffeine alone has been found to exhibit antidepressant effects in animal studies. It may also have antidepressant effects in humans. However, phytochemicals in tea other than caffeine may produce antidepressant effects in animals and humans. Among the phytochemicals in green tea are the flavonoids epigallocatechin-3-gallate (59% of total), epigallocatechin (19%), epicatechin-3-gallate (13%), and epicatechin (6.4%). It also contains the phenolic acids chlorogenic acid and caffeic acid, the flavonols kaemferol, myricetin, and quercetin; and the amino acid analogue, l-theanine.[3]

7.10.1 ANTIOXIDANT

The polyphenols in tea have potent antioxidant effects. Daily consumption of tea over several weeks decreases oxidative DNA damage, lipid peroxidation, and free radical generation in human blood samples.[4] In diabetic subjects, six cups a day of black tea significantly reduced oxidative damage to lymphocyte DNA as well as plasma levels of malondialdehyde, an indicator of lipid peroxidation.[5] Acute oral dosing with EGCG has also been shown to have neuroprotective effects against hippocampal neuronal damage following global ischemia in gerbils. This damage is believed to be due to oxygen free radical injury.[6]

7.10.2 ANTI-INFLAMMATORY

Many studies have demonstrated various anti-inflammatory effects of green tea and its constituent phytochemicals. Epigallocatechin-3-gallate, the primary tea polyphenol, inhibits UVB-induced infiltration of leukocytes in the skin of human subjects. This migration of leukocytes, part of the inflammatory response, is a source of reactive oxygen species and inflammatory prostaglandin metabolites.[7] In a mouse study, oral administration of epigallocatechin-3-gallate limited brain inflammation and neuronal damage in a model of protein-induced Autoimmune Encephalomyelitis. TNF-α was reduced as was proliferation of inflammatory T-cells. The results also show that at least some of the phytochemicals in tea cross from the bloodstream into the brain. In the same study, it was found that epigallocatechin-3-gallate inhibited inflammatory activity of human myelin-specific CD4 T-cells, in part through inhibition of NfKB.[8]

7.10.3 ANTIDIABETIC/ANTI-METABOLIC SYNDROME

Green tea has been shown to improve glucose tolerance and insulin sensitivity, primarily through the action of epigallocatechin-3-gallate.[9] In rats, replacing drinking water with green tea led to lower fasting serum levels of glucose, insulin, and triglycerides.[10] Not all studies have shown such encouraging results for tea. For example, in one study, four weeks of daily tea in diabetic patients showed no reduction in inflammatory markers, serum glucose, insulin resistance, or lipid values.[11]

7.10.4 PRECLINICAL ANTIDEPRESSANT-LIKE EFFECTS

Animal studies have both shown antidepressant-like effects of tea as well as suggested mechanisms of action. Mice orally administered green tea polyphenols for seven days showed improved performance in the forced swimming and tail suspension tests, suggestive of antidepressant effects. The polyphenols also reduced serum corticosterone and ACTH levels in mice subjected to the forced swimming test.[12]

In mice subjected to weeks of chronic unpredictable stress, concurrent oral administration of tea polyphenols decreased immobility in the forced swim and tail suspension tests. The polyphenols also normalized the stress-induced decreases in serotonin and norepinephrine in the hippocampus and prefrontal cortex. Activities of superoxide dismutase and catalase were normalized in those areas, and lipid peroxidation reduced.[13] L-Theanine extracted from green tea also reduced immobility of mice in the forced swim and tail suspension tests.[14]

7.10.5 HUMAN ANTIDEPRESSANT EFFECTS

In mice, L-theanine had antidepressant-like effects at doses as low as 1 mg/kg. It has been estimated that the average Japanese green tea drinker consumes about 20 mg of the substance. Thus, L-theanine was studied in human subjects suffering unremitted MDD. Subjects were given 250 mg a day in addition to their medication. After six weeks, there were significant improvements in symptoms of depression, anxiety, sleep disturbance, and cognitive impairments. This dose of theanine was beyond what could easily be obtained in drinking tea. Moreover, the open-label design leaves room for confound.[15] Interestingly, L-theanine has been found to act as a potent NMDA receptor antagonist,[16] as does ketamine. Thus, the L-theanine in green tea could be contributing antidepressant effect through that increasingly well-known mechanism.

A connection has been made between anhedonia, MDD, and deficits in learning enhanced by monetary rewards. In a randomized, double-blind, placebo-controlled study, green tea capsules or placebo were given to subjects who were not suffering depression. After five weeks, those receiving the green tea showed reduced reaction times in the incentivized learning paradigm. They also showed significant reductions in both the Montgomery-Asberg Depression Rating Scale and the 17-item Hamilton Rating Scale for Depression. Of course, since none of the subjects were actually suffering MDD at any time during the study, the significance of those findings is questionable.[17]

Although epidemiological evidence and animal studies strongly suggest that tea has mood-elevating effects and may lower the risk of developing MDD, there is as yet no clear and convincing evidence that tea or the phytochemicals it contains can be used to treat significant degrees of existing MDD.

7.10.6 DOSAGE

There are no dosage recommendations for tea. It is generally drunk *ad libitum*, and can be tailored to individual tastes. Nonetheless, a recommendation of four cups a day would be reasonable.

7.10.7 TOXICITY

Tea has quite low toxicity. For example, the essential oil of tea leaves, one of the more potent forms of the herbs, has been reported have an oral LD50 of 8560 mg/

kg in rats.[18] The oral LD50 of EGCG in rats was estimated to be between 186.8 mg/kg and 1,868 mg/kg.[19] The LD50 of tea polysaccharides was reported to be 4.19 g/kg.[20]

7.10.8 SAFETY IN PREGNANCY

The *Physicians' Desk Reference for Nonprescription Drugs and Dietary Supplements* recommends that pregnant women limit their intake of tea such that caffeine intake remains below 300 mg daily.[21]

7.10.9 DRUG INTERACTIONS

In the *Botanical Safety Handbook*, the warning of potential drug interaction with various forms of *Camellia sinensis* involves concerns over the caffeine they contain. Caffeine is metabolized primarily by the CYP1A2 enzyme. Thus, there can be interactions with medications also metabolized by that enzyme. It was found, for example, that co-administration of fluvoxamine can substantially increase serum caffeine levels.[22] Other medications largely metabolized by CYP1A2 with potential for interactions with caffeine from coffee intake include mexiletine, clozapine, psoralens, idrocilamide, phenylpropanolamine, furafylline, theophylline, and quinolones.[23]

REFERENCES

1. Pham NM, Nanri A, Kurotani K, et al. Green tea and coffee consumption is inversely associated with depressive symptoms in a Japanese working population. *Public Health Nutr.* 2013;17(3):625–633.
2. Hintikka J, Tolmunen T, Honkalampi K, et al. Daily tea drinking is associated with a low level of depressive symptoms in the Finnish general population. *Eur J Epidemiol.* 2005;20(4):359–363.
3. Cabrera C, Artacho R, Giménez R. Beneficial effects of green tea—A review. *J Am Coll Nutr.* 2006;25(2):79–99.
4. Klaunig JE, Xu Y, Han C, et al. The effect of tea consumption on oxidative stress in smokers and nonsmokers. *Proc Soc Exp Biol Med.* 1999;220(4):249–254.
5. Lean ME, Noroozi M, Kelly I, et al. Dietary flavonols protect diabetic human lymphocytes against oxidative damage to DNA. *Diabetes.* 1999;48(1):176–181.
6. Lee S-R, Suh SI, Kim SP. Protective effects of the green tea polyphenol (–)-epigallocatechin gallate against hippocampal neuronal damage after transient global ischemia in gerbils. *Neurosci Lett.* 2000;287(3):191–194.
7. Katiyar SK, Matsui MS, Elmets CA, Mukhtar H. Polyphenolic antioxidant (-)-epigallocatechin-3-gallate from green tea reduces UVB-Induced inflammatory responses and infiltration of leukocytes in human skin. *Photochem Photobiol.* 1999;69(2):148–153.
8. Aktas O, Prozorovski T, Smorodchenko A, et al. Green tea epigallocatechin-3-gallate mediates T cellular NF–B inhibition and exerts neuroprotection in autoimmune encephalomyelitis. *J Immunol.* 2004;173(9):5794–5800.
9. Anderson RA, Polansky MM. Tea enhances insulin activity. *J Agric Food Chem.* 2002;50(24):7182–7186.
10. Wu LY, Juan CC, Ho LT, Hsu YP, Hwang LS. Effect of green tea supplementation on insulin sensitivity in Sprague-Dawley rats. *J Agric Food Chem.* 2004;52(3):643–648.

11. Ryu OH, Lee J, Lee KW, et al. Effects of green tea consumption on inflammation, insulin resistance and pulse wave velocity in type 2 diabetes patients. *Diabetes Res Clin Pract.* 2006;71(3):356–358.

12. Zhu WL, Shi HS, Wei YM, et al. Green tea polyphenols produce antidepressant-like effects in adult mice. *Pharmacol Res.* 2012;65(1):74–80.

13. Liu Y, Jia G, Gou L, et al. Antidepressant-like effects of tea polyphenols on mouse model of chronic unpredictable mild stress. *Pharmacol Biochem Behav.* 2013;104:27–32.

14. Yin C, Gou L, Liu Y, et al. Antidepressant-like effects of L-theanine in the forced swim and tail suspension tests in mice. *Phytother Res.* 2011;25(11):1636–1639.

15. Hidese S, Ota M, Wakabayashi C, et al. Effects of chronic l-theanine administration in patients with major depressive disorder: An open-label study. *Acta Neuropsychiatr.* 2017;29(2):72–79.

16. Di X, Yan J, Zhao Y, et al. L-Theanine protects the APP (Swedish mutation) transgenic SH–SY5Y cell against glutamate-induced excitotoxicity via inhibition of the NMDA receptor pathway. *Neuroscience.* 2010;168(3):778–786.

17. Zhang Q, Yang H, Wang J, et al. Effect of green tea on reward learning in healthy individuals: A randomized, double-blind, placebo-controlled pilot study. *Nutr J.* 2013;12:84.

18. Ash M, Ash I. Handbook of preservatives. *Endicot NY Synapse Inf Resour.* 2004:317.

19. Bedrood Z, Rameshrad M, Hosseinzadeh H. Toxicological effects of *Camellia sinensis* (green tea): A review. *Phytother Res.* 2018;32(7):1163–1180.

20. Nie SP, Xie MY. A review on the isolation and structure of tea polysaccharides and their bioactivities. *Food Hydrocoll.* 2011;25(2):144–149.

21. PDR. *The Physicians' Desk Reference for Non-Prescription Drugs and Dietary Supplements.* 27th ed. Montvale, NJ: Medical Economics Co; 2006.

22. Jeppesen U, Loft S, Poulsen HE, Brśen K. A fluvoxamine-caffeine interaction study. *Pharmacogenetics.* 1996;6(3):213–222.

23. Carrillo JA, Benitez J. Clinically significant pharmacokinetic interactions between dietary caffeine and medications. *Clin Pharmacokinet.* 2000;39(2):127–153.

7.11 CANNABIS

Cannabis is a genus of plants in the family *Cannabaceae*. The genus includes three species, *sativa, indica,* and the lesser known *ruderalis.* Cannabis is thought to have originally come from central Asia, but it has been spread by humans throughout the temperate and tropical areas of the world. For thousands of years it has been used for fiber, recreation, and spiritual quests. It has also been used for medicinal purposes. Hippocrates did not speak of using cannabis. However, his fifth century BCE contemporary, Herodotus, described the use of cannabis seeds by the Scythians in their mourning rituals. The Scythians likely appreciated its mind-altering effects, as they were said to toss cannabis seeds onto red-hot rocks and inhale the vapors that were released – then howl. Pliny the Elder, the first century Roman historian, catalogued some medical uses of cannabis that included the drying up of leaking semen, ridding the ears of vermin, easing the pain of gout and arthritis, and treating the bellies of farm animals.[1] The Chinese used hemp for rope, cloth, and bowstrings. However, as far back as 4000 BCE, references were also made to its use as an anesthetic for surgical procedures. Cannabis was valued in the Ayurvedic Medicine of ancient India and in the Islamic medical tradition. The European Renaissance herbalist, John Gerard, recommended it as it "consumeth wind and drieth up seed." Nicholas Culpeper, in his *Complete Herbal*, recommended the decoction of the roots, as this "allayeth inflammations, easeth the pain of gout, tumours or knots of joints, pain of hips."[2] The medical use of cannabis has persisted, and among its current applications are the treatment of chronic pain, muscle spasticity, cachexia, nausea, inflammation, and seizures.[3] Most psychiatrists have heard the claims that, "marijuana helps me relax," "weed helps me sleep," or "I don't need an antidepressant because I smoke cannabis."

Cannabis contains a variety of unique, structurally related phytochemicals collectively known as cannabinoids. The best known of these substances is Δ9-tetrahydrocannabinol, or THC, as it is the substance most responsible for the characteristic "high" obtained from smoking the herb. Another increasingly well-recognized cannabinoid is cannabidiol, or CBD, as it is being found to convey many of the medicinal benefits of cannabis, but without producing the typical "high." However, there are scores of other cannabinoids of varying importance and concentration in cannabis. Among those are tetrahydrocannabinolic acid, Δ8-tetrahydrocannabinol, cannabigerol, cannabichromene, cannabinol, cannabicitran, cannabidiolic acid, cannabielsoin, cannflavin, and others.[4] The plant also contains a wide variety of flavonoids and terpenes that may contribute medicinal effects. These include some phytochemicals that are found in other medicinal plants, such as borneol, camphor, β-caryophyllene, geraniol, humulene, linalool, myrcene, pinene, terpineol, and others. It is likely these substances combine to produce the familiar aroma of cannabis.[5]

7.11.1 ANTIOXIDANT

Possibly because of its well-known intoxicating effects, it is rarely noted that cannabis also has antioxidant and anti-inflammatory effects that could contribute to antidepressant effects. For example, extract of cannabis was found to increase the reduced glutathione content in the liver and thus significantly decrease rates of lipid peroxidation.[6] Oil pressed from seed of *Cannabis sativa* is rich in antioxidant flavonoids that perform well as radical scavengers in the *in vitro* DPPH assay.[7]

7.11.2 ANTI-INFLAMMATORY

Orally administered CBD reduced inflammation in the carrageenan-induced rat paw edema test. The cannabinoid dampened COX activity and thus decreased plasma levels of the inflammatory prostaglandin-E2. It also demonstrated antioxidant effects in attenuating production of oxygen-derived free radicals and nitric oxide.[8]

CBD in particular shows many effects in the brain that are currently associated with antidepressant action. It dampens lipopolysaccharide-induced activation of NF-kβ and interferon in microglial cells.[9] It stimulates synthesis of BDNF, through activation of the PI3K/Akt/mTOR pathway and inhibition of GSK-3β.[10] It stimulates PPARγ, and thus enhances insulin sensitivity in brain tissue.[11]

7.11.3 ANTIDIABETIC/ANTI-METABOLIC SYNDROME

There are no reports in the literature of cannabis being used to ameliorate diabetes or symptoms of Metabolic Syndrome. Indeed, the well-known appetite stimulating effects of cannabis and the use of the cannabinoid antagonist rimonabant as a weight control agent would suggest that cannabis use is somewhat antithetical to such.

7.11.4 PRECLINICAL ANTIDEPRESSANT-LIKE EFFECTS

Preclinical studies using mice and rats have provided strong and uniform evidence that the cannabinoids in cannabis exert antidepressant-like effects. In mice, acute intraperitoneal administration of THC reduced immobility in the forced swim and tail suspension tests. The cannabinoids, CBD, and cannabichromene also showed this antidepressant-like effect, as did the synthetic CB1 agonist CP55,940. Several other cannabinoids that were tested, Δ8-tetrahydrocannabinol, cannabigerol, cannabicitran, cannabicyclol, and cannabinol, did not exhibit this effect.[12,13]

Others found a similar effect of acute CBD in decreasing immobility in the forced swim test, and further discovered that this effect was blocked by co-administration of the 5-HT1A receptor antagonist, WAY100635. This suggested that the acute effect of CBD was mediated in part by activation of 5-HT1A receptors. However, this CBD treatment did not increase levels of BDNF in the hippocampus.[14]

CBD also has an antidepressant-like effect in rats. In a study more consistent with common use of cannabis, sub-chronic administration of CBD over 14 days reduced immobility in the forced swim test. CBD was also found to increase BDNF in the amygdala, and both the behavioral and neurochemical effects of the phytochemical were similar to those produced by sub-chronic administration of the standard antidepressant medication, imipramine.[15]

Anxiety is a common comorbidity in MDD, and at times is a barrier to treatment of the latter illness. Thus, it is important to note that CBD was shown to exert anxiolytic-like effects in rats. Daily administration of CBD for a week following exposure to a feline predator increased movement in the elevated plus maze, a standard preclinical test of anxiolytic effects. This effect of CBD was associated with increases in 5-HT1A receptor RNA expression and blocked by the 5-HT1A receptor antagonist. Thus, the study confirmed that the effect of CBD was partially due to enhancement of activity at 5-HT1A receptors. However, no increases in BDNF were found in the brains of those animals.[16]

Sub-chronic administration of the cannabinoid type 1 receptor agonist, HU-210, reduces immobility of rats in the forced swim test,[17] as does the CB1 agonist, WIN55,212-2.[18] Other cannabinoidergic substances with such antidepressant-like effects in rats are oleamide, a natural cannabinoid agonist, and the synthetic endocannabinoid uptake inhibitor, AM404.[19]

In the bulbectomized mouse, CBD restores the preference for sucrose while enhancing serotonergic activity in the frontal cortex. As was seen in the aforementioned studies of CBD, the behavioral effect of the substance was blocked by the 5-HT1A antagonist, WAY100635.[20] In bulbectomized rats, chronic administration of THC attenuated the hyperactivity typical of such animals in the open field. Curiously, chronic administration of the cannabinoid receptor antagonist, rimonabant, also blocked this hyperactivity. In intact animals, chronic THC increased BDNF expression levels in the hippocampus and frontal cortex, but rimonabant had no effect.[21] In a consistent finding, chronic, but not acute, administration of the cannabinoid agonist, HU210, stimulated neurogenesis in the hippocampal dentate gyrus of adult rats, an effect thought to underlie the actions of many antidepressant medications.[22] Nonetheless, at least two studies have provided results contrary to the above. Both the cannabinoid antagonist rimonabant[23] and the inverse agonist AM251[24] have been shown to exert antidepressant-like effects in mice.

7.11.5 HUMAN ANTIDEPRESSANT EFFECTS

There are no specific studies of cannabis or its phytochemical constituents as treatments for MDD in human subjects. The only means by which the effect of cannabinoids on MDD can be ascertained are through epidemiological studies of cannabis users who may or may not co-morbidly suffer MDD, or by studying levels of endocannabinoids in subjects with or without MDD. In addition, there are the aforementioned studies of the effects of blocking human cannabinoid receptors with the antagonist rimonabant.

Studies of the natural disposition of the endocannabinoid system suggest changes in endocannabinoid activity in individuals suffering MDD. However, data are somewhat inconsistent, and it is not yet clear if such activity is increased or decreased in MDD. For example, a polymorphism that would result in reduced expression of the of the CB1 gene reduced risk of MDD in patients with Parkinson's disease.[25] Moreover, a post-mortem study revealed elevated levels of the CB1 receptor and CB1-receptor-mediated G-protein signaling, as well as higher levels of endocannabinoids (including anandamide) in the dorsolateral prefrontal cortex of alcoholic suicide victims compared with alcoholic non-suicide subjects.[26]

However, basal serum concentrations of the endocannabinoid ligands anandamide and 2-arachidonoylglycerol were significantly reduced in women with MDD relative to matched controls, indicating a deficit in peripheral endocannabinoid activity.[27] Furthermore, a recent study has shown that electroconvulsive therapy significantly increases levels of the endocannabinoids anandamine and 2-arachidonoylglycerol in the cerebrospinal fluid of patients with MDD.[28]

Epidemiological studies of depression in users of cannabis have provided little, if any, evidence that cannabis might be useful in the treatment of MDD. In an often-cited study, the risk of first episode of MDD was moderately associated with the number of occasions of cannabis use and with more advanced stages of cannabis use, though the strength of the association was deemed "modest at best."[29] In another influential epidemiological study published in the *American Journal of Psychiatry*, it was found that those with a diagnosis of cannabis abuse at the beginning of the study were four times more likely than those with no cannabis abuse diagnosis to have depressive symptoms at the follow-up assessment. Furthermore, participants who suffered MDD at the beginning of the study, but did not abuse cannabis, showed no greater likelihood than non-depressed participants of later starting to abuse cannabis. The latter result was seen as diminishing the possibility that people with depression drifted toward using cannabis as a means to self-treat their depression.[30] The perhaps not-so-subtle key to these results was abuse, rather than mere use, of cannabis, which may have skewed results.

Indeed, the literature has frequently associated heavy use of cannabis with increased risk of MDD and other psychiatric disorders.[31] As a caveat, in one such study associating cannabis dependence with psychiatric disorders, it was noted that most had developed psychiatric issues prior to becoming dependent on cannabis.[32] One of the most compelling studies, a study of twins, found that between monozygotic twins, the one who used cannabis frequently was more likely to report MDD and suicidal ideation than their identical twin who had used cannabis less frequently.[33]

Despite evidence that cannabis, particularly heavy use, can cause or exacerbate MDD, there is also evidence that blocking cannabis receptors with the cannabinoid antagonist rimonabant can precipitate anxiety and MDD of a severe nature. In fact, rimonabant was blocked from being marketed as a weight loss drug in the United States because of its severe psychiatric side effects.[34] Apparently, either over- or under-stimulation of cannabinoid receptors in the brain can disturb mood.

The final complication is that there are many cannabinoids, terpenoids, and other phytochemicals contained in the cannabis plant, and they may act in quite different fashion from one another. The composition of phytochemicals can vary from one strain of cannabis to another, with some containing high concentrations of THC and others containing high concentrations of cannabinoids such as CBD that buffer THC.[35] These different cannabinoids may also interact with each other through the so-called "entourage effect" and thus possibly create new and unexpected effects.[36] It is entirely possible that strains of cannabis that are high in CBD may offer benefits that are not seen in those that are high in THC. Indeed, psychiatrically, CBD appears to be the most helpful component of cannabis. Nonetheless, at the present time, cannabis cannot be recommended as a treatment for MDD.

7.11.6 DOSAGE

Because of the variance of strength of cannabis and the means by which it is commonly used – that is, by smoking until the subjective sense of a satisfying degree of intoxication – it is difficult to determine precisely what doses of Δ9-THC are required to produce specific effects. Therapeutic use of the FDA approved form of Δ9-THC, dronabinol, is commonly in doses of 2.5–40 mg daily, with a maximum approved dose of 150 mg daily.[37] Commercially available cannabidiol is often dispensed in doses of 10 mg taken several times a day. I note, however that in a study of treatment of schizophrenia, patients were started on 40 mg cannabidiol a day, and this was increased up to 1280 mg/day without adverse effects.[38]

7.11.7 TOXICITY

Cannabis appears to have very low acute toxicity. In rats, the intragastric LD50 Δ9-THC in oil vehicle was reported to be 800 mg/kg, whereas the intravenous and inhaled LD50s were roughly the same at 36–40 mg/kg.[39] In another study, the dried ethanolic extract of intact cannabis, which was 3% Δ9-THC, was gavaged into male albino rats. The LD50 of that extract was 1729.6 mg/kg, or about 50mg Δ9-THC/kg.[40] That finding compares well with the above findings for intravenous and inhaled

Δ9-THC. Extrapolated to an adult human of 75 kg, this would be a very large dose of about 3600 mg Δ9-THC.

In experienced male cannabis smokers, daily oral Δ9-THC doses of 10–30 mg every 4 h, escalating to 210 mg/day caused no obvious toxic effects, but did begin to cause tolerance in subjective effects and physiological signs after several weeks.[41]

7.11.8 SAFETY IN PREGNANCY

Linn et al. found no statistical difference in likelihood of adverse pregnancy outcomes among women who continued to smoke cannabis. However, they advised against its use in pregnancy due to lack of definitive information.[42] A later study by Fergusson et al. drew a similar conclusion.[43] Many women have reported cannabis to be very effective in relieving the nausea of morning sickness during pregnancy.[44] Ironically, there have also been reports of cannabis causing the so-called Cannabinoid Hyperemesis Syndrome during pregnancy.[45]

7.11.9 DRUG INTERACTIONS

CBD is metabolized primarily by the CYP3A4, CYP2C19, and CYP2C19 enzymes in the liver.[46] THC is metabolized primarily by CYP2C9 and CYP3A4.[47] Thus, cannabis can potentially alter blood levels of certain medications that are also metabolized by those enzymes.

Cannabis has also been reported to enhance the pain-relieving effects of opiates, which is of importance for treatment of both pain and substance abuse. This interaction is not due to competition for P450 enzymes in the liver, but rather to synergistic interactions between respective G-proteins within neurons.[48]

REFERENCES

1. Butrica J. The medicinal use of cannabis among the Greeks and Romans. In: Russo E, Grotenherman F, editors. *The Handbook of Cannabis Therapeutics: From Bench to Bedside.* New York: Haworth; 2006. p. 23–42.
2. Warf B. High points: An historical geography of cannabis. *Geogr Rev.* 2014;104(4):414–438.
3. Aggarwal SK, Carter GT, Sullivan MD, ZumBrunnen C, Morrill R, Mayer JD. Medicinal use of cannabis in the United States: Historical perspectives, current trends, and future directions. *J Opioid Manag.* 2009;5(3):153–168.
4. Choi YH, Hazekamp A, Peltenburg-Looman AMG, et al. NMR assignments of the major cannabinoids and cannabiflavonoids isolated from flowers of Cannabis sativa. *Phytochem Anal.* 2004;15(6):345–354.
5. Elzinga S, Fischedick J, Podkolinski R, Raber JC. Cannabinoids and terpenes as chemotaxonomic markers in cannabis. *Nat Prod Chem Res.* 2015;3:181.
6. Comelli F, Bettoni I, Colleoni M, Giagnoni G, Costa B. Beneficial effects of a *Cannabis sativa* extract treatment on diabetes-induced neuropathy and oxidative stress. *Phytother Res.* 2009;23(12):1678–1684.
7. Smeriglio A, Galati EM, Monforte MT, et al. Polyphenolic compounds and antioxidant activity of cold-pressed seed oil from Finola cultivar of *Cannabis sativa* L. *Phytotherapy.* 2016;30(8):1298–1307.

8. Costa B, Colleoni M, Conti S, et al. Oral anti-inflammatory activity of cannabidiol, a non-psychoactive constituent of cannabis, in acute carrageenan-induced inflammation in the rat paw. *Naunyn Schmiedebergs Arch Pharmacol.* 2004;369(3):294–299.

9. Kozela E, Pietr M, Juknat A, Rimmerman N, Levy R, Vogel Z. Cannabinoids delta(9)-tetrahydrocannabinol and cannabidiol differentially inhibit the lipopolysaccharide-activated NF-kappaB and interferon-beta/STAT proinflammatory pathways in BV-2 microglial cells. *J Biol Chem.* 2010;285(3):1616–1626.

10. Giacoppo S, Pollastro F, Grassi G, Bramanti P, Mazzon E. Target regulation of PI3K/Akt/mTOR pathway by cannabidiol in treatment of experimental multiple sclerosis. *Fitoterapia.* 2017;116:77–84.

11. Scuderi C, Steardo L, Esposito G. Cannabidiol promotes amyloid precursor protein ubiquitination and reduction of beta amyloid expression in SHSY5YAPP+ cells through PPARgamma involvement. *Phytother Res.* 2014;28(7):1007–1013.

12. El-Alfya AT, Ivey K, Robinson K, et al. Antidepressant-like effect of Δ9-tetrahydrocannabinol and other cannabinoids isolated from *Cannabis sativa* L. *Pharmacol Biochem Behav.* 2010;95(4):434–442.

13. Ivey KD, Ross SA, Ahmed S, et al. Evaluation of Delta9-Tetrahydrocannabinol and other cannabinoids for antidepressant-like actions in the mouse forced swim test. *Planta Med.* 2008;74:P-28.

14. Zanelati TV, Biojone C, Moreira FA, Guimarães FS, Joca SR. Antidepressant-like effects of cannabidiol in mice: Possible involvement of 5-HT1A receptors. *Br J Pharmacol.* 2010;159(1):122–128.

15. Réus GZ, Stringari RB, Ribeiro KF, et al. Administration of cannabidiol and imipramine induces antidepressant-like effects in the forced swimming test and increases brain-derived neurotrophic factor levels in the rat amygdala. *Acta Neuropsychiatr.* 2011;23(5):241–248.

16. Campos AC, Ferreira FR, Guimarães FS. Cannabidiol blocks long-lasting behavioral consequences of predator threat stress: Possible involvement of 5HT1A receptors. *J Psychiat Res.* 2012;46(11):1501–1510.

17. Morrish AC, Hill MN, Riebe CJN, Gorzalka BB. Protracted cannabinoid administration elicits antidepressant behavioral responses in rats: Role of gender and noradrenergic transmission. *Physiol Behav.* 2009;98(1–2):118–124.

18. Bambico FR, Katz N, Debonnel G, Gobbi G. Cannabinoids elicit antidepressant-like behavior and activate serotonergic neurons through the medial prefrontal cortex. *J Neurosci.* 2007;27(43):11700–11711.

19. Hill MN, Gorzalka BB. Pharmacological enhancement of cannabinoid CB1 receptor activity elicits an antidepressant-like response in the rat forced swim test. *Eur Neuropsychopharmacol.* 2005;15(6):593–599.

20. Linge R, Jiménez-Sánchez L, Campa L, et al. Cannabidiol induces rapid-acting antidepressant-like effects and enhances cortical 5-HT/glutamate neurotransmission: Role of 5-HT1A receptors. *Neuropharmacology.* 2016;103:16–26.

21. Elbatsh MM, Moklas MAA, Marsden CA, Kendall DA. Antidepressant-like effects of Δ9-tetrahydrocannabinol and Rimonabant in the olfactory bulbectomised rat model of depression. *Pharmacol Biochem Behav.* 2012;102(2):357–365.

22. Jiang W, Zhang Y, Xiao L, et al. Cannabinoids promote embryonic and adult hippocampus neurogenesis and produce anxiolytic- and antidepressant-like effects. *J Clin Invest.* 2005;115(11):3104–3116.

23. Griebel G, Stemmelin J, Scatton B. Effects of the cannabinoid CB1 receptor antagonist Rimonabant in models of emotional reactivity in rodents. *Biol Psychiatry.* 2005;57(3):261–267.

24. Shearman LP, Rosko KM, Fleischer R, et al. Antidepressant-like and anorectic effects of the cannabinoid CB1 receptor inverse agonist AM251 in mice. *Behav Pharmacol.* 2003;14(8):573–582.

25. Barrero FJ, Ampuero I, Morales B, et al. Depression in Parkinson's disease is related to a genetic polymorphism of the cannabinoid receptor gene (CNR1). *Pharmacogenomics J.* 2005;5(2):135–141.
26. Vinod KY, Arango V, Xie S, et al. Elevated levels of endocannabinoids and CB1 receptor-mediated G-protein signaling in the prefrontal cortex of alcoholic suicide victims. *Biol Psychiatry.* 2005;57(5):480–486.
27. Hill MN, Miller GE, Carrier EJ, Gorzalka BB, Hillard CJ. Circulating endocannabinoids and N-acyl ethanolamines are differentially regulated in major depression and following exposure to social stress. *Psychoneuroendocrinology.* 2009;34(8):1257–1262.
28. Kranaster L, Hoyer C, Aksay SS, et al. Electroconvulsive therapy enhances endocannabinoids in the cerebrospinal fluid of patients with major depression: a preliminary prospective study. *Eur Arch Psy Clin N.* 2017;267(8):781–786.
29. Chen CY, Wagner FA, Anthony JC. Marijuana use and the risk of major depressive episode: Epidemiological evidence from the United States National comorbidity Survey. *Soc Psych Psych Epid.* 2002;37(5):199–206.
30. Bovasso GB. Cannabis abuse as a risk factor for depressive symptoms. *Am J Psychiatry.* 2001;158(12):2033–2037.
31. Degenhardt L, Hall W, Lynskey M. Exploring the association between cannabis use and depression. *Addiction.* 2003;98(11):1493–1504.
32. Agosti V, Nunes E, Levin F. Rates of psychiatric comorbidity among U.S. residents with lifetime cannabis dependence. *Am J Drug Alcohol Abus.* 2002;28(4):643–652.
33. Agrawal A, Nelson EC, Bucholz KK, et al. Major depressive disorder, suicidal thoughts and behaviours, and cannabis involvement in discordant twins: A retrospective cohort study. *Lancet Psychiatry.* 2017;4(9):706–714.
34. Mitchell PB, Morris MJ. Depression and anxiety with Rimonabant. *Lancet.* 2007;370(9600):1671–1672.
35. De Petrocellis L, Di Marzo V. Non-CB1, non-CB2 receptors for endocannabinoids, plant cannabinoids, and synthetic cannabimimetics: focus on G-protein-coupled receptors and transient receptor potential channels. *J Neuroimmune Pharmacol.* 2010;5:103–121.
36. Russo EB. Taming THC: Potential cannabis synergy and phytocannabinoid-terpenoid entourage effects. *BJP. Br J Pharmacol.* 2011;163(7):1344–1364.
37. *Physicians Desk Reference.* 60th ed. Montvale, NJ: Thomson; 2006. p. 3334–3336.
38. Zuardi AW, Hallak JEC, Dursun SM, et al. Cannabidiol monotherapy for treatment-resistant schizophrenia. *J Psychopharmacol.* 2006;20(5):683–686.
39. Rosenkrantz H, Heyman IA, Braude MC. Inhalation, parenteral and oral LD50 values of Δ9-tetrahydrocannabinol in Fischer rats. *Toxicol Appl Pharmacol.* 1974;28(1): 18–27.
40. Yassa HA, Dawood AEWA, Shehata MM, Abdel Aal KM, Shehata MM. Subchronic toxicity of cannabis leaves on male albino rats. *Hum Exp Toxicol.* 2010;29(1):37–47.
41. Jones RT, Benowitz N, Bachman J. Clinical studies of cannabis tolerance and dependence. *Ann N Y Acad Sci.* 1976;282:221–239.
42. Linn S, Schoenbaum SC, Monson RR, Rosner RI, Stubblefield PC, Ryan KJ. The association of cannabis use with outcome of pregnancy. *Am J Public Health.* 1983;73(10):1161–1164.
43. Fergusson DM, Horwood LJ, Northstone K, ALSPAC Study Team. Maternal use of cannabis and pregnancy outcome. *BJOG.* 2002;109(1):21–27.
44. Westfall RE, Janssen PA, Lucas P, Capler R. Survey of medicinal cannabis use among childbearing women: Patterns of its use in pregnancy and retroactive self-assessment of its efficacy against 'morning sickness.' *Complement Ther Clin.* 2006;12(1):27–33.
45. Andrews KH, Bracero LA. Cannabinoid hyperemesis syndrome during pregnancy: A case report. *J Reprod Med.* 2015;60(9–10):430–432.

46. Jiang R, Yamaori S, Takeda S, Yamamoto I, Watanabe K. Identification of cytochrome P450 enzymes responsible for metabolism of cannabidiol by human liver microsomes. *Life Sci.* 2011;89(5–6):165–170.
47. Watanabe K, Yamaori S, Funahashi T, Kimura T, Yamamoto I. Cytochrome P450 enzymes involved in the metabolism of tetrahydrocannabinols and cannabinol by human hepatic microsomes. *Life Sci.* 2007;80(15):1415–1419.
48. Cichewicz DL. Synergistic interactions between cannabinoid and opioid analgesics. *Life Sci.* 2004;74(11):1317–1324.

7.12   CECROPIA

Cecropia is a genus of tropical trees of Central and South America. They are locally known as "embauba" or trumpet-trees, and have been used in folk medicine as treatments for asthma, cough, hypertension, diabetes, and inflammation.[1] Among the main phytochemicals contained in species of *Cecropia* are soorientin, orientin and isovitexin chlorogenic acid, apigenin, diosmetin, luteolin, and procyanidin, as well as various glycosylated flavonoids.[2]

7.12.1   ANTIOXIDANT

Extracts of *Cecropia pachystachya* are rich in flavonoids, and have been shown to exert potent antioxidant effects, *in vitro*, in the standard DPPH, β-carotene bleaching, and TBARS assays.[3]

Hydroethanolic extracts of *Cecropia glaziovii* were found to be effective, *in vivo*, in protecting liver tissue from the oxidative damage of CCl_4. It inhibited lipid peroxidation, as per the TBARS assay, and enhanced the activity of liver superoxide dismutase and catalase.[4] Similar antioxidant effects and protection of antioxidant enzymes are seen in brain tissue.

7.12.2   ANTI-INFLAMMATORY

The anti-inflammatory activity of extract of *Cecropia pachystachya* was evaluated by croton oil-induced ear edema test. When used orally, the anti-inflammatory effect of the extract was similar to that of indomethacin with 53% inhibition of the ear edema. Also, results on topical treatment were similar to that of dexamethasone, with 83% inhibition of the edema.[5]

Aqueous extract *Cecropia glaziovii* leaves attenuated the inflammatory effect of carrageenan injected into the pleural cavity of rats. The extract reduced concentrations of the inflammatory cytokines, TNF-α, IL-1β, and IL-6, and minimized cell infiltration in a manner similar to that of dexamethasone.[6]

7.12.3   ANTIDIABETIC/ANTI-METABOLIC SYNDROME

In alloxan-induced diabetic rats, extract of *Cecropia pachystachya* lowered serum glucose by 60%, which was no different than the effects of metformin or glibenclamide.[7] Ethanolic extract of *Cecropia glaziovii*, as well as well as two of its constituents, chlorogenic and caffeic acids, significantly lowered serum glucose levels in fasted alloxan-induced diabetic rats. It also improved glucose tolerance in these animals.[8] Chlorogenic acid also reduced memory deficits and anxiety in streptozotocin-induced diabetic rats, as well as reducing lipid peroxidation in diabetic brain tissue.[9] Thus, these effects are both peripheral and central.

7.12.4 PRECLINICAL ANTIDEPRESSANT-LIKE EFFECTS

There are several studies showing species of *Cecropia* to have antidepressant-like effects in rodents. Chronic treatment with the aqueous extract of *Cecropia glazioui* reduced the immobility of rats in the forced swim test. Biochemical analyses of the hippocampal neurotransmitters in these animals showed significant increases in monoamine levels. The conclusion was that the antidepressant-like effect of *C. glazioui* extract was most likely due to the blockade of the monoamine uptake in the CNS.[10] A similar antidepressant-like effect in the forced swim test was seen in rats after acute administration of extract of the closely related species, *Cecropia pachystachya*. In this case, further analyses showed prevention of oxidative damage as evidenced by reductions in carbonyl, TBARS, and nitrite/nitrate levels in some, but not all, areas of the brain. There were no alterations of superoxide dismutase or catalase activity in any brain structures, nor were there changes in levels of any inflammatory cytokines in any brain area. Of course, neither a single injection of extract nor a single stressful test procedure may have been enough to produce such changes.[11]

Other studies have explored effects of *Cecropia pachystachya* on rodents that had been subjected to chronic unpredictable stress, an animal model of MDD. In rats subjected to stress for 40 days, five days of daily administration of extract of *Cecropia pachystachya* reduced immobility in the forced swim test, as well as increased grooming behavior in the splash test, another test of antidepressant-like effect. Treatment also reversed increases in TBRS, myeloperoxidase activity, and nitrite/nitrate concentrations in some brain regions, all of which are indications of antioxidative and anti-nitrosative effects. The activities of the antioxidants superoxide dismutase and catalase were restored in some brain areas. Treatment increased the activity of mitochondrial complex IV throughout the brain, suggesting enhancement of energy utilization.[12]

Cecropia pachystachya also showed antidepressant-like effects in mice subjected to chronic unpredictable stress. Chronic treatment with the extract decreased immobility in the forced swim test. It also attenuated stress-induced increases in lipid peroxidation in the hippocampus and prefrontal cortex. However, no changes were observed in the activity of the antioxidant enzymes superoxide dismutase and catalase in the brain.[13]

Interestingly, behavioral tests of the related species, *Cecropia membranacea*[14] and *Cecropia* peltate,[15] showed anxiolytic-like, but not antidepressant-like, effects of those plants.

7.12.5 HUMAN ANTIDEPRESSANT EFFECTS

There are no published studies of the use of *Cecropia* for the treatment of MDD in human subjects.

7.12.6 Dosage

Uniform dosage recommendations for *Cecropia pachystachya* are not available. However, there is a published study of the hypoglycemic effects of the closely related *Cecropia obtusifolia* plant in human patients with diabetes type II. In that study, benefits without adverse effects were seen with aqueous extracts of the leaves prepared as a single dose by boiling 13.5 g of dried and milled leaves of *Cecropia obtusifolia* for 5 min in 1 l of water. This was said to be the recommendation of traditional healers.[16]

7.12.7 Toxicity

The LD50 of dried aqueous extract of was determined to be higher than 5.0 g/kg in rats or by extrapolation, higher than 375 g in a 75 kg human adult.

7.12.8 Safety in Pregnancy

There are no reports in the literature concerning effects of *Cecropia* species on pregnancy outcome. Pregnant female rats that received 1.0 g/kg of extract per day throughout pregnancy gave birth to normal pups. The weight and the physical development of both genders of pups were not affected, but the uprightness latency and the negative geotaxis reflexes were enhanced and the rearing frequency decreased. The toxicity to pregnant females and pups was deemed low.[17]

7.12.9 Drug Interactions

Review of the literature revealed no reports on interactions of *Cecropia pachystachya* with action or metabolism of medications.

REFERENCES

1. Lorenzi H, Matos FJ. Plantas medicinais do Brasil. In: *Nova Odessa*. 1st ed. São Paulo: Instituto Plantarum; 2012.
2. Costa GM, Ortmann CF, Schenkel EP, Reginatto FH. An HPLC-DAD method to quantification of main phenolic compounds from leaves of *Cecropia* species. *J Braz Chem Soc*. 2011;22(6):1096–1102.
3. Pacheco NR, Pinto NCC, da Silva JM, et al. *Cecropia pachystachya*: A species with expressive in vivo topical anti-inflammatory and in vitro antioxidant effects. *BioMed Res Int*. 2014; Article ID 301294.
4. Petronilho F, Dal-Pizzol F, Costa GM, et al. Hepatoprotective effects and HSV-1 activity of the hydroethanolic extract of *Cecropia glaziovii* (embaúba-vermelha) against acyclovir-resistant strain. *J Pharmaceut Biol*. 2012;50(7):7911–7918.
5. Aragão DM, Lima IV, da Silva JM, et al. Anti-inflammatory, antinociceptive and cytotoxic effects of the methanol extract of *Cecropia pachystachya* Trécul. *Phytother Res*. 2013;27(6):926–930.
6. Müller SD, Florentino D, Ortmann CF, et al. Anti-inflammatory and antioxidant activities of aqueous extract of *Cecropia glaziovii* leaves. *J Ethnopharmacol*. 2016;185:255–262.

7. Aragão DMO, Guarize L, Lanini J, da Costa JC, Garcia RM, Scio E. Hypoglycemic effects of *Cecropia pachystachya* in normal and alloxan-induced diabetic rats. *J Ethnopharmacol*. 2010;128(3):629–633.

8. Arend DP, Santos TCd, Cazarolli LH, et al. In vivo potential hypoglycemic and in vitro vasorelaxant effects of *Cecropia glaziovii* standardized extracts. *Rev Bras Farmacogn*. 2015;25(5):473–484.

9. Stefanello N, Schmatz R, Pereira LB, et al. Effects of chlorogenic acid, caffeine, and coffee on behavioral and biochemical parameters of diabetic rats. *Mol Cell Biochem*. 2014;388(1–2):277–286.

10. Rocha FF, Lima-Landman MTR, Souccar C, Tanae MM, De Lima TC, Lapa AJ. Antidepressant-like effect of *Cecropia glazioui* Sneth and its constituents: In vivo and in vitro characterization of the underlying mechanism. *Phytomedicine*. 2007;14(6):396–402.

11. Ortmann CF, Abelaira HM, Réus GZ, et al. LC/QTOF profile and preliminary stability studies of an enriched flavonoid fraction of *Cecropia pachystachya* Trécul leaves with potential antidepressant-like activity. *Biomed Chromatogr*. 2017;31(11):e3982.

12. Ortmann CF, Réus GZ, Ignácio ZM, et al. Enriched flavonoid fraction from *Cecropia pachystachya* Trécul Leaves exErts antidepressant-like behavior and protects brain Against oxidative stress in rats subjected to chronic mild stress. *Neurotox Res*. 2016;29(4):469–483.

13. Gazal M, Ortmann CF, Martins FA, et al. Antidepressant-like effects of aqueous extract from *Cecropia pachystachya* leaves in a mouse model of chronic unpredictable stress. *Brain Res Bull*. 2014;108:10–17.

14. Jaramill RA, Rincon J, Guerrero MF. Kind of activity antiabsence in mice produced by the methanolic extract from *Cecropia membranacea* Trécul. *Rev Fac Quim Farm*. 2008;15(2):267–273.

15. Chávez JO, Velanda JR, Pabón MG. Neuropharmacological profile of the butanolic fraction obtained from leaves of *Cecropia peltata* L. *Rev Colomb Cienc Quím Farm*. 2013;42(2):244–259.

16. Revilla-Monsalve MC, Andrade-Cetto A, Palomino-Garibay MA, Wiedenfeld H, Islas-Andrade S. Hypoglycemic effect of *Cecropia obtusifolia* Bertol aqueous extracts on type 2 diabetic patients. *J Ethnopharmacol*. 2007;111(3):636–640.

17. Gerenutti M, Prestes AF, Silva MG, et al. The effect of *Cecropia glazioui* Snethlage on the physical and neurobehavioral development of rats. *Pharmazie*. 2008;63(5): 398–404.

7.13   *CENTELLA ASIATICA* (GOTU KOLA)

Centella asiatica (Gotu Kola), is a perennial herbaceous creeper belonging to the family Umbellifere. It is found throughout India in moist places up to an altitude of 1800 m. It is found in swampy areas of other tropical and subtropical countries including Pakistan, Sri Lanka, Madagascar, South Africa, South Pacific, and Eastern Europe. Known as mandukparni, Indian pennywort, or jalbrahmi, it has been used as a medicine in the Ayurvedic tradition of India for thousands of years and was described in the ancient Indian medical text *Sushruta Samhita*. *Centella asiatica* was also known to the ancient Chinese herbalists, and was referred to as a "miracle elixir of life."[1] *Centella asiatica* has been used for the treatment of various skin conditions such as leprosy, lupus, varicose ulcers, eczema, psoriasis, diarrhea, fever, amenorrhea, diseases of the female genitourinary tract, and for relieving anxiety and improving cognition. In Ayurveda, it is one of the main herbs for revitalizing the nerves and brain cells. Eastern healers relied on CA to treat emotional disorders, such as depression.[2,3]

The primary active constituents of *Centella asiatica* are thought to be triterpenoids. The total extract also contains tannins, sterols, flavonoids, and other components including brahmosides, brahminosides, madecassoside, beta-chariophylen, trans-beta-pharnesen, germachrene D, campesterol, sitosterol, stigmasterol, chercetin, kempferol, hydrochotine, and vallerine.[4]

7.13.1 ANTIOXIDANT

Centella asiatica performed very well in a battery of *in vitro* antioxidant tests, including DPPH radical-scavenging activity, superoxide radical-scavenging activity in riboflavin/light/NBT system, and hydroxyl radical-scavenging activity, and inhibition of lipid peroxidation induced by FeSO4 in egg yolk. It was among the most effective in antioxidant action among 11 commonly used Indian plants.[5]

Centella asiatica has been shown to exert significant *in vivo* antioxidant effects. The extract of *Centella asiatica* prevented nitropropionic acid-induced oxidative stress and mitochondrial dysfunctions in the brains of mice. Markers of oxidative damage, malondialdehyde, ROS, and hydroperoxides, were substantially reduced in the striatum.[6]

7.13.2 ANTI-INFLAMMATORY

Centella asiatica has traditionally been used for the treatment of arthritis, and recent studies suggest such treatment may be effective. It was recently shown that madecassoside, an active phytochemical in *Centella asiatica*, alleviated infiltration of inflammatory cells, synovial hyperplasia, and joint destruction in the type II collagen-induced arthritis model of rheumatoid arthritis in mice.[7] An aqueous extract of *Centella asiatica* also reduced prostaglandin E2-induced paw edema, a common test of anti-inflammatory effects.[8] Aqueous extract of *Centella asiatica* in drinking water improved performance of old mice in the Morris water maze. These effects appeared to be due to induction of mitochondrial and neuronal antioxidant enzymes, including Nrf2 and hemeoxygenase-1, in the hippocampus and frontal cortex.[9]

7.13.3 ANTIDIABETIC/ANTI-METABOLIC SYNDROME

Centella asiatica exhibits antidiabetic effects. When administered to glucose-load mice, it lowered serum glucose in manner almost comparable to the diabetic medication, glibenclamide.[10] Asiatic acid purified from *Centella asiatica* ameliorates the development of fatty liver in mice consuming a high fat diet. It lowered plasma insulin secretion and enhanced insulin sensitivity, and also decreased serum levels of the inflammatory cytokines, interleukin (IL)-1β, IL-6, and tumor necrosis factor-α.[11] These reversals of metabolic syndrome might be expected to diminish the risk of MDD and possibly help reduce its symptoms in existing cases.

7.13.4 PRECLINICAL ANTIDEPRESSANT-LIKE EFFECTS

Several animal studies support the notion that *Centella asiatica* exerts antidepressant-like effects. Acute intraperitoneal treatment with asiatic acid purified from extracts of *Centella asiatica* improved performance of rats in both the forced swim and elevated plus-maze tests. This suggests both antidepressant and anxiolytic effects of the substance.[12] In olfactory bulbectomized rats, a classical animal model of MDD, two weeks of oral administration of extracts of *Centella asiatica* improved performance in the open-field and elevated plus-maze tests, further suggesting antidepressant and anxiolytic effects of the herb. Treatment also produced normalization of food intake and weight. These effects of *Centella asiatica* were similar to those of the standard antidepressants fluoxetine, desipramine, and imipramine.[13] Triterpenes isolated from *Centella asiatica* also decreased immobility time in mice subjected to the forced swim test, with results comparable to those of imipramine.[14]

7.13.5 HUMAN ANTIDEPRESSANT EFFECTS

In healthy human subjects randomly assigned to receive either a single dose of *Centella asiatica* or placebo, *Centella asiatica* significantly attenuated the peak acoustic startle response. This suggests an anxiolytic effect of the herb. The single dose of *Centella asiatica* had no effect on self-rated mood.[15] A more recent study found that *Centella asiatica* does have anxiolytic effects in individuals diagnosed with Generalized Anxiety Disorder per the Hamilton's Brief Psychiatric Rating Scale. *Centella asiatica* not only reduced anxiety over 60 days of treatment, but also significantly reduced reports of depression. Unfortunately, while the study provides the only statistically analyzed data concerning antidepressant effects of *Centella asiatica*, the study was not placebo-controlled.[16]

In a two month long, randomized, placebo-controlled, double-blind study in healthy individuals of an average of 65 years of age, *Centella asiatica* enhanced cognitive function in a standard computerized test battery. It also improved mood per the Bond-Lader visual analogue scales, a test designed to analyze the three mood factors: alertness, calmness, and contentedness. Significant improvements were seen in alertness and calmness. However, contentedness, ostensibly the factor most closely related to absence of depression, was unchanged by treatment.[17] In a similar three-month study, *Centella asiatica* improved physical functioning, but had no effects on subjective impressions of social functioning and general mental health, including depression, anxiety, behavioral-emotional control, and general positive affect.[18] Thus, it can be concluded that there are no published studies of the usefulness of *Centella asiatica* in treating MDD, and no indications from existing studies that *Centella asiatica* improves mood in healthy adults.

7.13.6 DOSAGE

Easley and Horne recommend 500 to 1000 mg capsules of dried leaf powder three times a day. Alternatively, they suggest tincture of fresh plant (1:95% alcohol) or dried plant (1:5, 40% alcohol) 1–4 ml three times a day.[19]

7.13.7 TOXICITY

The LD50 of *Centella asiatica* in mice is reported be over 4000 mg/kg, which is far beyond the typical human dosage.[20]

7.13.8 SAFETY IN PREGNANCY

It has been reported that chronic treatment may prevent women from becoming pregnant by causing spontaneous abortion.[21] Because there is little or no information regarding the safety of this herb during breastfeeding, nursing mothers are advised to refrain from taking this herb.

7.13.9 DRUG INTERACTIONS

A 2010 review of the pharmacological effects of *Centella asiatica* noted "no reports documenting negative interactions between *Centella asiatica* and medications to date." The only caveat was that high doses can be sedating, and that they could possibly enhance the effects of soporifics and sedating anxiolytics.[1] However, at least one study has shown extracts of *Centella asiatica* to exert non-competitive inhibitory effects on CYP3A4 and CYP2D6 enzymes in human liver microsomes.[22]

REFERENCES

1. Gohil KJ, Patel JA, Gajjar AK. Pharmacological review on *Centella asiatica*: A potential herbal cure-all. *Indian J Pharm Sci.* 2010;72(5):546–556.
2. *PDR for Herbal Medicine.* 1st ed. Montvale, NJ: Medical Economics Co; 1999. p. 729.
3. Hagemann RC, Burnham TH, Granick B, Neubauer D. Gotu kola. In: *The Lawrence Review of Natural Products: Facts and Comparisons.* St. Louis, MO: JB Lippincott; 1996. p. 41–42.
4. Srivastava R, Shukla YN, Kumar S. Chemistry and pharmacology of *Centella asiatica*: A review. *J Med Arom Plant Sci.* 1997;19:1049–1056.
5. Dasgupta N, De B. Antioxidant activity of some leafy vegetables of India: A comparative study. *Food Chem.* 2007;101(2):471–474.
6. Shinomol GK, Muralidhara. Prophylactic neuroprotective property of Centella asiatica against 3-nitropropionic acid induced oxidative stress and mitochondrial dysfunctions in brain regions of prepubertal mice. *NeuroToxicology.* 2008;29(6):948–957.
7. Liu M, Dai Y, Yao X, et al. Anti-rheumatoid arthritic effect of madecassoside on type II collagen-induced arthritis in mice. *Int Immunopharmacol.* 2008;8(11):1561–1566.
8. Somchit MN, Sulaiman MR, Zuraini A, et al. Antinociceptive and antiinflammatory effects of *Centella asiatica*. *Indian J Pharmacol.* 2004;36:377–380.
9. Gray NE, Harris CJ, Quinn JF, Soumyanath A. *Centella asiatica* modulates antioxidant and mitochondrial pathways and improves cognitive function in mice. *J Ethnopharmacol.* 2016;180:78–86.
10. Haque S, Naznine T, Ali M, et al. Antihyperglycemic activity of leaves of *Brassica oleracea*, *Centella asiatica* and *Zizyphus mauritiana*: Evaluation through oral glucose tolerance tests. *Adv Nat Appl Sci.* 2013;7(5):519–525.
11. Yan SL, Yang HT, Lee YJ, Lin CC, Chang MH, Yin MC. Asiatic acid ameliorates hepatic lipid accumulation and insulin resistance in mice consuming a high-fat diet. *J Agric Food Chem.* 2014;62(20):4625–4631.

12. Ceremuga TE, Valdivieso D, Kenner C, et al. Evaluation of the anxiolytic and antidepressant effects of Asiatic acid, a compound from gotu kola or *Centella asiatica*, in the male Sprague Dawley rat. *AANA J.* 2015;83(2):91–98.
13. Kalshetty P, Aswar U, Bodhankar S, Sinnathambi A, Mohan V, Thakurdesai P. Antidepressant effects of standardized extract of *Centella asiatica* L in olfactory bulbectomy model. *Biomed Aging Path.* 2012;2(2):48–53.
14. Chen Y, Han T, Qin L, Rui Y, Zheng H. Effect of total triterpenes from *Centella asiatica* on the depression behavior and concentration of amino acid in forced swimming mice. *J Chin Med Mater.* 2003;26(12):870–873.
15. Bradwejn J, Zhou Y, Koszycki D, Shlik J. A double-blind, placebo-controlled study on the effects of gotu kola (*Centella asiatica*) on acoustic startle response in healthy subjects. *J Clin Psychopharmacol.* 2000;20(6):680–684.
16. Jana U, Sur TK, Maity LN, Debnath PK, Bhattacharyya D. A clinical study on the management of generalized anxiety disorder with *Centella asiatica*. *Nepal Med Coll J.* 2010;12(1):8–11.
17. Wattanathorn J, Mator L, Muchimapura S, et al. Positive modulation of cognition and mood in the healthy elderly volunteer following the administration of *Centella asiatica*. *J Ethnopharmacol.* 2008;116(2):325–332.
18. Mato L, Wattanathorn J, Muchimapura S, et al. *Centella asiatica* improves physical performance and health-related quality of life in healthy elderly volunteer. *Evid Based Complement Altern Med.* 2011;579467.
19. Easley T, Horne S *The Modern Herbal Dispensary.* Berkeley, CA: North Atlantic Books; 2016. p. 242.
20. Chauhan PK, Singh V. Acute and Subacute Toxicity study of the acetone Leaf extract of Centella asiatica in Experimental Animal Models. *Asian Pac J Trop Biomed.* 2012;2(2):S511–S513.
21. Dutta T, Basu UP. Crude extract of Centella asiatica and products derived from its glycosides as oral antiferility agents. *Indian J Exp Biol.* 1968;6(3):181–182.
22. Savai J, Varghese A, Pandita N, Chintamaneni M. Investigation of CYP3A4 and CYP2D6 Interactions of Withania somnifera and Centella asiatica in Human Liver Microsomes. *Phytother Res.* 2015;29(5):785–790.

7.14 *CHRYSACTINIA MEXICANA*

Chrysactinia mexicana, with the common name Damianita daisy, is a species of flowering plants in the sunflower family, native to Mexico and to the southwestern United States. In traditional Mexican herbal medicine, the herb has been used to treat fever and rheumatism, and as a diuretic, anticonvulsant, stimulant, and tonic. Perhaps its major use has been as an aphrodisiac.[1] Indeed, in a study of male rats, acute administration of extract of *Chrysactinia mexicana* restored sexual activity to rats that had reached a state of sexual exhaustion. The treatment significantly increased the proportion of sexually exhausted rats that resumed copulation after ejaculation, and reduced subsequent ejaculation latencies.[2] There is also evidence that the extract of *Chrysactinia mexicana*, as well as some of its purified components phytochemicals have antidepressant effects.

Chemical analysis has shown the plant to contain 1,8-cineole, pinene, apigenin, caffeic acid, ferulic acid, coumaric acid, linalool, myrcene, limonene, and various sterols, acetylenes, thiophenes, monoterpenes, and oxygenated sesquiterpenes.[3,4] The main components of the essential oil are eucalyptol, piperitone, and linalyl acetate.[5]

7.14.1 ANTIOXIDANT

Methanolic extracts from flowers of *Chrysanctinia mexicana* exhibited strong antioxidant activity in a standard, DPPH free radical assay. Its effects were similar to those of the control antioxidant phytochemical quercetin.[6]

7.14.2 ANTI-INFLAMMATORY

In view of the herb's traditional use for treatment of rheumatic symptoms, anti-inflammatory effects of *Chrysactinia mexicana* would be expected. Chronic pretreatment with extract of *Chrysactinia mexicana* partially reversed the systemic inflammatory response to lipopolysaccharide injections in chickens. It also reduced local inflammatory effects of injection of phytohemagglutinin into the wing webs of the birds.[7]

7.14.3 ANTIDIABETIC/ANTI-METABOLIC SYNDROME

Several of the primary constituent flavonoids in *Chrysactinia mexicana*, including apigenin,[8] caffeic acid,[9] and ferulic acid[10] have been found to exert hypoglycemic effects in alloxan- or streptozocin-induced diabetic rodents.

7.14.4 PRECLINICAL ANTIDEPRESSANT-LIKE EFFECTS

There is only one study of antidepressant-like effects of crude extract of *Chrysactinia mexicana*. Treatment of mice with the extract decreased immobility in both the forced swim and tail suspension tests, and those effects were similar to those of the standard antidepressant, clomipramine.[11] However, several of the individual flavonoids identified in *Chrysactinia mexicana* are common among medicinal plants and have themselves been found to produce antidepressant effects. For example, acute treatment with ferulic acid, a common flavonoid also found in *Chrysactinia mexicana*, decreased immobility in the tail suspension and forced swim tests in mice. Those effects were thought due to increases in serotonergic and noradrenergic activities, as levels of those transmitters were increased in the hippocampus and frontal cortex.[12] Interestingly, ferulic acid, like ketamine, acts as an NMDA receptor antagonist and protects nerve cells against NMDA excitotoxic effects.[13]

Two weeks of pretreatment with linalool, a common terpene shared by *Chrysactinia mexicana*, protected rats from the depression-like effects of chronic restraint stress. When subjected to the forced swim test, treated rats exhibited significantly less immobility.[14] The caffeic acid[15] and apigenin[16] in the herb have also been reported to have antidepressant-like effects. Thus, it appears that multiple phytochemicals could work in concert to produce an antidepressant-like effect of *Chrysactinia mexicana*.

7.14.5 HUMAN ANTIDEPRESSANT EFFECTS

There are no studies of effects of *Chrysactinia mexicana* on MDD in human subjects.

7.14.6 Dosage

No dosage information is available.

7.14.7 Toxicity

In the above study of the antidepressant effects of *Chrysactinia mexicana* in mice, an attempt was also made to assess the toxicity of the herb. It was reported that no deaths or adverse effects in mice occurred with doses up to 5,000 mg/kg.

7.14.8 Safety in Pregnancy

Mexican women boil *Chrysactinia mexicana* and drink the liquid with the traditional belief that tea from the shrub is helpful during pregnancy. No other definitive information appears in the literature.[17]

7.14.9 Drug Interactions

Review of the literature reveals no reports on effects of *Chrysactinia mexicana* on drug action or metabolism.

REFERENCES

1. Goetz P. Traitement des troubles de la libido masculine. *Phytothérapie*. 2006;4(1):9–14.
2. Estrada-Reyes R, Ferreyra-Cruz OA, Jiménez-Rubio G, Hernández-Hernández OT, Martínez-Mota L. Prosexual effect of *Chrysactinia mexicana* A. Gray (*Asteraceae*), false damiana, in a model of male sexual behavior. *Bio Med Res Int*. 2016:Article ID 2987917.
3. Guerra-Boone L, Álvarez-Román R, Salazar-Aranda R, et al. Chemical compositions and antimicrobial and antioxidant activities of the essential oils from *Magnolia grandiflora*, *Chrysactinia mexicana*, and *Schinus molle* Found in Northeast Mexico. *Nat Prod Commun*. 2013;8(1):135–138.
4. Delgado G, Ríos MY. Monoterpenes from *Chrysactinia mexicana*. *Phytochemistry*. 1991;30(9):3129–3131.
5. Cárdenas-Ortega NC, Zavala-Sánchez MA, Aguirre-Rivera JR, Pérez-González C, Pérez-Gutiérrez S. Chemical composition and antifungal activity of essential oil of *Chrysactinia mexicana* Gray. *J Agric Food Chem*. 2005;53(11):4347–4349.
6. Salazar-Aranda R, Pérez-López LA, López-Arroyo J, Alanís-Garza BA, Waksman de Torres N. Antimicrobial and antioxidant activities of plants from northeast of Mexico. *Evid-Based Complement Altern Med*. 2011; Article ID 536139.
7. García-López JC, Álvarez-Fuentes G, Pinos-Rodríguez JM, et al. Anti-inflammatory effects of *Chrysactinia mexicana* Gray extract in growing chicks (*Gallus gallusdomesticus*) challenged with LPS and PHA. *Int J Curr Microbiol App Sci*. 2017;6(1):550–562.
8. Panda S, Kar A. Apigenin (4',5,7-trihydroxyflavone) regulates hyperglycaemia, thyroid dysfunction and lipid peroxidation in alloxan-induced diabetic mice. *J Pharm Pharmacol*. 2007;59(11):1543–1548.
9. Celik S, Erdogan S, Tuzcu M. Caffeic acid phenethyl ester (CAPE) exhibits significant potential as an antidiabetic and liver-protective agent in streptozotocin-induced diabetic rats. *Pharmacol Res*. 2009;60(4):270–276.

10. Ohnishi M. Antioxidant activity and hypoglycemic effect of ferulic acid in STZ-induced diabetic mice and KK-A^{y} mice. *BioFactors*. 2004;21(1–4):315–319.

11. Cassani J, Ferreyra-Cruz OA, Dorantes-Barrón AM, Villaseñor RM, Arrieta-Baez D, Estrada-Reyes R. Antidepressant-like and toxicological effects of a standardized aqueous extract of *Chrysactinia mexicana* A. Gray (*Asteraceae*) in mice. *J Ethnopharmacol*. 2015;171:295–306.

12. Chen J, Lin D, Zhang C, et al. Antidepressant-like effects of ferulic acid: Involvement of serotonergic and norepinergic systems. *Metab Brain Dis*. 2015;30(1):129–136.

13. Yu L, Zhang Y, Liao M, et al. Neurogenesis-enhancing effect of sodium ferulate and its role in repair following stress-induced neuronal damage. *WJNS*. 2011;1:9–18.

14. Saiyudthong S, Srijittapong D, Mekseepralard C. Subchronic administration of linalool decreases depressive-like behaviour in restrained rats. *J Pharm Pharmacol*. 2017;7:401–407.

15. Takeda H, Tsuji M, Inazu M, Egashira T, Matsumiya T. Rosmarinic acid and caffeic acid produce antidepressive-like effect in the forced swimming test in mice. *Eur J Pharmacol*. 2002;449(3):261–267.

16. Li R, Wang X, Qin T, Qu R, Ma S. Apigenin ameliorates chronic mild stress-induced depressive behavior by inhibiting interleukin-1β production and NLRP3 inflammasome activation in the rat brain. *Behav Brain Res*. 2016;296:318–325.

17. Zingg RM. Mexican folk remedies of Chihuahua, Mexico. *J Wash Acad Sci*. 1932;22:174–181.

7.15   *CIMICIFUGA RACEMOSA* (BLACK COHOSH)

Cimicifuga racemosa, commonly known as black cohosh, is a member of the buttercup family (*Ranunculaceae*). It is native to the forests of eastern North America from Georgia to Ontario. Native Americans used *Cimicifuga racemosa* for a variety of ailments including rheumatism, malaria, sore throats, and complications associated with childbirth.[1] The name "cohosh" itself comes from an Algonquian Indian word meaning "rough," which describes the hard, knotted rhizomes that contain the plant's medicinal properties.[2] Europeans have used this important medicinal plant to treat menopausal symptoms for over 40 years, and the herb has been approved by the German Commission E for premenstrual discomfort, dysmenorrhea, or menopausal neurovegetative ailments. There is also evidence that *Cimicifuga racemosa* may have antidepressant effects in the context of stages of reproductive life in women, and perhaps beyond this specific indication.

The *C. racemosa* rhizome contains numerous chemical components including triterpene glycosides (actein, 27-deoxyactein, cimicifugoside), phenolic acids (isoferulic acid, caffeic acid, ferulic acid, cinnamic acid, coumaric acid, fukinolic acid), flavonoids, volatile oils, and tannins.[3,4]

7.15.1   ANTIOXIDANT

A number of the phytochemical constituents of *Cimicifuga racemosa* have been found to have antioxidant effects. This has been established in standard biochemical tests, such as the *in vitro* scavenging of free radicals in the DPPH assay. A study also showed ability of specific phytochemicals purified from extracts of *Cimicifuga racemosa* to protect DNA from the destructive oxidizing effects of the toxic chemical, menadione. Several of the phytochemicals, some common to medicinal plants, such as caffeic acid and ferulic acid, and some rather unique to black cohosh, such as the cimicifugic acids and cimiracemates, are potent antioxidants.[5]

7.15.2 ANTI-INFLAMMATORY

Cimicifuga racemosa has also been found to exert potent anti-inflammatory effects in animals. This is consistent with its historic use to relieve the discomfort of rheumatic conditions. A single oral dose of methanolic extract of *Cimicifuga racemosa* attenuated inflammation in the carrageenan-induced paw edema model in rats. In fact, its anti-inflammatory effect was stronger than that of the standard anti-inflammatory medication indomethacin.[6] Cimiracemate A, a specific constituent of the extracts of the *Cimicifuga racemose* rhizome, can suppress the production of TNF-α from lipopolysaccharide stimulated macrophages by nearly half. This anti-inflammatory effect appears to be due to blocking the effects of both nuclear factor-kappaB and mitogen activated protein kinase.[7]

7.15.3 ANTIDIABETIC/ANTI-METABOLIC SYNDROME

A common effect of herbs with antidepressant properties is the ability to improve insulin sensitivity, diabetes, and Metabolic Syndrome. *Cimicifuga racemosa* has not had a reputation of being effective for treatment of diabetes. In an animal study, ovariectomized rats fed for six weeks with food laced with *Cimicifuga racemosa* extract had significantly lower fasting insulin levels than those of control animals. However, there were no differences in responses of the two groups to a glucose load.[8] In a study of menopausal women, three months of oral dosing of 160mg of *Cimicifuga racemosa* extract had no significant effects on total cholesterol, LDL, HDL, triglycerides, glucose, or fasting insulin.[9] Despite this unimpressive finding, it is likely that the herb could indirectly exercise attenuation of symptoms of Metabolic Syndrome through its established antioxidant and anti-inflammatory effects.

The apparent efficacy of *Cimicifuga racemosa* in reducing symptoms of menopause has led to suspicion of interactions with estrogen receptors. However, studies showing no effect of extracts on uterine weight or vaginal cellular cornification in female rats has cast doubt on that possibility. On the other hand, extracts have been found to displace ligands from various serotonin receptors, particularly the 5-HT1A and 5-HT7 receptors, where there is evidence of partial agonism. It has been suggested that effects of serotonin in the hypothalamus may mitigate some symptoms of menopause, such as hot flashes. This is likely to be the basis of the relief offered by SSRIs. However, both 5-HT1A and 5-HT7 receptors have been associated with mood effects and MDD.[10] A particular molecule, Nω-methylserotonin, has been isolated from extracts of the herb, and it showed extremely high, picomolar affinity for the 5-HT7 receptor and acts as a partial agonist.[11] There is compelling evidence that 5-HT7 receptor antagonists may have antidepressant properties. For example, it is thought that 5-HT7 antagonism may contribute to the antidepressant effects of the atypical antipsychotic agent, lurasidone.[12] What if any role partial agonist effects at 5-HT7 receptors play in potential antidepressant effects of *Cimicifuga racemosa* is unclear. Some component of *Cimicifuga racemosa* also acts as a partial agonist at human mu opiate receptors.[13]

7.15.4 PRECLINICAL ANTIDEPRESSANT-LIKE EFFECTS

Studies have demonstrated antidepressant-like effects of *Cimicifuga racemosa* in animals, and there is compelling evidence of such effects in humans. In a study of ovariectomized female rats, treatment with either estradiol or *Cimicifuga racemosa* reduced immobility in the forced swim test.[14] A single oral administration of *Cimicifuga racemosa* to mice also reversed some of the effects of immobilization stress, a treatment that results in depression-like changes in behavior and neurochemistry. The herb attenuated the stress-induced increases in corticosterone, as well as reversing changes in monoamine transmitter to metabolite ratios in the hypothalamus, hippocampus, and cortex of the mice.[15]

The closely related herb from the same genus, *Cimicifuga foetida*, has been used in Chinese medicine to treat the symptoms of menopause. A week of treating female mice with *Cimicifuga foetida* significantly reduced immobility in the forced swim and tail suspension tests. The herb also normalized the reduction of sucrose intake and stress-induced increases in serum corticosterone in rats that had been subjected to chronic unpredictable stress.[16]

7.15.5 HUMAN ANTIDEPRESSANT EFFECTS

Many studies have shown the ability of *Cimicifuga racemosa* to relieve common discomforts of menopause. In one of the earliest published human studies, extract of *Cimicifuga racemosa* improved the scores of menopausal women in a questionnaire for self-evaluation of depression and in the Hamilton Anxiety Scale. However, none of the subjects were formally diagnosed with MDD, and neither scale is a standard instrument for assessing MDD.[17]

Beyond that, there is a report of three cases of what appear to be MDD successfully treated with black cohosh.[18] In the first case, a 43-year old woman with constant abdominal pain of two years duration, and irregular painful menses, also reported feeling "very depressed." She added

> I can't focus, I flip from one thing to another. I'm too easily distracted. I feel such guilt for being so dysfunctional. I feel totally helpless and hopeless. Why get out of bed? There is nothing to look forward to. The despair is overwhelming.

After a month of *Cimicifuga racemosa* in tincture 30 drops three times a day, she reported great improvement of all symptoms of depression and pain. It was noted that she had taken Prozac in the past and the *Cimicifuga racemosa* worked better. In the second case, a woman struggled with episodes of depression in which a black mood would seem to descend on her out of nowhere and last for a few hours to days. During these times, she was more irritable and weepier and would struggle with a terrible despair and sense of doom. She was instructed to take 30 drops of tincture of *Cimicifuga racemosa* on those particular days, up to every two hours if needed.

"She reported that this made an enormous difference for her, often within one or two doses." In the final case, a fifty-year-old woman complained of depression and difficulty concentrating at work after a neighbor had cut down several old trees near her home. "The grief is overwhelming," she said, crying as she spoke. "I can't get out from under it. No matter how much we do to help the planet, there are still too many people who just don't care. It's too dark and hopeless. I can't believe people have the power to do these things." Two days after starting 30 drops of tincture of *Cimicifuga racemosa* three or four times a day, the patient called to report that she was much better. "The depression is gone!" she said. "I still feel grief at the loss of those beautiful trees, but I'm not overwhelmed by the feelings anymore and I'm more able to focus on my work again. That herb you gave me is amazing!"

7.15.6 DOSAGE

Pursell recommends decoction, 1–2 teaspoons per cup, drinking 1–3 cups per day. He also notes use of tincture, 3–60 drops up to three times a day.[19] An ethanolic extract of the rhizome of this plant standardized to contain 1 mg of triterpenes calculated as 27-deoxyacteine per 20-mg tablet (trade name Remifemin) is widely marketed for the relief of menopausal disorders, including hot flashes and profuse sweating.

7.15.7 TOXICITY

The LD50 for intraperitoneal administration of *Cimicifuga racemosa* in mice was found to be greater than 500 mg/kg, and over 1,000 mg/kg by oral administration.[20] There is evidence from a study of rats that black cohosh, in doses similar to those used by women to relieve menopausal discomfort, may cause oxidative stress in the liver.[21]

7.15.8 SAFETY IN PREGNANCY

Pursell warns against its use during the first trimester of pregnancy. The *Botanical Safety Handbook* notes conflicting information in the literature, but cites modern literature showing *Cimicifuga racemosa* to be useful to prevent miscarriage.[22]

7.15.9 DRUG INTERACTIONS

The *Botanical Safety Handbook* notes no known interactions with drugs. However, at least study has found significant interactions of *Cimicifuga racemose* with CY2D6 and CY3A4 enzymes, even to the extent of increasing serum levels of tamoxefin.[23]

REFERENCES

1. Predny ML, De Angelis P, Chamberlain JL. *Black Cohosh (Actaea racemosa): An Annotated Bibliography*. Gen. Tech. Rep. SRS-97. Asheville, North Carolina: Department of Agriculture Forest Service, Southern Research Station. 2006.

2. Tetherow H. Black cohosh [Online]. 2001. Accessed at: http://www.altmed.creight on.edu/blackcohosh. Accessed January 17, 2019.

3. Borrelli F, Ernst E. *Cimicifuga racemosa*: A systematic review of its clinical efficacy. *Eur J Clin Pharmacol.* 2002;58(4):235–241.

4. Gödecke T, Nikolic D, Lankin DC, et al. Phytochemistry of Cimicifugic acids and associated bases in *Cimicifuga racemosa* Root Extracts. *Phytochem Anal.* 2009;20(2):120–133.

5. Burdette JE, Chen SN, Lu ZZ, et al. Black cohosh (*Cimicifuga racemosa* L.) protects against menadione-induced DNA damage through scavenging of reactive oxygen species: Bioassay-directed isolation and characterization of active principles. *J Agric Food Chem.* 2002;50(24):7022–7028.

6. Rathore R, Rahal A, Mandil R, et al. Comparison of the antiinflammatory activity of plant extracts from *Cimicifuga racemosa* and *Mimosa pudica* in a rat model. *Aust Vet Pract.* 2012;42(3):274–278.

7. Yang CLH, Chik SCC, Li JCB, et al. Identification of the bioactive constituent and its mechanisms of action in mediating the anti-inflammatory effects of black cohosh and related *Cimicifuga* species on human primary blood macrophages. *J Med Chem.* 2009;52(21):6707–6715.

8. Rachoń D, Vortherms T, Seidlová-Wuttke D, Wuttke W. Effects of black cohosh extract on body weight gain, intra-abdominal fat accumulation, plasma lipids and glucose tolerance in ovariectomized Sprague-Dawley rats. *Maturitas.* 2008;60(3-4):209–215.

9. Spangler L, Newton KM, Grothaus LC, et al. The effects of black cohosh therapies on lipids, fibrinogen, glucose and insulin. *Maturitas.* 2007;57(2):195–204.

10. Burdette JE, Liu J, Chen SN, et al. Black Cohosh Acts as a Mixed Competitive Ligand and Partial Agonist of the serotonin Receptor. *J Agric Food Chem.* 2003;51(19):5661–5670.

11. Powell SL, Gödecke T, Nikolic D, et al. In vitro serotonergic activity of black cohosh and identification of Nω-Methylserotonin as a potential active constituent. *J Agric Food Chem.* 2008;56(24):11718–11726.

12. Ishibashi T, Horisawa T, Tokuda K, et al. Pharmacological profile of lurasidone, a novel antipsychotic agent with potent 5-hydroxytryptamine 7 (5-HT7) and 5-HT1A receptor activity. *JPET.* 2010;334(1):171–181.

13. Rhyu MR, Lu J, Webster DE, et al. Black cohosh (*Actaea racemosa, Cimicifuga racemosa*) behaves as a mixed competitive ligand and partial agonist at the human μ opiate receptor. *J Agric Food Chem.* 2006;54(26):9852–9857.

14. Hu GM, Shen GM, Xu H. *Cimicifuga racemosa* alyets behavior and PVN C-fos expression in ovariectomized rats subjected to the forced swimming test. *Acta Endocrinol (Buc).* 2012;8(4):529–537.

15. Nadaoka I, Yasue M, Sami M, Kitagawa Y. Oral administration of *Cimicifuga racemosa* extract affects immobilization stress-induced changes in murine cerebral monoamine metabolism. *Biomed Res.* 2012;33(2):133–137.

16. Ye L, Hu Z, Du G, et al. Antidepressant-like effects of the extract from *Cimicifuga foetida* L. *J Ethnopharmacol.* 2012;144(3):683–691.

17. Warnecke G. Influence of a phytopharmaceutical on climacteric complaints. *Die Meizinisch Welt.* 1985;36:871–874.

18. Cimicifuga FD. *Cimicifuga* for depression. *Med Herbalism.* 2001;7(1-2):10–11.

19. Pursell JJ. *The Herbal Apothecary.* Portland, Oregon: Timber Press. p. 65. 2016.

20. Genazzani E, Sorrentino L. Vascular action of Actaea racemose L. *Nature.* 1962;194:544–545.

21. Brites L, Gilglioni EH, Garcia RF, et al. *Cimicifuga racemosa* impairs fatty acid β-oxidation and induces oxidative stress in livers of ovariectomized rats with renovascular hypertension. *Free Rad Biol Med.* 2012;53(4):680–689.

22. Gardner Z, McGuffin M, editors. *Botanical Safety Handbook, 3rd edition*. Boca Raton, FL: CRC Press. p. 17. 2013.
23. Li J, Gödecke T, Chen SN, et al. In vitro metabolic interactions between black cohosh (Cimicifuga racemosa) and tamoxifen via inhibition of cytochromes P450 2D6 and 3A4. *Xenobiotica*. 2011;41(12):1021–1030.

7.16 *CINNAMOMUM ZEYLANICUM* (CINNAMON)

Cinnamon is a spice obtained from the inner bark of several tree species from the genus *Cinnamomum*. Ceylon cinnamon (the source of its Latin name, *zeylanicum*) or "true cinnamon" is indigenous to Sri Lanka and southern parts of India. There is also *Cinnamomum cassia*, known as Chinese cinnamon. Cinnamon in its various forms was known by the ancients and highly prized. It has been used primarily as a food spice, but also has a long history of use as a medicinal herb. In Ayurvedic Medicine, cinnamon is considered a remedy for respiratory, digestive, and gynaecological ailments. Modern research has established that cinnamon has properties including anti-microbial, hypoglycemic, antihypertensive, anti-oxidant, and free-radical scavenging properties, inhibition of tau aggregation, and anti-nociception, anti-inflammatory, and hepato-protective effects.[1] Among the phytochemicals in the various cinnamons are cinnamaldehyde – the predominant molecule – as well as cinnamic acid, cinnamyl alcohol, coumarin, eugenol, linalool, benzyl benzoate, and δ-cadinene.[2] A significant difference between "true" Cinnamon and Chinese Cinnamon is that Chinese cinnamon can contain much higher, even potentially dangerous, levels of coumarin. As is commonly seen, the species of the genus *Cinnamomum* will here be generally referred to by the common name cinnamon.

7.16.1 ANTIOXIDANT

In the *in vitro* superoxide anion scavenging, DPPH free radical scavenging, and ABTS radical cation decolorization assays, methanolic extracts of cinnamon were potent antioxidants, with roughly two-thirds the effect of vitamin C.[3] Also *in vitro*, the essential oil of cinnamon and two of its main constituents, eugenol and cinnamaldehyde, potently reduced peroxynitrite-induced nitration and lipid peroxidation.[4] Thus, cinnamon counteracts both nitrosative and oxidative damage.

Cinnamon acts as a potent antioxidant in intact animals. In rats fed high fat diets, the addition of cinnamon increased hepatic and cardiac levels of glutathione and reduced markers of lipid peroxidation in those tissues.[5] In an interesting human study, the antioxidant effects of ten days of drinking of cinnamon tea were evaluated in hospital operating room workers who were exposed to anesthetic gases on a daily basis. Apparently, anesthetic gases generate free radicals and oxidative damage. The workers that drank the tea had significantly lower serum levels of biomarkers of lipid peroxidation, while retaining higher degrees of antioxidant capacity.[6]

Importantly, oral consumption of cinnamon has antioxidant effects in the brain. Four weeks of daily consumption of aqueous extract of cinnamon had protective effects on brain tissue in rats subjected to chronic cerebral ischemia. Cinnamon increased activity of superoxide dismutase and levels of BDNF and NGF, but decreased levels of malondialdehyde, a marker of lipid peroxidation. The Y-maze test showed that cinnamon also spared some cognitive function in those animals.[7]

7.16.2 ANTI-INFLAMMATORY

Along with antioxidant effects, cinnamon and its constituents also exert anti-inflammatory effects through various cell signaling pathways. Rats that become obese from chronic high fat diets exhibit the classical signs of diabetes type II in humans, that is, hyperglycemia, hyperlipidemia, oxidative stress, and non-alcoholic fatty liver. Addition of cinnamon polyphenols to the high fat diet decreased body weight, visceral fat, serum glucose and insulin concentrations, liver antioxidant enzymes, and serum and liver concentrations of the oxidation marker malondialdehyde. However, in the livers of the same animals, cinnamon also suppressed the inflammatory actor, NF-κB and enhanced expressions of PPAR-α, IRS-1, Nrf2, and HO-1.[8] All of these actions in brain tissue are associated with antidepressant effects.

7.16.3 ANTIDIABETIC/ANTI-METABOLIC SYNDROME

Studies have shown that cinnamon improves symptoms of diabetes in humans. For example, a 40 day course of 1 to 6 g of cinnamon a day reduced fasting serum glucose (18–29%), triglyceride (23–30%), LDL cholesterol (7–27%), and total cholesterol (12–26%) levels in patients with diabetes type II.[9] Cinnamon extract has also been shown to reduce insulin resistance in *in vitro* and *in vivo* studies by increasing phosphatidylinositol 3-kinase activity in the insulin signaling pathway and thus potentiating insulin action.[10] In a study of women suffering Polycystic Ovary Syndrome, a condition associated with both insulin resistance and MDD, eight weeks of treatment with cinnamon lowered fasting serum glucose levels, improved responses to glucose loads, and improved insulin sensitivity.[11]

Cinnamon can also improve insulin sensitivity in the brain. That in itself would be expected to have antidepressant effects. However, part of this effect may be through activation of the Akt and PI3K pathways and inhibition of GSK-3β, all of which would further contribute to such effects. Of particular relevance is a study of rats on high fat and high fructose diets that cause metabolic syndrome, insulin resistance, changes similar to those of human patients with diabetes type II. Rats given cinnamon along with the high fat and fructose diet showed restoration toward normal of brain levels of Akt and PI3k activity that had been diminished by the diet. Moreover, while the diet increased GSK-3β activity, cinnamon reduced GSK-3β activity, as is the case with ketamine, and many antidepressant treatments. Cinnamon also restored normal brain mitochondrial responses to restraint stress in animals on the diet.[12]

Central activity of cinnamon treatment is also shown in streptozotocin-induced diabetic rats whose diabetic symptoms were further exacerbated by high fat diets. Three weeks of treatment with cinnamaldehyde, the main active phytochemical in cinnamon, reduced serum glucose and improved cognitive function in the Morris water maze, open field, and elevated plus maze tests. This treatment also decreased elevated levels of the inflammatory cytokines, TNF-α and IL-6 in the hippocampus and cortex of those animals.[13]

Cinnamon has also been found to disaggregate tau protein tangles in animal models of Alzheimer's disease. *In vitro*, HEK293 cells exposed to cinnamon extract for 48 hours produce less phosphorylated tau, which is the form that results in tangles. This was found to be due to decreased activity of GSK-3β, which is the enzyme that phosphorylates the tau protein in brain tissue.[14]

7.16.4 PRECLINICAL ANTIDEPRESSANT-LIKE EFFECTS

Several studies have established that cinnamon has antidepressant-like effects in rodents. Administration over several days, though not acute, single dose treatment, reduces immobility in the forced swim and tail suspension tests in mice.[15,16] Methanolic extract of *Cinnamomum cassia*, also has an antidepressant-like effect in mice by its decreasing immobility in the tail suspension test. The extract was also noted to enhance 5-HTP-induced head twitches, suggesting enhancement of serotonergic activity. In fact, both the antidepressant-like effect and enhancement of head twitch caused by the extract were similar to those of fluoxetine.[17]

In one study, two weeks of daily oral administration of the essential oil of cinnamon to mice decreased of immobility in both the forced swim and tail suspension tests. The treatment also increased time in the open arms of the elevated plus maze, which suggests anxiolytic effects as well. Further study showed the most predominant phytochemical in the essential oil to be trans-cinnamaldehyde.[18] Purified methyl-eugenol, another major component of the essential oil of cinnamon, also showed antidepressant and anxiolytic-like effects in mice.[19] In yet another study, antidepressant-like effects of eugenol were seen in mice, and further study showed that the phytochemical increased levels of BDNF in the hippocampi of those animals.[20]

Extract of cinnamon has also been shown to reverse the effects of treatments that cause depression-like behavior in mice and rats, suggesting that the herb may be useful in treating, as well as preventing, depression. Reserpine is well-known to cause depression in humans, and depression-like behavior in animals. Mice that had received an injection of reserpine 18 hours before exhibited increased immobility in the forced swim test. This immobility was not observed in mice that received an acute injection of extract of cinnamon an hour before the test.[21]

Chronic oral administration of extract of *Cinnamomum zeylanicum* was found to attenuate the anxiety- and depression-like behaviors of rats that had been chronically administered lead acetate. Lead-induced immobility in the forced swim test was reduced by cinnamon.[22]

7.16.5 HUMAN ANTIDEPRESSANT EFFECTS

The only study evaluating the effects of cinnamon in patients with MDD concerned the ability of the herb to enhance effects of the standard antidepressant, fluoxetine. The individuals had not responded to fluoxetine alone, but addition of essential oil of cinnamon over eight weeks significantly improved response to treatment. Beck Depression Index scores at the end of eight weeks were 26.9 in the fluoxetine group and 18.36 in the group that received both fluoxetine and cinnamon. There were no differences between groups in sexual and gastrointestinal complications or weight changes.[23]

7.16.6 DOSAGE

Easley and Horne suggest doses to be 0.5–2 g of cinnamon bark three times a day, 30 to 60 drops of tincture (1:5, 60% alcohol) up to three times daily, or 0.05–0.2 g essential oil per day.[24]

7.16.7 TOXICITY

The LD50 in mice of essential oil is 2.65–3.5 g/kg.[25] In tests of the toxicity of dried aqueous extracts of cinnamon in rats, it was found that kidney damage began to appear at doses as low as 0.1 g/kg. There were no animal deaths at doses up to 2 g/kg. However, there were signs of serious liver and kidney toxicity at that dose.[26] However, those doses are above typical doses of aqueous cinnamon in human studies, where 500 mg of dried aqueous extract per day have been used. Moreover, Anderson noted "In all the human studies involving cinnamon, or aqueous extracts of cinnamon, there have been no reported adverse events."[27]

7.16.8 SAFETY IN PREGNANCY

Traditional Chinese Medical texts advise not to use cinnamon during pregnancy.[28,29] In animal studies high doses of cinnamaldehyde isolated from cinnamon were found to increase rates of fetal malformations.[30]

7.16.9 DRUG INTERACTIONS

The *Botanical Safety Handbook* notes no known interactions with drugs.[31] However, one review has suggested caution when using with medications metabolized by P450 enzymes CY1A2 and CY2E1, as *Cinnamomum* species may interfere with their metabolism.[32]

REFERENCES

1. Ranasinghe P, Pigera S, Premakumara GAS, Galappaththy P, Constantine GR, Katulanda P. Medicinal properties of 'true' cinnamon (*Cinnamomum zeylanicum*): A systematic review. *BMC Complement Altern Med*. 2013;13:275.
2. He ZD, Qiao CF, Han QB, et al. Authentication and quantitative analysis on the chemical profile of cassia bark (*cortex cinnamomi*) by high-pressure liquid chromatography. *J Agric Food Chem*. 2005;53(7):2424–2428.
3. Mathew S, Abraham TE. Studies on the antioxidant activities of cinnamon (*Cinnamomum verum*) bark extracts, through various in vitro models. *Food Chem*. 2006;94(4):520–528.
4. Chericoni S, Prieto JM, Iacopini P, Cioni P, Morelli I. In vitro activity of the essential oil of *Cinnamomum zeylanicum* and eugenol in peroxynitrite-induced oxidative processes. *J Agric Food Chem*. 2005;53(12):4762–4765.

5. Dhuley J. Anti-oxidant effects of cinnamon (*Cinnamomum verum*) bark and greater cardamom (*Amomum subulatum*) seeds in rats fed high fat diet. *J Pharmacol.* 1999;4(51):99–106.
6. Ranjbar A, Ghaseminejhad S, Takalu H, Baiaty A, Rahimi F, Abdollahi M. Antioxidative stress potential of cinnamon (*Cinnamomum zeylanicum*) in operating room personnel; a before/after cross sectional clinical trial. *Int J Pharmacol.* 2007;3(6):482–486.
7. Zhang W. Effect of cinnamon intervention on oxidative stress and nerve factor expression in rats with chronic cerebral ischemia. *J Tradit Chin Med.* 2010-07.
8. Tuzcu Z, Orhan C, Sahin N, Juturu V, Sahin K. Cinnamon polyphenol extract inhibits hyperlipidemia and inflammation by modulation of transcription factors in high-fat diet-Fed Rats. *Oxid Med Cell Longev.* 2017;2017:Article ID 1583098.
9. Khan A, Safdar M, Ali Khan MMA, Khattak KN, Anderson RA. Cinnamon improves glucose and lipids of people with Type 2 diabetes. *Diabetes Care.* 2003;26(12):3215–3218.
10. Kirkham S, Akilen R, Sharma S, Tsiami A. The potential of cinnamon to reduce blood glucose levels in patients with type 2 diabetes and insulin resistance. *Diabetes Obes Metab.* 2009;11(12):1100–1113.
11. Wang JG, Anderson RA, Graham GM, et al. The effect of cinnamon extract on insulin resistance parameters in polycystic ovary syndrome: A pilot study. *Fertil Steril.* 2007;88(1):240–243.
12. Batandier C, Poulet L, Canini F, et al. Effects of a high fat/High fructose diet on brain mitochondria are counteracted by cinnamon in stressed rats. *FASEB.* 2013;27(1suppl).
13. Jawale A, Datusalia AK, Bishnoi M, Sharma SS. Reversal of diabetes-induced behavioral and neurochemical deficits by cinnamaldehyde. *Phytomedicine.* 2016;23(9):923–930.
14. Donley J, Hurt W, Stockert A. Effects of cinnamon extract on glycogen synthase kinase 3 phosphorylation of tau. *FASEB J.* 2016;30:814.
15. Reyhaneh S. Antidepressant-like activity of *Cinnamon verum*: An acute and sub-acute study. *Int Cong Compliment Altern Med.* 2015;1.
16. Emamghoreishi M, Ghasemi F. Antidepressant Effect of Aqueous and hydroalcoholic extracts of *Cinnamon zeylanicum* in the Forced Swimming Test. *Asian J Psychiatr.* 2011;4(1):S44.
17. Zada W, Zeeshan S, Bhatti HA, Mahmood W, Rauf K, Abbas G. *Cinnamomum cassia*: An implication of serotonin reuptake inhibition in animal models of depression. *Nat Prod Res.* 2016;30(10):1212–1214.
18. Sohrabi R, Pazgoohan N, Seresht HR, Amin B. Repeated systemic administration of the cinnamon essential oil possesses anti-anxiety and anti-depressant activities in mice. *Iran J Basic Med Sci.* 2017;20(6):708–714.
19. Norte MCB, Cosentino RM, Lazarini CA. Effects of methyl-eugenol administration on behavioral models related to depression and anxiety, in rats. *Phytomedicine.* 2005;12(4):294–298.
20. Irie Y, Itokazu N, Anjiki N, Ishige A, Watanabe K, Keung WM. Eugenol exhibits antidepressant-like activity in mice and induces expression of metallothionein-III in the hippocampus. *Brain Res.* 2004;1011(2):243–246.
21. Ghaderi H, Rafieian M, Nezhad HR. Effect of hydroalcoholic *Cinnamomum zeylanicum* extract on reserpine-induced depression symptoms in mice. *Pharmacophore.* 2018;9(2):35–44.
22. Fadaei S, Asle-Rousta M. Anxiolytic and antidepressant effects of cinnamon (*Cinnamomum verum*) extract in rats receiving lead acetate. *Sci J Kurdistan Univ Med Sci.* 2017;22(6):e31–e39.
23. Ghaderi H, Nikan R, Rafieian-Kopaei M, Biyabani E. The effect of *Cinnamon zeylanicum* essential oil on treatment of patients with unipolar nonpsychotic Major Depression Disorder treated with fluoxetine. *Pharmacophore.* 2017;8(3):24–31.

24. Easley T, Horne S. *The Modern Herbal Dispensary*. Berkeley, CA: North Atlantic Books; 2016. p. 212.
25. Price L, Price S. *Aromatherapy for Health Professionals E-Book*. London: Churchill, Livingstone; 2012.
26. Ahmad RA, Serati-Nouri H, Majid FAA, Sarmidi MR, Aziz RA. Assessment of potential toxicological effects of cinnamon bark aqueous extract in rats. *Int J Biosci Biochem Bioinform*. 2015;5(1):36–44.
27. Anderson RA. Chromium and polyphenols from cinnamon improve insulin sensitivity. *Proc Nutr Soc*. 2008;67(1):48–53.
28. Bensky D, Clavey S, Stoger E. *Chinese Herbal Medicine: Materia Medica*. 3rd ed. Seattle, WA: Eastland Press; 2004.
29. Chen J, Lin D, Zhang C, et al. Antidepressant-like effects of ferulic acid: Involvement of serotonergic and norepinergic systems. *Metab Brain Dis*. 2015;30(1):129–136.
30. Mantovani A, Stazi AV, Macri C, Ricciardi C, Piccioni A, Badellino E. Prenatal (segment II) toxicity study of cinnamic aldehyde in the Sprague-Dawley rat. *Food Chem Toxicol*. 1989;27(12):781–786.
31. Gardner Z, McGuffin M, editors. *Botanical Safety Handbook*. 3rd ed. Boca Raton, FL: CRC Press; 2013. p. 216.
32. Ulbricht C, Seamon E, Windsor RC, et al. An evidence-based systematic review of cinnamon (Cinnamomum spp.) by the Natural Standard Research Collaboration. *J Diet Suppl*. 2011;8(4):378–454.

7.17   *COFFEA ARABICA* (COFFEE)

Coffee is the beverage produced from roasted seeds, or beans, from trees of the genus *Coffea*. There are as many as 100 species of *Coffea*, but the beverage comes primarily from two of them, *arabica* and *robusta*. *Coffea arabica* trees generate approximately 70% of the world's coffee production. The beans are flatter and more elongated than *Robusta* and are lower in caffeine. *Robusta* trees, while producing a less flavorful bean, are more resistant to disease and extremes of temperature. Thus, *robusta* trees are easier to cultivate and the beans are less costly. The herb will here be referred to by its most commonly consumed form, i.e., coffee.

Coffee is widely used as a richly flavored stimulant deriving its effects primarily from caffeine. Caffeine (1,3,7-trimethylxanthine) is the world's most frequently ingested psychoactive substance, with approximately 80% consumed in the form of coffee. Along with caffeine, coffee contains the related methylxanthines theobromine and theophylline that contribute mild stimulant effects. However, coffee contains a complex mix of phytochemicals that produce effects beyond those of the methylxanthines. Some of these phytochemicals, e.g., chlorogenic acid, quinic acid, caffeic acid, ferulic acid, trigonelline, and coumaric acid, are natural to green coffee beans and others, e.g., melanoidins, N-methylpyridinium, and acrylamide, are introduced by the roasting process.[1]

7.17.1   ANTIOXIDANT

Caffeine protects against cellular damage caused by free radicals and peroxides, and exhibits antioxidant activity comparable to that of the endogenous antioxidant glutathione.[2] Caffeine and its metabolites theobromine and xanthine have been shown to decrease lipid peroxidation.[3]

Chronic treatment with orally administered caffeic acid blocks the oxidative damage produced by subsequent intrastriatal injection of the excitotoxin quinolinic acid. Caffeic acid decreased lipid peroxidation and nitrite concentration, prevented depletion of SOD and catalase, and mitigated impairment in mitochondrial activities in *ex vivo* striatum.[4] Its antioxidant activities were equal to those of α-tocopherol.[5]

7.17.2   ANTI-INFLAMMATORY

Anti-inflammatory effects of caffeine are seen in its ability to inhibit the production of tumor necrosis factor alpha (TNF-α) in LPS-stimulated human whole blood.[6] Anti-inflammatory properties of caffeic acid have been shown in a number of studies. For example, caffeic acid suppressed LPS-induced signaling pathways in Raw 264.7 macrophage cells, namely p38 MAPK, JNK1/2, and NF-κB. It also decreased nitrite concentration in LPS stimulated macrophages.[7]

Pretreatment of mice with trigonelline purified from coffee protected brain tissue against the oxidative and inflammatory effects of exposure to lipopolysaccharide. It significantly decreased signs of oxidative stress in the hippocampus and cortex by normalizing levels of superoxide dismutase and glutathione and reducing lipid peroxidation. It also countered the lipopolysaccharide-induced rise in the inflammatory cytokines TNF-α and IL-6, and restored levels of BDNF.[8] Trigonelline also stimulates neurite outgrowth *in vitro* in human neuroblastoma SK-N-SH cells.[9]

7.17.3   ANTIDIABETIC/ANTI-METABOLIC SYNDROME

In rats made diabetic by streptozotocin and a high carbohydrate/high fat diet, trigonelline decreased blood glucose, and normalized insulin level, insulin sensitivity, and the activities of superoxide dismutase, catalase, glutathione, and inducible nitric oxide synthase.[10]

Chlorogenic acid, one of the more abundant polyphenols in coffee, exhibits antioxidant, anti-inflammatory, and insulin-enhancing effects. Chlorogenic acid protected neurons *in vitro* against H_2O_2-induced oxidative stress and prevented neuronal death by reducing the accumulation of intracellular reactive oxygen species.[11] In rats, it blocked the inflammatory effect induced by carrageenan.[12] In genetically diabetic mice – due to the Leprdb/db mutation – chronic administration of chlorogenic acid attenuated hepatic steatosis and improved lipid profiles and skeletal muscle glucose uptake, which in turn improved fasting glucose level, glucose tolerance, insulin sensitivity, and dyslipidemia.[13]

7.17.4   PRECLINICAL ANTIDEPRESSANT-LIKE EFFECTS

A number of animal studies have both confirmed antidepressant efficacy of coffee and its constituent phytochemicals and revealed the mechanisms by which these effects might occur. Mice chronically administered caffeine withstood chronic unpredictable stress in a manner comparable to those treated with the antidepressant desipramine. Caffeine also maintained hippocampal serotonin and dopamine levels in a manner similar to that of the antidepressant.[14] Caffeine enhanced the antidepressant effects of bupropion and duloxetine in mice.[15]

The primary mechanism by which caffeine affects the brain is the blocking of adenosine type 2 receptors. These receptors mediate some of the detrimental effects of stress on mood and cognitive function. Treatment with caffeine produced an antidepressant effect in mice in the forced swim and tail suspension tests. It also increased preference for sucrose and improved memory in the modified Y-maze. These effects of caffeine were mimicked by both the potent adenosine antagonist SCH58261 and by the use of genetically modified, adenosine receptor knock-out mice for the procedures.[16]

The caffeic acid in coffee reduces the duration of immobility and freezing in the forced swim test in mice, which is indicative of antidepressant activity.[17] Studies have also found the coffee phytochemical ferulic acid to have antidepressant effects. Ferulic acid improved performance in the tail suspension and forced swim tests. These effects were reversed by co-administration of 5-HT1A and 5-HT2 receptor antagonists, suggesting serotonergic mediation of the effect.[18] Ferulic acid also reduced immobility time in the tail suspension and forced swim tests in animals that had been treated with the depression-inducing drug reserpine. The effects were comparable to those of imipramine. Ferulic acid normalized monoamine levels in the hippocampus and frontal cortex, and mitigated oxidative and nitrosative stress and inflammation. Levels of nitrite, LPO, IL-1β, TNF-α, substance P, NF-κβ, p65, and caspase-3 were reduced in the frontal cortex and hippocampus, whereas activities of glutathione and superoxide dismutase were upregulated toward normal.[19]

Ferulic acid ameliorated the stress-induced depression-like behavior of mice treated with high doses of the stress hormone corticosterone, and increased levels of BDNF mRNA level in the hippocampi of those animals. This in turn led to proliferation of neural stem cells and neural progenitor cells in that area.[20] It should be noted that ferulic acid has exhibited antidepressant effects at concentrations higher than those generally found in coffee.[21]

7.17.5 HUMAN ANTIDEPRESSANT EFFECTS

There is at least one major review of the literature concerning the effects of coffee intake on MDD in humans.[22] In that review it was concluded that that there is an inverse relationship between consumption of coffee and incidence of depression. It was further suggested that caffeine was mainly responsible for the apparent antidepressant effects of coffee. However, in one study the ingestion of pure caffeine was found not to have significant effects on mood.[23] Moreover, in an epidemiological study,[24] coffee consumption (decaffeinated and/or caffeinated) reduced risk of depression, whereas consumption of either caffeinated and/or decaffeinated soft drinks increased risk of depression. Thus, other phytochemicals, perhaps along with caffeine, may have roles in the salutary effect of chronic coffee consumption on mood.

7.17.6 DOSAGE

In the above study showing reduction of risk with either caffeinated or decaffeinated coffee, the difference seen was between those that drank no coffee and those that drank four or more cups of coffee. It appears that no further advantage is gained with more than four cups per day. Moreover, given the fact that some individuals experience anxiety or even heart rhythm irregularities with high caffeine intake, a limit of four cups a day may seem reasonable.[25]

7.17.7 TOXICITY

The acute toxicity of coffee oil, arguably the most concentrated form of coffee phytochemicals, is low in toxicity with an oral LD50 in mice of 5 g/kg.[26] The LD50 of orally administered caffeine in male albino rats is reported to be 367 mg/kg.[27] This would extrapolate to about 25 g for a 75 kg human adult. Note that a strong cup of coffee contains about 125mg of caffeine.

7.17.8 SAFETY IN PREGNANCY

There has been found to be no dangers posed by moderate use of coffee during pregnancy.[28]

7.17.9 DRUG INTERACTIONS

Caffeine is metabolized primarily by the CYP1A2 enzyme, thus, there can be interactions with medications also metabolized by that enzymes. It was found, for example, that co-administration of fluvoxamine can substantially increase serum caffeine levels.[29] Other medications largely metabolized by CYP1A2 with potential for interactions with caffeine from coffee intake include mexiletine, clozapine, psoralens, idrocilamide, phenylpropanolamine, furafylline, theophylline, and quinolones.[30]

REFERENCES

1. Bøhn SK, Blomhoff R, Paur I. Coffee and cancer risk, epidemiological evidence, and molecular mechanisms. *Mol Nutr Food Res.* 2014;58(5):915–930.
2. Devasagayam TPA, Kamat JP, Mohan H, Kesavan PC. Caffeine as an antioxidant: Inhibition of lipid peroxidation induced by reactive oxygen species. *Biochim Biophys Acta Biomembr.* 1996;1282(1):63–70.
3. Azam S, Hadi N, Khan NU, Hadi SM. Antioxidant and pro-oxidant properties of caffeine, theobromine and xanthine. *Med Sci Monit.* 2003;9(9):BR325–BR330.
4. Kalonia H, Kumar P, Kumar A, Nehru B. Effect of caffeic acid and rofecoxib and their combination against intrastriatal quinolinic acid induced oxidative damage, mitochondrial and histological alterations in rats. *Inflammopharmacology.* 2009;17(4):211–219.
5. Chen JH, Ho CT. Antioxidant activities of caffeic acid and its related hydroxycinnamic acid compounds. *J Agric Food Chem.* 1997;45(7):2374–2378.
6. Horrigan LA, Kelly JP, Connor TJ. Caffeine suppresses TNF-α production via activation of the cyclic AMP/protein kinase A pathway. *Int Immunopharmacol.* 2004;4(10–11):1409–1417.
7. Búfalo MC, Ferreira I, Costa G, et al. Propolis and its constituent caffeic acid suppress LPS-stimulated pro-inflammatory response by blocking NF-κB and MAPK activation in macrophages. *J Ethnopharmacol.* 2013;149(1):84–92.

8. Chowdhury AA, Gawali NB, Munshi R, Juvekar AR. Trigonelline insulates against oxidative stress, proinflammatory cytokines and restores BDNF levels in lipopolysaccharide induced cognitive impairment in adult mice. *Metab Brain Dis.* 2018;33(3):681–691.

9. Tohda C, Nakamura N, Komatsu K, Hattori M. Trigonelline-induced neurite outgrowth in human neuroblastoma SK-N-SH cells. *Biol Pharm Bull.* 1999;22(7):679–682.

10. Zhou J, Zhou S, Zeng S. Experimental diabetes treated with trigonelline: Effect on β cell and pancreatic oxidative parameters. *Fundam Clin Pharmacol.* 2013;27(3):279–287.

11. Cho ES, Jang YJ, Hwang MK, Kang NJ, Lee KW, Lee HJ. Attenuation of oxidative neuronal cell death by coffee phenolic phytochemicals. *Mutat Res.* 2009;661(1–2):18–24.

12. Dos Santos MD, Almeida MC, Lopes NP, de Souza GE. Evaluation of the anti-inflammatory, analgesic and antipyretic activities of the natural polyphenol chlorogenic acid. *Biol Pharm Bull.* 2006;29(11):2236–2240.

13. Ong KW, Hsu A, Tan BKH. Anti-diabetic and anti-lipidemic effects of chlorogenic acid are mediated by ampk activation. *Biochem Pharmacol.* 2013;85(9):1341–1351.

14. Pechlivanova DM, Tchekalarova JD, Alova LH, Petkov VV, Nikolov RP, Yakimova KS. Effect of long-term caffeine administration on depressive-like behavior in rats exposed to chronic unpredictable stress. *Behav Pharmacol.* 2012;23(4):339–347.

15. Kale PP, Addepalli V. Augmentation of antidepressant effects of duloxetine and bupropion by caffeine in mice. *Pharmacol Biochem Behav.* 2014;124:238–244.

16. Kaster MP, Machado NJ, Silva HB, et al. Caffeine acts through neuronal adenosine A2a receptors to prevent mood and memory dysfunction triggered by chronic stress. *Proc Natl Acad Sci USA.* 2015;112(25):7833–7838.

17. Takeda H, Tsuji M, Inazu M, Egashira T, Matsumiya T. Rosmarinic acid and caffeic acid produce antidepressive-like effect in the forced swimming test in mice. *Eur J Pharmacol.* 2002;449(3):261–267.

18. Zeni ALB, Zomkowski ADE, Maraschin M, Rodrigues ALS and Tasca CI. Ferulic acid exerts antidepressant-like effect in the tail suspension test in mice: Evidence for the involvement of the serotonergic system. *Eur J Pharmacol.* 2012;679(1–3):68–74.

19. Xu Y, Zhang L, Shao T, et al. Ferulic acid increases pain threshold and ameliorates depression-like behaviors in reserpine-treated mice: Behavioral and neurobiological analyses. *Metab Brain Dis.* 2013;28(4):571–583.

20. Yabe T, Hirahara H, Harada N, et al. Ferulic acid induces neural progenitor cell proliferation in vitro and in vivo. *Neuroscience.* 2010;165(2):515–524.

21. Hall S, Desbrow B, Anoopkumar-Dukie S, et al. A review of the bioactivity of coffee, caffeine and key coffee constituents on inflammatory responses linked to depression. *Food Res Int.* 2015;76(3):626–636.

22. Tenore GC, Daglia M, Orlando V, et al. Coffee and depression: A short review of literature. *Curr Pharm Des.* 2015;21(34):5034–5040.

23. Loke WH, Meliska CJ. Effects of caffeine use and ingestion on a protracted visual vigilance task. *Psychopharmacology.* 1984;84(1):54–57.

24. Guo X, Park Y, Freedman ND, et al. Sweetened beverages, coffee, and tea and depression risk among older US adults. *PLoS One.* 2014;9(4):e94715.

25. Wang HR, Woo YS, Bahk WM. Caffeine-induced psychiatric manifestations: A review. *Int Clin Psychopharmacol.* 2015;30(4):179–182.

26. Viani R. Physiologically active substances in coffee. In: Clarke RJ, Macrae R, editors. Coffee. Vol 3. Physiology. London: Elsevier Applied Science; 1988. p. 1–31.

27. Adamson RH. The acute lethal dose 50 (LD50) of caffeine in albino rats. *Regul Toxicol Pharmacol.* 2016;80:274–276.

28. Linn S, Schoenbaum SC, Monson RR, Rosner B, Stubblefield PG, Ryan KJ. No association between coffee consumption and adverse outcomes of pregnancy. *N Engl J Med.* 1982;306(3):141–145.

29. Jeppesen U, Loft S, Poulsen HE, Brśen K. A fluvoxamine-caffeine interaction study. *Pharmacogenetics.* 1996;6(3):213–222.

30. Carrillo JA, Benitez J. Clinically significant pharmacokinetic interactions Between dietary caffeine and medications. *Clin Pharmacokinet.* 2000;39(2):127–153.

7.18   *CORIANDRUM SATIVUM* (CORIANDER)

Coriandrum sativum (coriander) is an annual herb of the family *Umbelliferae*. The family includes a variety of medical herbs, including angelica, anise, celery, and fennel. *Coriandrum sativum* is an ancient herb that has been cultivated for at least the last 7,000 years, primarily for spice. Indeed, the plant supplies two spices, *Coriandrum sativum* seed and its flavorful leaves that are called cilantro. It also has a long history of use as a medicinal plant. It is mentioned in Sanskrit literature as far back as 5000 BC and in the Greek Eber Papyrus as early as 1550 BC *Coriandrum sativum* was used in traditional Greek medicine by Hippocrates (ca. 460–377 BC). The Egyptians called it the "spice of happiness", likely because it was considered an aphrodisiac.[1] In Ayurvedic Medicine, *Coriandrum sativum* seeds have been used to treat digestive complaints, whereas in traditional Chinese Medicine, the seeds are administered to treat indigestion, anorexia, influenza with no sweating, bad breath, and unpleasant body odors.[2]

The major constituents of the essential oil from *Coriandrum sativum* seeds are linalool, α-pinene, γ-terpinene, geranyl acetate, and camphor. Alcohol and water-soluble compounds from the seeds include gallic, chlorogenic, caffeic, and ferulic acids. The primary constituents of the green leaves are aldehydes and alcohols, such as 2E-decenal, decanal, 2E-decen-1-ol, and decanol, as well as caffeic, ferulic, gallic, and chlorogenic acids.[3] Many of the constituent phytochemicals are found in other medical plants, including some with antidepressant effects.

7.18.1   ANTIOXIDANT

Extracts of *Coriandrum sativum* seed and, even more so, the leaves showed significant antioxidant capacity in several standard *in vitro* tests. The plant exhibited significant antioxidant effects in the DPPH radical scavenging and 15-lipoxygenase tests, with the aqueous and alcohol extracts being most effective.[4] *Ex vivo* treatment of human lymphocytes with extract of *Coriandrum sativum* seeds protected them from H_2O_2-induced oxidative stress. The extract increased activities of the antioxidant enzymes, superoxide dismutase, catalase, glutathione peroxidase, gluthatione reductase, and glutathione-S-transferase. It also increased levels of glutathione. Reductions in levels of thiobarbituric acid reactive substances in the lymphocytes demonstrated the ability of extract of *Coriandrum sativum* to reduce lipid peroxidation.[5]

Coriandrum sativum also showed antioxidant effects *in vivo*. Pretreatment of rats with extract of *Coriandrum sativum* attenuated CCl_4-induced oxidative damage in the liver. Levels of antioxidant enzymes were maintained in liver tissue, and lipid peroxidation prevented, as evidenced by reduced levels of thiobarbituric acid reactive substances.[6] Perhaps most significant is the demonstration of *in vivo* central neuroprotective effects of the herb. Fifteen days of oral pretreatment with dried methanolic extract of *Coriandrum sativum* leaves protected brain tissue of rats from the oxidative damage of ischemia and reperfusion. Lipid peroxidation in brain, per the thiobarbituric acid reactive substances assay, was reduced, and levels of gluta-thione, superoxide dismutase, and catalase were maintained.[7] Linalool, the primary monoterpene in *Coriandrum sativum* oil, also attenuated amyloid-beta-induced oxidative neurotoxicity in mice. Oxidative enzymes were preserved. In addition, the herb stimulated the Nrf2/HO-1 signaling pathway that has been associated with antidepressant effects.[8]

7.18.2 ANTI-INFLAMMATORY

Coriandrum sativum has long been used as an anti-inflammatory agent in Ayurvedic Medicine, and it is a primary ingredient in a traditional remedy in Sri Lanka known as *Maharasnadhi Quather*. Treatment over three months with that remedy was found to have antinociceptive and anti-inflammatory effects in humans suffering rheumatoid arthritis. In the same study *Maharasnadhi Quather* attenuated carrageenan-induced paw edema in rats, and apparently did so by inhibiting 5-lipoxygenase activity.[9] Topical application of *Coriandrum sativum* oil also significantly reduced ultraviolet-induced erythema of skin in human subjects, though it was less effective than hydrocortisone.[10]

7.18.3 ANTIDIABETIC/ANTI-METABOLIC SYNDROME

Coriandrum sativum has an antihyperglycemic effect in streptozotocin-diabetic mice.[11] Moreover, extract of *Coriandrum sativum* both stimulated release of insulin from pancreatic beta cells, and mimicked insulin in increasing 2-deoxyglucose transport, glucose oxidation, and incorporation of glucose into glycogen in isolated murine abdominal muscle.[12] Give that oral administration of *Coriandrum sativum* preparations has central effects, one might expect similar insulin-like effects in brain tissue.

7.18.4 PRECLINICAL ANTIDEPRESSANT-LIKE EFFECTS

An acute intraperitoneal injection of ethanolic extract of *Coriandrum sativum* seed reduced immobility in mice in the forced swim test.[13] Two weeks of daily oral dosing of either aqueous or oil extract of *Coriandrum sativum* seed also reduced immobility of mice in both the forced swim and tail suspension test. Brain homogenates from treated mice also showed a marked inhibition of MAO-B.[14] Linalool, the predominant component of the essential oil of *Coriandrum sativum*, has itself been found to have antidepressant effects.[15]

Intraperitoneal administration of the hydroalcoholic extract of *Coriandrum sativum* also exhibits substantial anxiolytic effect. In four different procedures that test for anxiolytic effects, the elevated plus maze, open field test, light and dark test, and social interaction test, the extract was nearly as effective as diazepam.[16] The same extract also stimulates feeding in rats.[17]

Interestingly, inhalation of the volatile oil of *Coriandrum sativum* produced both an antidepressant- and anxiolytic-like effect in rats in the beta-amyloid (1-42) model of Alzheimer's disease. It was further shown that the treatment reduced catalase and increased glutathione in the hippocampi of treated animals, indicating that volatile oil components are bioavailable by this route.[18]

7.18.5 HUMAN ANTIDEPRESSANT EFFECTS

There are no published studies of the use of *Coriandrum sativum* in the treatment of MDD in human subjects.

7.18.6 DOSAGE

The German E Commission recognizes *Coriandrum sativum* seed as a treatment of digestive complaints and loss of appetite, but not for mood disorders or other psychiatric indications. The recommended dose of powdered seed is three grams per day. No side-effects, contraindications, or interactions with medications are noted.[19]

7.18.7 TOXICITY

Coriandrum sativum appears quite safe in reasonable doses. *Coriandrum sativum* seed at a dose of 750 mg/kg caused no mortality in rats, and the LD50 for the oil was found to be 4.13 g/kg.[20]

7.18.8 SAFETY IN PREGNANCY

There are no definitive studies of the safety of *Coriandrum sativum* during pregnancy of women. However, animal studies suggest safety at fairly high doses. The "no observed adverse effect level" of oral oil of *Coriandrum sativum* for pregnant rats was 250 mg/kg/day, whereas for developmental effects, it was 500 mg/kg/day.[21]

7.18.9 DRUG INTERACTIONS

The *Botanical Safety Handbook* notes no known interactions with drugs.[22]

REFERENCES

1. Burdock GA, Carabin IG. Safety assessment of coriander (*Coriandrum sativum* L.) essential oil as a food ingredient. *Food Chem Toxicol.* 2009;47:22–34.
2. Sahib NG, Anwar F, Gilani AH, Hamid AA, Saari N, Alkharfy KM. Coriander (*Coriandrum sativum* L.): A potential source of high-value components for functional foods and nutraceuticals—A review. *Phytother Res.* 2013;27:1439–1456.
3. Singletary K. Coriander: Overview of potential health benefits. *Nutr Today.* 2016;51:151–161.
4. Wangensteen H, Samuelsen AB, Malterud KE. Antioxidant activity in extracts of coriander. *Food Chem.* 2004;88:293–297.
5. Hashim MS, Lincy S, Remya V, Teena M, Anila L. Effect of polyphenolic compounds from *Coriandrum sativum* on H2O2-induced oxidative stress in human lymphocytes. *Food Chem.* 2005;92:653–660.

6. Sreelatha S, Padma PR, Umadevi M. Protective effect of *Coriandrum sativum* extracts on carbon tetrachloride-induced hepatoxicity in rats. *Food Chem Toxicol.* 2009;47:702–708.

7. Vekaria RH, Patel MN, Bhalodiya PN, Patel V, Desai TR, Tirgar PR. Evaluation of neuroprotective effect of *Coriandrum sativum* Linn. against ischemic reperfusion insult in brain. *Int J Phytopharmacol.* 2012;3(2):186–193.

8. Xu P, Wang K, Lu C, et al. Protective effects of linalool against amyloid beta-induced cognitive deficits and damages in mice. *Life Sci.* 2017;174:21–27.

9. Thabrew MI, Dharmasiri MG, Senaratne LJ. Anti-inflammatory and analgesic activity in the polyherbal formulation Maharasnadhi Quathar. *Ethnopharmacology.* 2003;85(2–3):261–267.

10. Reuter J, Huyke C, Casetti F, et al. Anti-inflammatory potential of a lipolotion containing coriander oil in the ultraviolet test. *J Ger Soc Dermatol.* 2008;6:847–851.

11. Swanston-Flatt SK, Day C, Bailey CJ, Flatt PR. Traditional plant treatments for diabetes: Studies in normal and streptozotocin diabetic mice. *Diabetologia.* 1990;33:462–464.

12. Gray AM, Flatt PR. Insulin-releasing and insulin-like activity of the traditional antidiabetic plant *Coriandrum sativum* (coriander). *Br J Nutr.* 1999;81:203–209.

13. Pathan A, Alshahrani A, Al-Marshad F. Neurological assessment of seeds of *Coriandrum sativum* by using antidepressant and anxiolytic like activity on albino mice. *Ethnopharmacology.* 2015;2015(3):102–105.

14. Kharade SM, Gumate DS, Patil VM, Kokane SP, Naikwade NS. Behavioral and biochemical studies of seeds of Coriander sativum in various stress models of depression. *Int J Curr Res Rev.* 2011;3:4–8.

15. Saiyudthong S, Srijittapong D, Mekseepralard C. Subchronic administration of linalool decreases depressive-like behaviour in restrained rats. *J Pharm Pharmacol.* 2017;7:401–407.

16. Mahendra P, Bisht S. Anti-anxiety activity of *Coriandrum sativum* assessed using different experimental anxiety models. *Indian J Pharmacol.* 2011;43:574–577.

17. Nematy M, Kamgar M, Mohajeri SMR, et al. The effect of hydroalcoholic extract of *Coriandrum sativum* on rat appetite. *Avicenna J Phytomed.* 2013;3:91–97.

18. Cioanca O, Hritcu L, Mihasan M, Trifan A, Hancianu M. Inhalation of coriander volatile oil increased anxiolytic-antidepressant-like behaviors and decreased oxidative status in beta-amyloid (1–42) rat model of Alzheimer's disease. *Physiol Behav.* 2014;131:68–74.

19. Blumenthal M, Busse WR, Goldberg A, et al. *The Complete German Commission E Monograph: Therapeutic Guide to Herbal Medicines.* Austin, TX: American Botanical Council; 1998. p.117.

20. Önder A. Coriander and its phytoconstituents for the beneficial effects. In: El-Shemy H, editor. Potential of Essential Oils. London: Intech Open; 2018. p.165–185.

21. Burdock GA, Carabin IG. Safety assessment of coriander (*Coriandrum sativum* L.) essential oil as a food ingredient. *Food Chem Toxicol.* 2009;47:22–34.

22. Gardner Z, McGuffin M, editors *Botanical Safety Handbook.* 3rd ed. Boca Raton, FL: CRC Press; 2013. p. 264.

7.19 *CORYDALIS YANHUSUO*

Corydalis yanhusuo is a genus of about 470 species of herbaceous plants in the *Papaveraceae* family, native to the temperate zones of Asia and eastern Africa. Corydalis Rhizoma, named *Yuan Hu* in China, is the dried tuber of *Corydalis yanhusuo*. This species has been used for centuries in Chinese and Korean schools of medicine for pain relief. Recent studies utilizing modern experimental and statistical methods have confirmed the pain-relieving effects of corydalis. Indeed, the herb appears to be as effective as the NSAID diclofenac for osteoarthritis pain, with fewer adverse side effects.[1]

Some of the nociceptive effects of *Corydalis yanhusuo* are mediated by antagonism of dopamine type 2 (DA2) receptors. At least two alkaloids from corydalis, l-tetrahydropalmatine and dehydrocorybulbine, block DA2 receptors. DA2 receptor knock-out mice do not exhibit an anti-nociceptive response to *Corydalis yanhusuo* in models of neuropathic pain, though they obtain relief from pain due to inflammation.[2] Tetrahydropalmatine also antagonizes activity at 5-HT2 receptors, and thus it shares some important characteristics with the so-called atypical anti-psychotics that can enhance the effects of standard antidepressants.[3] Recent studies have also shown that alkaloids from *Corydalis yanhusuo* may be useful in the treatment of MDD.

7.19.1 ANTIOXIDANT

Alkaloids of *Corydalis yanhusuo* were shown to be efficient scavengers of oxygen radicals in the lecithin lipid peroxidation test per the TBARS assay.[4] *In vivo*, extract of *Corydalis yanhusuo* protected cerebral tissue of rats subjected to cerebral ischemia due to carotid artery ligation. This effect is generally seen as being due to prevention of oxidative damage from reperfusion.[5]

7.19.2 ANTI-INFLAMMATORY

Anti-inflammatory effects of *Corydalis yanhusuo* are seen in its inhibition of lipopolysaccharide- induced release of IL-8 from human monocytic THP-1 cells.[6] A similar effect was seen in cultured mouse macrophages. A methanolic extract of *Corydalis yanhusuo* root dampened lipopolysaccharide-stimulated release of NO and PGE2 due to inhibition of iNOS and COX-2 activity, respectively. It also attenuated release of the inflammatory cytokines, TNF-α, IL-1β, and IL-6.[7] These effects may have been secondary to prevention of the translocation of NF-κB that stimulates synthesis of those cytokines. In the same study, *Corydalis yanhusuo* was also found to inhibit carrageenan-induced paw edema, which is largely medicated by COX-2.

In intact mice, antinociceptive effects were seen after administration of dehydrocorydaline, another alkaloid isolated from corydalis. Those effects were partially blocked by naloxone, indicating that central opiate receptors mediate some of the pain-relieving effects of *Corydalis yanhusuo*. However, in the formalin test, dehydrocorydaline decreased the expression of caspase 6 (CASP6), TNF-α, IL-1β, and IL-6 proteins in mouse spinal cord, showing that much of the nociceptive effect of corydalis alkaloids are likely due to anti-inflammatory action.[8]

7.19.3 ANTIDIABETIC/ANTI-METABOLIC SYNDROME

Although not widely seen as an agent to treat diabetes, *Corydalis yanhusuo* has been found to decrease blood glucose in normal and diabetic mice and improve sugar tolerance in insulin-resistant mice.[9]

7.19.4 Preclinical Antidepressant-Like Effects

Several rodent studies have shown antidepressant effects of corydalis alkaloids. An alcohol extraction of *Corydalis yanhusuo* containing the alkaloids tetrahydropalmatine, corydaline, tetrahydroberberine, and protopine exhibited anti-depressant effects in significantly reducing the detrimental effects of chronic unpredictable mild stress in rats in the open field and sucrose preference tests. The effects were similar to those of treatment with venlafaxine.[10]

In another study, eight days of administration of tetrahydropalmatine to rats prior to a two-hour session of restraint stress significantly improved performance in the elevated plus maze, sucrose preference, and forced swim tests. The highest dose of tetrahydropalmatine ameliorated the effects of restraint stress in a manner similar to that of fluoxetine. Tetrahydropalmatine also reduced serum corticosterone levels in response to the restraint stress, and tended to normalize increased corticotrophin-releasing factor and decreased neuropeptide Y immunoreactivities in the periventricular hypothalamus.[11]

In mice, *Corydalis yanhusuo* was shown to improve performance in the forced swim test of antidepressant effect after subjection to chronic unpredictable mild stress. Oral administration for 14 days also reversed the elevated plasma corticosterone levels, body weight loss, decrease in proliferation of hippocampal precursor cells, and elevated rates of hippocampal cell apoptosis.[12]

7.19.5 Human Antidepressant Effects

There are no human trials of *Corydalis yanhusuo* in the treatment of MDD. Nonetheless, orally administered *Corydalis yanhusuo* alkaloids enter the human brain and produce pharmacological effects. For example, orally administered tetrahydropalmatine significantly attenuates opiate craving and increases rates of abstinence in human heroin addicts.[13]

7.19.6 Dosage

Easley and Horne recommend tincture (1:3 50% alcohol) 1–5 ml as needed, standard decoction 3–8 oz as needed, standardized extract containing tetrahydropalmatine at 100 mg to 200 mg daily, or tea powder concentrate 1–2g three times daily.[14]

7.19.7 Toxicity

The LD50 of the crude alkaloids of *Corydalis adunca* maxim in rats was reported to be 1.833g/kg.[15] Extrapolated to a 75 kg human adult, this would be approximately 128 g.

7.19.8 Safety in Pregnancy

Reference texts from Traditional Chinese Medicine advise against the use of *Corydalis yanhusuo* during pregnancy.[16,17]

7.19.9 Drug Interactions

The *Botanical Safety Handbook* notes no known interactions of *Corydalis yanhusuo* with medications.[18] Studies have, however, shown possible interactions. A total alkaloid extract of *Corydalis yanhusuo* upregulated levels of cytochromes P450 enzymes in rat liver,[19] whereas in another study, the alkaloid corydaline extracted from the herb potently inhibited CYP2C19 and CYP2C9 enzymes. To a lesser extent, it inhibited CYP3A.[20]

REFERENCES

1. Zuo C, Yin G, Cen XM, Xie QB. Controlled clinical study on compound Decumbent Corydalis Rhizome and diclofenac in treatment of knee osteoarthritis. *China J Chin Mater Med*. 2015;40(1):149–153.
2. Wang L, Zhang Y, Wang Z, et al. The antinociceptive properties of the *Corydalis yanhusuo* extract. *PLoS One*. 2016;11(9):e0162875.
3. Lin MT, Chueh FY, Hsieh MT, Chen CF. Antihypertensive effects of dl-tetrahydropalmatine: An active principle isolated from Corydalis. *Clin Exp Pharmacol Physiol*. 1996;23(8):738–742.
4. Bei YX, Dong X, Li GP, Gao YT. The general alkaloid of corydalis on active oxygens and its antioxidation in vitro. *Food Sci Technol*. 2008-09.
5. Tao YW, Wang Y, Lai BL, Tian GY, Feng JP, Su JX. Protective effect of polysaccharide from the root of Rhizoma Corydalis on focal ischemic cerebral infarct induced by middle cerebral artery occlusion in rats. *J Med Plants Res*. 2013;7(4):140–147.
6. Oh YC, Choi JG, Lee YS, et al. Tetrahydropalmatine inhibits pro-inflammatory mediators in lipopolysaccharide-stimulated THP-1 Cells. *J Med Food*. 2010;13(5):1125–1132.
7. Choi WY, Park SJ, Lee JR, et al. Modulation of inflammatory mediators by corydalis tuber in LPS-stimulated Raw264.7 cells and its inhibitory effects on carrageenan-stimulated paw edema in rats. *Orient Pharm Exper Med*. 2014;14(4):319–328.
8. Yin ZY, Li L, Chu SS, Sun Q, Ma ZL, Gu XP. Antinociceptive effects of dehydrocorydaline in mouse models of inflammatory pain involve the opioid receptor and inflammatory cytokines. *Sci Rep*. 2016;6:27129.
9. Yu XC. Predication of anti-diabetes effects of *Corydalis yanhusuo* alkaloids with pharmacological network technology and experimental validation in ICR mice. *J Chin Pharm Sci*. 2014;49(11):913–918.
10. Wu H, Wang P, Liu M, et al. A 1H-NMR-Based metabonomic study on the anti-depressive effect of the total alkaloid of corydalis rhizma. *Molecules*. 2015;20(6):10047–10064.
11. Lee B, Sur B, Yeom M, Shim I, Lee H, Hahm DH. L-Tetrahydropalmatine ameliorates development of anxiety and depression-related symptoms induced by single prolonged stress in rats. *Biomol Ther*. 2014;22(3):213–222.
12. Sun GG, Shih JH, Chiou SH, Hong CJ, Lu SW, Pao LH. Chinese herbal medicines promote hippocampal neuroproliferation, reduce stress hormone levels, inhibit apoptosis, and improve behavior in chronically stressed mice. *J Ethnopharmacol*. 2016;193(4):159–168.
13. Yang Z, Shao Y, Li S, et al. Medication of l-tetrahydropalmatine significantly ameliorates opiate craving and increases the abstinence rate in heroin users: A pilot study. *Acta Pharmacol Sin*. 2008;29(7):781–788.
14. Easley T, Horne S. *The Modern Herbal Dispensary*. Berkeley, CA: North Atlantic Books; 2016. p. 218.
15. Zhang L, Zhang F, Wang J, Shangjiu HU, Liu T. Study on anti-inflammatory and analgesic effects of alkaloids of corydalis adunca maxim. *China Pharm*. 2001;0(12).

16. Bensky D, Clavey S, Stoger E. *Chinese Herbal Medicine: Materia Medica*. 3rd ed. Seattle, WA: Eastland Press; 2004.
17. Chen JK, Chen TT. *Chinese Medical Herbology and Pharmacology*. City of Industry, CA: Art of Medicine Press; 2004.
18. Gardner Z, McGuffin M, editors. *Botanical Safety Handbook*. 3rd ed. Boca Raton, FL: CRC Press; 2013. p. 268.
19. Yan J, He X, Feng S, et al. Up-regulation on cytochromes P450 in rat mediated by total alkaloid extract from Corydalis yanhusuo. *BMC Complement Altern Med*. 2014:14–306.
20. Ji HY, Liu KH, Lee H, et al. Corydaline inhibits multiple cytochrome P450 and UDP-glucuronosyltransferase enzyme activities in human liver microsomes. *Molecules*. 2011;16(8):6591–6602.

7.20 *CROCUS SATIVA* (SAFFRON)

Saffron is the familiar spice made of the stigmata plucked from the flowers of the *Crocus sativa* plant. The plant is cultivated from the lands around the Mediterranean Sea and eastward across Iran, India, Tibet, and China. Nearly 80% of commercial saffron is grown in Iran, where much of the research into the herb's medicinal properties has been performed. It has traditionally been seen as an adaptogen, stimulant, aphrodisiac, and antidepressant.[1]

The principle phytochemical of saffron is safranal, which provides its characteristic flavor. Other active components are crocin and crocetin, which lend color to saffron, and the glucoside picrocrocin, which adds a slightly bitter flavor to the spice. Various antioxidant anthocyanins can be isolated from the blue flowers.[2] Saffron is not a unique source of crocin, as crocin is also a major contributor to the medicinal effects of *Gardenia jasminoides*.

7.20.1 ANTIOXIDANT

In vivo antioxidant effects of saffron have been demonstrated in several models. For example, both aqueous and ethanolic extracts of saffron protected mouse livers from the oxidative damage of CCl_4. The extracts significantly decreased serum levels of AST and ALT, and histopathological studies of liver tissue showed reductions in the incidence of CCl_4-induced lesions.[3] In human subjects, ingestion of 50 mg saffron in milk twice a day for 3–6 weeks was found to significantly decrease serum lipoprotein oxidation.[4]

Antioxidant effects were also seen in the brains of mice sub-chronically administered saffron, thus demonstrating central action. Saffron-treated mice exhibited significant improvements in learning and memory that was accompanied by reduced lipid peroxidation products, higher total brain antioxidant activity, and reduced caspase-3 activity in *ex vivo* brain homogenates. Saffron and its constituents crocetin and safranal also protected against H_2O_2-induced toxicity in human neuroblastoma cells. ROS production was dampened, caspase-3 activation reduced, and cell viability improved.[5]

7.20.2 ANTI-INFLAMMATORY

Saffron has anti-inflammatory effects that may be due in part to the potent antioxidant effects of crocetin and crocin. Indeed, part of many inflammatory conditions is the production of reactive oxygen species that are quenched by crocetin.[6] In the carrageenan-induced inflammation model in rats, both crocin and safranal reversed pain, edema, immobility, and neutrophil infiltration in a manner similar to that of the NSAID, diclofenac.[7]

Ethanolic and aqueous extracts of saffron also produced anti-inflammatory and antinociceptive effects in the chronic sciatic nerve constriction model of neuropathic pain in rats. Both extracts reduced levels of the inflammatory cytokines TNF-α, IL-1β, and IL-6. Levels of the oxidative stress marker malondialdehyde were decreased, whereas protective reduced glutathione level increased.[8]

Anti-inflammatory effects are also seen in brain cells. In cultured microglial cells from rat brain, crocin and crocetin inhibited lipopolysaccharide-induced NO release, and reduced release of TNF-α, IL-1β, and intracellular reactive oxygen species. The compounds also effectively reduced lipopolysaccharide-elicited NF-κB activation.[9]

7.20.3   ANTIDIABETIC/ANTI-METABOLIC SYNDROME

Extract of saffron has an anti-hyperglycemic effect in alloxan-induced diabetic rats.[10] Purified crocin and safranal from saffron also exhibit the full anti-hyperglycemic effect of the crude extract in such rats.[11]

In cultured mouse skeletal muscle cells, saffron enhanced insulin sensitivity and glucose uptake through the phosphorylation of AMP-activated protein kinase, acetyl-CoA carboxylase, and mitogen-activated protein kinases, but not activation of the PI 3-kinase/Akt pathway. Interestingly, co-treatment with saffron and insulin further improved the insulin sensitivity by both insulin-independent (AMPK/ACC and MAPKs) and insulin-dependent (PI 3-kinase/Akt/mTOR) pathways.[12]

Saffron also shows effects in the brains of diabetic animals. Aqueous extracts of saffron decreased serum glucose and TNF-α levels, and reversed cognitive deficits in streptozotocin-induced diabetic rats. In the hippocampus, it increased the antioxidants glutathione, superoxide dismutase, and catalase (CAT), and substantially decreased induced nitric oxide synthase activity.[13,14]

Several studies show that saffron and its constituents dampen the response to stress and lower glucocorticoid levels. Intraperitoneally injected aqueous extract of saffron reduced serum corticosterone levels in rats stressed by electrical shock. Stress-induced anorexia, increased sniffing, rearing, locomotion, and coping time were also reduced.[15] Daily injection of the active component crocin to rats subjected to chronic restraint stress spared the memory of a previously passive avoidance maneuver. Crocin also reduced to normal cortical and hippocampal concentrations of corticosterone.[16]

7.20.4   PRECLINICAL ANTIDEPRESSANT-LIKE EFFECTS

A large number of animal studies show efficacy of saffron and its constituent phytochemicals in the treatment of depression as well as providing indications of the mechanisms by which they work. For example, sub-chronic treatment of rats with aqueous extract of saffron reduced immobility time on the forced swim test in a manner similar to imipramine. The hippocampal levels of BDNF, CREB, and p-CREB were significantly increased in saffron treated rats. VGF protein expression was also increased, but not significantly.[17] Administration of crocin over 21 days also significantly reduced immobility time in rats subjected to the forced swim test. Treatment with crocin or imipramine increased levels of BDNF, phosphorylated CREB, and VGF in the hippocampus.[18]

The means by which saffron might influence BDNF, CREB, and VGF, and then in turn, neuronal plasticity is unclear. However, several studies have found evidence of glutamatergic mediation. Alcohol-extract of saffron blocked electrically stimulated NMDA and non-NMDA postsynaptic potentials mediated by glutamate. Saffron also blocked NMDA- and kainate-induced depolarizations, but not those induced by AMPA. Trans-crocetin extracted from saffron blocked NMDA depolarization, but not that of kainate.[19] Binding studies have also shown that saffron extracts displace NMDA receptor radioligands, though the binding affinity of the extract, in the low micromolar range, is substantially lower than that of ketamine.[20] Thus, the antidepressant effects of saffron may be mediated in part by a blockade of NMDA receptor activity in a manner similar to that of ketamine.

7.20.5 HUMAN ANTIDEPRESSANT EFFECTS

Enough clinical studies of the antidepressant effect of saffron in human subjects have been performed to warrant reviews of this effect. The conclusions have been that saffron supplementation can indeed improve symptoms of depression in adults with MDD.[21–23]

In one such study, six weeks of treatment with saffron significantly reduced symptoms of depression in adults with mild to moderately severe MDD as per the Hamilton Rating Scale. The saffron group suffered no significant side effects.[24] In a similar six-week study, saffron was as effective as imipramine in treating depression with fewer side effects.[25] A study has also shown a standardized extract of saffron to be safe and effective in treating mild to moderate depression in adolescents.[26]

In yet another study, the flower petal of *Crocus sativus* – not the stigmata that are the source of saffron spice – was compared against the antidepressant effects of fluoxetine. After eight weeks of treatment, petal of *Crocus sativus* was found to be as effective as fluoxetine in the treatment of mild to moderate depression. However, in both treatments, the remission rate was only 25%.[27]

7.20.6 DOSAGE

Easley and Horne state the dosage of saffron to be as tincture of dry herb (1:10, 40%) 5–20 drops three times daily.[28] Available commercially are saffron extract capsules of 88.25 mg, standardized to 0.3% safranal, to be taken twice a day.

7.20.7 TOXICITY

Saffron is considered non-toxic, with an LD50 in mice of 27 g/kg.[29]

7.20.8 SAFETY IN PREGNANCY

Saffron has been used in folk medicine as an abortifacient, and is thus not recommended during pregnancy.[30] The view in Chinese Traditional Medicine is also that saffron should be avoided in pregnancy.[31] Safety in lactation has not been established.

7.20.9 DRUG INTERACTIONS

The *Botanical Safety Handbook* notes no known interactions with commonly prescribed medications.[32] Nonetheless, crocin extracted from *Crocus sativa* significantly inhibited the activities of CYP3A, CYP2C11, CYP2B, and CYP2A enzymes in in rat liver microsomes, whereas safranal significantly enhanced the activities of CYP2B, CYP2C11, and CYP3A enzymes.[33]

REFERENCES

1. Schmidt M, Betti G, Hensel A. Saffron in phtyotherapy: Pharmacology and clinical uses. *Wein Wochenschr.* 2007;157(13–14):315–319.
2. Gohari AR, Saeidnia S, Mahmoodabadi MK. An overview on saffron, phytochemicals, and medicinal properties. *Pharmacogn Rev.* 2013;7(13):61–66.
3. Iranshahi M, Khoshangosht M, Mohammadkhani Z, Karimi G. Protective effects of aqueous and ethanolic extracts of saffron stigma and petal on liver toxicity induced by carbon tetrachloride in mice. *Pharmacol Line.* 2011;1:203–212.
4. Verma SK, Bordia A. Antioxidant property of Saffron in man. *Indian J Med Sci.* 1998;52(5):205–207.
5. Papandreou MA, Tsachaki M, Efthimiopoulos S, Cordopatis P, Lamari FN, Margarity M. Memory enhancing effects of saffron in aged mice are correlated with antioxidant protection. *Behav Brain Res.* 2011;219(2):197–204.
6. Pomaa A, Fontecchio G, Carlucci G, Chichiriccò G. Anti-inflammatory properties of drugs from saffron crocus. *Antiinflamm Antiallergy Agents Med Chem.* 2012;11(1):37–51.
7. Tamaddonfard E, Farshid AA, Eghdami K, Samadi F, Erfanparast A. Comparison of the effects of crocin, safranal and diclofenac on local inflammation and inflammatory pain responses induced by carrageenan in rats. *Pharmacol Rep.* 2013;65(5):1272–1280.
8. Amin B, Abnous K, Motamedshariaty V, Hosseinzadeh H. Attenuation of oxidative stress, inflammation and apoptosis by ethanolic and aqueous extracts of *Crocus sativus* L. stigma after chronic constriction injury of rats. Anais da Academia Brasileira de Ciencências. 2014;86(4):1821–1832.
9. Nam KN, Park YM, Jung HJ, et al. Anti-inflammatory effects of crocin and crocetin in rat brain microglial cells. *Eur J Pharmacol.* 2010;648(1–3):110–116.
10. Elgazar AF, Rezq AA, Bukhari HM. Anti-hyperglycemic effect of saffron extract in alloxan-induced diabetic rats. *Eur J Biol Sci.* 2013;5(1):14–22.
11. Kianbakht S, Hajiaghaee R. Anti-hyperglycemic effects of saffron and its active constituents, crocin and safranal, in alloxan-induced diabetic rats. *JMP.* 2011;3(39):82–89.
12. Kang C, Lee H, Jung ES, et al. Saffron (*Crocus sativus* L.) increases glucose uptake and insulin sensitivity in muscle cells via multipathway mechanisms. *Food Chem.* 2012;135(4):2350–2358.
13. Samarghandian S, Azimi-Nezhad M, Samini F. Ameliorative effect of saffron aqueous extract on hyperglycemia, hyperlipidemia, and oxidative stress on diabetic encephalopathy in streptozotocin induced experimental diabetes mellitus. *BioMed Res Int.* 2013;2013:417928

14. Georgiadou G, Tarantilis P, Pitsikas N. Effects of the active constituents of *Crocus sativus* L., crocins, in an animal model of obsessive-compulsive disorder. *Neurosci Lett.* 2012;528(1):27–30.

15. Hooshmandi Z, Rohani AH, Eidi A, Fatahi Z, Golmanesh L, Sahraei H. Reduction of metabolic and behavioral signs of acute stress in male Wistar rats by saffron water extract and its constituent safranal. *J Pharmaceut Biol.* 2011;49(9):947–954.

16. Dastgerdi AH, Radahmadi M, Pourshanazari AA, Dastgerdi HH. Effects of crocin on learning and memory in rats Under chronic restraint stress with special focus on the hippocampal and frontal cortex corticosterone levels. *Adv Biomed Res.* 2017;6:157.

17. Ghasemi T, Abnous K, Vahdati F, Mehri S, Razavi BM. Antidepressant effect of *Crocus sativus* aqueous extract and its effect on CREB, BDNF, and VGF transcript and protein levels in rat hippocampus. *Drug Res.* 2015;65(7):337–343.

18. Hassani FV, Naseri V, Razavi BM, Mehri S, Abnous K, Hosseinzadeh H. Antidepressant effects of crocin and its effects on transcript and protein levels of CREB, BDNF, and VGF in rat hippocampus. *DARU J Pharm Sci.* 2014;22:16.

19. Berger F, Hensel A, Nieber K. Saffron extract and trans-rocetin inhibit glutamatergic synaptic transmission in rat cortical brain slices. *Neuroscience.* 2011;180:238–247.

20. Lechtenberg M, Schepmann D, Niehues M, Hellenbrand N, Wünsch B, Hensel A. Quality and functionality of saffron: Quality control, species assortment and affinity of extract and isolated saffron compounds to NMDA and σ1 (sigma-1) receptors. *Planta Med.* 2008;74(7):764–772.

21. Hausenblas HA, Saha D, Dubyak PJ, Anton SD. Saffron (*Crocus sativus* L.) and major depressive disorder: A meta-analysis of randomized clinical trials. *J Integr Med.* 2013;11(6):377–383.

22. Lopresti AL, Drummond PD. Saffron (*Crocus sativus*) for depression: A systematic review of clinical studies and examination of underlying antidepressant mechanisms of action. *Hum Psychopharmacol.* 2014;29(6):517–527.

23. Kamalipour M, Jamshidi AH, Akhondzadeh S. Antidepressant effect of *Crocus sativus*: An evidence-based review. *J Med Plant.* 2010;9(6):35–38.

24. Akhondzadeh S, Tahmacebi-Pouri N, Noorbala AA, et al. *Crocus sativus* L. in the treatment of mild to moderate depression: A double-blind, randomized and placebo-controlled trial. *Phytother Res.* 2005;19(2):148–151.

25. Akhondzadeh S, Fallah-Pour H, Afkham K, Jamshidi AH, Khalighi-Cigaroudi F. Comparison of *Crocus sativus* L. and imipramine in the treatment of mild to moderate depression: A pilot double-blind randomized trial. *BMC Complement Altern Med.* 2004;4:12.

26. Lopresti AL, Drummond PD, Inarejos-García AM, Prodanov M. Saffron, a standardised extract from saffron (*Crocus sativus* L.) for the treatment of youth anxiety and depressive symptoms: A randomised, double-blind, placebo-controlled study. *J Affect Disord.* 2018;232:349–357.

27. Akhondzadeh A, Moshiri E, Noorbala AA, Jamshidi AH, Abbasi SH, Akhondzadeh S. Comparison of petal of *Crocus sativus* L. and fluoxetine in the treatment of depressed outpatients: A pilot double-blind randomized trial. *Prog Neuropsychopharmacol Biol Psychiatry.* 2007;31(2):439–442.

28. Easley T, Horne S. *The Modern Herbal Dispensary.* Berkeley, CA: North Atlantic Books; 2016. p. 297.

29. Abdullaev F. Biological properties and medicinal use of saffron (*Crocus sativus* L.). *Acta Hortic (ISHS).* 2007;739(739):339–345.

30. Wichtl M. *Herbal Drugs and Phytopharmaceuticals: A Handbook for Practice on a Scientific Basis.* 3rd ed. Boca Raton, FL: CRC Press; 2004.

31. Bensky D, Clavey S, Stoger E. *Chinese Herbal Medicine: Materia Medica.* 3rd ed. Seattle, WA: Eastland Press; 2004.

32. Gardner Z, McGuffin M, editors. *Botanical Safety Handbook*. 3rd ed. Boca Raton, FL: CRC Press; 2013. p. 274.

33. Dovrtelova G, Noskova K, Jurica J, Turjap M, Zendulka O. Can bioactive compounds of Crocus sativus L. influence the metabolic activity of selected CYP enzymes in the rat? *Physiol Res*. 2015;64(Suppl. 4):S453–S458.

7.21 *CURCUMA LONGA* (TURMERIC)

Curcuma longa, commonly known as turmeric, is a flowering plant of the ginger family that is native to Southeast Asia and the Indian subcontinent. All parts of the plant are used for medical purposes, but the greatest use comes from the dried rhizome. A number of sesquiterpenes and other complex phytochemicals have been isolated from turmeric rhizome, but it is richest in curcuminoids and furanodiene. The best studied curcuminoid is curcumin. Not surprisingly, turmeric shares some phytochemicals with ginger.[1] Among the many phytochemicals identified in turmeric are ferulic acid, coumaric acid, pinene, sabinene, myrcene, phellandrene, careen, terpinene, cymene, limonene, cineole, zingiberene, sesquiphellandrene, nerolidol, santalenone, turmerol, germacrone, turmerone, bisabolone, trans-alpha-atlantone, stigmasterol, and sitosterol.[2] Along with antioxidant and anti-nociceptive effects, both animal and human trials have shown turmeric and its phytochemicals to be of value in the treatment of MDD.

7.21.1 ANTIOXIDANT

Extracts of turmeric showed significant free radical scavenging and reducing power. Those antioxidant activities of turmeric were attributed primarily to curcumin, ferulic acid, and p-coumaric acid.[3] In the standard test of *in vivo* antioxidant capacity, extract of turmeric added to the diet of rats protected their livers from the oxidative damage of CCl_4. Serum levels of bilirubin, cholesterol, and transaminases were reduced in rats that received turmeric, though they did not reach normal levels.[4]

7.21.2 ANTI-INFLAMMATORY

Crude extracts of turmeric inhibited lipopolysaccharide-induced production of TNF-α and prostaglandin E2 by human leukemia cells. The seemingly more potent fractions containing curcumin also showed inhibition of COX-2.[5] Treatment of mice with curcumin prior to administration of lipopolysaccharide also attenuated inflammatory kidney damage. Expression of monocyte chemoattractant protein-1 and subsequent macrophage infiltration in renal tissue were reduced. The DNA-binding activity of NF-κB was inhibited, and production of IL-8 and IL-2 attenuated.[6]

Curcumin also has anti-inflammatory and neuroprotective effects in rat models of stroke. In rats pretreated with curcumin, brain levels of TNF-α and IL-6 were reduced after induced ischemia. Treatment also maintained mitochondrial function. The effects of curcumin were blocked by the SIRT1 antagonist sirtinol. Thus, curcumin acts by a mechanism similar to resveratrol in providing neuroprotection.[7]

7.21.3 Antidiabetic/Anti-Metabolic Syndrome

Curcumin enhances insulin sensitivity and normalizes serum adipocytokines, leptin, and adiponectin in human patients.[8,9]

7.21.4 Preclinical Antidepressant-Like Effects

In a study using rats, curcumin reversed the depression-like syndrome produced by chronic corticosterone treatment, and may have done so by increasing BDNF expression in the hippocampus and frontal cortex.[10] There is also evidence that curcumin reverses depression-like behaviors in mice by enhancing serotonergic and dopaminergic activity in the brain, due in part to inhibition of both MAO-A and -B.[11]

7.21.5 Human Antidepressant Effects

A 2014 study found curcumin to be effective in the treatment of MDD. After four weeks of treatment, curcumin was more effective than placebo in reducing symptoms of depression and anxiety as measured by the Inventory of Depressive Symptomology and Spielberger State-Trait Anxiety Inventory.[12] A similar subsequent study confirmed those results, although patients suffering so-called atypical depression benefited the most. The study showed the combination of curcumin and saffron to be particularly effective.[13]

Chronic supplementation with curcumin also enhances the efficacy of standard antidepressant treatment. Supplementation with curcumin 1000 mg twice a day significantly reduced scores on the Hamilton Depression and Montgomery-Asberg Depression Rating Scales in patients already receiving antidepressants. The addition of curcumin also reduced serum levels of inflammatory cytokines IL-1β and tumor necrosis factor α level, increases plasma brain-derived neurotrophic factor levels, and decreases salivary cortisol concentrations compared with placebo group.[14]

7.21.6 Dosage

The German E Commission gives the dosing of turmeric root as 1.5–3 g of root daily.[15] The bioavailability of curcumin is quite low, but its absorption from the gut can be substantially increased by adding piperine, a phytochemical from black pepper.[16]

7.21.7 Toxicity

In studies of the essential oil of turmeric using rats and mice, no acute toxicity or mortality was seen in any animals up to the maximum dose of 5,000 mg/kg body weight. Thus, the LD50 was deemed to be greater than 5,000 mg/kg, which extrapolated to a 160-pound adult would be approximately 375 g. Furthermore, repeated administration of the essential oil for 90 days in rats at a dose of 1,000 mg/kg body weight did not induce any observable toxic effects, compared with the corresponding control animals.[17]

7.21.8 Safety in Pregnancy

Reference texts from Traditional Chinese Medicine advise against use of either *Curcuma longa* or *Curcuma zedoaria* during pregnancy.[18,19]

7.21.9 Drug Interactions

The *Botanical Safety Handbook* notes no known drug interactions with either *Curcuma longa* or *Curcuma zedoaria*.[20] Curcumin did inhibit recombinant human CYP1A2, CYP3A4, CYP2D6, CYP2C9, and CYP2B6 enzymes. However, this occurred at mid-micromolar concentrations and, due to relatively low exposure of the liver to ingested curcumin, effects on drug metabolism by the liver may be minimal.[21]

REFERENCES

1. Lobo R, Prabhu KS, Shirwaikar A, Shirwaikar A. *Curcuma zedoaria* Rosc. (white turmeric): A review of its chemical, pharmacological and ethnomedicinal properties. *JPP* 2009;61(1):13–21.
2. Singh G, Kapoor IPS, Singh P, De Heluani CS, De Lampasona MP, Catalan CA. Comparative study of chemical composition and antioxidant activity of fresh and dry rhizomes of turmeric (*Curcuma longa* Linn.). *Food Chem Toxicol*. 2010;48(4):1026–1031.
3. Kumar GS, Nayaka H, Dharmesh SM, Salimath PV. Free and bound phenolic antioxidants in amla (*Emblica officinalis*) and turmeric (*Curcuma longa*). *J Food Compos Anal*. 2006;19(5):446–452.
4. Deshpande UR, Gadre SG, Raste AS, Pillai D, Bhide SV, Samuel AM. Protective effect of turmeric (*Curcuma longa* L.) extract on carbon tetrachloride-induced liver damage in rats. *Indian J Exp Biol*. 1998;36(6):573–577.
5. Lantz RC, Chen GJ, Solyom AM, Jolad SD, Timmermann BN. The effect of turmeric extracts on inflammatory mediator production. *Phytomedicine*. 2005;12(6–7):445–452.
6. Zhong F, Chen H, Han L, Jin Y, Wang W. Curcumin attenuates lipopolysaccharide-induced renal inflammation. *Biol Pharm Bull*. 2011;34(2):226–232.
7. Miao Y, Zhao S, Gao Y, et al. Curcumin pretreatment attenuates inflammation and mitochondrial dysfunction in experimental stroke: The possible role of Sirt1 signaling. *Brain Res Bull*. 2016;121:9–15.
8. Vieira de Melo IS, dos Santos AF, Bueno NB. Curcumin or combined curcuminoids are effective in lowering the fasting blood glucose concentrations of individuals with dysglycemia: Systematic review and meta-analysis of randomized controlled trials. *Pharmacol Res*. 2018;128:137–144.
9. Panahi Y, Hosseini MS, Khalili N, et al. Effects of supplementation with curcumin on serum adipokine concentrations: A randomized controlled trial. *Nutrition*. 2016;32(10):1116–1122.
10. Zhen Huang Z, Zhong XM, Li ZY, Feng CR, Pan AJ, Mao QQ. Curcumin reverses corticosterone-induced depressive-like behavior and decrease in brain BDNF levels in rats. *Neurosci Lett*. 2011;493(3):145–148.
11. Kulkarni SK, Bhutani MK, Bishnoi M. Antidepressant activity of curcumin: Involvement of serotonin and dopamine system. *Psychopharmacology*. 2008;201(3):435.
12. Lopresti AL, Maes M, Maker GL, Hood SD, Drummond PD. Curcumin for the treatment of major depression: A randomized, double-blind, placebo-controlled study. *J Affect Dis*. 2014;167:368–375.

13. Lopresti AL, Drummond PD. Efficacy of curcumin, and a saffron/curcumin combination for the treatment of major depression: A randomized, double-blind, placebo-controlled study. *J Affect Dis.* 2017;207(1):188–196.

14. Yu JJ, Pei LB, Zhang Y, Wen ZY, Yang JL. Chronic supplementation of curcumin enhances the efficacy of antidepressants in major depressive disorder: A randomized, double-blind, placebo-controlled pilot study. *J Clin Psychopharmacol.* 2015;35(4):406–410.

15. Blumenthal M, Busse WR, Goldberg A, et al. The Complete German Commission E Monographs. Austin, TX: American Botanical Council; 1998. p. 222.

16. Shoba G, Joy D, Joseph T, Majeed M, Rajendran R, Srinivas PS. Influence of piperine on the pharmacokinetics of curcumin in animals and human volunteers. *Planta Med.* 1998;64(4):353–356.

17. Aggarwal ML, Chacko KM, Kuruvilla BT. Systematic and comprehensive investigation of the toxicity of curcuminoid-essential oil complex: A bioavailable turmeric formulation. *Mol Med.* 2016;13(1):592–604.

18. Bensky D, Clavey S, Stoger E. *Chinese Herbal Medicine: Materia Medica.* 3rd ed. Seattle, WA: Eastland Press; 2004.

19. Chen JK, Chen TT. *Chinese Medical Herbology and Pharmacology.* City of Industry, CA: Art of Medicine Press; 2004.

20. Gardner Z, McGuffin M, editors. *Botanical Safety Handbook.* 3rd ed. Boca Raton, FL: CRC Press; 2013. p. 264.

21. Appiah-Opong R, Commandeur JNM, van Vugt-Lussenburg B, Vermeulen NP. Inhibition of human recombinant cytochrome P450s by curcumin and curcumin decomposition products. *Toxicology.* 2007;235(1–2):83–91.

7.22 *CYPERUS ROTUNDUS*

Cyperus rotundus, also known as purple nutsedge, or nutgrass, is a grass native to southern Asia, Africa, and southern Europe. It has been an important herb in Chinese and Ayurvedic Medicine. The various components of *Cyperus rotundus* are thought to have anticonvulsant, antidiabetic, anti-inflammatory, anti-obesity, antioxidant, neuroprotective, and nootropic effects. The dried root contains a variety of alkaloids, flavonoids, terpenoids, glycosides, and sesquiterpenes. Many are novel and pharmacologically active.[1] Among the specific phytochemicals that have been identified are: α-cyperone, α-rotunol, β-cyperone, β-pinene, β-rotunol, β-selinene, camphene, copaene, cyperene, cyperenone, cyperol, cyperolone, cyperotundone, copadiene, epoxyguaiene, cymene, isocyperol, isokobusone, kobusone, limonene, rotunduskone, myristic acid, oleanolic acid, oleanolic acid-3-oneohesperidoside, p-cymol, patchoulenone, rotundene, rotundenol, rotundone, selinatriene, sitosterol, sugeonol, and sugetriol. Some of the phytochemicals in *Cyperus rotundus* are found in other medical herbs. For example, α-cyperone, β-selinene, cyperene, sugeonol, isokobusone, and rotundone are phytochemicals that contribute the spicy aromas and flavor of black pepper.[2]

7.22.1 ANTIOXIDANT

In vitro pretreatment with *Cyperus rotundus* extracts increased production of BDNF and prevented hydrogen peroxide-induced damage to cultured human neurons.[3] The methanolic extract of *Cyperus rotundus* was also found to inhibit NO production in cultured mouse macrophages stimulated with interferon-γ and lipopolysaccharide. This was due to inhibition of inducible nitric oxide synthase. The extract also suppressed the production of superoxide by phorbol ester-stimulation of these cells.[4]

In intact rats, prior treatment with a methanolic extract of *Cyperus rotundus* partially protected liver tissue against the oxidative damage of CCl_4.[5] Pre-treatment with ethanolic extracts of *Cyperus rotundus* also protected against the behavioral and cognitive deficits caused by sodium nitrite-induced hypoxia.[6] Thus, it is clear that the herb crosses into the brain to have central nervous system effects.

7.22.2 ANTI-INFLAMMATORY

It has been found that α-cyperone isolated from alcohol extracts of *Cyperus rotundus* has anti-inflammatory effects in mouse macrophages stimulated by lipopolysaccharide. α-Cyperone suppressed production of the inflammatory prostaglandin PGE2 and dampened expression of inducible COX-2. α-Cyperone also downregulated the production of the inflammatory cytokine IL-6 and suppressed the transcriptional activity of the intracellular inflammation factor, NFκB.[7] Fulgidic acid isolated from *Cyperus rotundus* also has potent anti-inflammatory, antioxidative, and antinitrosative effects. Fulgidic acid reduced the production of nitric oxide, prostaglandin E2, TNF-α, and IL-6 in lipopolysaccharide-induced macrophages. It did so in part by suppressing inducible nitric oxide synthase and COX-2 at the protein level.[8]

In the carrageenan-induced edema and formaldehyde-induced arthritis tests in living mice, the active triterpenoids of *Cyperus rotundus* have a greater anti-inflammatory effect than hydrocortisone.[9]

7.22.3 ANTIDIABETIC/ANTI-METABOLIC SYNDROME

Cyperus rotundus is a traditional Ayurvedic treatment for diabetes, and at least one study has shown it to have potent hypoglycemic effects in rats with alloxan-induced diabetes.[10] *Cyperus rotundus* also protects against the protein oxidation and glycoxidation caused by persistently high glucose levels of diabetes.[11] Protein glycation and neuroinflammation from advanced glycation end-products are known to occur in the human brain where they contribute to various neuropathologies and degenerative diseases.[12]

7.22.4 PRECLINICAL ANTIDEPRESSANT-LIKE EFFECTS

There are several reports of extracts of *Cyperus rotundus* having antidepressant-like effects in rodents. Studies have found *Cyperus rotundus* to have antidepressant-like effects in the tail suspension and forced swim tests in mice.[13,14] Two specific phytochemicals, cyprotusides A and B, purified from *Cyperus rotundus* showed antidepressant-like effects in mice in the forced swim test.[15] Another report showed a phenolic glycoside from *Cyperus rotundus*, rotunduside F, to have antidepressant-like effects in mice similar to those of fluoxetine.[16]

Cyperus rotundus is also known to have sedative and anxiolytic properties, which can be of benefit in treating some individuals with MDD. In mice, ethanolic extract of *Cyperus rotundus* showed anxiolytic effects in the "light-dark box" paradigm. Mice that received either the extract or diazepam exhibited similar anxiolytic effects in spending more time in the lighted but exposed area of the apparatus.[17] Isocurcuminol, a sesquiterpine isolated from *Cyperus rotundus*, is a GABA-A receptor agonist and thus likely contributes to anxiolytic effects.[18]

7.22.5 HUMAN ANTIDEPRESSANT STUDIES

Yeung et al. has noted *Cyperus rotundus* to be among the most commonly used single ingredients in TCM herbal treatments of depression. However, there are no references in the literature to studies on the use of *Cyperus rotundus* alone in the treatment of MDD in human subjects.

7.22.6 DOSAGE

The literature does not include a specific dose for use of *Cyperus rotundus* as a monotherapy. However, in the TCM combination, *Chaihu shugansan*, the recipe calls for a decoction of a combination of herbs that includes 4.5 g of dried rhizome of *Cyperus rotundus*, with the decoction consumed twice daily.[19]

7.22.7 TOXICITY

Acute toxicological studies in rats showed no mortality or morbidity up to 2,000 mg/ kg body weight, which extrapolated to a 75 kg human adult would be approximately 140 g. Sub-chronic toxicity studies at that dose revealed no changes in food, water consumption, or body weight.[20]

7.22.8 SAFETY IN PREGNANCY

The *Botanical Safety Handbook* notes there is no definitive guidance from the literature concerning the safety of *Cyperus rotundus* during pregnancy.[21]

7.22.9 DRUG INTERACTIONS

The *Botanical Safety Handbook* notes no known interactions with drugs. Indeed, extract of *Cyperus rotundus* only weakly inhibits CYP1A2, CYP3A4, and CYP2D6 enzymes.[22]

REFERENCES

1. Singh N, Pandey BR, Verma P, Bhalla M, Gilca M. Phyto-pharmacotherapeutics of *Cyperus rotundus* Linn. (Motha): An overview. *Indian J Nat Prod Res.* 2012;3(4):467–476.
2. Siebert TE, Wood C, Elsey GM, Pollnitz AP. Determination of rotundone, the pepper aroma impact compound, in grapes and wine. *J Agric Food Chem.* 2008;56(10):3745–3748.

3. Kumar KH, Khanum F. Hydroalcoholic extract of *Cyperus rotundus* ameliorates H2O2-induced human neuronal cell damage via its anti-oxidative and anti-apoptotic machinery. *Cell Mol Neurobiol*. 2013;33(1):5–17.
4. Seo WG, Pae HO, Oh GS, et al. Inhibitory effects of methanol extract of *Cyperus rotundus* Linn. Linn. Rhizomes on nitric oxide and superoxide productions by murine macrophage cell line, RAW 264.7 cells. *J Ethnopharmacol*. 2001;76(1):59–64.
5. Rao YR, Rao MP, Rao VR, Suresh K, Rafi M, Rao TN. Preliminary phytochemical screening and hepatoprotective activity of methanolic leaves extract of *Cyperus rotundus* in CCl₄ induced hepatotoxicity in albino Wistar rats. *World J Pharm Pharm Sci*. 2014;3(3):787–795.
6. Jebasingh D, Jackson DD, Venkataraman S, Adeghate E, Emerald BS. The protective effects of *Cyperus rotundus* on behavior and cognitive function in a rat model of hypoxia injury. *Pharml Biol*. 2014;52(12):1558–1569.
7. Jung SH, Kim SJ, Jun BG, et al. α-Cyperone, isolated from the rhizomes of *Cyperus rotundus*, inhibits LPS-induced COX-2 expression and PGE2 production through the negative regulation of NFκB signalling in RAW 264.7 cells. *J Ethnopharmacol*. 2013;147(1,2):208–214.
8. Shin JS, Hong Y, Lee HH, et al. Fulgidic acid isolated from the rhizomes of *Cyperus rotundus* suppresses LPS-induced iNOS, COX-2, TNF-α, and IL-6 expression by AP-1 inactivation in RAW264.7 macrophages. *Biol Pharm Bull*. 2015;38(7):1081–1086.
9. Gupta MB, Palit TK, Singh N, Bhargava KP. Pharmacological study to isolate the active constituents of *Cyperus rotundus* responsible for anti-inflammatory, antipyretic and analgesic activity. *Indian J Med Res*. 1971;59(1):76–82.
10. Raut NA, Gaikwad NJ. Antidiabetic activity of hydro-ethanolic extract of *Cyperus rotundus* in alloxan induced diabetes in rats. *Fitoterapia*. 2006;77(7–8):585–588.
11. Ardestani A, Yazdanparast R. *Cyperus rotundus* suppresses AGE formation and protein oxidation in a model of fructose-mediated protein glycoxidation. *Int J Biol Macromol*. 2007;41(5):572–578.
12. Pamplona R, Dalfó E, Ayala V, et al. Proteins in Human Brain Cortex Are Modified by Oxidation, glycoxidation, and lipoxidation: Effects of Alzheimer's disease and identification of lipoxidation targets. *J Biol Chem*. in press.
13. Zhou ZL, Liu YH. Study on antidepressant effect and mechanism by *Cyperus rotundus* extracts. *Chin J Exp Tradit* Formulae. 2012-07.
14. Wang JM, Ma YX, Zhang B, Li QW, Cui Y. Antidepressant-like effects of extracts isolated from rhizomes of *Cyperus rotundus* L. *Lishizhen Med Mater Med Res*. 2013-04.
15. Zhou ZL, Lin SQ, Yin WQ. New cycloartane glycosides from the rhizomes of *Cyperus rotundus* and their antidepressant activity. *J Asian Nat Prod Res*. 2016;18(7):662–668.
16. Lin SQ, Zhou ZL, Zhang HL, Yin WQ. Phenolic glycosides from the rhizomes of *Cyperus rotundus* and their antidepressant activity. *J Korean Soc Appl Biol Chem*. 2015;58(5):685–691.
17. Sheik SH, Vedhaiyan N, Singaravel S. Evaluation of Central Nervous System Activities of *Cyperus rotundus* L. extract on rodents. *Curr Res Neurosci*. 2015;5(1):10–19.
18. Ha J, Lee KY, Choi HC, et al. Modulation of radioligand binding to the GABA A-benzodiazepine receptor complex by a new component from *Cyperus rotundus*. *Biol Pharm Bull*. 2002;25(1):128–130.
19. Yeung WF, Chung KF, Ng KY, Yu YM, Ziea ET, Ng BF. A systematic review on the efficacy, safety and types of Chinese herbal medicine for depression. *J Psychiatr Res*. 2014;57:165–175.
20. Jebasingh D, Jackson DD, Venkataraman S, Emerald BS. Physiochemical and toxicological studies of the medicinal plant *Cyperus rotundus* L (Cyperaceae). *Int J Appl Res Nat Prod*. 2013;5(4):1–8.

21. Gardner Z, McGuffin M, editors. *Botanical Safety Handbook*. 3rd ed. Boca Raton, FL: CRC Press; 2013. p. 296.

22. Essaidi I, Brahmi Z, Koubaier HBH, et al. Phenolic composition and antioxidant, antimicrobial and cytochrome P450 inhibition activities of Cyperus rotundus tubers. *Mediterr J Chem*. 2015;4(4):201–208.

7.23   *ECHIUM AMOENUM*

Echium amoenum is sometimes referred to as Iranian Borage, but it is distinct from *Borago officinalis*, which is sometimes called European Borage. *Echium amoenum* is from the *Echium* genus, grows in the Alborz hillsides, has not yet been domesticated, and cannot be cultivated. Among the phytochemicals that have been identified in *Echium amoenum* are: cadinene, amorphene, ledene, muurolene, caryophyllene, alloaromadendrene, viridiflorol, calacorene, rosmarinic acid, and spatullenol.[1] Most phytochemicals are not unique, but rather appear in a number of other families of plants.

7.23.1   Antioxidant

Echium amoenum exhibits antioxidant effects. A study of the effects in healthy human subjects showed reduction of blood lipid peroxidation levels, and increases in antioxidant capacity and thiol molecules. It was suspected that major components of these effects were rosmarinic acid and various constituent flavonoids.[2]

7.23.2   Anti-Inflammatory

Echium amoenum exhibited antioxidant and anti-inflammatory effects in mice injected serially with cerulein in a model of pancreatitis. Pretreatment with the herb significantly reduced myeloperoxidase activity and lipid peroxidation in the pancreas, as well as serum levels of inflammatory cytokines.[3] *Echium amoenum* also showed neuroprotective effects in rats subjected to carotid occlusion and reperfusion. The herb spared animals from neuronal loss, reduced myeloperoxidase, and improved cognitive function, ostensibly from antioxidant and anti-inflammatory effects.[4]

7.23.3   Antidiabetic/Anti-Metabolic Syndrome

No references to such effects noted in the literature.

7.23.4   Preclinical Antidepressant-Like Effects

Echium amoenum has been used by Iranian physicians and herbalist to treat anxiety. In a study of mice, the ethanolic extract of *Echium amoenum* flowers increased the percentage of time spent in the open arms of the elevated plus-maze, which is taken to indicate an anxiolytic effect.[5]

There are no animal studies of antidepressant-like effects of *Echium amoenum*. However, the closely related plant of the same genus, *Echium vulgare* – which is used interchangeably with *Echium amoenum* by folk healers of Iran for the treatment of mood disorders – was found to reduce immobility of mice in the forced swim test.[6]

7.23.5 HUMAN ANTIDEPRESSANT EFFECTS

There has been one small randomized controlled trial evaluating the effects of *Echium amoenum* on MDD human subjects. After four weeks of treatment, adults with mild to moderate major depressive disorder enjoyed significantly fewer symptoms of depression per the Hamilton Rating Scale for Depression. This improvement was no longer significant at six weeks.[7] There is also a report of modest efficacy of *Echium amoenum* decoction in reducing symptoms of Obsessive Compulsive Disorder as measured by the Yale-Brown Obsessive Compulsive (Y-BOCS) and Hamilton Rating Scale for Anxiety (HAM-A).[8] Thus, in patients with MDD and co-morbid OCD may benefit from *Echium amoenum* as a component of treatment.

7.23.6 DOSAGE

In the above noted study of effects of *Echium amoenum* on depression, human subjects received 375 mg of dried aqueous extract of *Echium amoenum* each day. There were no significant differences between placebo and drug treated groups in side effects observed.

7.23.7 TOXICITY

In the study of antioxidant effects in humans described above, subjects received doses of 7 mg air-dried flowers/kg twice daily for 14 days without adverse effects. In rats, the LD50 of intraperitoneally administered aqueous extract of *Echium amoenum* was found to be 2.03 g/kg, which extrapolates to approximately 140 g in a 75 kg human adult.[9]

7.23.8 SAFETY IN PREGNANCY

Review of the literature reveals no studies of the safety of the use of *Echium amoenum* in pregnancy of women. However, one study showed that the herb reduced the teratogenic effects of high doses of lamotrigine in mice without adding additional adverse effects.[10]

7.23.9 DRUG INTERACTIONS

Review of the literature revealed no studies of the possible interactions of *Echium amoenum* with commonly used medications.

REFERENCES

1. Ghassemi N, Sajjadi SE, Ghannadi A, Shams-Ardakani M, Mehrabani M. Volatile constituents of a medicinal plant of Iran, *Echium amoenum* Fisch, and C.A. Mey. *DARU*. 2003;11(1):32–33.
2. Ranjbar A, Khorami S, Safarabadi M, et al. Antioxidant activity of Iranian *Echium amoenum* Fisch & C.A. Mey flower decoction in humans: A cross-sectional before/after clinical trial. *eCAM*. 2006;3(4):469–473.
3. Abed A, Minaiyan M, Ghannadi A, Mahzouni P, Babavalian MR. Effect of *Echium amoenum* Fisch. et Mey a traditional Iranian herbal remedy in an experimental model of acute pancreatitis. *ISRN Gastroenterol*. 2012; ID 141548.
4. Safaeian L, Tameh AA, Ghannadi A, Naghani EA, Tavazoei H, Alavi SS. Protective effects of *Echium amoenum* Fisch. and C.A. Mey. against cerebral ischemia in the rats. *Adv Biomed Res*. 2015;4:107.
5. Rabbani M, Sajjadi SE, Vaseghi G, Jafarian A. Anxiolytic effects of *Echium amoenum* on the elevated plus-maze model of anxiety in mice. *Fitoterapia*. 2004;75(5):457–464.
6. Moallem SA, Hosseinzadeh H, Ghoncheh F. Evaluation of antidepressant effects of aerial parts of *Echium vulgare* on mice. *Iran J Basic Med Sci*. 2007;10(3):189–196.
7. Sayyaha M, Sayyah M, Kamalinejad M. A preliminary randomized double-blind clinical trial on the efficacy of aqueous extract of *Echium amoenum* in the treatment of mild to moderate major depression. *Prog Neuropsychopharmacol Biol Psychiatry*. 2006;30(1):166–169.
8. Sayyah M, Boostani H, Pakseresht S, Malaieri A. Efficacy of aqueous extract of *Echium amoenum* in treatment of obsessive-compulsive disorder. *Prog Neuropsychopharmacol Biol Psychiatry*. 2009;33(8):1513–1516.
9. Zamansoltani F, Nassiri-Asl M, Karimi R, Mamaghani-Rad P. Hepatotoxicity effects of aqueous extract of *Echium amoenum* in rats. *Pharmacologyonline*. 2008;1:432–438.
10. Mahsa K, Habibolah J. The antioxidant effect of *Echium amoenum* to prevent teratogenic effects of lamotrigine on the skeletal system and fetal growth in mice. *J Pharm Clin Res*. 2017;3(4):555617.

7.24 *ELEUTHEROCOCCUS SENTICOCCUS* (SIBERIAN GINSENG)

For a number of years, in regions of what had been the Soviet Union, there was pursuit of substances that non-specifically enhance an organism's ability to withstand stress and other adverse conditions. These substances, including various herbs, have been referred to as adaptogens. One of the best known and most studied of the adaptogens is *Eleutherococcus senticoccus*, or Siberian Ginseng.[1] The name Siberian Ginseng is misleading, as the plant is not related to *Panax ginseng*. *Eleutherococcus senticoccus* may also be referred to as *Acanthopanax senticosus*.

The many phytochemicals found in *Eleutherococcus senticosus* include sesamine, syringin, eleutherosides, eleutherans, isofraxidin, sinapaldehyde glucoside, coniferaldehyde glucoside, coniferin, caffeoylquinic acids, α-bergamotnen, δ-elemene, βelemene, γ-cadinene, and α-pinene. Among the effects attributed to *Eleutherococcus senticoccus* are anti-inflammatory, anti-oxidative, anti-carcinogenic, anti-fatigue, antidiabetic, hypolipidemic, immunoprotective, and immunoregulatory.[2]

In microglia, *Eleutherococcus senticosus* affects scores of genes involved in the adaptive stress response and adaptive stress-response signaling pathways, including corticotropin-releasing hormone, cAMP-mediated, protein kinase A, and CREB, nitric oxide synthase, MAPK, and other pathways involved in stress modulation and neuroinflammation.[3] *In vitro, Eleutherococcus senticoccus* is able to ameliorate effects of stress by dampening corticosterone-induced neurotoxicity. PC12 cells – a neuron-like cell line derived from a pheochromocytoma of the rat adrenal medulla – suffer damage when exposed to high concentrations of corticosterone, and this is taken as an *in vitro* model of depression. Treatment with extract of *Eleutherococcus senticoccus* increased cell viability, decreased lactate dehydrogenase release (a sign of loss of cell membrane integrity), suppressed the apoptosis of PC12 cells, attenuated the intracellular Ca2+ overloading, and upregulated the BDNF mRNA level and CREB protein expression compared with the corresponding corticosterone-treated group.[4]

7.24.1 ANTIOXIDANT

Aqueous extracts of *Eleutherococcus senticosus* have high total phenol contents and radical scavenging activity. Such extracts were shown to protect human lymphocytes from the oxidative DNA damage caused by exposure to H_2O_2.[5] Treatment with *Eleutherococcus senticosus* also increased levels of mitochondrial glutathione peroxidase and reduced glutathione, as well as attenuated conversion of NO to ONO2- by hydrogen peroxide.[6]

Seven days of oral treatment with *Eleutherococcus senticosus* gave substantial hepatoprotective effects to rats subjected to CCl_4. Treatment increased levels of the antioxidant enzymes, superoxide dismutase, catalase, and glutathione peroxidase in liver homogenates, as well as reduced degree of tissue damage.[7]

7.24.2 ANTI-INFLAMMATORY

Extracts of *Eleutherococcus senticosus* induce anti-neuroinflammatory and neuroprotective processes in cultured hippocampal and microglial cells through activation of Nrf2 and enhancement of HO-1 expression. These intracellular factors lie at the crux of oxidation and inflammation regulatory pathways. Nrf2 has a wide range of cytoprotective functions, including anti-oxidative stress, regulating glutathione synthesis, and anti-inflammatory effects. Heme oxygenase-1 serves as a powerful antioxidant, anti-inflammatory, and counter-agent to stresses of various types.[8] Activation of these factors by *Eleutherococcus senticosus* reduce lipopolysaccharide-induced inflammatory, nitrosative and oxidative damage in hippocampal and microglial cells.[9] Activation of the Nrf2/HO-1 signaling pathway in the brain is suspected of contributing to antidepressant effects.[10]

7.24.3 ANTIDIABETIC/ANTI-METABOLIC SYNDROME

Crude extract of *Eleutherococcus senticosus* reduces hyperglycemia in mice predisposed to Type II diabetes. This was found to be due, in part, to inhibition of α-glucosidase that in turn slows digestion of complex carbohydrates and absorption of glucose.[11]

Several specific phytochemicals purified from *Eleutherococcus senticosus* have also been shown to have antidiabetic and anti-Metabolic Syndrome effects. Syringin from *Eleutherococcus senticosus* decreased hyperglycemia in streptozotocin-induced diabetic rats. This was due in part to enhanced glycogen synthesis in the liver and stimulation of glucose uptake in muscle tissue.[12] Eleutheroside E purified from *Eleutherococcus senticosus* lowered insulin resistance in db/db mice that are genetically predisposed to Type 2 Diabetes. Eleutheroside E decreased blood glucose and serum insulin levels, as well as improved serum lipid profiles.[13]

7.24.4 PRECLINICAL ANTIDEPRESSANT-LIKE EFFECTS

Eleutherococcus senticosus shows antidepressant-like effects in intact animals. Of several so-called "adaptogens" used in Russian herbal medicine, *Eleutherococcus senticosus* was most effective in reducing immobility time in rats subjected to forced swim.[14] Others also observed this effect in rats in the forced swim test. They further isolated and identified phytochemicals in the extract of the plant, and determined that syringin was likely most responsible for the antidepressant-like effect.[15] Treatment with aqueous extracts of *Eleutherococcus senticosus* also significantly prolonged the swimming time of mice in the forced swim test, as well as inhibiting the reduction of NK activity and the corticosterone elevation induced by the test.[16]

7.24.5 HUMAN ANTIDEPRESSANT EFFECTS

There are limited human trials suggesting that *Eleutherococcus senticosus* can improve symptoms of depression. Subjects described as suffering "neurosis" had significantly improved sleep, well-being, appetite, stamina, cognitive function, and mood after four weeks of treatment with *Eleutherococcus senticosus*.[17] In a randomized study of *Eleutherococcus senticosus* on quality of life in elderly patients over 65 years of age, improvements in social functioning, mental health, and cognitive function were observed after four weeks of therapy, but these differences did not persist to the eight-week time point.[18]

Eleutherococcus senticosus in combination with lithium relieved symptoms in patients suffering bipolar depression. Its ability to relieve depression was similar to that of the combination of lithium and fluoxetine. However, unlike the fluoxetine, in which several patients switched into mania, no occurrences of mania were seen in those that received *Eleutherococcus senticosus*.[19]

7.24.6 DOSAGE

Doses of *Eleutherococcus senticosus* extract standardized to contain 2.24 mg eleutherosides improved fatigue in human subjects at doses of four 500 mg capsules per day.[20] In a test for enhancement of athletic performance, subjects received 2 ml (150 mg of the dried material) of a 33% ethanol extract of *Eleutherococcus senticosus* twice daily for eight days. This treatment improved performance and work capacity.[21]

7.24.7 TOXICITY

The oral LD50 of the 33% ethanolic extract is estimated to be 14.5 g/kg, which extrapolated to a 75 kg human adult would be approximately 1,000 g. The extract is not considered to be teratogenic in mice at 10 mg/kg.[22]

7.24.8 SAFETY IN PREGNANCY

The *Botanical Safety Handbook* notes no compelling evidence of adverse effects from using this herb during pregnancy. Animals studies are benign.[23]

7.24.9 DRUG INTERACTIONS

The *Botanical Safety Handbook* notes no known interactions with drugs. A study using normal human volunteers showed that standardized extracts of *Eleutherococcus senticosus* at generally recommended doses are unlikely to alter the metabolism of medications dependent on the CYP2D6 or CYP3A4 pathways for elimination.[24]

REFERENCES

1. Brekhman II, Dardymov IV. New substances of plant origin which increase nonspecific resistance. *Annu Rev Pharmacolog* 1969;9:419–430.
2. Sun Y-L, Lin-De L, Soon-Kwan H. *Eleutherococcus senticosus* as a crude medicine: Review of biological and pharmacological effects. *J Med Plants Res* 2011;5(25):5946–5952.
3. Panossian A, Seo E-J, Efferth T. Novel molecular mechanisms for the adaptogenic effects of herbal extracts on isolated brain cells using systems biology. *Phytomedicine* 2018;50:257–284.
4. Wu F, Li H, Zhao L, et al. Protective effects of aqueous extract from *Acanthopanax senticosus* against corticosterone-induced neurotoxicity in PC12 cells. *J Ethnopharmacol* 2013;148(3):861–868.
5. Park H-R, Park E, Rim A-R, et al. Antioxidant activity of extracts from *Acanthopanax senticosus*. *Afr J Biotechnol* 2006;5(23):2388–2396.
6. Vaško L, Vašková J, Fejerčáková A, et al. Comparison of some antioxidant properties of plant extracts from *Origanum vulgare, Salvia officinalis, Eleutherococcus senticosus* and *Stevia rebaudiana. In vitro Cell Dev-An* 2014;50:614–622.
7. Lee S, Son D, Ryu J, et al. Anti-oxidant activities of *Acanthopanax senticosus* stems and their lignan components. *Arch Pharm Res* 2004;27(1):106–110.
8. Jiang L-J, Zhang SM, Li CW, et al. Roles of the Nrf2/HO–1 pathway in the anti-oxidative stress response to ischemia reperfusion brain injury in rats. *Eur Rev Pharmacol* 2017;21(7):1532–1540.
9. Jin ML, Park SY, Kim YH, et al. *Acanthopanax senticosus* exerts neuroprotective effects through HO–1 signaling in hippocampal and microglial cells. *Environ Toxicol Pharmacol* 2013;35(2):335–346.
10. Zhang F, Fu Y-Y, Zhou X, et al. Depression-like behaviors and heme oxygenase-1 are regulated by lycopene in lipopolysaccharide-induced neuroinflammation. *J Neuroimmunol* 2016;298:1–8.
11. Watanabe K, Kamata K, Sato J, Takahashi T. Fundamental studies on the inhibitory action of *Acanthopanax senticosus* Harms on glucose absorption. *J Ethnopharmacol* 2010;132(1):193–199.

12. Niu H-S, Liu I-M, Cheng J-T, et al. Hypoglycemic effect of syringin from *Eleutherococcus senticosus* in streptozotocin-induced diabetic rats. *Planta Med* 2008;74(2):109–113.

13. Ahn J, Um MY, Lee H, et al. Eleutheroside E, an Active Component of *Eleutherococcus senticosus*, Ameliorates insulin Resistance in Type 2 Diabetic db/db Mice. *Evid-Based Compl Alt* 2013:Article ID 934183.

14. Kurkin VA, Dubishchev AV, Ezhkov VN, et al. Medicinal Plants: Antidepressant activity of some phytopharmaceuticals and phenylpropanoids. *Khimfarm Z* 2006;40(11):33–38.

15. Nishibe S, Kinoshita H, Takeda H, Okano G. Phenolic compounds from stem bark of *Acanthopanax senticosus* and their pharmacological effect in chronic swimming stressed rats. *Chem Pharmaceut Bull* 1990;38(6):1763–1765.

16. Kimura Y, Sumiyoshi M. Effects of various *Eleutherococcus senticosus* cortex on swimming time, natural killer activity and corticosterone level in forced swimming stressed mice. *J Ethnopharmacol* 2004;95(2–3):447–453.

17. Panossian AG. Adaptogens in mental and behavioral disorders. *Psychiat Clin N Am* 2013;36(1):49–64.

18. Cicero AFG, Derosa G, Brillante R, et al. Effects of Siberian ginseng (*Eleutherococcus senticosus*) on elderly quality of life: A randomized trial. *Arch Gerontol Geriatr*; Suppl. 2004;9:69–73.

19. Weng S-H, Tang J, Wang G, et al. Comparison of the addition of Siberian ginseng (*Acanthopanax senticosus*) Versus fluoxetine to lithium for the treatment of bipolar disorder in adolescents: A randomized, double-blind trial. *Curr Ther Res* 2007;68(4):280–290.

20. Hartz AJ, Bentler S, Noyes R, et al. Randomized controlled trial of Siberian ginseng for chronic fatigue. *Psychol Med* 2004;34(1):51–61.

21. Asano K, Takahashi T, Miyashita M, et al. Effect of *Eleutherococcus senticosus* extract on human physical working capacity. *Planta Med* 1986;3(3):175–177.

22. Halstead BW, Hood LL. *Eleutherococcus senticosus/Siberian Ginseng: An Introduction to the Concept of Adaptogenic Medicine*. Long Beach, CA: Oriental Healing Arts Institute; 1984:65.

23. Gardner Z, McGuffin M, editors. *Botanical Safety Handbook, 3rd edition*. Boca Raton, FL: CRC Press; 2013:321.

24. Donovan JL, DeVane CL, Chavin KD, et al. Siberian ginseng (*Eleutheroccus senticosus*) effects on CYP2D6 and CYP3A4 activity in normal volunteers. *Drug Metab Dispos* 2003;31(5):519–522.

7.25 *EPIMEDIUM BREVICORNUM* (HORNY GOAT WEED)

Epimedium brevicornum, often referred to as horny goat weed, is an important medicinal plant that has been used in Western herbalism and in various traditional Chinese formulations for thousands of years. Hundreds of phytochemicals, mostly flavonoids, have been identified in the genus *Epimedium*, with icariin being the most abundant. Icariin is pharmacologically active and has been used to strengthen bones, nervous system, and cardiovascular function. It is thought to possess anti-cancer, anti-inflammatory, and immunoprotective effects. *Epimedium brevicornum* is particularly well known in regards to its alleged ability to restore libido and sexual performance, hence the reference, horny goat weed.[1]

7.25.1 ANTIOXIDANT

Extracts of *Epimedium brevicornum* were shown to exert significant antioxidant effects in three *in vitro* assays, the DPPH, phosphomolybdenum, and ferric reducing assays. These effects correlated with its high phenolic content.[2] In *ex vivo* testing of liver tissue from animals that had been chronically treated with methanolic extract of *Epimedium brevicornum*, the extract significantly preserved glutathione levels after challenge with the oxidizing agent tert-butylhydroperoxide.[3]

7.25.2 Anti-Inflammatory

Epimedium brevicornum has been shown to have anti-inflammatory properties that could contribute to antidepressant effects. A flavone, 2″-hydroxy-3″-en-anhydroicaritin, isolated from the herb prevents lipopolysaccharide-induced production of nitric oxide and prostaglandin E2 and inhibits COX-2 in a mouse macrophage line. This effect is mediated by blocking NF-κB and MAPK signaling.[4] Icariin, another flavone of *Epimedium brevicornum*, also acts as an anti-inflammatory agent in cultured human nucleus pulposus cells. Icariin blocks IL-1beta-induction of COX-2 and iNOS and increases production of prostaglandin E2 and nitric oxide. Icariin suppressed IL-1β-induced activation of the inflammatory MAPK and NF-κB signaling pathways.[5]

7.25.3 Antidiabetic/Anti-Metabolic Syndrome

Chronic stress decreases insulin sensitivity in the hypothalamic arcuate nucleus of rats. It appears to do so by enhancing the corticotrophin releasing hormone that activates SOCS3, which in turn blocks the insulin signaling pathway through the suppression of both insulin receptor substrate 2 (IRS2) and phosphatidylinositol-3-kinase (PI3-K) activation in the arcuate nucleus hypothalamic ARC. Both icariin and fluoxetine dampen CRF system hyperactivity and thus restore insulin signaling in the hypothalamus. Overall glucose tolerance was also improved.[6]

The icariin metabolites icaritin and icariside II from the closely related herb, *Epimedium koreanum*, were evaluated for their ability to inhibit protein tyrosine phosphatase 1B (PTP1B) and α-glucosidase. Inhibitors of these enzymes are known to enhance insulin sensitivity.[7,8] Results showed potent inhibitory activities with 50% inhibition concentrations in the low micromolar range.[9]

Chronic unpredictable stress increases corticotropin releasing hormone expression levels in rat hypothalamus, and decreases glucocorticoid and 5-HT1A receptors in the hippocampus and frontal cortex. Behaviorally, it reduces sucrose intake. Icariin reversed those behavioral and neurochemical effects of stress.[10] In a similar study, icariin attenuated increases in serum corticotropin-releasing hormone and corticosterone levels in chronically stressed rats, as well as reversing serum elevations of IL-6 and TNF-a.[11]

Icariin also protects brain cells from the damaging effects of glucocorticoid. Exposure of cultured rat hypothalamic cells to high concentrations of corticosterone causes cell death due to decreases in mitochondrial membrane potential, increase in caspase-3 activity, elevation of intracellular reactive oxygen species, and decreased superoxide dismutase activity. Pretreatment of cells with icariin suppresses these corticosterone-induced events, and this effect of icariin appears to be mediated by activation of the PI3-K/Akt pathway.[12]

Another effect of icariin that could improve mood and combat depression is its ability to enhance serum testosterone levels.[13] One study showed, along with increased serum testosterone, increased hypothalamic DA and 5-HT levels, and eNOS, and increased activity of the PI3K/ AKT pathway in penile tissue.[14] Interestingly, icariin likely enhanced erectile function in these animals through increasing endothelial nitric oxide, which itself has been linked to an antidepressant effect.[15]

7.25.4 PRECLINICAL ANTIDEPRESSANT-LIKE EFFECTS

A variety of studies have found icariin to be an effective antidepressant in animal models of depression. Chronic unpredictable stress produces a depression-like syndrome in rats that is mediated in part by activation of the hippocampal NLRP3 inflammasome. Rats exposed to unpredictable showed decreases sucrose preference test and increased immobility in the forced swim test that are taken as rodent analogues of human depression. Icariin reversed these behavioral indications of depression. These animals also showed activation of the inflammasome as indicated by increases in oxidative-nitrosative stress markers, tumor necrosis factor-alpha, interleukin-1β (IL-1β), activation of the nuclear factor kappa B (NF-κB) signaling pathway, and increased inducible nitric oxide synthase (iNOS) mRNA expression in the hippocampus. Icariin freely crossed the blood-brain barrier, downregulated the activation of hippocampal nod-like receptor protein 3 (NLRP3) inflammasome/caspase-1/IL-1β axis, and thus reversed the above-noted neurochemical effects of stress.[16]

Icariin is also effective in the animal model of glucocorticoid-induced depression. Rats subjected to 21 daily injections of corticosterone exhibited reduced sucrose intake and increased immobility time in the forced swim test (FST). These animals also had reduced hippocampal levels of brain-derived neurotrophic factor. Icariin significantly increased sucrose intake and decreased the immobility time in the forced swim test, suggesting its potent antidepressant activity. Decreased hippocampal BDNF levels were restored.[17]

Murine studies have shown antidepressant efficacy of *Epimedium brevicornum*. Pretreatment with extracts of *Epimedium brevicornum* significantly reduced the duration of murine immobility in subsequent tail suspension and forced swim tests. Treatment was also found to inhibit MAO-A and MAO-B activities in the brains and livers of the mice.[18]

7.25.5 HUMAN ANTIDEPRESSANT EFFECTS

There is only one report on antidepressant effects of *Epimedium brevicornum* or its phytochemical constituents in humans, and this is in regard to bipolar depression. In this small study, individuals suffering alcoholism as well as depression from either bipolar type I or type II were treated daily with icariin for eight weeks. By the end of the study, participants exhibited significant reductions in scores in the Hamilton Rating Scale for Depression, Quick Inventory of Depressive Symptomatology-Self Report (QIDS-SR), and Hamilton Rating Scale for Anxiety (HAMA). Heavy drinking days also decreased significantly. Icariin was well tolerated and no participants withdrew due to side-effects.[19]

7.25.6 Dosage

Pursell recommends infusions of 1–2 teaspoons of dried herb per cup, taking 1–2 cups per day. It may also be taken as tincture 5–30 drops 1–3 times a day.[20]

7.25.7 Toxicity

Epimedium brevicornum is a traditional Chinese medicine, and no adverse reactions were reported from clinical practice. In animal studies, the median lethal dose (LD50) of epimedium water extract is more than 80 g/kg (or over 5,6000g for a 75 kg human adult), and all the results, including mouse bone marrow micronucleus test, Ames test, and TK Gene mutation test were negative.[21]

7.25.8 Safety in Pregnancy

The *Botanical Safety Handbook* notes there is no definitive guidance from the literature concerning the safety of *Epimedium brevicornum* during pregnancy.[22]

7.25.9 Drug Interactions

The *Botanical Safety Handbook* notes no known interactions with drugs. *Epimedium brevicornum* extracts inhibited activities of CYP2C19, CYP2E1, CYP2C9, CYP3A4, CYP2D6, and CYP1A2. However, fairly high concentrations were required, making it unlikely to be clinically significant. Nonetheless, the authors of the paper in which the data were reported did not rule out that possibility.[23]

REFERENCES

1. Li C, Li Q, Mei Q, Lu T. Pharmacological effects and pharmacokinetic properties of icariin, the major bioactive component in Herba Epimedii. *Life Sci.* 2015;126:57–68.
2. Asif N, Amjad S, Hussain K, et al. Physicochemical characterization and antioxidant activity of extract of Epimedium grandiflorum. *J Pharm Chem.* 2016;10(2):11–16.
3. Yim TK, Ko KM. Antioxidant and immunomodulatory activities of Chinese tonifying herbs. *Pharm Biol.* 2002;40(5):329–335.
4. Ci X, Liang X, Luo G, et al. Regulation of inflammatory mediators in lipopolysaccharide-stimulated RAW 264.7 cells by 2″-hydroxy-3″-en-anhydroicaritin involves down-regulation of NF-κB and MAPK expression. *Int Immunopharmacol.* 2010;10(9):995–1002.
5. Hua W, Zhang Y, Wu X, et al. Icariin attenuates interleukin-1β-induced inflammatory response in human nucleus pulposus cells. *Curr Pharm Des.* 2017;23(39):6071–6078.
6. Pan Y, Hong Y, Zhang QY, Kong LD. Impaired hypothalamic insulin signaling in CUMS rats: Restored by icariin and fluoxetine through inhibiting CRF system. *Psychoneuroendocrinology.* 2013;38(1):122–134.
7. Través PG, Pardo V, Pimentel-Santillana M, et al. Pivotal role of protein tyrosine phosphatase 1B (PTP1B) in the macrophage response to pro-inflammatory and anti-inflammatory challenge. *Cell Death Dis.* 2014;5:e1125.
8. van de Laar F, Lucassen PL, Akkermans RP, Van De Lisdonk EH, Rutten GE, Van Weel C. α-glucosidase inhibitors for patients with type 2 diabetes. *Diabetes Care.* 2005;28(1):154–163.

9. Kim DH, Jung HA, Sohn HS, Kim JW, Choi JS. Potential of icariin metabolites from Epimedium koreanum Nakai as antidiabetic therapeutic agents. *Molecules.* 2017;22(6):986–1000.

10. Pan Y, Wang FM, Qiang LQ, Zhang DM, Kong LD. Icariin attenuates chronic mild stress-induced dysregulation of the LHPA stress circuit in rats. *Psychoneuroendocrinology.* 2010;35(2):272–283.

11. Pan Y, Zhang WY, Xia X, Kong LD. Effects of icariin on hypothalamic-pituitary-adrenal axis action and cytokine levels in stressed Sprague-Dawley rats. *Biol Pharm Bull.* 2006;29(12):2399–2403.

12. Zhang H, Liu B, Wu J, et al. Icariin inhibits corticosterone-induced apoptosis in hypothalamic neurons via the PI3-K/Akt signaling pathway. *Mol Med Rep.* 2012;6(5):967–972.

13. Zhang Z-B, Yang QT. The testosterone mimetic properties of icariin. *Asian J Androl.* 2006;8(5):601–605.

14. Ding J, Tang Y, Tang Z, et al. Icariin improves the sexual function of male mice through the PI3K/AKT/eNOS/NO signaling pathway. *Andrologia.* 2018;50(1):e12802.

15. Dhir A, Kulkarni K. Nitric oxide and major depression. *Nitric Oxide.* 2011;24(3):125–131.

16. Liu B, Xu C, Wu X, et al. Icariin exerts an antidepressant effect in an unpredictable chronic mild stress model of depression in rats and is associated with the regulation of hippocampal neuroinflammation. *Neuroscience.* 2015;294:193–205.

17. Gonga MJ, Han B, Wang SM, Liang SW, Zou ZJ. Icariin reverses corticosterone-induced depression-like behavior, decrease in hippocampal brain-derived neurotrophic factor (BDNF) and metabolic network disturbances revealed by NMR-based metabonomics in rats. *J Pharm Biomed.* 2016;123:63–73.

18. Zhong H-B, Pan Y, Kong L. Antidepressant effect of *Epimedium brevicornum* extracts. *Chin Tradit Herb Drugs.* 2005;36:1506–1510.

19. Xiao H, Wignall N, Brown ES. An open-label pilot study of icariin for co-morbid bipolar and alcohol use disorder. *Am J Drug Alcohol Abus.* 2016;42(2):162–167.

20. Pursell JJ. *The Herbal Apothecary.* Portland, OR: Timber Press; 2016. p. 116.

21. Sui HX, Gao P, Xu HB. The safety evaluation of *Herba Epimedii* water extract. *Carcinog Teratog Mutagen.* 2006;18:439.

22. Gardner Z, McGuffin M, editors. *Botanical Safety Handbook.* 3rd ed. Boca Raton, FL: CRC Press; 2013. p. 328.

23. Liu KH, Kim MJ, Jeon BH, et al. Inhibition of human cytochrome P450 isoforms and NADPH-CYP reductase in vitro by 15 herbal medicines, including Epimedii herba. *J Clin Pharm Ther.* 2006;31(1):83–91.

7.26 *FOENICULUM VULGARE* (FENNEL)

Foeniculum vulgare (fennel) is a perennial plant in the carrot family. It is native to the shores of the Mediterranean, but has become widely cultivated in many parts of the world. Its leaves and seeds are used as spice, whereas its celery-like base is consumed as a vegetable. It has also been used as one of the main flavorings in the alcoholic drink, absinthe. The highly aromatic plant is rich in phytochemicals, among which are fenchone, estragole, eugenol, p-anisaldehyde, α-phellandrene, trans-anethole, quercetin, kaemferol, caffeoylquinic acid, limonene, linalool, cineol, thujene, myrcene, pinene, rosmarinic acid, luteolin, and many others.[1] Many of those have significant medicinal effects and have been shown to exert antifungal, antibacterial, antioxidant, antithrombotic, and hepatoprotective activities, and it has been widely used as carminative, digestive, lactogogue, diuretic, and in treating various respiratory and gastrointestinal disorders.[2] There is compelling data showing antidepressant effects of *Foeniculum vulgare* in both preclinical and clinical studies.

7.26.1 ANTIOXIDANT

The phytochemicals in extract of *Foeniculum vulgare* include potent antioxidants as demonstrated by the *in vitro* DPPH scavenging and FRAP reducing assays. The methanolic extract of *Foeniculum vulgare* demonstrated a hydroxyl group scavenging capacity that was 76% that of ascorbic acid.[3] The extract also exhibited potent antioxidant effect in inhibition of lipid peroxidation in both rat liver and brain homogenates. Those effects exceeded the antioxidant effects of ascorbic acid.[4] In mice, extract of *Foeniculum vulgare* reversed lead-induced decreases of the antioxidant enzymes superoxide dismutase and peroxiredoxin 6 in in the cortex and hippocampus.[5]

7.26.2 ANTI-INFLAMMATORY

Foeniculum vulgare and its chemical constituents are known to exert anti-inflammatory effects. This appears to occur at a deep, cellular level. Acute administration of extract of *Foeniculum vulgare* reduced damage in mice that had developed acute inflammatory lung injury after injection of lipopolysaccharide. *Foeniculum vulgare* reduced numbers of activated immune cells and suppressed production of the inflammatory cytokines IL-6 and TNF-alpha. Activity of the proinflammatory mediator matrix metalloproteinase 9 was reduced, as was that of the immune modulator nitric oxide. *Foeniculum vulgare* also significantly decreased the phosphorylation of ERK, that in turn dampened activity of the inflammatory actor, NF-κB.[6] In a very similar study, one of the main constituents of the essential oil of *Foeniculum vulgare*, trans-anethole, was administered to mice that had been treated with lipopolysaccharide. Again, inflammatory lung injury was diminished. Trans-anethole decreased levels of the inflammatory cytokine, IL-17, but increased levels of the anti-inflammatory IL-10. Numbers of macrophages and eosinophils were also reduced.[7]

   Foeniculum vulgare also attenuated two inflammatory processes that are mediated by cyclooxygenase and prostaglandins. Acute oral administration of *Foeniculum vulgare* extract reduced both carrageenan-induced paw edema and arachidonic acid-induced ear edema in mice. The effects were equal to those of the standard nonsteroidal anti-inflammatory drug indomethacin.[8]

7.26.3 ANTIDIABETIC/ANTI-METABOLIC SYNDROME

Foeniculum vulgare exhibits some insulin-enhancing, antidiabetic effects, though these are not prominent effects of the herb. Treatment with extract of *Foeniculum vulgare* reduced hyperglycemia in streptozocin-induced diabetic rats. Curiously, it did not affect glucose levels in control animals.[9] In rats with obesity induced by high fat diets, a model of Metabolic Syndrome, methanolic extract of *Foeniculum vulgare* reduced hyperlidemia, hyperinsulinemia, hyperglycemia, and hyperleptinemia.[10] Leptin-resistance, like insulin resistance, may contribute to MDD.[11]

7.26.4 PRECLINICAL ANTIDEPRESSANT-LIKE EFFECTS

Studies in rodents have shown that *Foeniculum vulgare* has antidepressant-like effects. Acute intraperitoneal administration of ethanolic extract of *Foeniculum vulgare* decreased immobility of rats in the forced swim and tail suspension test. The antidepressant effect of *Foeniculum vulgare* was slightly less than that of acutely administered fluoxetine.[12] Acute oral administration of methanolic extract of *Foeniculum vulgare* also decreased immobility of rats in the forced swim test. In the same study, brain tissue of mice treated with the extract of *Foeniculum vulgare* showed increases in glutathione levels, but decreases in markers of nitration and lipid oxidation.[13] Three weeks of oral administration of *Foeniculum vulgare* oil also decreased immobility of rats in the forced swim test in a manner similar to that of fluoxetine.[14] Chronic administration of *Foeniculum vulgare* extract has been found affect concentrations of neurotransmitters in various areas of rat brain. Monoamines were uniformly increased throughout the brain, as was acetylcholine. In fact, the extract was found to significantly inhibit acetylcholinesterase.[15]

7.26.5 HUMAN ANTIDEPRESSANT EFFECTS

There are no studies of effects of *Foeniculum vulgare* in human subjects diagnosed with MDD. However, there are reports of effects of *Foeniculum vulgare* in women experiencing discomforts of menopause and premenstrual syndrome. Menopausal women given 200 mg of *Foeniculum vulgare* oil daily for eight weeks experienced significant improvements in self-reported general quality of life, including psychological and sexual aspects, in comparison to baseline and control values.[16] In a similar study, menopausal women chronically treated with *Foeniculum vulgare* showed improvements in anxiety and depression scores, but these apparent improvements did not reach statistical significance.[17]

Reports on effects of *Foeniculum vulgare* on mood in premenstrual syndrome have been quite inconsistent. In one study, twice-daily treatment with *Foeniculum vulgare* for a month provided significant improvement in mood.[18] In another study, treatment with *Foeniculum vulgare* significantly improved symptoms of premenstrual syndrome that included mood components, but the results did not include which specific components of the syndrome improved.[19] In yet another study, modest improvements in mood and anxiety were noted, but they did not reach significance.[20] *Foeniculum vulgare* also failed to improve mood when used in women with dysmenorrhea.[21]

7.26.6 DOSAGE

The German E Commission notes the daily dose of *Foeniculum vulgare* oil to be 0.1–0.6 ml, which is equal to 0.1–0.6 g of herb. Their suggested dose for *Foeniculum vulgare* seed is 5–7 g.[22]

7.26.7 TOXICITY

The LD50 of essential oil of *Foeniculum vulgare* in mice is reported to be 1.038 ml/kg.[23] This would be extrapolated to approximately 70 ml for a 75 kg adult human.

7.26.8 SAFETY IN PREGNANCY

The editors of the *Botanical Safety Handbook* stated their belief that *Foeniculum vulgare* is safe for use during pregnancy and lactation.[24]

7.26.9 DRUG INTERACTIONS

The *Botanical Safety Handbook* notes no known interactions of *Foeniculum vulgare* with commonly used drugs in human subjects. However, extract of seed of *Foeniculum vulgare* has been found to inhibit the CY3A4 enzyme from human liver microsomes.[25]

REFERENCES

1. Badgujar SB, Patel VV, Bandivdekar AH. *Foeniculum vulgare* Mill: A review of its botany, phytochemistry, pharmacology, contemporary application, and toxicology. *BioMed Res Int.* 2014:1–32.
2. Rather MA, Dar BA, Sofi SN, Bhat BA, Qurishi MA. Foeniculum vulgare: A comprehensive review of its traditional use, phytochemistry, pharmacology, and safety. *Arab J Chem.* 2016;9(Suppl 2):S1574–S1583.
3. Chatterjee S, Goswami N, Bhatnagar P. Estimation of phenolic components and in vitro antioxidant activity of fennel (Foeniculum vulgare) and ajwain (Trachyspermum ammi) seeds. *Adv Biores.* 2012;3(2):109 –118.
4. Koppula S, Kumar H. *Foeniculum vulgare* Mill (*Umbelliferae*) attenuates stress and improves memory in wister rats. *Trop J Pharm Res.* 2013;12(4):553–558.
5. Bhatti S, Ali Shah SA, Ahmed T, Zahid S. Neuroprotective effects of *Foeniculum vulgare* seeds extract on lead-induced neurotoxicity in mice brain. *Drug Chem Toxicol.* 2018;41(4):4399–4407.
6. Lee HS, Kang P, Kim KY, Seol GH. *Foeniculum vulgare* Mill. protects against lipopolysaccharide-induced acute lung injury in mice through ERK-dependent NF-κB activation. *Korean J Physiol Pharmacol.* 2015 Mar;19(2):183–189.
7. Zhang S, Chen X, Devshilt I, et al. Fennel main constituent, trans-anethole treatment against LPS-induced acute lung injury by regulation of Th17/Treg function. *Mol Med Rep.* 2018;18(18):21369–21376.
8. Choi EM, Hwang JK. Antiinflammatory, analgesic and antioxidant activities of the fruit of Foeniculum vulgare. *Fitoterapia.* 2004;75(6):557–565.
9. Mhaidat NM, Abu-zaiton AS, Alzoubi KH, Alzoubi W, Alazab RS. Antihyperglycemic properties of *Foeniculum vulgare* extract in streptozocin-induced diabetes in rats. *Int J Pharmacol.* 2015;11(1):72–75.
10. Shahat AA, Ahmed HH, Hammouda FM, Ghaleb H. Regulation of obesity and lipid disorders by *Foeniculum vulgare* extracts and Plantago ovata in high-fat diet-induced obese rats. *Am J Food Tech.* 2012;7(10):622–632.
11. Mendelson SD. *Metabolic Syndrome and Psychiatric Illness: Interactions, Pathophysiology, Assessment and Treatment.* London: Academic Press; 2008.
12. Josephine IG, Elizabeth AA, Muniappan M, Muthiah NS. Antidepressant activity of *Foeniculum vulgare* in forced swimming and tail suspension test. *Res J Pharm Biol Chem Sci.* 2014;5(2):448–454.
13. Singh JN, Sunil K, Rana AC. Antidepressant activity of methanolic extract of *Foeniculum vulgare* (fennel) fruits in experimental animal models. *J App Pharm Sci.* 2013;3(09):65–70.

14. Perveen T, Emad S, Ahmad S, et al. Fennel oil treatment mimics the anti-depressive and anxiolytic effects of fluoxetine without altering the serum cholesterol levels in rats. *Pak J Zool.* 2017;49(6):2291–2297.
15. Mona AR, Seham M, Amira AB. Study on the effect of fennel extract on some neurotransmitters and their related ions in different brain areas of adult male albino rats. *Bull Fac Sci Cairo Univ.* 2010;78:1–29.
16. Kian RF, Bekhradi R, Rahimi R, Golzareh P, Mehran A. Evaluating the effect of fennel soft capsules on the quality of life and its different aspects in menopausal women: A randomized clinical trial. *Nurs Pract Today.* 2017;4(2):87–95.
17. Ghazanfarpour M, Mohammadzadeh F, Shokrollahi P, et al. Effect of *Foeniculum vulgare* (fennel) on symptoms of depression and anxiety in postmenopausal women: A double-blind randomised controlled trial. *J Obstet Gynaecol.* 2018;38(1):121–126.
18. Omidali F. The effect of pilates exercise and consuming fennel on pre-menstrual syndrome symptoms in non-athletic girls. *CMJA.* 2015;5(2):1203–1213.
19. Pazoki H, Bolouri G, Farokhi F, Azerbayjani MA. Comparing the effects of aerobic exercise and *Foeniculum vulgare* on pre-menstrual syndrome. *Middle East Fertil Soc J.* 2016;21(1):61–64.
20. Delaram M, Kheiri S, Hodjati MR. Comparing the effects of echinophora-platyloba, fennel and placebo on pre-menstrual syndrome. *J Reprod Infertil.* 2011;12(3):221–226.
21. Khorshidi N, Ostad SN, Mosaddegh M, Soodi M. Clinical effects of fennel essential oil on primary dysmenorrhea. *Iran J Pharm Res.* 2003;6(2):89–93.
22. Blumenthal M, Busse WR, Goldberg A, et al. *The Complete German Commission E Monograph: Therapeutic Guide to Herbal Medicines.* Austin, TX: American Botanical Council. 1998; p. 128–129.
23. Özbek H. Investigation of the level of the lethal dose 50 and the hypoglycemic effect of *Foeniculum vulgare* Mill. Fruit essential oil extract in healthy and diabetic mice. *Van Tip Derg.* 2002;9(4):98–103.
24. Gardner Z, McGuffin M, editors. *Botanical Safety Handbook.* 3rd ed. Boca Raton, FL: CRC Press. 2013; p. 361.
25. Zaidi SFH, Kadota S, Tezuka Y, Tezuka Y. Inhibition on human liver cytochrome P450 3A4 by constituents of fennel (Foeniculum vulgare): Identification and characterization of a mechanism-based inactivator. *J Agric Food Chem.* 2007;55(25):10162–10167.

7.27 GINKGO BILOBA

Ginkgo biloba, also known as the maidenhair tree, is one of the oldest species of tree in the world. Indeed, the tree is the only living species in the division *Ginkgophyta*, all others being extinct. Native to China, the tree is widely cultivated, and was cultivated early in human history. The trees can grow more than 130 feet tall and can live for over 1,000 years. Some trees in China are said to be over 2,500 years old. The seeds of ginkgo can be eaten, but it is the leaves that serve as a source of medicine.

A variety of health benefits have been attributed to *Ginkgo biloba*, and among the many uses are treating poor memory, dementia, dizziness, depression, cerebrovascular and cardiovascular insufficiency, oxidative damage, low cellular energy levels, inflammation, migraine, allergies, and asthma.[1]

The leaves of *Ginkgo biloba* contain a variety of flavonol glycosides, biflavones, proanthocyanidins, alkylphenols, simple phenolic acids, and polyprenols. Some of these substances, such as quercetin, isorhamnetin, myricetin, rutin, luteolin, and kaempferol, are common in medicinal plants. *Ginkgo biloba* also contains a number of unique phytochemicals, e.g., the terpene trilactones, ginkgolides A, B, C, J, and bilobalide.[2]

7.27.1 ANTIOXIDANT

In vitro, extracts of ginkgo leaves were shown to have substantial antioxidant capacity in the DPPH radical scavenging assay. These effects were attributed to its content of protocatechuic and p-coumaric acids, quercetin, rutin, isoginkgetin, catechin, and other phenolics.[3] This property allowed *Ginkgo biloba* to protect cultured rat cerebellar neurons from the oxidative damage of hydrogen peroxide.[4]

In vivo, administration of extract of *Ginkgo biloba* to rats prior to and during subchronic injection of CCl_4 protected their livers from oxidative damage. Treatment reduced serum levels of transaminases and levels of markers of lipid peroxidation in liver homogenates. At the same time, treatment increased glutathione and activities of the antioxidant enzymes superoxide dismutase, catalase, glutathione peroxidase, and glutathione reductase in liver.[5]

In gerbils subjected to cerebral ischemia, pretreatment with extract of *Ginkgo biloba* reduced oxidative damage in brain tissue. Levels of malondialdehyde were reduced. Moreover, the increases of nitrite and nitrate observed in the hippocampus after cerebral ischemia were dose-dependently reduced in animals pretreated with Ginkgo.[6]

7.27.2 ANTI-INFLAMMATORY

Treatment with *Ginkgo biloba* dampened carrageenan-induced paw edema in rats. This inflammatory response is mediated by NF-κB and subsequent stimulation of COX, NOS, IL-6, and other inflammatory cytokines.[7] The inflammatory effects of lipopolysaccharide in lung tissue of rats were also attenuated by *Ginkgo biloba* extract. Neutrophil infiltration and increases in myeloperoxidase activity were decreased, as was lipid peroxidation. Ginkgo prevented the activation and translocation of NF-κB that initiates much of the cellular inflammatory response.[8]

7.27.3 ANTIDIABETIC/ANTI-METABOLIC SYNDROME

Along with antioxidant and anti-inflammatory effects, ginkgo also exerts antidiabetic effects. In streptozocin-induced diabetic rats, chronic treatment with extract of *Ginkgo biloba* normalized serum glucose and improved lipid profiles.[9] The herb also appears to enhance insulin sensitivity in cultured human liver cells. Addition of ginkgo extract to culture medium containing normal hepatocytes increased mRNA levels of PPARc, insulin receptor substrate, and glucose-6-phosphatase, all of which enhance the effects of insulin. In cultures of insulin resistant hepatocytes, ginkgo extract improved glucose consumption, much in the manner of the diabetic medication rosiglitazone.[10]

7.27.4 PRECLINICAL ANTIDEPRESSANT-LIKE EFFECTS

Preclinical studies have provided compelling evidence that *Ginkgo biloba* has antidepressant effects. Subchronic administration of extracts of *Ginkgo biloba* reduced immobility of rats in the forced swimming test and of mice in the tail suspension test. Imipramine produced stronger effects in these models.[11] Repeated oral administration of an extract of *Ginkgo Biloba* (EGB 761) to rats and mice also helped prevent so-called "learned helplessness" caused by unavoidable shock. However, it did not reverse that depression-like condition after it was established.[12] Acute administration of EGb 761 extract also partially blocked the depression-like reduction of sucrose intake induced by lipopolysaccharide.[13]

Several studies have found antidepressant-like effects of *Ginkgo biloba* and obtained neurochemical data to help determine the mechanisms by which those effects occur. In one such study, mice were treated with EGb761 for 17 days. After these animals were found to exhibit the expected antidepressant effect of that treatment – i.e., reduced immobility in the forced swim test – they were sacrificed and their brains studied. The antidepressant effects of *Ginkgo biloba* were associated with reductions in lipid peroxidation and superoxide radical production in the midbrain, hippocampus, and prefrontal cortex. Treatment also increased dopaminergic and serotonergic neurotransmission in those areas.[14] Daily treatment with diterpene ginkgolides purified from *Ginkgo biloba* also decreased immobility in the tail suspension test and increased sucrose consumption. Subsequent neurochemical analysis of hippocampal tissue showed enhanced neurotransmitter metabolism, reduction in oxidative stress, increased glutathione metabolism, improved lipid and glucose metabolism, and reductions in the neurotoxic metabolite kynurenic acid.[15]

7.27.5 HUMAN ANTIDEPRESSANT EFFECTS

Ginkgo biloba appears to improve cognitive function in humans, particularly those suffering various forms of dementia. In a prospective, randomized, double-blind, placebo-controlled study, the *Ginkgo biloba* extract EGb 761 significantly improved cognitive performance in patients diagnosed with Alzheimer's or multi-infarct dementias.[16] A review noted that virtually all studies of effects of chronic treatment with ginkgo extract on cognitive function in patients with cerebral insufficiency showed positive effects. They saw no marked differences between the effects of *Gingko biloba* and those of Hydergine (though the benefits of the latter drug are themselves questionable).[17] In a more recent review, it was concluded that "EGb761 at 240 mg/day is able to stabilize or slow decline in cognition, function, behavior, and global change at 22-26 weeks in cognitive impairment and dementia, especially for patients with neuropsychiatric symptoms."[18]

In healthy young human volunteers, a single dose of *Ginkgo biloba* reduced increases in blood pressure and cortisol release in response to physical and mental stresses. Such an effect could help provide resiliency against the development of MDD.[19] However, while there have been studies of effects of *Ginkgo biloba* on mood in humans, these have been in the context of improving mood in individuals with some form of dementia. For example, in a double-blind study, patients suffering from chronic cerebrovascular insufficiency one month of treatment with extract of *Ginkgo biloba* improved motor activity, speech comprehension and production, and mood.[20] In a similar study of elderly patients with mild cognitive impairment, *Ginkgo biloba* increased motivation and interest in activities of daily living.[21] In elderly patients with cerebral insufficiency but also formal diagnoses of MDD, the addition of *Ginkgo biloba* extract improved response to standard antidepressant treatment.[22] Beyond such studies, there have been no formal studies of the ability of *Ginkgo biloba* to treat otherwise healthy individuals in whom MDD is the primary diagnosis.

7.27.6 DOSAGE

The German E Commission recommends 120–240 mg of dried extract 2–3 times a day.[23] A number of authors have cautioned that the quality of various extracts if *Ginkgo biloba* can vary considerably. For this reason, standardized extracts such as the well-known EGb 761 have been developed and marketed. In such standardized extracts, high concentrations of ginkgolides A, B, C, J, and bilobalide are maintained while the mildly toxic ginkgolic acids are kept in concentrations less than 5 mg/kg (5 ppm).

7.27.7 TOXICITY

Ginkgo biloba appears to be quite non-toxic. In a study to determine LD50 of intra-peritoneal injection of *Ginkgo biloba* in rats, no animal deaths were seen until the dose of 8 g/kg. This dose is well beyond typical human therapeutic doses.[5]

7.27.8 SAFETY IN PREGNANCY

The *Botanical Safety Handbook* notes there to be no definitive guidance from the literature concerning the safety of *Ginkgo biloba* during pregnancy, as evidence is inconsistent.[24]

7.27.9 DRUG INTERACTIONS

Ginkgo biloba has been suspected of potentially dangerous interactions with medications that have included anticoagulants, MAO inhibitors, alprazolam, and haloperidol.[25] A study also revealed significant inductive effect of ginkgo on CYP2C19 activity. In one case, this may have resulted in sub-therapeutic blood levels of depakote and dilantin with a resulting fatal breakthrough seizure.[26]

REFERENCES

1. Chan P-O, Xia Q, Fu PP. *Ginkgo biloba* leave extract: Biological, medicinal, and toxicological effects. *J Environ Sci Health C*. 2007;25(3):211–244.
2. van Beek TA, Montoro P. Chemical analysis and quality control of *Ginkgo biloba* leaves, extracts, and phytopharmaceuticals. *J Chromatogr A*. 2009;1216(11):2002–2032.
3. Ellnain-Wojtaszek M, Kruczyński Z, Kasprzak J. Investigation of the free radical scavenging activity of *Ginkgo biloba* L. leaves. *Fitoterapia*. 2003;74(1–2):1–6.
4. Oyama Y, Chikahisa L, Ueha T, Kanemaru K, Noda K. *Ginkgo biloba* extract protects brain neurons against oxidative stress induced by hydrogen peroxide. *Brain Res*. 1996;712(2):349–352.

5. Naik SR, Panda VS. Antioxidant and hepatoprotective effects of *Ginkgo biloba* phytosomes in carbon tetrachloride-induced liver injury in rodents. *Liver Int.* 2007;27(3):393–399.

6. Calapai G, Crupi A, Firenzuoli F, et al. Neuroprotective effects of *Ginkgo biloba* extract in brain ischemia are mediated by inhibition of nitric oxide synthesis. *Life Sci.* 2000;67(22):2673–2683.

7. Hedayat I, Abdel Salam OME, Baiuomy AR. Effect of *Ginkgo biloba* extract on carrageenan-induced acute local inflammation in gamma irradiated rats. *Pharmazie.* 2005;60(8):614–619.

8. Huang CH, Yang ML, Tsai CH, Li YC, Lin YJ, Kuan YH. *Ginkgo biloba* leaves extract (EGb 761) attenuates lipopolysaccharide-induced acute lung injury via inhibition of oxidative stress and NF-κB-dependent matrix metalloproteinase-9 pathway. *Phytomedicine.* 2013;20(3–4):303–309.

9. Cheng D, Liang B, Li Y. Antihyperglycemic effect of *Ginkgo biloba* extract in Streptozosin-induced diabetes in rats. *BioMed Res Int.* 2013; ID 162724.

10. Zhou L, Meng Q, Qian T, Yang Z. *Ginkgo biloba* extract enhances glucose tolerance in hyperinsulinism-induced hepatic cells. *J Nat Med.* 2011;65(1):50–56.

11. Sakakibara H, Ishida K, Grundmann O, et al. Antidepressant effect of extracts from *Ginkgo biloba* Leaves in behavioral models. *Biol Pharm Bull.* 2006;29(8):1767–1770.

12. Porsalt RD, Martin P, Lenègre A, Fromage S, Drieu K. Effects of an extract of *Ginkgo biloba* (EGB 761) on "learned helplessness" and other models of stress in rodents. *Pharmacol Biochem Behav.* 1990;36(4):963–971.

13. Yeh K-Y, Shou SS, Lin YX, Chen CC, Chiang CY, Yeh CY. Effect of *Ginkgo biloba* extract on lipopolysaccharide-induced anhedonic depressive-like behavior in male rats. *Phytother Res.* 2015;29(2):260–266.

14. Rojasa P, Serrano-García N, Medina-Camposet ON, Pedraza-Chaverri J, Ögren SO, Rojas C. Antidepressant-like effect of a *Ginkgo biloba* extract (EGb761) in the mouse forced swimming test: Role of oxidative stress. *Neurochem Int.* 2011;59(5):628–636.

15. Liang Z, Bai S, Shen P, et al. GC–MS-based metabolomic study on the antidepressant-like effects of diterpene ginkgolides in mouse hippocampus. *Behav Brain Res.* 2016;314(1):116–124.

16. Kanowski S, Herrmann WM, Stephan K, Wierich W, Hörr R. Proof of efficacy of the *Ginkgo biloba* special extract EGb 761 in outpatients suffering from mild to moderate primary degenerative dementia of the Alzheimer type or multi-infarct dementia. *Pharmacopsychiatry.* 1996;29(2):47–56.

17. Kleijnen J, Knipschild P. *Ginkgo biloba* for cerebral insufficiency. *Br J Clin Pharmacol.* 1992;34(4):352–358.

18. Tana M-S, Yu JT, Tan CC, et al. Efficacy and adverse effects of *Ginkgo biloba* for cognitive impairment and dementia: A systematic review and meta-analysis. *J Alzheimers Dis.* 2015;43(2):589–603.

19. Jezova D, Duncko R, Lassanova M, Kriska M, Moncek F. Reduction of rise of in blood pressure and cortisol release during stress by *Ginkgo biloba* extract (EGB 761) in healthy volunteers. *J Physiol Pharmacol.* 2002;53(3):337–348.

20. Eckmann F, Schlag H. Kontrollierte Doppelbind-Studie zum Wirksamkeitsnachweis von tebonin forte bei Patienten mit zerebrovaskulärer Insuffizienz. *Fortschr Med.* 1982;100(31–32):1474–1478.

21. Wesnes K, Simmons D, Rook M, Simpson P. A double-blind placebo-controlled trial of Tanakan in the treatment of idiopathic cognitive impairment in elderly. *Hum Psychopharmacol Clin Exp.* 1987;2(3):159–169.

22. Schubert H, Halama P. Depressive episode primarily unresponsive to therapy in elderly patients: Efficacy of *Ginkgo biloba* extract EGb 761 in combination with antidepressants. *Geriatr Forsch* 1993;1:45–53.

23. Blumenthal M, Busse WR, Goldberg A, et al. *The Complete German Commission E Monograph: Therapeutic Guide to Herbal Medicines*. Austin, TX: American Botanical Council; 1998. p.138.
24. Gardner Z, McGuffin M, editors *Botanical Safety Handbook*. 3rd ed. Boca Raton, FL: CRC Press; 2013. p. 148.
25. Diamond BJ, Bailey MR. *Ginkgo biloba*: Indications, mechanisms, and safety. *Psychiatr Clin*. 2013;36(1):73–83.
26. Kupiec T, Raj V. Fatal seizures due to potential herb-drug interactions with *Ginkgo biloba*. *J Anal Toxicol*. 2005;29(7):755–758.

7.28   *GLYCYRRHIZA* (LICORICE)

Glycyrrhiza, or licorice, is a commonly used Chinese herbal medicine, derived from the dried roots and rhizomes of *Glycyrrhiza uralensis, glabra*, and *inflata*. Glycyrrhizin is the principal phytochemical in licorice root and makes up roughly 20% of licorice root extract.[1] Aside from glycyrrhizin, the *Glycyrrhiza* species also contain various triterpene saponins, flavonoids, coumarins, and other phenolics. Among these are glycyrrhizic acid, liquiritigenin, liquiritin, isoliquiritigenin, licoflavonol, licoricone, gancaonin, glabrone, glabridin, glycycoumarin, and others. The *Glycyrrhiza* have been shown to have a number of pharmacological properties including anti-inflammatory, antioxidative, antidiabetic, immunomodulatory, and neuroprotective effects.[2]

7.28.1   ANTIOXIDANT

In a battery of *in vitro* antioxidant assays, such as DPPH, ABTS, O2, and DMPD scavenging activities, and Fe3+-Fe2+ and Cu2+-Cu+ reducing abilities, extracts of licorice roots and aerial parts inhibited lipid peroxidation of linoleic acid emulsion by up to 87%. This was more effective than α-tocopherol or the commercial antioxidant, trolox.[3]

In vivo prior treatment with extract of *Glycyrrhiza* protected liver cells against the oxidative damage of CCl_4. Treatment reduced malondialdehyde, but increased glutathione in liver tissue. The activities of the enzymes, inducible nitric oxide synthase, and cyclooxygenase-2, were dampened in liver, and serum levels of TNF- α were reduced.[4]

An intracellular protein that is activated by aqueous extracts of *Glycyrrhiza inflate* is Nuclear factor (erythroid-derived 2)-like 2, or Nrf2. Nrf2 mobilizes the cell against oxidative stress, and its activation is associated with antidepressant effects.[5]

7.28.2   ANTI-INFLAMMATORY

Glycyrrhizin purified from *Glycyrrhiza* root inhibits the production of NO, PGE2, TNF-α, IL-6, and IL-1β, reduces the expression of iNOS and COX-2 genes, and blocks activation of the transcription factor NF-kβ in macrophages stimulated by lipopolysaccharide.[6] Glycyrrhizin also attenuates inflammatory responses induced by activators of the family of toll-like receptors in macrophages, a property it shares with a number of SSRI antidepressants.[7] *Glycyrrhiza* species also exhibit anti-inflammatory effects in the inhibition of prostaglandin E2 and leukotriene B production in murine macrophages and human neutrophils.[8]

7.28.3 Antidiabetic/Anti-Metabolic Syndrome

Glycyrrhiza also has significant antidiabetic and insulin-enhancing effects. Glycyrrhizin significantly improved blood glucose levels and enhanced insulin sensitivity in rats made diabetic with streptozotocin.[9] Phenolic phytochemicals extracted from *Glycyrrhiza glabra* species stimulate PPARγ,[10] whereas glycybenzofuran and glisoflavone from *Glycyrrhiza uralensis* inhibit protein tyrosine phosphatase 1B.[11] Protein tyrosine phosphatase 1B is a negative regulator of insulin and the leptin signaling pathway.

7.28.4 Preclinical Antidepressant-Like Effects

There are a number of rodent studies showing antidepressant effects of *Glycyrrhiza* and its extracts. One such study showed that treatment with liquiritigenin ameliorated the depression-like effects of chronic unpredictable stress in mice. Moreover, these antidepressant effects were accompanied by up-regulation of PI3K, Akt, mTOR, TrkB, and BDNF.[12] Of particular interest, isoliquiritigenin extracted from *Glycyrrhiza* root binds to NMDA receptors in rat cortical neurons and inhibits the cytotoxic glutamate-induced increase in Ca2+ influx.[13] Thus, like ketamine, isoliquiritigenin acts as an NMDA receptor antagonist.

Mitochondrial abnormalities play a role in MDD. Glycyrrhizin and licochalcone A extracted from *Glycyrrhiza inflata* upregulate proliferator-activated receptor γ coactivator 1α (PGC-1a), a critical factor in mitochondrial biogenesis.[14] PGC-1 also inhibits proinflammatory cytokine production.[15] It is suspected that increases in PGC-1 activity is what in part mediates the beneficial effects of exercise on MDD.[16]

7.28.5 Human Antidepressant Effects

Although *Glycyrrhiza* species are common components of Traditional Chinese Medicine's herbal combinations for the treatment of depression-like illnesses, a thorough review of the literature revealed no human clinical trials of *Glycyrrhiza* or any of its constituent phytochemicals used alone for the treatment of MDD.

7.28.6 Dosage

Easley and Horne recommend 500–1,000 mg of powdered root, tincture of dried root (1:5 40% alcohol) 1–5 ml, or decoction of 4–6 ounces of dried root up to four times daily.[17]

7.28.7 Toxicity

Toxicological studies in mice showed no mortality at doses up to 5,000 mg/kg. Thus, the LD50 was deemed to be above that dose.[18] Nonetheless, it has long been known

that high intake of licorice can cause hypermineralocorticoidism with sodium retention and potassium loss, edema, increased blood pressure, and depression of the
renin-angiotensin-aldosterone system. This is due to intake of glycyrrhizic acid.
There has been found to be individual variation in the susceptibility to glycyrrhizic
acid. In the most sensitive individuals, a regular daily intake of no more than about
100 mg glycyrrhizic acid, which corresponds to 50 g licorice sweets (assuming a
content of 0.2% glycyrrhizic acid), seems to be enough to produce adverse effects.
Most individuals who consume 400 mg glycyrrhizic acid daily experience adverse
effects. Considering that a regular intake of 100 mg glycyrrhizic acid/day is the
lowest-observed-adverse-effect level and using a safety factor of 10, a daily intake of
10 mg glycyrrhizic acid would represent a safe dose for most healthy adults.[19]

7.28.8　SAFETY IN PREGNANCY

The literature includes serious warnings about the use of the medicinal species of
Glycyrrhiza during pregnancy. Heavy consumption of forms of licorice that contain
glycyrrhizin are associated with early births[20] and adverse developmental and psychiatric effects.[21]

7.28.9　DRUG INTERACTIONS

The *Botanical Safety Handbook* notes that no interactions are expected at standard
therapeutic doses of *Glycyrrhiza* species. However, they advise that high doses
and prolonged use can potentiate potassium depletion of high-ceiling loop diuretics and thiazide diuretics, and steroids. It may also potentiate the effects of cardiac
glycosides.[22]

REFERENCES

1. Isbrucker RA, Burdock GA. Risk and safety assessment on the consumption of
Licorice root (*Glycyrrhiza* sp.), its extract and powder as a food ingredient, with empha-
sis on the pharmacology and toxicology of glycyrrhizin. *Regul Toxicol Pharmacol.*
2006;46(3):167–192.
2. Hosseinzadeh H, Nassiri-Asl M. Pharmacological effects of *Glycyrrhiza* spp. and its
bioactive constituents: Update and review. *Phytother Res.* 2015;29(12):1868–1886.
3. Tohma HS, Gulçin I. Antioxidant and radical scavenging activity of aerial parts and
roots of Turkish liquorice (*Glycyrrhiza glabra* L.). *Int J Food Prop.* 2010;13(4):657–671.
4. Lee C-H, Park SW, Kim YS, Kang SS, Kim JA, Lee SH. Protective mechanism of
glycyrrhizin on acute liver injury induced by carbon tetrachloride in mice. *Biol Pharm
Bull.* 2007;30(10):1898–1904.
5. Maes M, Fišar Z, Medina M, Scapagnini G, Nowak G, Berk M. New drug targets in
depression: inflammatory, cell-mediated immune, oxidative and nitrosative stress, mitochondrial, antioxidant, and neuroprogressive pathways. And new drug candidates—
Nrf2 activators and GSK-3 inhibitors. *Inflammopharmacology.* 2012;20(3):127–150.
6. Wang CY, Kao TC, Lo WH, Yen GC. Glycyrrhizic acid and 18β-glycyrrhetinic acid
modulate lipopolysaccharide-induced inflammatory response by suppression of NF-κB
through PI3K p110δ and p110γ inhibitions. *J Agric Food Chem.* 2011;59(14):7726–7733.

7. Schröfelbauer B, Raffetseder J, Hauner M, Wolkerstorfer A, Ernst W, Szolar OH. Glycyrrhizin, the main active compound in liquorice, attenuates pro-inflammatory responses by interfering with membrane-dependent receptor signalling. *Biochem J.* 2009;421(3):473–482.
8. Chandrasekaran CV, Deepak HB, Thiyagarajan P, et al. Dual inhibitory effect of *Glycyrrhiza glabra* (GutGard™) on COX and LOX products. *Phytomedicine.* 2011;18(4):278–284.
9. Sen S, Roy M, Chakrabort AS. Ameliorative effects of glycyrrhizin on streptozotocin-induced diabetes in rats. *J Pharm Pharmacol.* 2011;63(2):287–296.
10. Kuroda M, Mimaki Y, Honda S, Tanaka H, Yokota S, Mae T. Phenolics from *Glycyrrhiza glabra* roots and their PPAR-gamma ligand-binding activity. *Bioorg Med Chem.* 2010;18(2):962–970.
11. Yoon G, Lee W, Kim SN, Cheon SH. Inhibitory effect of chalcones and their derivatives from *Glycyrrhiza inflata* on protein tyrosine phosphatase 1B. *Bioorg Med Chem Lett.* 2009;19(17):5155–5157.
12. Tao W-W, Dong Y, Su Q, et al. Liquiritigenin reverses depression-like behavior in unpredictable chronic mild stress-induced mice by regulating PI3K/Akt/mTOR mediated BDNF/TrkB pathway. *Behav Brain Res.* 2016;308:177–186.
13. Kawakami Z, Ikarashi Y, Kase Y. Isoliquiritigenin is a novel NMDA receptor antagonist in Kampo medicine yokukansan. *Cell Mol Neurobiol.* 2011;31(8):1203–1212.
14. Chen CM, Weng YT, Chen WL, et al. Aqueous extract of *Glycyrrhiza inflata* inhibits aggregation by upregulating PPARGC1A and NFE2L2-ARE pathways in cell models of spinocerebellar ataxia 3. *Free Radic Biol Med.* 2014;71:339–350.
15. Vats D, Mukundan L, Odegaard JL, et al. Oxidative metabolism and PGC-1β attenuate macrophage-mediated inflammation. *Cell Metab.* 2006;4(1):13–24.
16. Farshbaf MJ, Ghaedi K, Megraw TL, et al. Does PGC1α/FNDC5/BDNF elicit the beneficial effects of exercise on neurodegenerative disorders? *NeuroMol Med.* 2016;18(1):1–15.
17. Easley T, Horne S. *The Modern Herbal Dispensary.* Berkeley, CA: North Atlantic Books; 2016. p. 261.
18. Pravin MA, Viswanathan V, Pharande RR. Acute toxicity studies of nano-formulations of *Glycyrrhiza glabra* extract in Swiss albino mice. *World J Pharm Pharm Sci.* 2017;6(6):820–829.
19. Størmer FC, Reistad R, Alexander J. Glycyrrhizic acid in liquorice—Evaluation of health hazard. *Food Chem Toxicol.* 1993;31(4):303–312.
20. Strandberg TE, Andersson S, Järvenpää AL, McKeigue PM. Preterm birth and licorice consumption during pregnancy. *Am J Epidemiol.* 2002;156(9):803–805.
21. Räikkönen K, Pesonen AK, Heinonen K, et al. Maternal licorice consumption and detrimental cognitive and psychiatric outcomes in children. *Am J Epidemiol.* 2009;170(9):1137–1146.
22. Gardner Z, McGuffin M, editors. *Botanical Safety Handbook.* 3rd ed. Boca Raton, FL: CRC Press; 2013. p. 409.

7.29 HEDYOSMUM BRASILIENSE

Hedyosmum brasiliense is a South American plant native to the Atlantic forest region. Its peppermint-like flavor is prized, and it is a popular beverage in Brazil and Bolivia. It is commonly known as "cidrão" or "Cha De Soldado," and in folk medicine this aromatic species is widely used as a calmative, tranquilizer, and hypnotic. Among the phytochemicals contained in *Hedyosmum brasiliense* are α-terpineol, curzerene, pinocarvone, β-thujene, podoandin, 1,2-epoxy-10α-hydroxy-podoandin, 1-hydroxy-10,15-methylenepodoandin, 15-acetoxy-isogermafurenolide, 8α/β,9α-hydroxy-onoseriolide, guaianolide podoandin, onoseriolide, scopoletin, vanillin, vanillic acid, protocatechuic aldehyde, and ethyl caffeate.[1,2]

7.29.1 ANTIOXIDANT/ANTI-INFLAMMATORY

Hedyosmum brasiliense has antioxidant and anti-inflammatory effects that manifest in the brains of intact animals. In mice intracerebroventricularly injected with Aβ1-42 peptide, a model of Alzheimer's disease, simultaneous injection of 15-acetoxy-isogermafurenolide, 15-hydroxy-isogermafurenolide, podoandin, 1,2-epoxy-10α-hydroxy-podoandin, 13-hydroxy-8,9-dehydroshizukanolide and aromadendrane-4β,10α-diol extracted from *Hedyosmum brasiliense* gave neuroprotective effects. Memory impairment was reduced, GSH activity was increased, and thiobarbituric acid reactive substance levels, indicative of lipid peroxidation and inflammation, were decreased by treatment with these compounds.[3]

7.29.2 ANTIDIABETIC/ANTI-METABOLIC SYNDROME

There are no reports in the literature of effects of the herb on diabetes or Metabolic Syndrome.

7.29.3 PRECLINICAL ANTIDEPRESSANT-LIKE EFFECTS

Ethanol extract of the leaves of *Hedyosmum brasiliense* improves performance in the tail suspension and forced swimming tests in mice. Further studies have shown that the sesquiterpene lactone podoandin, purified from the extract, also shows such antidepressant-like effects.[4]

In another study of mice, ethanolic extract of *Hedyosmum brasiliense* showed antidepressant-like effects in the forced swim test similar to those of Imipramine. Treatment with the extract also produced anxiolytic effects in the open field test. Those effects were reversed by flumazenil, suggesting that they were mediated by a benzodiazepine-like effect on GABA-A receptors. Finally, the extract increased barbiturate-induced sleeping time, an indication of enhancing sleep.[5]

7.29.4 HUMAN ANTIDEPRESSANT EFFECTS

With antidepressant, anxiolytic, and hypnotic effects, *Hedyosmum brasiliense* may be a useful herb for human psychiatric conditions. However, there are no human clinical trials testing the herb as a treatment for MDD.

7.29.5 DOSAGE

There are no formal studies from which to suggest doses to treat human ailments.

7.29.6 Toxicity

The herb has low toxicity with an LD50 in mice of 1.99 g/kg.[6] This would extrapolate to about 140 g for a 160-pound human adult.

7.29.7 Safety in Pregnancy

Review of the literature reveals no reports on the safety of using *Hedyosmum brasiliense* during pregnancy.

7.29.8 Drug Interaction

Review of the literature reveals no reports on interactions of *Hedyosmum brasiliense* with common medications.

REFERENCES

1. Kirchner K, Wisniewski A, Cruz AB, Biavatti MW, Netz DJ. Chemical composition and antimicrobial activity of *Hedyosmum brasiliense* Miq., Chloranthaceae, essential oil. *Rev Bras Farmacogn.* 2010;20(5):692–699.
2. Amoah SKS, de Oliveira FL, da Cruz ACH, et al. Sesquiterpene lactones from the leaves of *Hedyosmum brasiliense (Chloranthaceae)*. *Phytochemistry.* 2013;87:126–132.
3. Amoah SKS, Vecchia MTD, Pedrini B, et al. Inhibitory effect of sesquiterpene lactones and the sesquiterpene alcohol aromadendrane-4β,10α-diol on memory impairment in a mouse model of Alzheimer. *Eur J Pharmacol.* 2015;769:195–202.
4. Goncalves AE, Bürger C, Amoah SK, Tolardo R, Biavatti MW, de Souza MM. The antidepressant-like effect of *Hedyosmum brasiliense* and its sesquiterpene lactone, podoandin in mice: Evidence for the involvement of adrenergic, dopaminergic and serotonergic systems. *Eur J Pharmacol.* 2012;674(2–3):307–314.
5. Tolardo R, Zetterman L, Bitencourtt DR, et al. Evaluation of behavioral and pharmacological effects of *Hedyosmum brasiliense* and isolated sesquiterpene lactones in rodents. *J Ethnopharmacol.* 2010;128(1–2):63–70.
6. García-Zebadúa JC, Alfonso MGG, Marius MM, et al. Inhibition of HIV-1 reverse transcriptase, toxicological and chemical profile of Calophyllum brasiliense extracts from Chiapas, Mexico. *Fitoterapia.* 2010;82(7):1027–1034.

7.30 *HEMEROCALLIS CITRINA* (DAYLILY)

Hemerocallis citrina, or daylily, is a plant widely grown in East Asia. Its roots, flowers, and leaves have been used as foods and medicines for thousands of years. The plant has been used in the treatment of inflammation and liver disease. Given the traditional view of the liver playing a major role in MDD, it is not surprising that daylilies have also been used to treat sleep and mood disorders. Indeed, in Japan and China it has been known as the "forget one's sorrow plant."[1]

Hemerocallis citrina contains a wide variety of polyphenols, carotenoids, flavonoids, and other phytochemical forms. Many are shared with other medicinal plants. These phytochemicals include kaempferol, quercetin, isorhamnetin, phenethyl β-d-glucopyranoside, orcinol β-d-glucopyranoside, phloretin 2'-O-β-d-glucopyranoside, phloretin 2'O-β-d-xylopyranosyl-(1→6)-β-d-glucopyranoside, stelladerol, neo-xanthin, violaxanthin, violeoxanthin, lutein-5,6-epoxide, lutein, zeaxanthin, β-cryptoxanthin, pinnatanine, roseoside, phlomuroside, and lariciresinol. Many have been found to exhibit potent antioxidant and anti-inflammatory effects.[2–4]

7.30.1 ANTIOXIDANT

Extracts of *Hemerocallis citrina* showed high scavenging capacity for DPPH radicals, hydrogen peroxide, superoxide anion, and hydroxyl radicals.[5] In cultured liver cells, extract of *Hemerocallis citrina* protected against the oxidative damage of exposure to tert-butyl hydroperoxide. Treatment reduced levels of the enzymes ALT, AST, and LDH in the cell culture supernatant, thus indicating sparing of cell integrity. The extract also enhanced the activity of antioxidant enzymes, superoxide dismutase, and catalase in the cultured cells, as well as increasing levels of glutathione.[6]

In intact mice, 60 days of extract of *Hemerocallis citrina* administered by gavage resulted in reduced levels of malondialdehyde and increased levels of superoxide dismutase activity in liver tissue. Animals had received no other treatments, although the gavage process itself must be seen as a form of chronic stress.[7]

7.30.2 ANTI-INFLAMMATORY

In rats that had been subjected to chronic unpredictable stress, concurrent daily administration of daylily inhibited expression of the inflammatory cytokines, IL-1β, IL-6, and TNF- in frontal cortex and hippocampus. It also prevented activation of frontal cortex indoleamine 2,3-dioygenase, the stress-activated enzyme that depletes serotonin and increases activity in the excitotoxic kynurenine pathway. In the same animals, this treatment prevented the decreased preference for sucrose typically seen after chronic stress.[8]

7.30.3 ANTIDIABETIC/ANTI-METABOLIC SYNDROME

There are no specific references in the literature to antidiabetic effects of *Hemerocallis citrina*. However, several of the plant's primary phytochemical components, including quercetin,[9] kaempferol,[10] and isorhamnetin[11] have been to exhibit significant antidiabetic and anti-Metabolic Syndrome activities.

7.30.4 PRECLINICAL ANTIDEPRESSANT-LIKE EFFECTS

A number of animal studies have demonstrated antidepressant-like effects of *Hemerocallis citrina*. In a study of mice, *Hemerocallis citrina* significantly reversed the reduction of sucrose preference that followed administration of lipopolysaccharide. This was taken as evidence of an antidepressant-like effect. It also reduced the inflammatory activation of nuclear factor-κB (NF-κB), inducible nitric oxide synthase (iNOS), and COX-2 expression in the prefrontal cortex.[12] Another mouse study found similar antidepressant-like effects of *Hemerocallis citrina* in the forced-swim, tail suspension, and open field locomotor activity tests. *Hemerocallis citrina* enhanced 5-HT and NA levels in the frontal cortex and hippocampus as well as elevating DA levels in the frontal cortex. This gave evidence of monoaminergic contribution to the antidepressant-like effect.[13]

The ethanol extract of *Hemerocallis citrina* flowers also improved performance of rats in the forced swimming test. Rutin, one of the more prominent phytochemicals in *Hemerocallis citrina*, produced effects similar to those of the full ethanol extract, suggesting that rutin could be largely responsible for these antidepressant-like effects. As in mice, *Hemerocallis citrina* increased concentrations of serotonin, norepinephrine, and dopamine in the frontal cortex, hippocampus, striatum, and amygdala of the rats, suggesting that these effects could be due in part to enhancement of central monoaminergic activity.[14]

Two studies in rats have shown that *Hemerocallis citrina* can mitigate the damaging, depression-like effects of chronic stress and the primary stress hormone, corticosterone. Treatment with extract of *Hemerocallis citrina* enhanced performance in the open field and Morris Water maze tests in rats that had been subjected to chronic, unpredictable stress. It also restored sucrose preference. As shown in other paradigms, *Hemerocallis citrina* increased the concentrations of 5-HT, DA, and NE in the hippocampus and frontal cortex compared with untreated CUMS rats. Moreover, the increases in serum corticosterone and decreases in hippocampal BDNF levels were reversed.[15] *Hemerocallis citrina* also normalized sucrose preference and performance in the forced swim test in rats chronically administered corticosterone. It also restored normal levels of BDNF and TrkB receptors in the frontal cortex and hippocampus of those animals.[16]

7.30.5 HUMAN ANTIDEPRESSANT EFFECTS

Studies suggest likely antidepressant activity of *Hemerocallis citrina* through reduction of neuroinflammation, mitigation of damaging effects of stress, enhancement of BDNF, and maintenance of monoaminergic activity. However, despite the long history of use of *Hemerocallis citrina* in Chinese Traditional Medicine, there are no human studies of its use in the treatment MDD, at least, none that are accessible in the West.

7.30.6 DOSAGE

There are no reliable human dosage recommendations in the literature.

7.30.7 TOXICITY

In mice, no deaths or adverse effects were noted at orally administered doses of dried ethanolic extract up to 5,000 mg/kg, which would extrapolate to approximately 350 g for a 75 kg human adult. In that same study, significant antidepressant effects were noted at the 400 mg/kg dose.[17]

7.30.8 SAFETY IN PREGNANCY

Review of the literature revealed no studies of the safety of *Hemerocallis citrina* during human or animal pregnancy.

7.30.9 DRUG INTERACTIONS

Review of the literature revealed no studies of the possible interactions of *Hemerocallis citrina* with commonly used medications.

REFERENCES

1. Rodriguez-Enriquez MJ, Grant-Downton RT. A new day dawning: Hemerocallis (day-lily) as a future model organism. *AoB Plants.* 2012;pls055. doi:10.1093/aobpla/pls055
2. Cichewicz RH, Nair MG. Isolation and characterization of Stelladerol, a new anti-oxidant naphthalene glycoside, and other antioxidant glycosides from edible daylily (*Hemerocallis*) flowers. *J Agric Food Chem.* 2002;50(1):87–91.
3. Zhang Y, Cichewicz RH, Nair MG. Lipid peroxidation inhibitory compounds from daylily (*Hemerocallis fulva*) leaves. *Life Sci.* 2004;75(6):753–763.
4. Tai C-Y, Chen BH. Analysis and stability of carotenoids in the flowers of day-lily (*Hemerocallis disticha*) as affected by various treatments. *J Agric Food Chem.* 2000;48(12):5962–5968.
5. Wang Y-C. In vitro antioxidant capacity of daylily (*Hemerocallis disticha*) flowers cul-tivated in Taiwan. *Life Sci J.* 2013;10(3):1524–1527.
6. Wang J, Hu D, Hou J, et al. Ethyl acetate fraction of *Hemerocallis citrina* baroni decreases tert-butyl hydroperoxide-induced oxidative stress damage in BRL-3A cells. *Oxid Med Cell Longev.* 2018;2018: Article ID 1526125.
7. Fei Que F, Mao LC, Zheng XJ. In vitro and vivo antioxidant activities of daylily flow-ers and the involvement of phenolic compounds. *Asia Pac J Clin Nutr.* 2007;16 (Suppl 1):196–203.
8. Liu X-L, Luo L, Liu BB, et al. Ethanol extracts from *Hemerocallis citrina* attenuate the upregulation of proinflammatory cytokines and indoleamine 2,3-dioxygenase in rats. *J Ethnopharmacol.* 2014;153(2):484–490.
9. Nuraliev IN, Avezov GA. The efficacy of quercetin in alloxan diabetes. The efficacy of quercetin in alloxan diabetes. *Eksperimental'naia i Klinicheskaia Farmakologiia.* 1992;55(1):42–44.
10. Zhang Y, Liu D. Flavonol kaempferol improves chronic hyperglycemia-impaired pancreatic beta-cell viability and insulin secretory function. *Eur J Pharmacol.* 2011;670(1):325–332.
11. Lee J, Jung E, Lee J, et al. Isorhamnetin represses adipogenesis in 3T3-L1 cells. *Obesity.* 2009;17(2):226–232.
12. Li C-F, Chen XQ, Chen SM, et al. Evaluation of the toxicological properties and anti-inflammatory mechanism of *Hemerocallis citrina* in LPS-induced depressive-like mice. *Biomed Pharmacother.* 2017;91:167–173.
13. Gu L, Liu Y-J, Wang Y-B, Yi L-T. Role for monoaminergic systems in the antide-pressant-like effect of ethanol extracts from *Hemerocallis citrina*. *J Ethnopharmacol.* 2012;139(3):780–787.
14. Lin SH, Chang HC, Chenet PJ, Hsieh CL, Su KP, Sheen LY. The antidepressant-like effect of ethanol extract of daylily flowers (金針花 jīn zhēn Huā) in rats. *J Tradit Complement Med.* 2013;3(1):53–61.
15. Xu P, Wang KZ, Lu C, et al. Antidepressant-like effects and cognitive enhancement of the total phenols extract of *Hemerocallis citrina* Baroni in chronic unpredictable mild stress rats and its related mechanism. *J Ethnopharmacol.* 2016;194:819–826.
16. Yi LT, Li J, Li HC, et al. Ethanol extracts from *Hemerocallis citrina* attenuate the decreases of brain-derived neurotrophic factor, TrkB levels in rat induced by corticos-terone administration. *J Ethnopharmacol.* 2012;144(2):328–334.

17. Du B, Tang X, Liu F, et al. Antidepressant-like effects of the hydroalcoholic extracts of *Hemerocallis citrina* and its potential active components. *BMC Complement Altern Med.* 2014;14:326.

7.31   *HERICIUM ERINACEUS* (LION'S MANE)

Hericium erinaceus is an edible and medicinal mushroom, known in the United States and Europe as "Lion's Mane," in Japan as "*Yamabushitake*," and in China as "*Hou Tou Gu.*" *Hericium erinaceus* is known to exert a number of important biological effects, including induction of nerve growth factor, inhibition of the cytotoxicity of β-amyloid peptide, neuroprotection, reduction of blood glucose, and normalization of blood lipids.[1] Its most promising medicinal application is likely the treatment of Alzheimer's and other dementias. There is also data, albeit mostly from preclinical studies, that suggest potential use in the treatment of MDD.

A number of complex terpenoids, the hericenones and erinacines, appear unique to *Hericium erinaceus*. These molecules are believed to hold the mushroom's ability to stimulate neuronal growth.[2] Other substances isolated from *Hericium erinaceus* include ergosterol peroxide, cerevisterol, inoterpene, astradoric acid, betulin, oleanolic acid, ursolic acid, and hemisceramide.[3] Flavonoids common to many medicinal herbs, 4-hydroxybenzoic acid, syringic acid, 4-coumaric acid, and ferulic acid, have also been identified in *Hericium erinaceus*.[4]

7.31.1   ANTIOXIDANT

Hericium erinaceus has substantial activity in the in vitro DPPH assay, which is a widely accepted model to assess free radical-scavenging activity.[5] The mushroom also exhibits strong antioxidant effects *in vivo*. In streptozotocin-induced diabetic rats, chronic administration of aqueous extract of *Hericium erinaceus* not only reduced serum glucose, but also increased the activities of the antioxidant enzymes catalase, superoxide dismutase, and glutathione peroxidase in the livers of those animals. It also increased hepatic glutathione and reduced levels of malondialdehyde, a biomarker of oxidative stress.[6] Extract of *Hericium erinaceus* also protected the liver from the oxidative damage of CCl_4, while sparing antioxidant enzymes.[7]

7.31.2   ANTI-INFLAMMATORY

A variety of studies have demonstrated the anti-inflammatory effects of *Hericium erinaceus*. Treatment of cultured murine macrophages with extract of *Hericium erinaceus* attenuated the lipopolysaccharise-induced rise in levels of iNOS and cyclooxygenase-2, with consequent decreases in nitric oxide, prostaglandin E2, and reactive oxygen species. Treatment also inhibited the translocation of nuclear factor (NF)-κB p65 subunit, phosphorylation of I-κB, extracellular signal-regulated kinase (ERK), and c-Jun N-terminal kinase in a dose-dependent manner. The latter steps are necessary to mount the inflammatory response to lipopolysaccharide.[8]

Hericium erinaceus also exhibits potent anti-inflammatory effects *in vivo*. In rats subjected to experimental cerebral ischemia, the herb exerted an anti-inflammatory, neuroprotective effect. The highest dose of lion's mane extract reduced the area of brain infarct by 44% and reduced tissue levels of the inflammatory cytokines, TNF-α, IL1β, and IL-6. Treatment also suppressed reactive nitrogen species and downregulated inducible NO synthase. Similar effects were seen with the purified phytochemical from the mushroom erinacine A.[9]

7.31.3 ANTIDIABETIC/ANTI-METABOLIC SYNDROME

A methanolic extract of *Hericium erinaceus* rich in e D-threitol, D-arabinitol, and palmitic acid significantly reduced serum glucose in streptozotocin-induced diabetic rats. It also countered elevated triglyceride and total cholesterol levels in these animals.[10] This effect of *Hericium erinaceus* may have been due in part to activation of PPARα and PPARγ. A number of ergostane-type sterol fatty acid esters found in methanol extractions of the mushroom activate the transcriptional activity of these PPARs at low micromolar concentrations.[11]

Hericium erinaceus also reversed signs of Metabolic Syndrome in mice fed high fat diets. Chronic treatment with either hot water or ethanolic extracts of *Hericium erinaceus* normalized serum glucose, lowered elevated levels of triglyceride and cholesterol, decreased extent of mesenteric adipose tissue, and reduced the size of fatty livers.[12]

7.31.4 PRECLINICAL ANTIDEPRESSANT-LIKE EFFECTS

Several studies in mice have shown antidepressant effects of chronic administration of *Hericium erinaceus*. Oral administration of extract of *Hericium erinaceus* for four weeks decreased immobility of mice in the forced swim and tail suspension tests. Further evaluation of the brains of these animals showed the treatment stimulated neurogenesis in the hippocampus.[13] Another study showed that four weeks of alcoholic extract of *Hericium erinaceus* also reversed the depression-like effects of two weeks of restraint stress in mice. Immobility in the tail suspension and forced swim tests were again decreased. Anxiolytic-like effects were also noted, with increases in the number of entries and time spent in the open arm of the elevated plus maze. Chemical analyses of the brains of these mice further showed attenuation of the decreases in brain levels of norepinephrine, dopamine, and serotonin in the stressed mice. Treatment also diminished the stress-induced increases in levels of IL-6 and TNF-α. Finally, chronic treatment with *Hericium erinaceus* activated the PI3K/Akt pathway, and thus inhibited GSK-3β and activated BDNF and TrkB. This finding would help explain the stimulation of neurogenesis noted above. On the other hand, NF-κB signals were blocked, which would help explain the anti-inflammatory effects in the brain.[14]

Chronic treatment with amycenone, a patented standardized extract of *Hericium erinaceus*, was also effective, counteracting the depression-like effects of treatment with the inflammation-inducing agent lipopolysaccharide. Immobility in the forced swim and tail suspension tests were reduced. Treatment with amycenone attenuated the lipopolysaccharide-induced increases in serum TNF-α, as well as restoring diminished levels of the anti-inflammatory cytokine, IL-10.[15]

7.31.5 HUMAN ANTIDEPRESSANT EFFECTS

In the only published study of effects of *Hericium erinaceus* on mood in human subjects, perimenopausal women who were not diagnosed with MDD or any other psychiatric illness were given either *Hericium erinaceus* or placebo. After four weeks, scores on several test batteries were compared with those at baseline. In the Center for Epidemiologic Studies Depression Scale, scores in both groups declined, and there were no significances between the two. In the Keio Indefinite Complaints Index, two factors showed small but significant differences between the *Hericium erinaceus* and placebo group, those being motivation and palpitations. Differences in that instrument approached significance in irritability, anxiety, and concentration.[16] However, given the small differences and the overall number of comparisons made, those results should be taken with reservation.

Aside from the rather weak study above, there is also a case report concerning the use of *Hericium erinaceus* in an 86-year-old man with a 40-year history of severe recurrent MDD with multiple hospitalizations. Medications had included clomipramine, trimipramine, amitriptyline, setiptiline, fluvoxamine, and paroxetine. All had been only modestly effective and had caused side effects to the point that medications were stopped. When he was 86, due to persisting depression, he was started on mirtazapine. He benefited, but began to experience cognitive impairment. At that time, *Hericium erinaceus* was started for that impairment, but after one month he and his wife noted substantial improvement in both mood and cognitive function. This improvement continued, and after five months, he felt so well that he stopped the mirtazapine and continued the *Hericium erinaceus* alone. Thereafter he remained well, and better than he had been for many years, on just the *Hericium erinaceus*.[17]

There are several studies on the use of *Hericium erinaceus* in the treatment of dementia in human subjects. In one case, elderly Japanese men and women diagnosed with mild cognitive impairment by the Revised Hasegawa Dementia Scale were given a gram of dry powdered mushroom or placebo three times a day. After four months, cognitive function was significantly improved. This improvement was lost four weeks after the treatment was stopped.[18] In a study of elderly patients with more severe impairments, daily administration of five grams of lyophilized mushroom for six months resulted in significant improvements in cognitive function and abilities to perform activities of daily living per the Functional Independence Measure. It was noted that three of the patients diagnosed with Alzheimer's disease that had been bedridden were again able to get up for meals after the administration.[19]

7.31.6 DOSAGE

Hericium erinaceus is commercially available through several well-established brands in standardized dosing of about 1 g of fruiting extract per day.

7.31.7 TOXICITY

The LD50 of a dried, ethanol extract of *Hericium erinaceus* in rats was reported to higher than 300 mg/kg at a single dose.[20] This would extrapolate to approximately

21 g for a 75 kg human adult. In the human trials noted above, doses of 2–5 grams of dried powder per day have been given with benefits but without adverse effects.

7.31.8 SAFETY IN PREGNANCY

The literature includes no specific studies on the effects of *Hericium erinaceus* on pregnant human females. There are references to the use of mushrooms to control gestational diabetes, but these are not supported by scientific study. One study found chronic, supratherapeutic doses of *Hericium erinaceus* had no adverse acute and developmental toxicity in pregnant female rats of their offspring.[21]

7.31.9 DRUG INTERACTIONS

Review of the literature revealed no studies of the possible interactions of *Hericium erinaceus* with commonly used medications.

REFERENCES

1. Kawagishi H, Zhuang C. Compounds for dementia from Hericium erinaceum. *Drug Future.* 2008;33(2):149–155.
2. Kawagishi H, Ando M, Shinba K, et al. Chromans, hericenones F, G, and H from the mushroom Hericium erinaceum. *Phytochemistry.* 1992;32(1):175–178.
3. Zhang C-C, Yin X, Cao CY, Wei J, Zhang Q, Gao JM. Chemical constituents from *Hericium erinaceus* and their ability to stimulate NGF-mediated neurite outgrowth on PC12 cells. *Bioorg Med Chem Lett.* 2015;25(22):5078–5082.
4. Sol HL, Park S, Moon BK, et al. Targeted phenolic analysis in Hericium erinaceum and its antioxidant activities. *Food Sci Biotechnol.* 2012;21(3):881–888.
5. Ren L, Hemar Y, Perera CO, Lewis G, Krissansen GW, Buchanan PK. Antibacterial and antioxidant activities of aqueous extracts of eight edible mushrooms. *Bioact Carbohydr Diet Fibre.* 2014;3(2):41–51.
6. Liang B, Guo Z, Xie F, Zhao A. Antihyperglycemic and antihyperlipidemic activities of aqueous extract of *Hericium erinaceus* in experimental diabetic rats. *BMC Complement Altern Med.* 2013;13:253.
7. Cho W-S, Kim CJ, Park BS, et al. Inhibitory effect on proliferation of vascular smooth muscle cells and protective effect on CCl_4-induced hepatic damage of HEAI extract. *J Ethnopharmacol.* 2005;100(1–2):176–179.
8. Kim Y-O, Lee SW, Oh CH, Rhee YH. *Hericium erinaceus* suppresses LPS-induced pro-inflammation gene activation in RAW264.7 macrophages. *J Immunopharmacol Immunotoxicol.* 2012;34(3):504–512.
9. Lee K-F, Chen JH, Teng CC, et al. Protective effects of *Hericium erinaceus* mycelium and its isolated erinacine A against ischemia-injury-induced neuronal cell death via the inhibition of iNOS/p38 MAPK and nitrotyrosine. *Int J Mol Sci.* 2014;15(9):15073–15089.
10. Wang JC, Hu SH, Wanget JT, Chen KS, Chia YC. Hypoglycemic effect of extract of *Hericium erinaceus. J Sci Food Agric.* 2004;85(4):641–646.
11. Li W, Zhou W, Song SB, Shim SH, Kim YH. Sterol fatty acid esters from the mushroom Hericium erinaceum and their PPAR transactivational effects. *J Nat Prod.* 2014;77(12):2611–2618.
12. Hiwatashi K, Kosaka Y, Suzuki N, et al. *Yamabushitake* mushroom (*Hericium erinaceus*) improved lipid metabolism in mice fed a high-fat diet. *Biosci Biotechnol Biochem.* 2010;74(7):1447–1451.

13. Ryu S, Kim HG, Kim JY, Kim SY, Cho KO. *Hericium erinaceus* extract reduces anxiety and depressive behaviors by promoting hippocampal neurogenesis in the adult mouse brain. *J Med Food.* 2018;21(2):174–180.
14. Chiu C-H, Chyau CC, Chen CC, et al. Erinacine A-enriched *Hericium erinaceus* mycelium produces antidepressant-like effects through modulating BDNF/PI3K/Akt/GSK-3β signaling in mice. *Int J Mol Sci.* 2018;19(2):341–362.
15. Yao W, Zhang J, Dong C, et al. Effects of amycenone on serum levels of tumor necrosis factor-α, interleukin-10, and depression-like behavior in mice after lipopolysaccharide administration. *Pharmacol Biochem Behav.* 2015;136:7–12.
16. Nagano M, Shimizu K, Kondo R, et al. Reduction of depression and anxiety by 4 weeks *Hericium erinaceus* intake. *Biomed Res.* 2010;31(4):231–237.
17. Inanaga K. Marked improvement of neurocognitive impairment after treatment with compounds from Hericium erinaceum: A case study of recurrent depressive disorder. *Pers Med Universe.* 2014;3:46–48.
18. Mori K, Inatomi S, Ouchi K, Azumi Y, Tuchida T. Improving effects of the mushroom *Yamabushitake (Hericium erinaceus)* on mild cognitive impairment: A double-blind placebo-controlled clinical trial. *Phytother Res.* 2009;23(3):367–372.
19. Kasahara K, Kaneko N, Shimizu K. Effects of Hericium erinaceum on aged patients with impairment. *Gunma Med Suppl.* 2001;76:77–78.
20. Zhuang C, Kawagishi H, Zhang L, Anzai H. (2009). Anti-dementia substance from Hericium erinaceum and method of extraction. US Patent 20090274720.
21. Li I-C, Chen W-P, Chen Y-P, et al. Acute and developmental toxicity assessment of erincine A-enriched *Hericium erinaceus* mycelia in Sprague-Dawley rats. *Drug Chem Toxicol.* 2018;41(4):459–464.

7.32 *HIBISCUS ROSA-SINENSIS* (HIBISCUS)

Hibiscus rosa-sinensis, commonly referred to as simply hibiscus, is a flowering plant of the *Malvaceae* family native to East Asia. The flowers are best known for their beauty. However, the flowers are both edible and contain medicinal phytochemicals. They are rich in flavonoids, including quercetin, cyanidin, kaempferol, myricetin, pelargonidin, peonidin, delphinidin, petunidin, and malvidin, though the content varies across different colors and breeds.[1] Roots, leaves, and stems are also used. The genus *Hibiscus* contains approximately 200 species distributed throughout tropical and subtropical regions. Various closely related species of *Hibiscus* are used medicinally, and are believed to exert anticonvulsant, antioxidant, anti-inflammatory, antimicrobial, laxative, sedative, and antidepressant depressant effects.[2]

7.32.1 ANTIOXIDANT

Hibiscus has potent antioxidant capacity. *In vitro*, methanol extract of hibiscus effectively scavenged radicals in the DPPH assay.[3] Also *in vitro*, extract of hibiscus protected against lipid peroxidation and scavenged O_2-, H_2O_2, and NO in a manner similar to that of the commercial antioxidant BHA and almost as effectively as vitamin C.[4] *In vivo*, methanolic extract of hibiscus attenuated the oxidative damage to brain tissue caused by reserpine. The extract decreased lipid peroxidation, and increased levels of the antioxidant enzymes superoxide dismutase, catalase, and glutathione reductase.[5] Extract of hibiscus also prevented the oxidative damage consequent to cerebral ischemia and reperfusion in mouse brain. It similarly preserved activities of superoxide dismutase, catalase, and glutathione reductase.[6]

7.32.2 ANTI-INFLAMMATORY

Methanolic extract of hibiscus leaves attenuated both carrageenan- and dextran-induced paw edema in rats. It did so by inhibiting synthesis of prostaglandins.[7] In streptozotocin-induced diabetic rats, alterations of blood glucose, glycated hemoglobin, and antioxidant status were normalized by chronic administration extract of hibiscus. In liver tissue, the extract stimulated AKT and PI3K, and enhanced the expressions of Nrf2. Moreover, the inflammatory agents NF-κB and MAPK were dampened.[8] Thus, hibiscus exerts a range of actions including antioxidant, anti-inflammatory, and insulin-enhancing effects. Such effects in the brain would be expected to exert antidepressant effects.

7.32.3 ANTIDIABETIC/ANTI-METABOLIC SYNDROME

A similar study showed similar antidiabetic effects of hibiscus in streptozotocin-induced diabetic rats. Moreover, histopathological examination of liver found reduced fatty deposition, less lipid peroxidation, and decreases in inflammatory cell infiltration in the streptozotocin-treated animals that also received hibiscus extract.[9]

7.32.4 PRECLINICAL ANTIDEPRESSANT-LIKE EFFECTS

Many studies in rodents have demonstrated antidepressant-like effects of hibiscus. For example, acute administration of a methanol extracts of hibiscus to mice decreased immobility in the forced swim and tail suspension tests.[10] Methanol extracts of the closely related *Hibiscus tiliaceus*[11] and seed from *Hibiscus esculentus*[12] also decreased immobility in those tests.

In yet another study, ethanolic extract of hibiscus showed the antidepressant-like effect in mice in the forced swim and tail suspension tests. Also, effects of the extract on MAO-A and MAO-B were evaluated in mitochondria-rich homogenates of mouse brain. The extracts contained components that substantially reduced activity of MAO-A, though not the MAO-B enzyme.[13]

Daily administration of extract of the root bark of *Hibiscus syriacus* prevented depression-like effects of chronic restraint stress in mice as evidenced by decreases in immobility in the forced swim and tail suspension tests. It also restored appetite for sucrose in the restrained animals. Treatment reduced stress-induced levels of corticosterone, and in the brain, it increased the phosphorylation of CREB protein and the expression of BDNF. *In vitro*, the extract protected cultured human neuroblastoma cells from glucocorticoid activation oxidative damage.[14]

7.32.5 HUMAN ANTIDEPRESSANT EFFECTS

It is not uncommon for herbalists to recommend hibiscus as part of a treatment regimen for depression.[15] Nonetheless, there are no clinical studies of the use of hibiscus to treat MDD in human subjects. The German Commission E states that there are no known risks in using hibiscus. However, at the same time they say that there are no established therapeutic uses of the herb.[16]

7.32.6 Dosage

As an example of dose, an uncontrolled clinical trial using the ethanolic extract of *Hibiscus rosa-sinensis* flowers was carried out on 21 women of reproductive age by administering 750 mg/day in three divided doses.[17]

7.32.7 Toxicity

The LD50 of intraperitoneally administered dried ethanolic extract of dried hibiscus leaves in mice was reported to be 1.533 g/kg.[18] This would extrapolate to approximately 105 g for a 75 kg human adult.

7.32.8 Safety in Pregnancy

The *Botanical Safety Handbook* notes no known adverse effects of the use of hibiscus during pregnancy.[19]

7.32.9 Drug Interactions

The *Botanical Safety Handbook* notes no known interactions with drugs.

REFERENCES

1. Puckhaber LS, Stipanovic RD, Bost GA. Analyses for flavonoid aglycones in fresh and preserved Hibiscus flowers. In: Janick J, Whipkey A, editors. *Trends in New Crops and New Uses.* Alexandria, VA: ASHS Press; 2002. p. 556–563.
2. Panesar G, Kumar A, Sharma A. In vivo antianxiety and antidepressant activity of *Hibiscus sabdariffa* calyx extracts. *J Pharm Res.* 2017;11(8):962–966.
3. Prasad MP. In vitro phytochemical analysis and antioxidant studies of *Hibiscus* species. *Int J Pure Appl Biosci.* 2014;2(3):83–88.
4. Ghaffar FRA, El-Elaimy IA. In vitro, antioxidant and scavenging activities of *Hibiscus rosa sinensis* crude extract. *J Appl Pharm Sci.* 2012;2(2):51–58.
5. Nade VS, Dwivedi S, Kawale LA, Upasani CD, Yadav AV. Effect of *Hibiscus rosa sinensis* on reserpine-induced neurobehavioral and biochemical alterations in rats. *Indian J Exp Biol.* 2009;47(7):559–563.
6. Nade VS, Kawale LA, Dwivedi S, Yadav AV. Neuroprotective effect of *Hibiscus rosa sinensis* in an oxidative stress model of cerebral post-ischemic reperfusion injury in rats. *Pharm Biol.* 2010;48(7):822–827.
7. Tomar V, Kannojia P, Jain KN, Dubey KS. Anti-nociceptive and antiinflammatory activity of leaves of *Hibiscus-Rosa sinensis. Int J Res Ayurveda Pharm.* 2010;1(1): 201–205.
8. Pillai SS, Min S. Polyphenols rich *Hibiscus rosa sinensis* Linn. petals modulate diabetic stress signaling pathways in streptozotocin-induced experimental diabetic rats. *J Funct Food.* 2016;20:31–42.

9. Sankaran M, Vadivel A. Antioxidant and antidiabetic effect of *Hibiscus rosa sinensis* flower extract on streptozotocin-induced experimental rats-a dose response study. *Not Sci Biol.* 2011;3(4):13–21.
10. Shewale PB, Patil RA, Hiray YA. Antidepressant-like activity of anthocyanidins from *Hibiscus rosa-sinensis* flowers in tail suspension test and forced swim test. *Indian J Pharmacol.* 2012;44(4):454–457.
11. Vanzella C, Bianchetti P, Sbaraini S, et al. Antidepressant-like effects of methanol extract of *Hibiscus tiliaceus* flowers in mice. *BMC Complement Altern Med.* 2012;12:41.
12. Ebrahimzadeh MA, Nabavi SM, Nabavi SF. Antidepressant activity of *Hibiscus esculentus* L. *Eur Rev Pharm.* 2013;17(19):2609–2612.
13. Khalid L, Rizwani GH, Sultana V, Zahid H, Khursheed R, Shareef H. Antidepressant activity of ethanolic extract of *Hibiscus rosa sinensis* Linn. *Pak J Pharm Sci.* 2014;27(5):1327–1331.
14. Kim YH, Im A, Park BK, et al. Antidepressant-like and neuroprotective effects of ethanol extract from the root bark of *Hibiscus syriacus* L. *BioMed Res Int.* 2018; Article ID 7383869.
15. Yarnell E, Abascal K. Botanical treatments for depression: Part 2—Herbal corrections for mood imbalances. *Altern Complement Ther.* 2001;7(3):138–143.
16. Blumenthal M, Busse WR, Goldberg A, et al. *The Complete German Commission E Monograph: Therapeutic Guide to Herbal Medicines.* Austin, TX: American Botanical Council; 1998. p. 336.
17. Jadhav VM, Thorat RM, Kadam VJ, Sathe NS. *Hibiscus rosa sinensis* Linn—"Rudrapuspa": A review. *J Pharm Res.* 2009;2(7):1168–1173.
18. Singh NR, Nath AK, Agarwal RP. A pharmacological investigation of some indigenous drugs of plants origin for evaluation of their antipyretic, analgesic and anti-inflammatory activities. *J Res Indian Yoga Homeopath.* 1978;13:58–62.
19. Gardner Z, McGuffin M, editors *Botanical Safety Handbook.* 3rd ed. Boca Raton, FL: CRC Press; 2013. p. 433.

7.33 *HUMULUS LUPULUS* (HOPS)

Humulus lupulus, the flowers of which are known as hops, is a perennial plant in the hemp family (Cannabinaceae) that is native to the temperate zones of Europe, Asia, and the Americas. It is best known for its use in the brewing of beer. It lends a hearty, bitter flavor to the beverage and serves as a preservative owing to its antimicrobial properties. However, *Humulus lupulus* has also been used for centuries as a medicinal herb. Reports on the medicinal uses of *Humulus lupulus* date back to the Middle Ages. The oldest report may be an eleventh century book by the Arabic physician Mesue, who described anti-inflammatory properties of *Humulus lupulus*. In the thirteenth century, the Arabic botanist Ibn Al-Baytar noted soothing effects of the herb. Among the phytochemicals identified in *Humulus lupulus* are xanthohumol, humulone, cohumulone, adhumulone, lupulone, colupulone, adlupulone, catechin, quercetin, kaempferol, desmethylxanthohumol, 6-prenylnaringenin, 8-prenylnaringenin, ferulic acid, and resveratrol.[1]

7.33.1 ANTIOXIDANT

Research has shown *Humulus lupulus* and its constituent phytochemicals to have antioxidant, anti-inflammatory and antidiabetic effects. Xanthohumol (XN), extracted from *Humulus lupulus* and beer, has antioxidant activity higher than that of α-tocopherol (vitamin E) in inhibiting *in vitro* oxidation of human low-density lipoprotein.[2] Elderly women whose diets were supplemented for 45 days with two daily servings of alcohol-free beer containing *Humulus lupulus* had reductions in serum levels of oxidized low-density lipoproteins, thiobarbituric acid-reactive substances, and plasma carbonyl groups. Serum levels of α-tocopherol and erythrocytic glutathione were significantly increased.[3]

7.33.2   ANTI-INFLAMMATORY

Extract of *Humulus lupulus* blocked the production of inflammatory prostaglandin E2 in lipopolysaccharide (LPS)-stimulated human peripheral blood mononuclear cells. This was due to a selective inhibition of COX-2.[4] Similarly, humulone from *Humulus lupulus* inhibited phorbol ester-induced COX-2 expression in mouse skin by blocking activation of NF-κB.[5]

Both antioxidant and anti-inflammatory effects are likely responsible for neuroprotective effects of xanthohumol against hypoxic damage to brain tissue. Intraperitoneal administration of xanthohumol prior to middle cerebral artery occlusion in rats attenuated focal cerebral ischemia and improved subsequent neurobehavioral deficits. Levels of hypoxia-inducible factor-1α, TNF-α, inducible nitric oxide synthase, and active caspase-3 protein expressions were significantly reduced in ischemic regions.[6]   •

7.33.3   ANTIDIABETIC/ANTI-METABOLIC SYNDROME

Humulus lupulus has antidiabetic effects, enhances insulin sensitivity, and lowers the risk of Metabolic Syndrome. Administration of tetrahydro iso-alpha acids extracted from *Humulus lupulus* to high fat diet-fed obese and diabetic mice for eight weeks reduced body weight gain, glucose intolerance, and fasting hyperinsulinemia, and normalized insulin sensitivity markers. The treatment also increased plasma levels of the anti-inflammatory cytokine interleukin-10 and decreased levels of the pro-inflammatory cytokine granulocyte colony-stimulating factor.[7] A modified hop extract, Rhoisoalpha acids (RIAA) increased lipid accumulation in murine 3T3-L1 adipocyte by 225%, compared to 140% for troglitazone and 125% for rosiglitazone. The extract also improved insulin signaling through stimulation of the PI3K/Akt pathway, inhibition of GSK-3β, and reduction of IRS-1pS307, a phosphorylated form of the insulin receptor that is partially deactivated.[8] Isohumulones extracted from *Humulus lupulus* reduce insulin resistance by activation of peroxisome proliferator-activated receptors α and γ.[9]

7.33.4   PRECLINICAL ANTIDEPRESSANT-LIKE EFFECTS

Preclinical studies have provided evidence of antidepressant-like effects of *Humulus lupulus*. Extract of *Humulus lupulus* reduced immobility time of rats in the forced swim test.[10] Treatment with *Humulus lupulus*, in the form of non-alcoholic beer, similarly reversed the depression-like behavioral deficits in rats caused the pro-oxidant and inflammatory substance, aluminum nitrate. *Humulus lupulus* normalized performance in the forced swim test and reversed the oxidative and inflammatory damage in the brain.[11]

7.33.5 Human Antidepressant Effects

Humulus lupulus has long been used in folk medicine for insomnia, often a component of MDD. The German Commission E has approved *Humulus lupulus* for the treatment of restlessness, anxiety, mood disturbances, and sleep disturbances.[12] In an open-labeled study, nightly consumption of non-alcoholic beer, a good source of *Humulus lupulus*, significantly improved sleep and eased anxiety.[13]

There is only one published report on the effects of *Humulus lupulus* on MDD in human subjects. In a small, but randomized, placebo-controlled, double-blind study, the effects of *Humulus lupulus* were tested in otherwise healthy young adults with evidence of mild depression or anxiety per the Depression Anxiety Stress Scale-21. After four weeks of treatment, *Humulus lupulus* produced significant improvements in depression and anxiety scores. AM cortisol levels were also measured, but no differences were noted between treatment and control groups.[14]

The 8-prenylnaringenin found in *Humulus lupulus* is a potent phytoestrogen.[15] In a prospective, randomized, double-blind, placebo-controlled study, extracts of *Humulus lupulus* standardized to contain 8-prenylnaringenin (100 μg or 250 μg per dose) relieved symptoms in menopausal women. The responses were determined by means of a modified Kupperman index (KI) and a patients' questionnaire. Unfortunately, specific mood data was not included.[16]

7.33.6 Dosage

Marciano and Vizniak recommend dosage of 1 cup three times a day of infusion from 1 teaspoon of *Humulus lupulus* per cup. Alternatively, they suggest tincture (1:5, 60%) 2–3ml three times a day.[17]

7.33.7 Toxicity

Toxicological studies in animals stated that LD50 for orally administered hop extract in mice ranges from 500–3,500 mg/kg.[18] Extrapolated to a 75 kg human adult, these doses would be approximately 35–245 g.

7.33.8 Safety in Pregnancy

The *Botanical Safety Handbook* notes there is no definitive guidance from the literature concerning the safety of this herb during pregnancy.[19]

7.33.9 Drug Interactions

The *Botanical Safety Handbook* notes no known interactions with drugs. Extracts of *Humulus lupulus* do significantly inhibit a variety of cytochrome P450 enzymes, including CYP2C8, CYP2C9, CYP2C19, and CYP1A2.[20]

REFERENCES

1. Biendl M, Pinzl C. *Hops and Health. Uses, Effects, History.* Wolznach: German Hop Museum Wolznach; 2008.
2. Miranda CL, Stevens JF, Ivanov V, et al. Antioxidant and prooxidant actions of prenylated and nonprenylated chalcones and flavanones in vitro. *J Agric Food Chem.* 2000;48(9):3876–3884.
3. Alvarez JRM, Bellés VV, López-Jaén AB, Marín AV, Codoñer-Franch P. Effects of alcohol-free beer on lipid profile and parameters of oxidative stress and inflammation in elderly women. *Nutrition.* 2009;25(2):182–187.
4. Hougee S, Faber J, Sanders A, et al. Selective inhibition of COX-2 by a standardized CO2 extract of *Humulus lupulus* in vitro and its activity in a mouse model of zymosan-induced arthritis. *Planta Med.* 2006;72(3):228–233.
5. Lee JC, Kundu JK, Hwang DM, Na HK, Surh YJ. Humulone inhibits phorbol ester-induced COX-2 expression in mouse skin by blocking activation of NF-κB and AP-1: IκB kinase and c-JunN-terminal kinase as respective potential upstream targets. *Carcinogenesis.* 2007;28(7):1491–1498.
6. Yen T-L, Hsu CK, Lu WJ, et al. Neuroprotective effects of xanthohumol, a prenylated flavonoid from hops (*Humulus lupulus*), in ischemic stroke of rats. *J Agric Food Chem.* 2012;60(8):1937–1944.
7. Everard A, Geurts L, Van Roye M, Delzenne NM, Cani PD. Tetrahydro iso-alpha Acids from Hops Improve glucose homeostasis and reduce body weight gain and metabolic endotoxemia in high-fat diet-fed mice. *PLoS One.* 2012;7(3):e33858.
8. Tripp ML, Pacioretty L, Konda VR, et al. Selective kinase response modulators (SKRMS) from *Humulus lupulus* and Acacia nilotica modulate multiple kinases and improve insulin sensitivity in vitro and in vivo. *FASEB J.* 2007;21(5).
9. Yajima H, Ikeshima E, Shiraki M, et al. Isohumulones, bitter acids derived from hops, activate both peroxisome proliferator-activated receptor α and γ and reduce insulin resistance. *J Biol Chem.* 2004;279(32):33456–33462.
10. Zanoli P, Rivasi M, Zavatti M, Brusiani F, Baraldi M. New insight in the neuro-pharmacological activity of *Humulus lupulus* L. *J Ethnopharmacol.* 2005;102(1): 102–106.
11. Merino P, Santos-López JA, Mateos CJ, et al. Can nonalcoholic beer, silicon and hops reduce the brain damage and behavioral changes induced by aluminum nitrate in young male Wistar rats? *Food Chem Toxicol.* 2018;118:784–794.
12. Blumenthal M, Busse WR, Goldberg A, et al. The Complete German Commission E Monograph: Therapeutic Guide to Herbal Medicines. Austin, TX: American Botanical Council; 1998. p. 147.
13. Franco L, Sánchez C, Bravo R, et al. The sedative effect of non-alcoholic beer in healthy female nurses. *PLoS One.* 2012;7(7):e37290.
14. Kyrou I, Christou A, Panagiotakos D, et al. Effects of a hops (*Humulus lupulus* L.) dry extract supplement on self-reported depression, anxiety and stress levels in apparently healthy young adults: A randomized, placebo-controlled, double-blind, crossover pilot study. *Hormones.* 2017;16(2):171–180.
15. Milligan S, Kalita J, Pocock V, et al. Oestrogenic activity of the hop phyto-oestrogen, 8-prenylnaringenin. *Reproduction.* 2002;123(2):235–242.
16. Heyericka A, Vervarcke S, Depypere H, Bracke M, De Keukeleire D. A first prospective, randomized, double-blind, placebo-controlled study on the use of a standardized hop extract to alleviate menopausal discomforts. *Maturitas.* 2006;54(2):164–175.
17. Marciano M, Vizniak NA. *Botanical Medicine.* Toronto, ON: Prohealth; 2016. p. 225.
18. Hänsel R, Keller K, Rimpler H, Schneider G. *Hagers Handbuch der Pharmazeutische Praxis, Hrsg.* Berlin: Springer Verlag; 1993. p. 447–458.

19. Gardner Z, McGuffin M, editors. *Botanical Safety Handbook*. 3rd ed. Boca Raton, FL: CRC Press; 2013. p. 439.
20. Yuan Y, Qiu X, Nikolić D, et al. Inhibition of human cytochrome P450 enzymes by hops (*Humulus lupulus*) and hop prenylphenols. *Eur J Pharm Sci*. 2014;53(12):55–61.

7.34   *HUPERZIA SERRATA*

Huperzia serrata is a species of the *Huperziaceae* family. The plant is also known as "*qian ceng ta*" or Chinese Club Moss, and tea made from it has long been used in Traditional Chinese Medicinal system to enhance memory and treat fever and inflammatory conditions. *Huperzia serrata* contains a group of alkaloids known as Lycopodium alkaloids. The best known are huperzine A and huperzine B. Huperzine A, in particular, has been focused upon as it has potent cholinesterase inhibiting effects that make it a potentially important treatment of dementia. However, there is a vast array of other lycopodium alkaloids contained in *Huperzia serrata*. For example, the huperzines extend from huperzine A through V. There are also various serratidines, phlegmariurines, obscurines, and fawcettimines.[1] Moreover, along with the lycopodium alkaloids, there are other more common phytochemicals, such as flavonoids, lignins, phenolic acids, and terpenes, that have been isolated from *Huperzia serrata*. These have included p-hydroxybenzoic, vanillic, p-coumaric, and ferulic acids; p-hydroxybenzaldehyde, vanillin, acetovanillone, vanilloyl methyl ketone, α-hydroxy- and α-ethoxypropiovanillone.[2] Some of these phytochemicals have themselves been associated with antidepressant effects.

7.34.1   Antioxidant/Anti-Inflammatory Effects

Huperzine A was found to protect rat pheochromocytoma cells from the oxidative damage of H_2O_2. Pretreatment of the cells with huperzine A prior to H_2O_2 exposure significantly elevated antioxidant enzyme activities and decreased the level of lipid peroxidation.[3] The less studied alkaloid of *Huperzia serrata*, huperzine B, was later found to similarly protect against H_2O_2 with the same antioxidant properties.[4]

In infarcted rat heart tissue, huperzine A elevated activities of superoxide dismutase and glutathione peroxidase. It also reduced levels of the marker of oxidative damage, malondialdehyde, and increased concentration of glutathione. In the same study, huperzine A markedly inhibited the expression of the inflammatory cytokines, NF-κB, TNF-α, and IL-1β.[5]

In mice with induced experimental autoimmune encephalomyelitis, intraperitoneal injections of huperzine A suppressed the expression of monocyte chemoattractant protein-1, IL-6, TNF-α, and IL-1β in the spinal cord. Huperzine A also blocked release of monocyte chemoattractant protein-1 from cultured mouse astrocytes exposed to lipopolysaccharide, though it had no effect on production of IL-6, TNF-α, and IL-1β in those cells.[6]

7.34.2   Antidiabetic/Anti-Metabolic Syndrome

There are no references in the literature to such effects.

7.34.3 PRECLINICAL ANTIDEPRESSANT-LIKE EFFECTS

The active alkaloid of *Huperzia serrata*, huperzine A, has been used to treat memory loss and post-stroke changes. It was found to have both neuroprotective and antidepressant effects in a study of rats that had been subjected to middle cerebral artery occlusion followed by 18 days of chronic unpredictable mild stress. Administration of huperzine A over four weeks reversed some of the cognitive and memory deficits, as shown by improved performance in the water maze test. However, it also restored appetite for sucrose and reduced immobility in the forced swim test, which are taken to be indications of antidepressant effects. Chemical analyses revealed that the treatment upregulated hippocampal expression of 5-HT1A receptors, phosphorylated CREB and BDNF. It also increased levels of norepinephrine, serotonin, and dopamine in the hippocampus and prefrontal cortex.[7]

Stimulation of neurogenesis is thought to play a major role in mechanism of action of antidepressants. In a study of mice, huperzine A not only promoted the proliferation of cultured mouse embryonic hippocampal neural stem cells (NSCs), but also increased the newly generated cells in the subgranular zone of the hippocampus in adult mice. Huperzine A acted at least in part through stimulation of the mitogen-activated protein kinase/extracellular signal-regulated kinase (MAPK/ERK) signaling pathway, which was a well-known regulator of cell proliferation and differentiation. This effect of huperzine was blocked by an inhibitor of ERK.[8] Huperzine A also inhibits GSK3α/β activity, which is associated with antidepressant effects, and enhances β-catenin activity in mouse brain.[9] Higher beta-catenin signaling increases resiliency under stress, whereas defective beta-catenin signaling is associated with depression.[10]

7.34.4 HUMAN ANTIDEPRESSANT EFFECTS

All of the research into the effects of huperzine A on mood and cognitive function in humans has been performed in China. Thankfully, an English language paper published in the *Shanghai Archives of Psychiatry* describes some of the most recent studies in the use of huperzine to augment antidepressants in the treatment of MDD.[11] In this paper, Zheng describes three such studies in which the effects of an antidepressant alone were compared against those of the antidepressant plus huperzine A. All three studies were open label, and thus their overall quality was judged as poor. It should also be noted that in these studies, the ages of the subjects ranged from 16 to 60 years of age. Thus, it is not clear if the main target of the huperzine was cognitive loss and dementia, or if more general effects of the alkaloid were being explored. Nonetheless, the studies are instructive. In one study, the addition of huperzine to venlafaxine significantly improved scores on the Hamilton Depression Scale.[12] In another study, in which huperzine A was added to fluoxetine,[13] there was no significant difference between the groups in Hamilton scores. However, quality of life, as measured by the World Health Organization Quality of Life assessment, was significantly better in individuals who received combined treatment of fluoxetine and huperzine A. In the third study, in which huperzine A again augmented fluoxetine, improvements in cognitive function were noted, but there were no differences in the Hamilton scores.[14]

It is important to state that in a large Phase II trial, huperzine A dosed at 200 μg twice daily was found not to be effective in the treatment of mild to moderate Alzheimer's disease. Although small differences from placebo were noted, these were not significant.[15] However, in at least one other clinical trial, twice daily dosing of 200 μg huperzine A was found to be quite effective in the treatment of this condition.[16] Research also appears to have shifted toward ZT-1, a synthetic derivative of huperzine A.

7.34.5 DOSAGE

In 1997, Huperzine A was classified by the Food and Drug Administration as a dietary supplement, and it has since been marketed in the United States as a component of powdered *Huperzia serrata* in a twice-daily tablet containing 200–400 micrograms of huperzine A alkaloid. [17]

7.34.6 TOXICITY

A study in rats indicated that the LD50 of huperzine A, the most active component of *Huperzia serrata*, is 2–4 mg/kg, which is well above the typical human dose. Thus, the therapeutic window for *Huperzia serrata* is quite wide.[18]

7.34.7 SAFETY IN PREGNANCY

Review of the literature revealed no studies of the safety of *Huperzia serrata* during human or animal pregnancy. However, in a major review it was advised that the lack of information should dissuade pregnant women from using the herb.[19]

7.34.8 DRUG INTERACTIONS

Review of the literature revealed no studies of the possible interactions of *Huperzia serrata* with commonly used medications. However, it has been reported that Huperzine A, the primary active component of *Huperzia serrata*, is not metabolized by the human liver, but rather is excreted virtually unchanged by the kidneys. Thus, it is unlikely to cause clinically relevant drug–drug interactions when co-administrated with drugs that are metabolized by the cytochrome P450 isoenzyme systems.[20]

REFERENCES

1. Ma X-Q, Tan C, Zhu D, Gang DR, Xiao P. Huperzine A from *Huperzia* species—An ethnopharmacolgical review. *J Ethnopharmacol.* 2007;113(1):15–34.
2. Towers GHN, Maas WSG. Phenolic acids and lignins in the Lycopodiales. *Phytochemistry.* 1965;4(1):57–66.
3. Xiao XQ, Yang JW, Tang XC. Huperzine A protects rat pheochromocytoma cells against hydrogen peroxide-induced injury. *Neurosci Lett.* 1999;275(2):73–76.
4. Zhang HY, Tang XC. Huperzine B, a novel acetylcholinesterase inhibitor, attenuates hydrogen peroxide induced injury in PC12 cells. *Neurosci Lett.* 2000;292(1):41–44.

5. Sui X, Gao C. Huperzine A ameliorates damage induced by acute myocardial infarction in rats through antioxidant, anti-apoptotic and anti-inflammatory mechanisms. *Int J Mol Med*. 2014;33(1):227–233.

6. Tian GX, Zhu XQ, Chen Y, Wu GC, Wang J. Huperzine A inhibits CCL2 production in experimental autoimmune encephalomyelitis mice and in cultured astrocyte. *Int J Immunopathol Pharmacol*. 2013;26(3):757–764.

7. Duac Y, Liang H, Zhang L, Zhang L, Fu F. Administration of huperzine A exerts antidepressant-like activity in a rat model of post-stroke depression. *Pharmacol Biochem Behav*. 2017;158:32–38.

8. Ma T, Gong K, Yan Y, et al. Huperzine A promotes hippocampal neurogenesis in vitro and in vivo. *Brain Res*. 2013;1506:35–43.

9. Wang CY, Zheng W, Wang T, et al. Huperzine A activates Wnt/β-catenin signaling and enhances the nonamyloidogenic pathway in an Alzheimer transgenic mouse model. *Neuropsychopharmacology*. 2011;36(5):1073–1089.

10. Dias C, Feng J, Sun H, et al. β-catenin mediates behavioral resilience through Dicer1/microRNA regulation. *Nature*. 2014;516(7529):51–55.

11. Zheng W, Xiang YQ, Ungvari GS, et al. Huperzine A for treatment of cognitive impairment in major depressive disorder: A systematic review of randomized controlled trials. *Shanghai Arch Psychiatry*. 2016;28(2):64–71.

12. Liu SZ, Wang PJ, Yin AJ, Dang XJ, Guang H. Effects of huperzine A combined with venlafaxine for patients with depression. *Zhongguo Shi Yong Yi Yao*. 2010;5(11):151–152.

13. Gao YF, Li J, Meng H, et al. Effects of huperzine on cognition, function and life quality of patients with depression. *Chongqing Yi Xue*. 2007;36(6):483–485.

14. Yang ZB, Deng XM, Zhang GX, Yu XR. The study of huperzine combined with fluoxetine on cognition function of patients with depression. *Lin Chuang Jing Shen Yi Xue Za Zhi*. 2010;20(6):418–419.

15. Rafii MS, Walsh S, Little JT, et al. A phase II trial of huperzine A in mild to moderate Alzheimer disease. *Neurology*. 2011;76(16):1389–1394.

16. Zhang Z, Wang X, Chen Q, Shu L, Wang J, Shan G. Clinical efficacy and safety of huperzine alpha in treatment of mild to moderate Alzheimer disease, a placebo-controlled, double-blind, randomized trial. *Zhonghua Yi Xue Za Zhi*. 2002;82(14):941–944.

17. Ferreira A, Rodrigues M, Fortuna A, Falcão A, Alves G. Huperzine A from *Huperzia serrata*: A review of its sources, chemistry, pharmacology and toxicology. *Phytochem Rev*. 2016;15(1):51–85.

18. Ha GT, Wong RK, Zhang Y. Huperzine A as potential treatment of Alzheimer's disease: An assessment on chemistry, pharmacology, and clinical studies. *Chem Biodivers*. 2011;7(7):1189–1204.

19. Jacknowitz AI, Tracy TS. Huperzine. In: Cupp MJ, Tracy TS, editors. *Forensic Science: Dietary Supplements: Toxicology and Clinical Pharmacology*. Totowa, NJ: Hurnana Press Inc.; 2003.

20. Lin P-P, Li X-N, Yuan F, Chen WL, Yang MJ, Xu HR. Evaluation of the in vitro and in vivo metabolic pathway and cytochrome P450 inhibition/induction profile of huperzine A. *Biochem Biophys Res Commun*. 2016; 480(2):248–253.

7.35 *HYPERICUM PERFORATUM* (ST JOHN'S WORT)

Hypericum perforatum, widely known as St John's wort, is an herbaceous perennial plant native to Europe and Asia. It has been introduced into the United States where it has naturalized.[1] It has been used widely in both Chinese and Western herbal medicine. In Chinese it is *Guan Ye Lian Qiao* and, along with Siberian ginseng, it is an ingredient in the Chinese herbal treatment for depression *Shuganjieyu*. It was also known by Hippocrates and recommended by the great ancient Greek herbalist, Dioscorides.

Hypericum perforatum has been used traditionally for the treatment of agitation, neuralgia, fibrositis, sciatica, menopausal neurosis, anxiety, depression, and as a nerve tonic.[2] The herb has been very popular for the treatment of MDD, both in the United States and Europe. In 1984, the German Commission E designated *Hypericum perforatum* as an approved herb, and its safety and effectiveness are reevaluated periodically. *Hypericum perforatum* is likely the best studied herbal antidepressant, with both preclinical and clinical studies lending support for its efficacy.

The primary active components in *Hypericum perforatum* are hyperforin and hypericin. However, the plant also contains a variety of phenolics, many of which are known to have salutary effects in the brain, for example, the flavonoids kaempferol, quercetin, and luteolin.[3]

7.35.1 ANTIOXIDANT

Hypericum perforatum has potent antioxidant effects. Extracts of the herb and its constituent phenolics were shown to exert free radical scavenging capacity in the *in vitro* DPPH and APH assays, and antioxidant potential in the ascorbate/iron lipid peroxidation test. Extract of *Hypericum perforatum* also effectively scavenged NO.[4] Extract of *Hypericum perforatum* also protected rats from the oxidative hepatoxic damage from CCl_4. Pretreatment with crude alcoholic extract significantly reduced serum transaminase levels. In liver homogenates, levels of lipid peroxidation were also decreased, whereas glutathione levels were increased by the extract.[5]

7.35.2 ANTI-INFLAMMATORY

The anti-inflammatory effects of St John's wort are well established. Tincture of *Hypericum perforatum* reduced by 48% the anti-inflammatory effect of injecting formalin beneath the plantar aponeurosis of albino rats. This effect was comparable to that of hydrocortisone.[6] The anti-inflammatory effect of *Hypericum perforatum* is sufficient to be effective in treating atopic dermatitis in human subjects.[7] Hyperforin exhibits significant anti-inflammatory effects due in part to suppression of PGE2 biosynthesis by inhibiting activity of microsomal PGE2 synthase in rats.[8] In fact, *Hypericum perforatum* inhibits inflammatory responses by a variety of mechanisms. For example, *Hypericum perforatum* inhibits Substance P-induced synthesis of IL-6 in human astrocytoma cells.[9] Extract of *Hypericum perforatum* reduces inflammatory iNOS expression in human cell lines by reducing Janus kinase 2 activity and, in turn, STAT-1 α tyrosine phosphorylation. The more common pharmacological inhibition of iNOS is through blocking of NF-kβ.[10]

7.35.3 ANTIDIABETIC/ANTI-METABOLIC SYNDROME

Reports on the effects of *Hypericum perforatum* on diabetes and insulin sensitivity are mixed. However, unlike reports on many herbs used in the treatment of MDD, the predominance of data suggests the herb worsens insulin resistance and glucose tolerance. For example, it was found that long-term treatment with *Hypericum perforatum* impairs glucose tolerance by reducing insulin secretion in young, healthy men.[11] In an *in vitro* study, *Hypericum perforatum* inhibited differentiation of adipocytes and decreased insulin sensitivity in mature adipocytes.[12] In yet another study, *Hypericum perforatum* was found to attenuate insulin-sensitive glucose uptake in human adipocytes, apparently by blocking IRS-1 tyrosine phosphorylation, which is similar to the effect of the inflammatory cytokine, TNF-α, Interestingly, neither of those effects were due to hyperforin or hypericin.[13] These data suggest both that the apparent antidepressant effects of *Hypericum perforatum* are due to effects other than enhancing insulin sensitivity, and that use would be imprudent in individuals with diabetes type II or significant metabolic syndrome.

7.35.4   PRECLINICAL ANTIDEPRESSANT-LIKE EFFECTS

Pre-clinical studies have verified antidepressant effects of *Hypericum perforatum* as well as helped elucidate the mechanisms by which it acts in the brain. In a study using mice, doses of extract of *Hypericum perforatum* reversed depression-like behavioral effects of chronic administration of high doses of corticosterone. The herb also reversed corticosterone-induced decreases in hippocampal cell proliferation and reduction in axonal spine density in hippocampal neurons.[14] Somewhat consistent with that finding, hyperforin stimulated the expression of TRPC6 channels and TrkB and CREB activity in cultured mouse cortical neurons. Curiously, *in vivo*, hyperforin augmented the expression of TrkB in the cortex, but not in the hippocampus where neurogenesis remained unchanged.[15]

In another report, hypericin inhibited release of glutamate from rat cerebrocortical synaptosomes.[16] That is, in part, the mechanism by which the anticonvulsant lamotrigine imparts its mood-stabilizing effect.[17] One study has shown that hyperoside, a galactoside of quercetin that appears in small quantity in *Hypericum perforatum*, activates the PI3K/Akt cascade, spares mitochondria, and prevents activation of a mitochondrial pathway of apoptosis.[18]

Hypericum perforatum appears to have multiple mechanisms of action in producing antidepressant effects. *Hypericum perforatum*, particularly hyperforin, inhibits neuronal uptake not only of serotonin and noradrenaline – the common mechanisms of action of standard antidepressant medications – but also of dopamine, gamma-aminobutyric acid, and l-glutamate. Moreover, hyperforin does not act as a competitive inhibitor at the transmitter binding sites of the transporter proteins, but rather it affects the sodium gradient that in turn leads to an inhibition of transmitter uptake.[19] In a 2014 paper by Sell et al.,[20] it was admitted that we still do not understand the mechanism of action of hyperforin, the primary antidepressant component of *Hypericum perforatum*. Among the possibilities noted were: inhibition of 5-lipoxygenase; high affinity binding to the pregnane X receptor; release of Ca2+ and/or Zn2+ from intracellular stores; as well as the prominent hypothesis of diminishing reuptake of serotonin, dopamine, norepinephrine, acetylcholine, GABA, and glutamate. An earlier study had purported to show that hyperforin reduces uptake of the variety of neurotransmitters through activation of the somewhat mysterious neuronal channel protein, TRPC6. The activation of TRPC6 by hyperforin was also said to induce neuronal axonal sprouting, such as would be seen after the addition of various nerve growth factors.[21] However, Sell et al. found that the large hyperforin molecule might act as a protonophore independently of TRPC6, thereby altering local membrane electrical currents and affecting neurotransmitter uptake. In any case, the mechanism of action of *Hypericum perforatum* is clearly unusual, and its full elucidation may point the way to development of unique antidepressants.

Some of the antidepressant effects of *Hypericum perforatum* may be due to its amelioration of the effects of stress. Subchronic treatment with *Hypericum perforatum* decreases levels of corticosterone in rat brain, though not in serum.[22] *Hypericum perforatum* also buffers against chronic stress. In rats subjected to chronic restraint stress, treatment with *Hypericum perforatum* reduced plasma levels of ACTH, corticosterone, and TNF-α. It also prevented atrophy of thymus and spleen in stressed animals. These effects of *Hypericum perforatum* were similar to those of fluoxetine.[23]

7.35.5 HUMAN ANTIDEPRESSANT EFFECTS

Hypericum perforatum is perhaps the most thoroughly researched herbal treatment of MDD. In a 2008 review written for the well-respected *Cochrane Review*, authors analyzed the results of 29 studies, including comparisons with placebo and comparisons with standard antidepressant medications in adults suffering from mild to moderately severe symptoms. It was concluded that "hypericum extracts tested in the included trials are superior to placebo in patients with MDD; are similarly effective as standard antidepressants; and have fewer side effects than standard antidepressants."[24] In studies that included severely depressed patients, *Hypericum perforatum* has not been so effective.[25]

There is a lack of randomized controlled studies of *Hypericum perforatum* in the treatment of MDD in children and adolescents. However, in an open study of children with mild to moderate depressive symptoms, *Hypericum perforatum* was judged by patents and physicians to be both safe and effective in up to 100% of patients after six weeks. Tolerability was deemed good and no adverse events were reported. The primary caveat was that nearly 25% of the 101 young patients failed to complete the study.[26] There are also few studies of *Hypericum perforatum* for the treatment of MDD in geriatric populations. However, in one randomized, double-blind study of elderly patients with mild or moderate depressive episodes, daily treatment with *Hypericum perforatum* extract was equivalent in antidepressant effect to 20 mg per day dosing of fluoxetine.[27]

Some herbs have been used in addition to standard antidepressants to augment their effects in treatment resistant patients. However, it is important to warn against such use of *Hypericum perforatum*. There are no formal studies of effects of combining the herb with standard antidepressants. However, there have been a number of case reports about the development of serotonin syndrome in individuals that added *Hypericum perforatum* to various antidepressants.[28] It is not clear if this is due to effects of the herb on P450 enzymes, or to more direct serotonergic activation by the herb.

7.35.6 Dosage

The German E Commission recommends 2–4 g of herb per day, or 0.2–1.0 mg of hypericin from standardized extract per day.[29]

7.35.7 Toxicity

The LD50 of intraperitoneally administered dried methanolic extract of *Hypericum perforatum* is about 450 mg/kg in mice.[30] Extrapolated to a 75 kg human adult, this would be approximately 32 g.

7.35.8 Safety in Pregnancy

The *Botanical Safety Handbook* notes little basis for concern about the use of *Hypericum perforatum* during pregnancy. Animals studies have proven the herb to be benign, whereas a limited number of case reports noted no adverse effects on mother or newborn.[31]

7.35.9 Drug Interactions

A number of clinically significant interactions with *Hypericum perforatum* have been identified with prescribed medicines including warfarin, phenprocoumon, cyclosporin, HIV protease inhibitors, theophylline, digoxin, and oral contraceptives resulting in a decrease in concentration or effect of the medicines. These interactions are probably due to the induction of cytochrome P450 isoenzymes CYP2D6, CYP3A4, CYP2C9, CYP1A2, and the transport protein P-glycoprotein.[32]

Hypericum perforatum is known to affect drug metabolism. Crude extracts of the herb demonstrated inhibition of the enzymes CYP2D6, CYP2C9, and CYP3A4, and to a lesser extent CYP1A2 and CYP2C19.[33] Many psychiatric medications are metabolized by these enzymes.[34] Thus, *Hypericum perforatum* might be expected to increase blood levels of such medications. This may explain some reports of serotonin syndrome in geriatric patients adding *Hypericum perforatum* to prescribed antidepressants.

REFERENCES

1. Barnes J, Anderson LA, Phillipson JD. St John's wort (*Hypericum perforatum* L.): A review of its chemistry, pharmacology and clinical properties. *JPP.* 2001;53:583–600.
2. Newall CA, Anderson LA, Phillipson JD. *Herbal Medicines. A Guide for Health-Care Professionals.* 1st ed. London: Pharmaceutical Press; 1996.
3. Bombardelli E, Morazzoni P. *Hypericum perforatum. Fitoterapia.* 1995;66:43–68.
4. Silva BA, Malva JO, Dias ACP. St. John's wort (*Hypericum perforatum*) extracts and isolated phenolic compounds are effective antioxidants in several in vitro models of oxidative stress. *Food Chem.* 2008;110(3):611–619.
5. Popovic M, Jakovljevic V, Mimica-Dukic N, Kaurinovic B, Cebovic T. Effects of different extracts of *Hypericum perforatum* L. on the CCl_4-induced hepatotoxicity in rats. *Oxid Commun.* 2002;25(2):273–278.

6. Varma PN, Kumar S, Lohar DR, Chaturvedi D, Gaur GD. A chemo-pharmacological study of *Hypericum perforatum* L.: Anti-inflammatory action on albino rats. *Br Homoeopath J.* 1988;77(1):27–29.

7. Schempp CM, Windeck T, Hezel S, Simon JC. Topical treatment of atopic dermatitis with St. John's wort cream—A randomized, placebo controlled, double blind half-side comparison. *Phytomedicine.* 2003;10(Suppl IV):31–37.

8. Koeberle A, Rossi A, Bauer J, et al. Hyperforin, an anti-inflammatory constituent from St. John's wort, inhibits microsomal prostaglandin E2 synthase-1 and suppresses prostaglandin E2 formation in vivo. *Front Pharmacol.* 2011;2(7):1–10.

9. Fiebich BL, Höllig A, Lieb K. Inhibition of substance P-induced cytokine synthesis by St. John's wort extracts. *Pharmacopsychiatry.* 2001;34(Suppl1):26–28.

10. Tedeschi E, Menegazzi M, Margotto D, Suzuki H, Förstermann U, Kleinert H. Anti-inflammatory actions of St. John's wort: Inhibition of human inducible nitric-oxide synthase expression by down-regulating signal transducer and activator of transcription-1α (STAT-1α) activation. *JPET.* 2003;307:254–261.

11. Stage TB, Damkier P, Christensen MMH, Nielsen LBK, Højlund K, Brøsen K. Impaired glucose tolerance in healthy men treated with St. John's wort. *BCPT.* 2016;118(3):219–224.

12. Amini ZJ, Boyd B, Doucet J, Ribnicky DM, Stephens JM. St. John's wort inhibits adipocyte differentiation and induces insulin resistance in adipocytes. *Biochem Biophys Res Commun.* 2009 Oct 9;388(1):146–149.

13. Richard AJ, Amini ZJ, Ribnicky DM, Stephens JM. St. John's wort inhibits insulin signaling in murine and human adipocytes. *Biochim Biophys Acta.* 2012;1822(4): 557–563.

14. Crupi R, Mazzon E, Marino A, et al. *Hypericum perforatum* treatment: Effect on behaviour and neurogenesis in a chronic stress model in mice. *BMC Complement Altern Med.* 2011;11:7.

15. Gibon J, Deloulme JC, Chevallier T, Ladevèze E, Abrous DN, Bouron A. The antidepressant hyperforin increases the phosphorylation of CREB and the expression of TrkB in a tissue-specific manner. *Int J Neuropsychopharmacol.* 2013;16(1):189–198.

16. Chang Y, Wang SJ. Hypericin, the active component of St. John's wort, inhibits glutamate release in the rat cerebrocortical synaptosomes via a mitogen-activated protein kinase-dependent pathway. *Eur J Pharmacol.* 2010;634(1–3):53–61.

17. Ketter TA, Manji HK, Post RM. Potential mechanisms of action of lamotrigine in the treatment of bipolar disorders. *J Clin Psychopharmacol.* 2003;23(5):484–495.

18. Zeng KW, Wang XM, Ko H, Kwon HC, Cha JW, Yang HO. Hyperoside protects primary rat cortical neurons from neurotoxicity induced by amyloid β-protein via the PI3K/Akt/Bad/BclXL-regulated mitochondrial apoptotic pathway. *Eur J Pharmacol.* 2011;672(1–3):45–55.

19. Müller WE. Current St. John's wort research from mode of action to clinical efficacy. *Pharmacol Res.* 2003;47(2):101–109.

20. Sell TS, Belkacemi T, Flockerzi V, Beck A. Protonophore properties of hyperforin are essential for its pharmacological activity. *Sci Rep.* 2014;4:7500.

21. Leuner K, Kazanski V, Müller M, et al. Hyperforin—A key constituent of St. John's wort specifically activates TRPC6 channels. *FASEB* 2007;21(14):4101–4111.

22. Franklin M, Reed A, Murck H. Sub-chronic treatment with an extract of *Hypericum perforatum* (St John's wort) significantly reduces cortisol and corticosterone in the rat brain. *Eur Neuropsychopharmacol.* 2004;14(1):7–10.

23. Grundmanna O, Lv Y, Kelber O, Butterweck V. Mechanism of St. John's wort extract (STW3-VI) during chronic restraint stress is mediated by the interrelationship of the immune, oxidative defense, and neuroendocrine system. *Neuropharmacology.* 2010;58(4–5):767–773.

24. Linde K, Berner MM, Kriston L. St John's wort for major depression. *Cochrane DB Syst Rev.* 2008;(4): Art. No.: CD000448.
25. Shelton RC, Keller MB, Gelenberg A, et al. Effectiveness of St John's wort in major depression: A randomized controlled trial. *JAMA.* 2001;285(15):1978–1986.
26. Hubner W-D, Kirste T. Experience with St John's wort (*Hypericum perforatum*) in children under 12 years with symptoms of depression and Psychovegetative disturbances. *Phytother Res.* 2001;15(4):367–370.
27. Harrer G, Schmidt U, Kuhn U, Biller A. Comparison of equivalence between the St. John's wort extract LoHyp-57 and fluoxetine. *Arzneimittelforschung.* 1999;49(4):289–296.
28. Lantz MS, Buchalter E, Giambanco V. St. John's wort and antidepressant drug interactions in the elderly. *J Geriatr Psychiatry Neurol.* 1999;12(1):7–10.
29. Blumenthal M, Busse WR, Goldberg A, et al. The Complete German Commission E Monographs. Austin, TX: American Botanical Council; 1998. p.215.
30. Bukahri I, Dar A, Khan R. Antinociceptive activity of methanolic extracts of St. John's wort (*Hypericum perforatum*) preparation. *Pak J Pharmaceut Sci.* 2004;17(2):13–19.
31. Gardner Z, McGuffin M, editors. *Botanical Safety Handbook.* 3rd ed. Boca Raton, FL: CRC Press. p. 449; 2013.
32. Henderson L, Yue QY, Bergquist C, Gerden B, Arlett P. St John's wort (*Hypericum perforatum*): Drug interactions and clinical outcomes. *BJCP.* 2002;54(4):349–356.
33. Obach RS. Inhibition of human cytochrome P450 enzymes by constituents of St. John's wort, an herbal preparation used in the treatment of depression. *JPET.* 2000;294(1):88–95.
34. Rau T, Wohlleben G, Wuttke H, et al. CYP2D6 genotype: Impact on adverse effects and nonresponse during treatment with antidepressants—A pilot study. *Clin Pharmacol Ther.* 2004;75(5):386–393.

7.36   *ILEX PARAGUARIENSIS* (YERBA MATE)

Yerba mate is a tea brewed from the *Ilex paraguariensis* plant. The drink is widely consumed in South America, especially in Argentina, Brazil, Uruguay, and Paraguay. Indeed, the most common human use of *Ilex paraguariensis* is as this decoction, yerba mate. Thus, in this discussion the herb will generally be referred to by this name.

Yerba mate has a range of biological activities that are attributed to its high polyphenol content. In South America the yerba mate drink is used as a pick-me-up, similar to the way coffee and tea are used in North America and Europe. Indeed, yerba mate, coffee, and tea all contain mildly stimulating methylxanthines. The literature has shown that the hydroalcoholic extract of yerba mate can improve the cognitive function of rats, possibly through antagonism of adenosine receptors by caffeine and theobromine.[1]

Yerba mate contains a number of other flavonoids and phenolic acids that are common in food plants and medicinal herbs. Among the phytochemicals that have been identified in yerba mate are gallic acid, chlorogenic acid, caffeic acid, catechin, quercetin, rutin, and kaempferol. Many exhibit anti-oxidant and anti-inflammatory effects.[2]

7.36.1   ANTIOXIDANT

Phytochemical components of yerba mate appear to cross the blood–brain barrier to produce central effects. For example, chronic treatment with orally administered yerba mate reduced damage from pentylenetetrazol-induced seizures in the cerebellum, cerebral cortex, and hippocampus of rats. The oxidative damage in lipids and proteins and nitric oxide levels were reduced in those parts of the brain, while the activities of the antioxidant enzymes superoxide dismutase (Sod) and catalase (Cat) were maintained at normal levels.[3] In view of the roles now believed to be played by oxidative and nitrosative damage in the etiology of MDD, these effects would be expected to be protective against initiation, persistence, and progression of the illness.[4]

7.36.2 ANTI-INFLAMMATORY

Yerba mate has significant anti-inflammatory effects mediated by phytochemicals such as chlorogenic acid, caffeine, quercetin, oleanolic acid, and its various polyphenolic saponins. Crude extract of yerba mate somewhat inhibited PGE2/COX-2 pathway. However, the most potent component is quercetin, which in combination with saponins greatly attenuated lipopolysaccharide-induced production of PGE2 and Cox2 by mouse macrophages. It also suppressed IL-6 and IL-1β production, likely due to reduction of lipopolysaccharide-induced nuclear translocation of NF-κB.[5]

Oral administration of aqueous yerba mate extract attenuated carrageenan-induced edema in mice, which is a standard model of anti-inflammatory action. This effect appeared to be mediated by inhibition of COX-2 and inducible nitric oxide synthase.[6]

7.36.3 ANTIDIABETIC/ANTI-METABOLIC SYNDROME

Chronic consumption of yerba mate tea has been found to be useful against obesity and in improving the lipid parameters in humans and animal models. Animal studies have shown that yerba mate modulates signaling pathways that regulate adipogenesis, antioxidant, anti-inflammatory, and insulin signaling responses.[7] Such a combination of effects would likely be useful in the treatment of depression. Water extracts of yerba mate have been shown to improve markers of insulin resistance and inflammation in mice made obese with high fat diets. Treatment reduced glucose blood level and improved insulin sensitivity in liver and soleus muscle. In liver tissue, insulin substrate receptor (IRS)-1 and AKT phosphorylation were restored. Treatment also reduced serum levels of the inflammatory cytokines TNF-α and IL-6.[8] In a study of human subjects that were pre-diabetic, consumption of yerba mate tea promoted a significant increase of GSH concentration and a decrease of serum lipid hydroperoxides.[9]

7.36.4 PRECLINICAL ANTIDEPRESSANT-LIKE EFFECTS

At least two studies have evaluated antidepressant effects of yerba mate in rodents. Animals given yerba mate tea in place of water for four weeks showed evidence of antidepressant and anxiolytic effects in the elevated plus maze, open field, and forced swim tests.[10] In a similar study, hydroalcoholic extract of yerba mate produced an antidepressant-like effect in mice subjected to the tail suspension test.[11] The effects were similar to the antidepressant-like effect of ketamine in that they depended on the N-methyl-d-aspartate receptor and l-arginine-nitric oxide pathway, and the effect was also observed by combining otherwise subtherapeutic doses of yerba mate and ketamine. Anxiolytic, stimulant, and neuroprotective effects of yerba mate were demonstrated in another study of mice.[12]

7.36.5 HUMAN ANTIDEPRESSANT EFFECTS

There are no clinical studies of effects of yerba mate on depression in human patients. However, its phytochemical constituents are similar to those of coffee and tea, both of which have been found to reduce the risk of MDD.

7.36.6 DOSAGE

Like coffee and tea, yerba mate is generally drunk *ad libitum*. However, in the above-noted study of prediabetic human subjects, significant benefits were observed drinking 1 l per day of standard infusions.

7.36.7 TOXICITY

The oral LD50 of extract of yerba mate in mice was deemed to be greater than 5,000 mg/kg.[13] This would extrapolate to approximately 350 g for a 75 kg human adult.

7.36.8 SAFETY IN PREGNANCY

In a Brazilian study, it was found that consumption of yerba mate at least once a week during pregnancy caused neither prematurity nor smallness for gestational age. No other adverse effects were reported.[14] The *Physicians' Desk Reference for Nonprescription Drugs and Dietary Supplements* recommends that pregnant women limit their intake of mate such that caffeine intake remains below 300 mg daily.[15]

7.36.9 DRUG INTERACTIONS

In the *Botanical Safety Handbook*, the warning of potential drug interaction with yerba mate involves concerns over the caffeine it contains. Caffeine is metabolized primarily by the CYP1A2 enzyme. Thus, there can be interactions with medications also metabolized by that enzymes. It was found, for example, that co-administration of fluvoxamine can substantially increase serum caffeine levels.[16] Other medications largely metabolized by CYP1A2 with potential for interactions with caffeine from coffee intake include mexiletine, clozapine, psoralens, idrocilamide, phenylpropanolamine, furafylline, theophylline, and quinolones.[17]

REFERENCES

1. Prediger RDS, Fernandes MS, Rial D, et al. Effects of acute administration of the hydroalcoholic extract of mate tea leaves (*Ilex paraguariensis*) in animal models of learning and memory. *J Ethnopharmacol.* 2008;120(3):465–473.

2. Heck CI, De Mejia EG. Yerba mate tea (*Ilex paraguariensis*): A comprehensive review on chemistry, health implications, and technological considerations. *J Food Sci.* 2007;72(9):138–151.

3. dos Santos Branco C, Scola G, Rodrigues AD, et al. Anticonvulsant, neuroprotective and behavioral effects of organic and conventional yerba mate (*Ilex paraguariensis* St. Hil.) on pentylenetetrazol-induced seizures in Wistar rats. *Brain Res Bull.* 2013;92:60–68.

4. Maes M, Galecki P, Chang YS, Berk M. A review on the oxidative and nitrosative stress (O&NS) pathways in major depression and their possible contribution to the (neuro) degenerative processes in that illness. *Prog Neuropsychopharmacol Biol Psychiatry.* 2011;35(3):676–692.

5. Puangpraphant S, Gonzalez de Mejia E. Saponins in yerba mate tea (Ilex paraguariensis A. St.-Hil) and quercetin synergistically inhibit iNOS and COX-2 in lipopolysaccharide-induced macrophages through NFκB pathways. *J Agric Food Chem.* 2009;57(19):8873–8883.

6. Guillermo Schinella G, Neyret E, Cónsole G, et al. An aqueous extract of *Ilex paraguariensis* reduces carrageenan-induced edema and inhibits the expression of cyclooxygenase-2 and inducible nitric oxide synthase in animal models of inflammation. *Planta Med.* 2014;80(12):961–968.

7. Gambero A, Ribeiro ML. The positive effects of yerba maté (*Ilex paraguariensis*) in obesity. *Nutrients.* 2015;7(2):730–750.

8. Arçar DP, Bartchewsky W, Dos Santos TW, et al. Anti-inflammatory effects of yerba maté extract (*Ilex paraguariensis*) ameliorate insulin resistance in mice with high fat diet-induced obesity. *Mol Cell Endocrinol.* 2011;335(2):110–115.

9. Boaventura BCB, Di Pietro PF, Klein GA, et al. Antioxidant potential of mate tea (*Ilex paraguariensis*) in type 2 diabetic mellitus and pre-diabetic individuals. *J Funct Food.* 2013;5(3):1057–1064.

10. De Moraes Reis E, Neto FWS, Cattani VB, et al. Antidepressant-like effect of *Ilex paraguariensis* in rats. *BioMed Res Int.* 2014: Article ID 958209.

11. Ludka FK, Tandler LF, Kuminek G, Olescowicz G, Jacobsen J, Molz S. *Ilex paraguariensis* hydroalcoholic extract exerts antidepressant-like and neuroprotective effects: Involvement of the NMDA receptor and the l-arginine-NO pathway. *Behav Pharmacol.* 2016;27(4):384–392.

12. Santos DCS, Bicca MA, Blum-Silva CH, et al. Anxiolytic-like, stimulant and neuroprotective effects of *Ilex paraguariensis* extracts in mice. *Neuroscience.* 2015;292(30):13–21.

13. Jang SH, Hossain M, Lee JS, et al. Hepatoprotective effects of *Ilex paraguariensis* St. Hilaire (yerba mate) extract in rats. *IJTK.* 2018;17(4):707–715.

14. Santos IS, Matijasevich A, Valle NCJ. Maté drinking during pregnancy and risk of preterm and small for gestational age birth. *J Nutr.* 2005;135(5):1120–1123.

15. PDR. *The Physicians Desk References: For Non-Prescription Drugs and Dietary Supplements.* 27th ed. Montvale, NJ: Medical Economics Co; 2006.

16. Jeppesen U, Loft S, Poulsen HE, Brśen K. A fluvoxamine-caffeine interaction study. *Pharmacogenetics.* 1996;6(3):213–222.

17. Carrillo JA, Benitez J. Clinically significant pharmacokinetic interactions Between dietary caffeine and medications. *Clin Pharmacokinet.* 2000;39(2):127–153.

7.37   *LAVANDULA* (LAVENDER)

There are 32 recognized species of the genus *Lavandula*. These plants grow naturally in a distribution that stretches from the Canary Islands and Cape Verde east across the Mediterranean, North Africa, South West Asia, and the Arabian Peninsula. *Lavandula* has been since ancient times, and the name is derived from the Latin word *lavare*, "to wash," as the Romans used the plant to perfume their bath water. The species most commonly used in herbal medicines are *L. angustiflora*, *L. latifolia*, and *L. stoechas*.[1]

Lavender oil is obtained from steam distillation of the stem, leaves, and flowers of the plant. The primary phytochemical components of the essential oil are linalool, linalyl acetate, ocimene, cineole, terpinen-4-ol, and camphor. Also found are pinene, borneol, myrcene, farnescene, beta-caryophyllene, geraniol, and limonene.[2] For hundreds, if not thousands, of years, preparations of *Lavandula* have been used to treat anxiety, insomnia, arthritis, inflammation, indigestion, and headaches.

7.37.1 ANTIOXIDANT

In vitro, phenolics extracted from cultured *Lavandula vera* cells exhibited potent antioxidant effects, both in protecting sunflower oil from lipid peroxidation and in the DPPH radical assay.[3] When chronically administered to alloxan-induced diabetic rats, essential oil of lavender not only reduced hyperglycemia, but also buffered against oxidative damage in the liver. Levels of malondialdehyde were decreased, whereas activities of both superoxide dismutase and catalase were increased in livers of treated animals.[4] Extracts of lavender also protected cultured liver cells against the oxidative damage of alcohol.[5]

7.37.2 ANTI-INFLAMMATORY

Two of the main constituents of lavender oil, linalool, and linalyl acetate, exhibit anti-inflammatory effects in the carrageenan-induced edema model of inflammation in rats.[6] Linalool also dampened the inflammatory effects of LPS in cultured murine microglia cells. It inhibited LPS-induced release of TNF-α, IL-1β, NO, and PGE2, as well as LPS-induced NF-κB activation. Treatment of linalool also induced nuclear translocation of Nrf2 and expression of HO-1. The role played by Nrf2 was confirmed by the fact that the anti-inflammatory effect of linalool was attenuated by transfection with Nrf2 siRNA.[7]

Orally administered linalool also exerts anti-inflammatory effects in the brains of living animals. Linalool was administered every 48 hours for three months to aged mice with a triple transgenic model of Alzheimer's disease. This *Lavandula* terpene improved learning and spatial memory and increased motor behavior in the elevated plus maze. Hippocampi and amygdalae from the linalool-treated mice exhibited a significant reduction in extracellular β-amyloidosis, tauopathy, astrogliosis, and microgliosis. Of particular significance, linalool reduced levels of the pro-inflammatory markers p38 MAPK, NOS2, COX2, and IL-1β those areas of the brain.[8]

7.37.3 ANTIDIABETIC/ANTI-METABOLIC SYNDROME

Components of *Lavandula* may also act to reduce metabolic syndrome that predisposes to MDD. For example, linalool has been found to be protective in rats made diabetic by streptozotocin. It restored glucose-metabolizing enzymes, collagen content, and glucose transporter type 1 expression in kidney tissue. It also rescued tissue from oxidative stress and inflammation by decreasing the expression of NF-κB.[9] The combination of linalool and limonene, both of which are found in lavender, was particularly effective in lowering of blood glucose and glycated hemoglobin, improving the lipid profile in streptozotocin-induced diabetic rats.[10] Linalool is also an activator of PPARα,[11] which is increasingly being seen as a target of antidepressant action.[12]

Lavender oil can reduce components of the stress response, and this can be seen after either ingestion or inhalation. Inhalation of various components of lavender oil produce relaxation and measurable physiological changes. In a study of rats, inhaled lavender oil had anxiolytic effects similar to those of chlordiazepoxide as shown by behavior in open field tests.[13] This may be due in part to volatile components entering the brain.[14] Inhalation of linalool has also been found to reduce various parameters of the stress response in human subjects, including reduction in salivary cortisol.[15] One proposed mechanism of anxiolytic action of lavender oil is inhibition of voltage dependent calcium channels. Lavender oil acts at these receptors in primary hippocampal neurons within the same nanomolar range of concentration as does pregabalin.[16]

7.37.4 PRECLINICAL ANTIDEPRESSANT-LIKE EFFECTS

Preclinical studies lend compelling evidence of antidepressant effects of *Lavandula* and its phytochemicals. Two weeks of pretreatment with linalool protected rats from the depression-like effects of chronic restraint stress. When subjected to the forced swim test, treated rats exhibited significantly less immobility.[17] Curiously, no effects on BDNF were observed. Both aqueous and alcohol extras of *Lavandula* exhibited antidepressant effects in mice. Pretreatment with extract of *Lavandula officinalis* increased swimming time in the forced swim test and decreased immobility in the tail suspension test. Results were similar to those of fluoxetine.[18] A study of *Lavandula angustifolia* similarly found that whole extract significantly reduced immobility in mice subjected to the forced swim test. In evaluating fractions for activity, it was found that two specific constituents, 3-(3,4-dihydroxyphenyl)lactic acid and rosmarinic acid, have antidepressant effects.[19] Rosmarinic acid has also been found to have antidepressant effects in rats, that are mediated in part by increases in hippocampal BDNF.[20]

Geraniol, a monoterpenoid found in species of *Lavandula* – as well as in rose, ginger, and citrus – also has significant antidepressant effects in animal models. In mice exposed to chronic unpredictable stress, treatment with geraniol restored decreased sucrose preference and shortened immobile time in both the forced swimming and tail suspension tests. These antidepressant effects appeared to be mediated by dampening of neuroinflammation. Geraniol reduced stress-induced production of IL-1β in the brain by inhibiting nuclear factor kappa B (NF-κB) pathway activation and subsequent activity of the (NLRP3) inflammasome.[21]

7.37.5 HUMAN ANTIDEPRESSANT EFFECTS

Several clinical trials have provided evidence of antidepressant effects of lavandula in humans. In a 70-day, randomized, placebo-controlled study of mixed anxiety and depressive disorder (ICD-10 F41.2), Silexan (a proprietary lavender oil preparation) significantly reduced scores on the Montgomery Åsberg Depression Rating Scale. It also significantly reduced anxiety per the Hamilton Anxiety Rating Scale. Compared to placebo, the patients treated with Silexan had a better overall clinical outcome and showed more pronounced improvements of impaired daily living skills and health related quality of life.[22]

In another study, tincture of *Lavandula angustifolia* alone was ineffective. However, addition of the tincture to imipramine had a significantly greater antidepressant effect than imipramine alone, suggesting an advantage in using *Lavandula* as an augmentation strategy.[23] In a similar study, lavender significantly enhanced the antidepressant effects of citalopram.[24]

7.37.6 DOSAGE

Hoffman gives the dosing of preparations of Lavandula as 1–2 teaspoons of flower per cup of water taken three times a day. It was advised that lavender oil not be taken internally, but rather inhaled or rubbed on the skin.[25] Easley and Horne note tincture of dried flower and leaf (1:5,75% alcohol) 1–3 ml, three times daily.[26]

7.37.7 TOXICITY

The LD50 of essential oil of lavender in rats is reported to be 3.55 g/kg.[27] Extrapolated to a 75 kg human adult, this dose would be approximately 250 g.

7.37.8 SAFETY IN PREGNANCY

An expert review noted a lack of sufficient data about the use of *Lavandula angustifolia* by pregnant women. Yet, it did advise against its use during pregnancy due to its purported properties as an emmenagogue.[28]

7.37.9 DRUG INTERACTIONS

Essential oil of lavender increases pentobarbital sleep time in rats, suggesting competition for P450 enzymes.[29] However, in healthy human volunteers, daily administrations of a proprietary formulation of lavender oil had no clinically relevant inhibitory or inducing effects on the CYP1A2, CYP2C9, CYP2C19, CYP2D6, or CYP3A4 enzymes *in vivo*.[30]

REFERENCES

1. Upson T. The taxonomy of the genus *Lavandula* L. In: Lis-Balchin M, editor. *Lavender, the Genus Lavandula. Medicinal and Aromatic Plants: Industrial Profiles.* London: Taylor and Francis; 2002. p. 2–34.
2. Ghelardini C, Galeotti N, Salvatore G, Mazzanti G. Local anaesthetic activity of the essential oil of Lavandula angustifolia. *Planta Med.* 1999;65:700–703.
3. Kovatcheva EG, Koleva II, Ilieva M, Pavlov A, Mincheva M, Konushlieva M. Antioxidant activity of extracts from *Lavandula vera* MM cell cultures. *Food Chem.* 2001;72(3):295–300.

4. Sebai H, Selmi S, Rtibi K, Souli A, Gharbi N, Sakly M. Lavender (*Lavandula stoechas* L. essential oils attenuate hyperglycemia and protect against oxidative stress in alloxan-induced diabetic rats. *Lipids Health Dis.* 2013;12:189.

5. Farshori NN, Al-Sheddi ES, Al-Oqail MM, et al. Hepatoprotective potential of *Lavandula coronopifolia* extracts against ethanol induced oxidative stress-mediated cytotoxicity in HepG2 cells. *Toxicol Ind Health.* 2015;31(8):727–737.

6. Peana AT, D'Aquila PS, Panin F, Serra G, Pippia P, Moretti M. Anti-inflammatory activity of linalool and linalyl acetate constituents of essential oils. *Phytomedicine.* 2002;9(8):721–726.

7. Li Y, Lv O, Zhou F, Li Q, Wu Z, Zheng Y. Linalool inhibits LPS-induced inflammation in BV2 microglia cells by activating Nrf2. *Neurochem Res.* 2015;40(7):1520–1525.

8. Sabogal-Guáqueta AM, Osorio E, Cardona-Gómez GP. Linalool reverses neuropathological and behavioral impairments in old triple transgenic Alzheimer's mice. *Neuropharmacology.* 2016;102:111–120.

9. Deepa B, Anuradha CV. Effects of linalool on inflammation, matrix accumulation and podocyte loss in kidney of streptozotocin-induced diabetic rats. *Toxicol Mech Methods.* 2013;23(4):223–234.

10. More TA, Kulkarni BR, Nalawade ML, Arvindekar AU. Antidiabetic activity of linalool and limonene in streptozotocin-induced diabetic rat: A combinatorial therapy approach. *Int J Pharm Pharm Sci.* 2014;6:159.e163.

11. Jun H-J, Lee JH, Kim J, et al. Linalool is a PPARα ligand that reduces plasma TG levels and rewires the hepatic transcriptome and plasma metabolome[S]. *J Lipid Res.* 2014;55(6):1098–1110.

12. Song L, Wang H, Wang YJ, et al. Hippocampal PPARα is a novel therapeutic target for depression and mediates the antidepressant actions of fluoxetine in mice. *BJP.* 2018;175(14):2968–2987.

13. Shaw D, Annett JM, Doherty B, Leslie JC. Anxiolytic effects of lavender oil inhalation on open-field behaviour in rats. *Phytomedicine.* 2007;14(9):613–620.

14. Satou T, Takahashi M, Kasuya H, et al. Organ accumulation in mice After inhalation of single or mixed essential oil compounds. *Phytother Res.* 2013;27(2):306–311.

15. Höferl M, Krist S, Buchbauer G. Chirality influences the effects of linalool on physiological parameters of stress. *Planta Med.* 2006;72(13):1188–1192.

16. Schuwald AM, Nöldner M, Wilmes T, Klugbauer N, Leuner K, Müller WE. Lavender oil-potent anxiolytic properties via modulating voltage dependent calcium channels. *PLoS.* 2013;8(4):e59998.

17. Saiyudthong S, Srijittapong D, Mekseepralard C. Subchronic administration of linalool decreases depressive-like behaviour in restrained rats. *J Pharm Pharmacol.* 2017;7:401–407.

18. Abbasi Maleki S, Bekhradi R, Asgharpanah J, Abbasi Maleki F, Maleki A. Antidepressant effect of aqueous and hydroalcoholic extracts of *Lavandula officinalis* in forced swim test and tail suspension test in male mice. *AMUJ.* 2013;16(78):65–75.

19. Ueno T, Matsui Y, Masuda H, et al. Antidepressant-like effects of 3-(3,4-dihydroxyphenyl)lactic acid isolated from lavender (*Lavandula angustifolia*) flowers in mice. *Food Sci Technol Res.* 2014;20(6):1213–1219.

20. Jin X, Liu P, Yang F, Zhang YH, Miao D. Rosmarinic acid ameliorates depressive-like behaviors in a rat model of CUS and up-regulates BDNF levels in the hippocampus and hippocampal-derived astrocytes. *Neurochem Res.* 2013;38(9):1828–1837.

21. Deng X-Y, Xue JS, Li HY, et al. Geraniol produces antidepressant-like effects in a chronic unpredictable mild stress mice model. *Physiol Behav.* 2015;152(A):264–271.

22. Kasper S, Volz HP, Dienel A, Schläfke S. Efficacy of Silexan in mixed anxiety-depression - A randomized, placebo-controlled trial. *Eur Neuropsychopharm.* 2016;26(2):331–340.

23. Akhondzade S, Kashani L, Fotouhi A, et al. Comparison of *Lavandula angustifo-lia* Mill. tincture and imipramine in the treatment of mild to moderate depression: A double-blind, randomized trial. *Prog Neuropsychopharmacol Biol Psychiatry.* 2003;27(1):123–127.

24. Nikfarjam M, Parvin N, Assarzadegan N, Asghari S. The Effects of *Lavandula angustifolia* Mill Infusion on Depression in Patients Using citalopram: A comparison Study. *Iran Red Crescent Med J.* 2013;15(8):734–739.

25. Hoffmann D. *Medical Herbalism: The Science and Practice of Herbal Medicine.* Rochester, VT: Healing Arts Press; 2003. p. 562.

26. Easley T, Horne S *The Modern Herbal Dispensary.* Berkeley, CA: North Atlantic Books; 2016. p. 258.

27. Da Silva GL, Luft C, Lunardelli A, et al. Antioxidant, analgesic and anti-inflammatory effects of lavender essential oil. *Anais da Academia Brasileira de Ciências.* 2015;87(2 Suppl.):1397–1408.

28. Basch E, Foppa I, Liebowitz R, et al. Lavender (*Lavandula angustifolia* Miller). *J Herb Pharmacother.* 2004;4(2):63–78.

29. Guillemain J, Rousseau A, Delaveau P. Neurodepressive effects of the essential oil of Lavandula angustifolia Mill. *Ann Pharm Fr.* 1989;47(6):337–343.

30. Doroshyenko O, Rokitta D, Zadoyan G, et al. Drug cocktail interaction Study on the effect of the orally administered lavender oil preparation Silexan on cytochrome P450 enzymes in healthy volunteers. *Drug Metab Dispos.* 2013;41(5):987–993.

7.38   *LIGUSTICUM CHUANXIONG*

Ligusticum chuanxiong is a plant from the carrot family that is native to India, Kashmir, Nepal, and the Yunnan area of southern China. It is one of the herbs Yeung et al. report to be among the most commonly used in traditional Chinese herbal combinations to treat MDD-like illnesses. The herb contains alkaloids, phenolic acids, phthalide lactones, with ligustilide, tetramethylpyrazine, ferulic acid, and senkyunolide A being most prominent.[1]

Ligusticum chuanxiong has traditionally been used in China in the treatment of stroke. Ferulic acid, which is present in significant concentration in the herb, can reduce cerebral infarct area and neurological deficit score. The mechanism has been attributed to inhibition of superoxide radicals, intercellular adhesion molecule-1, and NF-κB expression in transient middle cerebral artery occlusion rats. Thus, the benefits are likely due to both antioxidant and anti-inflammatory effects.[2] This also speaks to its central effects.

7.38.1   ANTIOXIDANT

Tetramethylpyrazine from *Ligusticum chuanxiong* has significant antioxidant and neuroprotective effects. It scavenges free radicals generated by the neurotoxin kainic acid and thus can attenuate kainate-induced excitotoxicity in cultured rat hippocampal neurons.[3] Tetramethylpyrazine also protects the hippocampus from kainate-induced toxicity in intact rats. It partially alleviates kainate-induced status epilepticus in rats and prevents neuronal loss in the CA3 region. It acts, at least in part, by quenching ROS, blocking lipid peroxidation, and protecting activity of the antioxidant enzymes glutathione peroxidase and glutathione reductase.[4]

7.38.2 ANTI-INFLAMMATORY

The z-ligustilide and senkyunolide A in *Ligusticum chuanxiong* have anti-inflammatory effects. They inhibit lipopolysaccharide-induced TNF-α production in monocytes and suppress TNF-α-mediated NF-κB activation in reporter gene assays. *Ligusticum chuanxiong* has long been thought to have neuroprotective effects, and it has been found that Z-ligustilide and senkyunolide A inhibit the production of proinflammatory mediators in lipopolysaccharide-stimulated mouse microglial cells as well as in human macrophages.[5]

7.38.3 ANTIDIABETIC/ANTI-METABOLIC SYNDROME

There are no references in the literature to such effects.

7.38.4 PRECLINICAL ANTIDEPRESSANT-LIKE EFFECTS

Tetramethylpyrazine has an antidepressant effect in rodent models of depression.[6] It reduced the duration of immobility during forced swim tests and tail suspension test in rats and mice in a manner similar to clomipramine and fluoxetine. Unlike fluoxetine, it did not enhance serotonergic activity. Persistent exposure to corticosterone is useful for both cellular and behavioral models of depression. Senkyunolide A potently inhibits corticosterone-induced apoptosis that injures hippocampal tissue. It appears to do so by blocking corticosterone-induced overexpression of α-synuclein.[7]

7.38.5 HUMAN ANTIDEPRESSANT EFFECTS

There are no published reports of the use of *Ligusticum chuanxiong* in the treatment of MDD in human subjects. It is worth noting that among the phytochemicals *Ligusticum chuanxiong* contains are the non-steroidal phytoprogestins 3,8-dihydrodiligustilide and riligustilide.[8] Between the two, 3,8-Dihydrodiligustilide is the far more potent progesterone agonist, with nanomolar affinity for the progesterone receptor. Those same and similar phytoprogestins are also found in *Angelica sinensis*. The role these actions may play in the effects of *Ligusticum chuanxiong* on mood, particularly in women and sufferers of premenstrual dysphoric disorder, is unclear. However, buffering effects on natural progesterone activity may be therapeutic.

7.38.6 DOSAGE

The recommended dose of ligusticum is 3–6 grams daily, taken as part of a decoction. Some practitioners recommend a higher maximum dose of up to 10 grams per day. When used as a powder, a lower dose (1–1.5 grams) is administered. Ligusticum is available as a powder or pill, and is occasionally combined with water as part of a decoction.[9]

7.38.7 Toxicity

In mice, the LD50 of the essential oil of root and stem of *Ligusticum chuanxiong* was 328 mg/kg by intraperitoneal injection and 2982.37 mg/kg by oral administration.[10] Extrapolated to a 75 kg human adult, this dose would be approximately 210 g.

7.38.8 Safety in Pregnancy

Reference texts from Traditional Chinese Medicine advise against use of *Ligusticum chuanxiong* during pregnancy.[11,12]

7.38.9 Drug Interactions

The *Botanical Safety Handbook* notes no known interactions with drugs.[13] However, studies have shown that *Ligusticum chuanxiong* has inhibitory effects on the CYP2D and CYP3A families of cytochrome P450 enzymes.[14]

REFERENCES

1. Ran X, Ma L, Peng C, Zhang H, Qin LP. Ligusticum chuanxiong Hort: a review of chemistry and pharmacology. *Pharm Biol*. 2011;49(11):1180–1189.
2. Cheng CY, Ho TY, Lee EJ, Su SY, Tang NY, Hsieh CL. Ferulic acid reduces cerebral infarct through its antioxidative and anti-inflammatory effects following transient focal cerebral ischemia in rats. *Am J Chin Med*. 2008;36:1105–1119.
3. Shih YH, Wu SL, Chiou WF, Ku HH, Ko TL, Fu YS. Protective effects of tetramethylpyrazine on kainate-induced excitotoxicity in hippocampal culture. *Neuroreport*. 2002;13:515–519.
4. Li SY, Jia YH, Sun WG, et al. Stabilization of mitochondrial function by tetramethylpyrazine protects against kainate-induced oxidative lesions in the rat hippocampus. *Free Radic Biol Med*. 2010;48:597–608.
5. Or TCT, Yang CLH, Law AHY, Li JC, Lau AS. Isolation and identification of anti-inflammatory constituents from *Ligusticum chuanxiong* and their underlying mechanisms of action on microglia. *Neuropharmacology*. 2011;60(6):823–831.
6. Yu L, Jiang X, Liao M, Ma R, Yu T. Antidepressant-like effect of tetramethylpyrazine in mice and rats. *Neurosci Med*. 2011;2:142–148.
7. Gong S, Zhang J, Guo Z, Fu W. Senkyunolide A protects neural cells against corticosterone-induced apoptosis by modulating protein phosphatase 2A and α-synuclein signaling. *Drug Des Devel Ther*. 2018;12:1865–1879.
8. Lim LS, Shen P, Gong YH, Yong EL. Dimeric progestins from rhizomes of *Ligusticum chuanxiong*. *Phytochemistry*. 2006;67(7):728–734
9. Editorial Committee of Chinese Materia Medica. State Drug Administration of China. In: *Chinese Materia Medica*. Shanghai: Science and Technolgy Press; 1998.
10. Hou J, He X. Research development on volatile oil from chuanxiong rhizome. *J Med Plant Res*. 2012;6(12):2240–2248.
11. Bensky D, Clavey S, Stoger E. *Chinese Herbal Medicine: Materia Medica*. 3rd ed. Seattle, WA: Eastland Press; 2004.
12. Chen JK, Chen TT. *Chinese Medical Herbology and Pharmacology*. City of Industry, CA: Art of Medicine Press; 2004.
13. Gardner Z, McGuffin M, editors. *Botanical Safety Handbook*. 3rd ed. Boca Raton, FL: CRC Press; 2013. p. 510.

14. Tang JC, Zhang JN, Wu YT, Li ZX. Effect of the water extract and ethanol extract from traditional Chinese medicines *Angelica sinensis* (Oliv.) Diels, *Ligusticum chuanxiong* Hort. and *Rheum palmatum* L. on rat liver cytochrome P450 activity. *Phytother Res.* 2006;20(12):1046–1051.

7.39   *MAGNOLIA OFFICINALIS*

Magnolia officinalis is a species of magnolia tree native to the mountains and valleys of China. It has long been used in Traditional Chinese Medicine and is an essential component of at least two classical herbal combinations, *banxia houpu* and *zhi-zi-hou-po*, that are used to treat symptoms of MDD.[1] Modern studies have shown aqueous and methanolic extracts of Magnolia to exhibit antioxidative, anti-inflammatory, anti-tumorigenic, antidiabetic, anti-microbial, anti-nociceptive, anti-neurodegenerative, and antidepressant properties.[2]

The bark of the *Magnolia officinalis* tree is the major source of its medicinal phytochemicals. Among these chemicals, magnolol and its isomer, honokiol, are the most important. Other substances isolated from the bark are magnolianone, erythro-honokitriol, threo-honokitriol, magnaldehyde, magnatriol, randaiol, obovatol, magnolignan B, p-hydroxylbenzaldehyde, coniferaldehyde, coniferol alcohol, syringaldehyde, syringaresinol, and acteoside.[3]

7.39.1   ANTIOXIDANT

Magnolol and honokiol from magnolia are potent antioxidants. They inhibit lipid peroxidation in the mitochondria of rat heart and liver with a potency approximately 1,000 times greater than that of α-tocopherol.[4] Both magnolol and honokiol protected cultured rat cerebellar granule cells from the oxidative effects of hydrogen peroxide, glucose deprivation, and glutamatergic excitotoxicity, with honokiol being the more potent of the two.[5] Magnolol also showed neuroprotective antioxidant effects in intact animals. Intravenous administration of magnolol to rats attenuated heatstroke-induced neuronal damage due to increased free radical formation and lipid peroxidation in the brain.[6]

7.39.2   ANTI-INFLAMMATORY

Magnolia officinalis was found to be the most potent anti-inflammatory herb among four herbs commonly used in Traditional Chinese Medicine to treat inflammatory pain. It minimized secretion of the inflammatory cytokines IL-6 and IL-8 from lipopolysaccharide-stimulated human gingival fibroblasts and monocytes.[7]

Magnolol also blocks activation of the major inflammation mediator, the TLR4 receptor. Addition of magnolol to culture medium reversed the lipopolysaccharide-induced upregulation of TLR4 receptors in cultured mouse macrophages. This in turn attenuated TLR4-mediated activation of NF-κB and MAPK signaling and the subsequent release of pro-inflammatory cytokines TNF-α, IL-6, and IL-1β.[8] It may be recalled that the TLR4 is coming to be seen as an important bridge between the immune and neuroendocrine systems, and the neuropathology of MDD.[9]

Magnolol has exerted both antioxidant and ant-inflammatory effects in the brains of intact rats sustaining cerebral injury through carotid artery occlusion and reperfusion. It reduced levels of acute inflammatory cytokines, including IL-1β, TNF-α, and IL-6, as well as activity of inducible NO synthase. These effects were seen as following suppression of reactive oxygen species production and the upregulation of p-Akt and inhibition of GSK-3β.[10]

7.39.3   ANTIDIABETIC/ANTI-METABOLIC SYNDROME

Various magnolia species have been used to treat diabetes. *In vitro*, honokiol and magnolol each enhanced insulin-stimulated glucose uptake by mouse and human adipocytes, and interacted to even further enhance uptake. In insulin-resistant adipocytes, they enhanced glucose uptake nearly to the extent of the diabetes medication rosiglitazone. When administered together, the two phytochemicals were more effective than rosiglitazone. Inhibitors of the insulin-signaling pathway abolished the honokiol and magnolol-induced glucose uptake, suggesting that their antidiabetic effects are mediated by this signaling pathway.[11]

Chronic treatment with 4-O-methylhonokiol, a phytochemical in *Magnolia officinalis*, ameliorated some of the damaging effects of a high fat diet in intact mice. It mildly decreased body weight and fat mass, ameliorated hepatic steatosis, and enhanced insulin sensitivity. Interestingly, it also diminished the infiltration of mast cells into adipose tissue, which in humans serves as a source of inflammatory cytokines in Metabolic Syndrome.[12]

7.39.4   PRECLINICAL ANTIDEPRESSANT-LIKE EFFECTS

A substantial number of studies have shown antidepressant-like effects of extract of magnolia and its constituent phytochemicals. Sub-chronic pretreatment of mice with extract of magnolia attenuated the depression-like effects of several days of restraint stress. Mice treated with magnolia exhibited less immobility in both the forced swim and tail suspension tests. Treatment also attenuated the stress-induced decreases in serum levels of the antioxidant enzymes, superoxide dismutase, and glutathione peroxidase. Finally, whereas stress caused loss of serotonergic neurons in the dorsal raphe nucleus, treatment with extract of magnolia spared those neurons.[13]

Most evaluations of antidepressant effects of magnolia have focused on its two primary bioactive phytochemicals, magnolol and honokiol. When extract of *Magnolia officinalis* are administered to mice and humans, magnolol and other metabolites are recovered in urine of mice and humans. Two of those substances, magnolol and dihydroxydihydromagnolol, decreased immobility of mice in the forced swim test.[14] Mice treated with magnolol that showed reduced immobility in the forced swim test were also found to have been spared reductions of dopaminergic and serotonergic activities in the amygdala, as well as showing increases in noradrenergic activities in the amygdala and frontal cortex.[15]

Several studies have shown that magnolol and honokiol also attenuate the depression-like behaviors induced by stress or by administration of corticosterone that mimics stress-induced hormone release. Magnolol attenuated the behavioral and

neuroinflammatory effects of chronic unpredictable stress in mice. It restored preference for sucrose and reduced immobility in the forced swim test. It reduced levels of
the pro-inflammatory cytokines, IL-1β, IL-6, and TNF-α in the prefrontal cortex, and
also dampened oxidative effects by reducing lipid peroxidation and enhancing superoxide dismutase and glutathione peroxidase activity in that area.[16] Similar reversals of
depression-like behaviors were seen with magnolol in mice chronically administered
the stress hormone, corticosterone, and levels of BDNF, serotonin, and norepinephrine
were returned toward normal levels in the hippocampus.[17] Virtually the same effects
were seen with honokiol in mice chronically administered corticosterone.[18] These phytochemicals also reversed the behavioral and neurochemical depression-like effects of
inflammation-mediated reaction to lipopolysaccharide[19] and olfactory bulbectomy[20] in
mice. Of note, in the latter study, magnolol was found to stimulate the IP3K/Akt system
that in turn inhibits GSK-3β, as well as to stimulate neurogenesis in the hippocampus.

Both magnolol and honokiol reversed the depression-like effects of chronic unpredictable stress in rats. In one case, magnolol increased levels of glial fibrillary acidic
protein, an astrocyte marker, in the hippocampus and prefrontal cortex that reflected
sparing of astrocytes in those areas.[21] It normalized serotonergic activity in brain
tissue.[22] Honokiol restored toward normal serum levels of corticotrophin-releasing
hormone, adrenocorticotropic hormone, and corticosterone, as well as BDNF and
density of glucocorticoid receptors in the hippocampus.[23]

7.39.5   HUMAN ANTIDEPRESSANT EFFECTS

There are no published studies of effects of *Magnolia officinallis* alone on mood in
human subjects, and certainly none that formally address effects on MDD. However,
there are two studies examining effects of the combination of *Magnolia officinallis*
and *Phellodendron amurense* on general mood, cortisol, and stress. The earliest
study showed that six weeks of administration of this herbal combination to over-
weight women that complained of "stress eating" had only minor effects. It reduced
transitory episodes of anxiety but had no effect on long-standing feelings of anxiety
or depression as measured using the Spielberger TRAIT questionnaire. It also had no
effects on salivary cortisol, appetite, body morphology, or sleep. Part of these disappointing results may have been a very large dropout rate of nearly 50%.[24] In a later
study, four weeks of supplementation of the magnolia and phellodendron extract
reduced salivary cortisol as well as significantly reducing overall stress, tension,
depression, anger, fatigue, and confusion.[25]

7.39.6   DOSAGE

Powdered bark of *Magnolia officinalis* is commercially available under several
brand names in 400 mg capsules to be taken once daily.

7.39.7   TOXICITY

Toxicological studies of hydro-ethanolic extracts of the closely related plant,
Magnolia grandiflora were without deaths or obvious adverse effects in doses up

to 5.7 g/kg.[26] Extrapolated to a 75 kg human adult, that dose would be approximately 40 g.

7.39.8 SAFETY IN PREGNANCY

Reference texts from Traditional Chinese Medicine advise against use of *Magnolia* species during pregnancy.[27,28]

7.39.9 DRUG INTERACTIONS

The *Botanical Safety Handbook* notes no known interactions of *Magnolia* species with drugs.[29] Honokiol and magnolol, the primary bioactive components of magnolia officinalis, both potently inhibited the CYP1A2 isoenzyme, but had little effect on the other cytochrome P450 enzymes.[30]

REFERENCES

1. Liu Y, Wang D, Yang G, Shi Q, Feng F. Comparative pharmacokinetics and brain distribution of magnolol and honokiol after oral administration of *Magnolia officinalis* cortex extract and its compatibility with other herbal medicines in *Zhi-Zi-Hou-Po* Decoction to rats. *Biomed Chromatogr.* 2016;30(3):369–375.
2. Chen YH, Huang PH, Lin FY, et al. Magnolol: a multifunctional compound isolated from the Chinese medicinal plant *Magnolia officinalis. Eur J Integr Med.* 2011;3: 311–318.
3. Shen CC, Ni CL, Shen YC, et al. Phenolic constituents from the stem bark of *Magnolia officinalis. J Nat Prod.* 2009;72(1):168–171.
4. Shen JL, Man KM, Huang PH, et al. Honokiol and magnolol as multifunctional antioxidative molecules for dermatologic disorders. *Molecules.* 2010;15:6452–6465.
5. Lin YR, Chen HH, Ko CH, Chan MH. Neuroprotective activity of honokiol and magnolol in cerebellar granule cell damage. *Eur J Pharmacol.* 2006;537:64–69.
6. Chang CP, Hsu YC, Lin MT. Magnolol protects against cerebral ischaemic injury of rat heatstroke. *Clin Exp Pharmacol Physiol.* 2003;30(5–6):387–392.
7. Walker JM, Maitra A, Walker J, Ehrnhoefer-Ressler MM, Inui T, Somoza V. Identification of *Magnolia officinalis* L. bark extract as the most potent anti-inflammatory of four plant extracts. *Am J Chin Med.* 2013;41(3):531–544.
8. Fu Y, Liu B, Zhang N, et al. Magnolol inhibits lipopolysaccharide-induced inflammatory response by interfering with TLR4 mediated NF-κB and MAPKs signaling pathways. *J Ethnopharmacol.* 2013;145(1):193–199.
9. Liu JJ, Buisman-Pijlman F, Hutchinson MR. Toll-like receptor 4: innate immune regulator of neuroimmune and neuroendocrine interactions in stress and major depressive disorder. *Front Neurosci.* 2014;8:309.
10. Chen JH, Kuo HC, Lee KF, Tsai TH. Magnolol protects neurons against ischemia injury via the downregulation of p38/MAPK, CHOP and nitrotyrosine. *Toxicol Appl Pharmacol.* 2014;279(3):294–302.
11. Alonso-Castro AJ, Zapata-Bustos R, Domínguez F, García-Carrancá A, Salazar-Olivo LA. *Magnolia dealbata* Zucc and its active principles honokiol and magnolol stimulate glucose uptake in murine and human adipocytes using the insulin-signaling pathway. *Phytomedicine.* 2011;18(11):926–933.

12. Zhang Z, Chen J, Jiang X, et al. The magnolia bioactive constituent 4-O-methylhonokiol protects against high-fat diet-induced obesity and systemic insulin resistance in mice. *Oxid Med Cell Longev.* 2014;2014:Article ID 965954.

13. You JY, Woo C, Jeong H, Choi JH, Lee UJ. Experimental study on the antidepressant effects of *Magnolia Officinalis,* extracts. *J Intern Korean Med.* 2013;34(3):256–266.

14. Nakazawa T, Yasuda T, Ohsawa K. Metabolites of orally administered *Magnolia officinalis* extract in rats and man and its antidepressant-like effects in mice. *J Pharm Pharmacol.* 2003;55(11):1583–1591.

15. Nakazawa T, Yasuda T, Ohsawa K. Antidepressant-like effects of magnolol from *Magnolia officinalis* in the forced swimming test. *Nat Med.* 2003;57(6):221–226.

16. Cheng J, Dong S, Yi L, Geng D, Liu Q. Magnolol abrogates chronic mild stress-induced depressive-like behaviors by inhibiting neuroinflammation and oxidative stress in the prefrontal cortex of mice. *Int Immunopharmacol.* 2018;59:61–67.

17. Baia YT, Song L, Dai G, et al. Antidepressant effects of magnolol in a mouse model of depression induced by chronic corticosterone injection. *Steroids.* 2018;135: 73–78.

18. Pittaa S, Augustine BB, Kasala ER, et al. Honokiol reverses depressive-like behavior and decrease in brain BDNF levels induced by chronic corticosterone injections in mice. *Phcog. J.* 2013;5(5):211–215.

19. Sulakhiyaa K, Kumar P, Jangra A, et al. Honokiol abrogates lipopolysaccharide-induced depressive like behavior by impeding neuroinflammation and oxido-nitrosative stress in mice. *Eur J Pharmacol.* 2014;744:124–131.

20. Matsui N, Akae H, Hirashima N, et al. Magnolol enhances hippocampal neurogenesis and exerts antidepressant-like effects in olfactory bulbectomized mice. *Phytother Res.* 2016;30(11):1856–1861.

21. Li LF, Yang J, Ma SP, Qu R. Magnolol treatment reversed the glial pathology in an unpredictable chronic mild stress-induced rat model of depression. *Eur J Pharmacol.* 2013;711(1–3):42–49.

22. Li LF, Lu J, Li XM, et al. Antidepressant-like effect of magnolol on BDNF up-regulation and serotonergic system activity in unpredictable chronic mild stress treated rats. *Phytotherapy.* 2012;26(8):1189–1194.

23. Wang CM, Gan D, Wu J, Liao M, Liao X, Ai W. Honokiol exerts antidepressant effects in rats exposed to chronic unpredictable mild stress by regulating brain derived neurotrophic factor level and hypothalamus-pituitary-adrenal axis activity. *Neurochem Res.* 2018;43(8):1519–1528.

24. Kalman DS, Feldman S, Feldman R, Schwartz HI, Krieger DR, Garrison R. Effect of a proprietary Magnolia and Phellodendron extract on stress levels in healthy women: a pilot, double-blind, placebo-controlled clinical trial. *Nutr J.* 2008;7:11.

25. Talbott SM, Talbott JA, Pugh M. Effect of *Magnolia officinalis* and *Phellodendron amurense* (Relora®) on cortisol and psychological mood state in moderately stressed subjects. *J Int Soc Sports Nutr.* 2013;10:37.

26. Sokkar NM, Rabeh MA, Ghazal G, Slem AM. Determination of flavonoids in stamen, gynoecium, and petals of *Magnolia grandiflora* L. and their associated antioxidant and hepatoprotection activities. *Quim Nova.* 2014;37(4):667–671.

27. Bensky D, Clavey S, Stoger E. *Chinese Herbal Medicine: Materia Medica.* 3rd ed. Seattle, WA: Eastland Press; 2004.

28. Chen JK, Chen TT. *Chinese Medical Herbology and Pharmacology.* City of Industry, CA: Art of Medicine Press; 2004.

29. Gardner Z, McGuffin M, editors. *Botanical Safety Handbook.* 3rd ed. Boca Raton, FL: CRC Press; 2013. p. 531–533.

30. Joo J, Liu KH. Inhibitory effect of honokiol and magnolol on cytochrome P450 enzyme activities in human liver microsomes. *Mass Spec Lett.* 2013;4(2):34–37.

7.40 *MATRICARIA RECUTITA* (CHAMOMILE)

Matricaria recutita, commonly known as chamomile, is a member of the daisy family native to Europe and western Asia. There are two herbs that go by this name, German and Roman Chamomile. German chamomile (*Matricaria recutita* or *Matricaria chamomilla*) is the most commonly used, best studied, and more potent of the two, and from here on will be the species to which I am referring. The primary active phytochemicals in *Matricaria recutita* are apigenin and bisabolol. It also contains lesser amounts of luteolin, quercetin, rutin, and naringenin, all of which are known to be bioactive. The herb has long been used as a sedative, anxiolytic, antispasmodic, and as a treatment for mild skin inflammation.[1,2]

7.40.1 ANTIOXIDANT

Both water extracts and essential oil of *Matricaria recutita* have significant antioxidant effects. The water extracts of *Matricaria recutita* flowers showed higher antioxidant activity than the commercial antioxidant, butylated hydroxyanisole, in the linoleic acid and liposome models.[3] The essential oil of *Matricaria recutita* showed high capacity for scavenging of radicals in the DPPH test, and this was though due to its high content of the phenolic compounds chlorogenic acid, caffeic acid, p-coumaric acid, and ferulic acid.[4]

Extract of chamomile also demonstrated significant antioxidant effects *in vivo* in rats treated with CCl_4. Oral administration of the extract reduced serum levels of the liver transaminase enzymes AST and ALT and whole blood levels of malondialdehyde. At the same time, it spared blood levels of glutathione, superoxide dismutase, glutathione peroxidase, and catalase.[5]

7.40.2 ANTI-INFLAMMATORY

Recent studies have verified the traditional view of *Matricaria recutita* as having anti-inflammatory effects. *Matricaria recutita* treatment inhibited the release of lipopolysaccharide-induced prostaglandin E2 in murine macrophages. This effect was due to inhibition of COX-2 enzyme activity by *Matricaria recutita*. The NSAID, sulindac, had a similar effect.[6] When incubated with THP1 macrophages, *Matricaria recutita* extract reduced concentrations of IL-6 and TNF-α. The isolated polyphenolic compounds in *Matricaria recutita*, apigenin and quercetin, also reduced concentrations of those inflammatory cytokines. The Comet assay, used to study the protective effect of the isolated phenols against oxidative damage, showed positive results for apigenin and quercetin.[7] Apigenin also reduces the release of IL-6 and TNF-α from murine macrophages exposed to lipopolysaccharide in vitro.[8] Apigenin significantly reduces lipopolysaccharide-induced upregulation of NO and release of TNF-α in cultured murine microglia.[9]

Apigenin and luteolin may also protect from inflammation and oxidative damage by activating the Nrf2-antioxidant response element. Both apigenin and luteolin induced the Nrf2-ARE in a hepatic cell line, and this was blocked by exposure to the PI3K inhibitor LY294002. The two flavonoids also significantly decreased the production of NO, iNOS, and cytosolic phospholipase A2 (cPLA2), which were induced by exposure of the hepatocytes to lipopolysaccharide.[10] It has recently been suggested that activation of Nrf2 *per se* may have antidepressant effects.[11]

7.40.3 ANTIDIABETIC/ANTI-METABOLIC SYNDROME

High calorie diets, both high fat and high fructose, promote oxidative stress and chronic low-grade inflammation that predispose to brain dysfunction and neurodegeneration. They lead to insulin resistance and Metabolic Syndrome with increased risk of MDD. Apigenin appears to reduce the inflammation associated with Metabolic Syndrome. In rats fed high fat diets for 16 weeks, apigenin lowered plasma levels of pro-inflammatory cytokines, including MCP-1, IFN-γ, TNF-α, and IL-6. It also lowered fasting blood glucose, which was due in part to decreased insulin resistance, lower serum insulin, and downregulation of hepatic gluconeogenic enzymes' activities.[12]

In another study of rats fed high calorie diets, apigenin significantly improved the antioxidant machinery in the hippocampus, reduced ROS levels, and blocked the activation of the stress kinases, inhibitor of kappa B kinase beta, and c-Jun NH2 terminal kinase. It prevented the nuclear translocation and activation of NF-kβ in hippocampal neurons. The plasma levels of glucose, insulin, TNF-a, and IL-6 were also reduced.[13]

Simple decoction of *Matricaria recutita* appears to have antidiabetic effects in rats. It inhibits sucrase, and in high doses suppresses hyperglycemia after sucrose loading. In rats with streptozotocin-induced diabetes, it suppressed blood glucose levels and increased liver synthesis of glycogen. In human subjects, *Matricaria recutita* decoction three times a day also significantly decreased concentration of HbA1C, serum insulin levels, homeostatic model assessment for insulin resistance, total cholesterol, triglyceride, and low-density lipoprotein cholesterol compared with the control group.[14]

Some studies show that, at least in some circumstances, apigenin can inhibit the akt/mTOR pathway. Indeed, an important anti-inflammatory effect of apigenin may be mediated by reducing the TLR4-dependent activation of Akt, mTOR, and NF-κB pathways, and activation of JNK and p38-MAPK.[15] Some SSRIs, including fluoxetine, also block the TLR4 receptor. Inhibition of the Akt/mTOR pathway might be expected to disinhibit GSK-3β, an effect contrary to the effects of many antidepressant treatments. However, there has been a report that apigenin can inhibit GSK-3β by directly acting on the enzyme.[16]

7.40.4 PRECLINICAL ANTIDEPRESSANT-LIKE EFFECTS

Numerous studies have demonstrated antidepressant-like effects of apigenin in rodents. In rats subjected to chronic unpredictable stress, apigenin reduced depression-like behavior, such as diminished sucrose intake and less locomotor behavior in the open field. This effect of apigenin was due in part to upregulation of PPARγ expression and downregulation of the expression of the neuronal NLRP3 inflammasome. PPARγ, itself a recent target for antidepressant effects, regulates adipogenesis, lipid metabolism, cell proliferation, inflammation, and insulin sensitization. PPARγ-mediated inhibition of the inflammasome in turn reduced the production of IL-1β, which contributes to depression-like behavior due to inflammatory effects. This is so-called "sickness behavior." Treatment with GW9662, a selective PPARγ inhibitor, diminished the inhibitory effects of apigenin on the NLRP3 inflammasome and diminished its antidepressant effects.[17]

Apigenin also reversed the depression-like behaviors of mice treated with lipopolysaccharide. Pre-treatment with apigenin reduced immobility time in the tail suspension test, and increased sucrose preference in a manner similar to treatment with fluoxetine. Both apigenin and fluoxetine attenuated lipopolysaccharide-induced pro-inflammatory cytokines IL-1β and TNF-α in the prefrontal cortex. Both also suppressed iNOS and cyclooxygenase-2 activity in the prefrontal cortex through dampening activation of NF-κβ activation. The contribution of brain TNF-α – and ostensibly other inflammatory factors – in the depression-like behaviors of mice exposed to lipopolysaccharide was demonstrated in showing that intracerebroventricular administration of TNF-α produced the depression-like behavior, and that this also was reversed by pretreatment with apigenin or fluoxetine.[18] Along with ameliorating inflammation in the brain, apigenin also inhibits oxidative stress, and restores activity in the ERK/CREB/BDNF pathway in the brain.[19] Perhaps most significantly, apigenin stimulates neurogenesis in the adult rat brain.[20]

7.40.5 Human Antidepressant Effects

The mild anxiolytic effects of *Matricaria recutita* in human subjects are well known. Daily treatment with *Matricaria recutita* extract for eight weeks restores normal diurnal pattern of cortisol secretion in patients with Generalized Anxiety Disorder.[21] However, there is more recent evidence of antidepressant effects in humans as well. A 2012 study evaluated effects of *Matricaria recutita* in subjects that comorbidly suffered anxiety and depression. The analysis of results performed with the use of the Hamilton Depression scale revealed that a slight but significant decrease in depression symptoms in the *Matricaria recutita* group in comparison with the control group.[22] In a test of post-partum women with sleep difficulty, *Matricaria recutita* significantly improved insomnia, fatigue, and symptoms of depression, as per the Postpartum Sleep Quality Scale, Postpartum Fatigue Scale, and Edinburgh Postnatal Depression Scale, respectively. The difference from control disappeared after four weeks.[23]

Most recently, the anxiolytic effects of *Matricaria recutita* were evaluated, but the analysis included scales to evaluate mood and general well-being. All the subscales on the Psychological General Well Being Index improved, including the depression scale. However, significant improvement was seen in only the anxiety and well-being scales.[24]

7.40.6 Dosage

Hoffman recommends infusion of 2–3 teaspoons of herb in 1 cup hot water, to be drunk up to four times a day. Alternatively, 3–10ml of tincture (1:5 in 50% ethanol) may be taken three times a day.[25]

7.40.7 Toxicity

In rats the LD50 of dried decoction of *Matricaria recutita* flowers was above 3,200 mg/kg, as up to that dose, no toxic effects were observed.[26] In mice, the LD50 of the essential oil of *Matricaria recutita* was noted to be above 5,000 mg/kg, as up to and including that dose, no animal deaths were noted.[27]

7.40.8 Safety in Pregnancy

In a study of Palestinian women, the use of *Matricaria recutita* during pregnancy was quite common. Nearly half of pregnant women used the herb. Although the sample, was relatively small, i.e., 300 women, there was not a statistically significance in pregnancy outcomes between those that used *Matricaria recutita* and those that did not.[28] The *Botanical Safety Handbook* noted no concerns for use, but also noted no firm evidence of safety.[29]

7.40.9 Drug Interaction

The crude essential oil of *Matricaria recutita* significantly inhibited the activity of the CYP1A2 isoenzyme, and to lesser extents, the CYP2C9, CYP2D6, and CYP3A4 isoforms.[30] However, the Botanical Safety Handbook notes no known interactions with drugs.

REFERENCES

1. Murti K, Panchal MA, Gajera V, Solanki J. Pharmacological properties of *Matricaria recutita*: a review. *Pharmacologia.* 2012;3(8):348–351.
2. Singh O, Khanam Z, Misra N, Srivastava MK. Chamomile (*Matricaria chamomilla* L.): an overview. *Pharmacogn Rev.* 2011;5(9):82–95.
3. Al-Ismail KH, Aburjai T. Antioxidant activity of water and alcohol extracts of chamomile flowers, anise seeds and dill seeds. *J Sci Food Agric.* 2004;84(2):173–178.
4. Roby MHH, Sarhan MA, Selim KAH, Khalel KI. Antioxidant and antimicrobial activities of essential oil and extracts of fennel (*Foeniculum vulgare* L.) and chamomile (*Matricaria chamomilla* L.). *Ind Crop Prod.* 2013;44:437–445.
5. Aksoy L, Sözbilir NB. Effects of *Matricaria chamomilla* L. on lipid peroxidation, antioxidant enzyme systems, and key liver enzymes in CCl_4-treated rats. *J Toxicol Environ Chem.* 2012;94(9):1780–1788.
6. Srivastava JK, Pandey M, Gupta S. Chamomile, a novel and selective COX-2 inhibitor with anti-inflammatory activity. *Life Sci.* 2009;85(19–20):663–669.
7. Drummond EM, Harbourne N, Marete E, et al. Inhibition of proinflammatory biomarkers in THP1 macrophages by polyphenols derived from chamomile, meadowsweet and willow bark. *Phytother Res.* 2013;27(4):588–594.
8. Smolinski AT, Pestka JJ. Modulation of lipopolysaccharide-induced proinflammatory cytokine production in vitro and in vivo by the herbal constituents apigenin (chamomile), ginsenoside Rb1 (ginseng) and parthenolide (feverfew). *Food Chem Toxicol.* 2003;41(10):1381–1390.
9. Shanmugam K, Holmquist L, Steele M, et al. Plant-derived polyphenols attenuate lipopolysaccharide-induced nitric oxide and tumour necrosis factor production in murine microglia and macrophages. *Mol Nutr Food Res.* 2008;52(4):427–438.

10. Paredes-Gonzalez X, Fuentes F, Jeffery S, et al. Induction of NRF2-mediated gene expression by dietary phytochemical flavones apigenin and luteolin. *Biopharm Drug Dispos.* 2015;36(7):440–451.
11. Maes M, Fišar Z, Medina M, Scapagnini G, Nowak G, Berk M. New drug targets in depression: inflammatory, cell-mediated immune, oxidative and nitrosative stress, mitochondrial, antioxidant, and neuroprogressive pathways. And new drug candidates—Nrf2 activators and GSK-3 inhibitors. *Inflammopharmacology.* 2012;20(3):127–150.
12. Jung UJ, Cho YY, Choi MS. Apigenin ameliorates dyslipidemia, hepatic steatosis and insulin resistance by modulating metabolic and transcriptional profiles in the liver of high-fat diet-induced obese mice. *Nutrients.* 2016;8(5)305.
13. Jagan K, Chandrasekaran SP, Kalaivanan K, Ramachandran V, Carani Venkatraman A. Apigenin attenuates hippocampal oxidative events, inflammation and pathological alterations in rats fed high fat, fructose diet. *Biomed Pharmacother.* 2017;89:323–331.
14. Rafraf M, Zemestani M, Asghari-Jafarabadi M. Effectiveness of chamomile tea on glycemic control and serum lipid profile in patients with type 2 diabetes. *J Endocrinol Invest.* 2015;38(2):163–170.
15. Kim A, Lee CS. Apigenin reduces the Toll-like receptor-4-dependent activation of NF-κB by suppressing the Akt, mTOR, JNK, and p38-MAPK. *Naunyn Schmiedebergs Arch Pharmacol.* 2018;391(3):271–283.
16. Johnson JL, Rupasinghe SG, Stefani F, Schuler MA, Gonzalez de Mejia E. Citrus flavonoids luteolin, apigenin, and quercetin inhibit glycogen synthase kinase-3beta enzymatic activity by lowering the interaction energy within the binding cavity. *J Med Food.* 2011;14:325–333.
17. Li R, Wang X, Qin T, Qu R, Ma S. Apigenin ameliorates chronic mild stress-induced depressive behavior by inhibiting interleukin-1β production and NLRP3 inflammasome activation in the rat brain. *Behav Brain Res.* 2016;296:318–325.
18. Li R, Zhao D, Qu R, Fu Q, Ma S. The effects of apigenin on lipopolysaccharide-induced depressive-like behavior in mice. *Neurosci Lett.* 2015;594:17–22.
19. Zhao L, Wang JL, Liu R, Li XX, Li JF, Zhang L. Neuroprotective, anti-amyloidogenic and neurotrophic effects of apigenin in an Alzheimer's disease mouse model. *Molecules.* 2013;18(8):9949–9965.
20. Taupin P. Apigenin and related compounds stimulate adult neurogenesis. *Expert Opin Ther Pat.* 2009;19(4):523–527.
21. Keefe JR, Guo W, Li QS, Amsterdam JD, Mao JJ. An exploratory study of salivary cortisol changes during chamomile extract therapy of moderate to severe generalized anxiety disorder. *J Psychiatr Res.* 2018;96:189–195.
22. Amsterdam JD, Shults J, Soeller I, Mao JJ, Rockwell K, Newberg AB. Chamomile (*Matricaria recutita*) may provide antidepressant activity in anxious, depressed humans: an exploratory study. *Altern Ther Health Med.* 2012;18(5):44–49.
23. Chang, SM, Chen CH. Effects of an intervention with drinking chamomile tea on sleep quality and depression in sleep disturbed postnatal women: a randomized controlled trial. *J Adv Nurs.* 2016;72(2):306–315.
24. Keefe JR, Mao JJ, Soeller I, Li QS, Amsterdam JD. Short-term open-label chamomile (*Matricaria chamomilla* L.) therapy of moderate to severe generalized anxiety disorder. *Phytomedicine.* 2016;23(14):1699–1705.
25. Hoffmann D. *Medical Herbalism: The Science and Practice of Herbal Medicine.* Rochester, VT: Healing Arts Press; 2003. p. 566.
26. Sebai H, Jabri MA, Souli A, et al. Antidiarrheal and antioxidant activities of chamomile (*Matricaria recutita* L.) decoction extract in rats. *J Ethnopharmacol.* 2014;152:327–332.
27. Hajjaj G, Bounihi A, Tajani M, Cherrah YA, Zellou AM. Evaluation of CNS activities of *Matricaria chamomilla* L. essemtial oil in experimental animals from Morocco. *Int J Pharm Pharm Sci.* 2013;5(2):530–534.

28. Al-Ramahi R, Jaradat N, Adawi D. Use of herbal medicines during pregnancy in a group of Palestinian women. *J Ethnopharmacol*. 2013;150(1):79–84.
29. Gardner Z, McGuffin M, editors. *Botanical Safety Handbook*. 3rd ed. Boca Raton, FL: CRC Press; 2013. p. 541.
30. Ganzera, M, Schneider P, Stuppner H. Inhibitory effects of the essential oil of chamomile (*Matricaria recutita* L.) and its major constituents on human cytochrome P450 enzymes. *Life Sci*. 2006;78(8):856–861.

7.41 *MELISSA OFFICINALIS* (LEMON BALM)

Melissa officinalis, commonly known as lemon balm, is a perennial plant of the mint family that is native to south-central Europe, the Mediterranean Basin, Iran, and Central Asia. As noted by Maud Grieve, in her famous work *A Modern Herbal*,[1] "It was highly esteemed by Paracelsus, who believed it would completely revivify a man. It was formerly esteemed of great use in all complaints supposed to proceed from a disordered state of the nervous system." She further noted "The London Dispensary (1696) says: 'An essence of Balm, given in Canary wine, every morning will renew youth, strengthen the brain, relieve languishing nature and prevent baldness.'" The herb has long been used for a wide range of human complaints, including depression, psychosis, hysteria, insomnia, epilepsy, headaches, vertigo, syncope, malaise, flatulence, indigestion, colic, nausea, anemia, asthma, bronchitis, amenorrhea, rheumatism, ulcers, and wounds.[2] It is rich in polyphenolic phytochemicals, many of which have been identified as active components of extracts of various other herbs. Among the substances it contains are rosmarinic acid, caffeic acids, chlorogenic acid, metrilic acid, tannins, luteolin, apigenin, monoterpene glycosides, β-caryophyllene, germacrene, triterpenes, citronellal, citrals, ocimene, citronellol, geraniol, nerol, linalool, and ethric oil.[3]

7.41.1 ANTIOXIDANT

The herb also produces anti-inflammatory and antioxidative effects both peripherally and centrally. *In vitro*, extracts of *Melissa officinalis* had neuroprotective effects against Aβ-induced cytotoxicity and oxidative stress in cultured PC12 cells, a neuron-like cell line derived from a pheochromocytoma of the rat adrenal medulla. Production of reactive oxygen species was reduced, as were levels of the lipid peroxidation biomarkers, malondialdehyde, and thiobarbituric acid reactive substances. Glutathione peroxidase activity was increased.[4] Extracts of *Melissa officinalis* also exert anti-oxidative and anti-inflammatory neuroprotective effects *in vivo*. In rats subjected to transient hippocampal ischemia, oil of *Melissa officinalis* inhibited generation of malondialdehyde and reduced concentrations of thiobarbituric acid reactive substances in the hippocampus. It also significantly increased the antioxidant enzyme caspase and suppressed expression of the hypoxiainducible factor 1-alpha gene, which is induced under conditions of hypoxia.[5]

7.41.2 ANTI-INFLAMMATORY

Melissa officinalis was effective in two standard animal models of inflammation. Oral administration of aqueous extracts reduced swelling and pain in both the histamine- and carrageenan-induced paw edema tests in rats. The effects of *Melissa officinalis* were comparable to those of the potent, non-steroidal anti-inflammatory drug, indomethacin. The herbal treatment also countered the oxidative effects secondary to the inflammatory process. It increased glutathione and reduced whole blood levels of thiobarbituric acid reactive substances and malondialdehyde, both of which are markers of lipid peroxidation.[6]

7.41.3 ANTIDIABETIC/ANTI-METABOLIC SYNDROME

Melissa officinalis exhibits antidiabetic effects and acts to reverse components of the Metabolic Syndrome. Four weeks of treatment with the herb normalized weight and reduced serum glucose in streptozotocin (STZ)-induced diabetic rats.[7] Enhancement of insulin sensitivity and improvement in dyslipidemia were also observed in mice. In cultured human adipocytes, extract of *Melissa officinalis* were found to induce peroxisome proliferator-activated receptor target gene expression, which was suspected to be at least partially responsible for the effects observed in the intact mice.[8] The protective, antidiabetic effects of *Melissa officinalis* are further enhanced by its apparent ability to inhibit the formation of advanced glycation end products, which is yet another source of inflammation and oxidative stress.[9]

7.41.4 PRECLINICAL ANTIDEPRESSANT-LIKE EFFECTS

Several studies have shown antidepressant-like effects of *Melissa officinalis* in mice and rats. Ten days of treatment with an ethanolic extract of *Melissa officinalis* improved performance of rats in the forced swim test. Rats treated with *Melissa officinalis* exhibited less immobility, which is interpreted as an antidepressant-like effect. The effect was similar to that in rats treated with fluoxetine. Treatment also produced an anxiolytic effect as demonstrated by increases of open arm entries and time spent in open arm times of the elevated plus-maze. In fact, rats treated with extract of *Melissa officinalis* showed less anxiety than those of the vehicle-treated rats, and exhibited behaviors similar to those of rats treated with diazepam.[10] In another study with rats, treatment with extract of *Melissa officinalis* and rosmarinic acid purified from the extract both decreased immobility time in the forced swim test.[11] Both the aqueous extract and essential oil of *Melissa officinalis* decreased immobility in the forced swimming test in mice. Those effects were comparable to those of the antidepressants, fluoxetine and imipramine.[12]

In mice treated with extract of *Melissa officinalis* for three weeks, increases in hippocampal cell proliferation and neuroblast differentiation were observed through bromodeoxyuridine and Calbindin D-28k labeling, respectively. These increases were up to 245.2% of the vehicle-treated group. This was suspected to be due in part to reductions in serum corticosterone that were also observed in those animals.[13]

7.41.5 HUMAN ANTIDEPRESSANT EFFECTS

There are several studies of the effects of *Melissa officinalis* on mood in humans, though none are formal evaluations of treatment effects on MDD. In a randomized,

placebo-controlled, double-blind, balanced-crossover study of healthy young participants, acute administration of *Melissa officinalis* improved "calmness" as assessed
by the Bond-Lader mood scales. Although calmness was reported following the lowest dose, reduction in alertness was reported after the highest dose, suggesting sedation. No change was seen in the "content" component of the Bond-Lader mood scale,
suggesting no effect on emotion, arguably the *sine qua non* of an antidepressant.[14]

In a randomized trial in young women suffering premenstrual dysphoric disorder, chronic treatment with *Melissa officinalis* significantly reduced psychosomatic
symptoms, anxiety, insomnia, and social function disorder per the General Health
Standard Questionnaire.[15]

In a study of treatment of insomnia with the combination of the herbs, *Melissa
officinalis* and *Nepeta menthoides*, there was significant improvement of scores on
the Beck Depression index.[16] This was despite the fact that none of the subjects were
formally diagnosed as suffering MDD. Thus, while animal studies suggest antidepressant effects of *Melissa officinalis*, and potential pharmacological mechanisms
for such have been described, there is as yet no compelling evidence that *Melissa
officinalis* alone has efficacy in the treatment of MDD in human subjects.

7.41.6 DOSAGE

Hoffman suggests dosing of tincture (1:5, 40%) to be 2–6ml three times a day.
Alternatively, one may take as an infusion of 2–3 teaspoons of dried herb in one cup
of water 2–3 times a day.[17]

7.41.7 TOXICITY

The LD50 of the dried alcoholic extract of *Melissa officinalis* in mice is reported to
be 4.5 g/kg.[18] This would extrapolate to be approximately 315 g for a 75 kg human
adult.

7.41.8 SAFETY IN PREGNANCY

The *Botanical Safety Handbook* notes there to be no definitive guidance from the
literature concerning the safety of this herb during pregnancy.[19]

7.41.9 DRUG INTERACTIONS

The *Botanical Safety Handbook* notes no known interactions with drugs.

REFERENCES

1. Grieve MA. *Modern Herbal*. New York, NY: Harcourt, Brace & Company; 1931.
2. Miraj S, Azizi N, Kiani S. A review of chemical components and pharmacological
   effects of *Melissa officinalis* L. *Der Pharm Lett*. 2016;8(6):229–237.
3. Carnat AP, Carnat A, Fraisse D, Lamaison JL. The aromatic and polyphenolic composition of lemon balm (*Melissa officinalis* L. subsp. officinalis) tea. *Pharm Acta Helv*.
   1998;72:301–305.

4. Sepand MR, Soodi M, Hajimehdipoor H, Soleimani M, Sahraei E. Comparison of neuroprotective effects of *Melissa officinalis* total extract and its acidic and non-acidic fractions against A β-induced toxicity. *Iran J Pharm Res.* 2013;12(2):415–423.

5. Bayat M, Tameh AA, Ghahremani MH, et al. Neuroprotective properties of *Melissa officinalis* after hypoxic-ischemic injury both in vitro and in vivo. *DARU.* 2012;20:42

6. Birdane YO, Buyukokuroglu ME, Birdane FM, et al. Anti-inflammatory and antinociceptive effects of *Melissa Officinalis* L. in rodents. *Revue Méd Vét.* 2007;158(2):75–81.

7. Hasanein P, Riahi H. Antinociceptive and antihyperglycemic effects of *Melissa officinalis* essential oil in an experimental model of diabetes. *Med Princ Pract.* 2015;24:47–52.

8. Weidner C, Wowro SJ, Freiwald A, et al. Lemon balm extract causes potent antihyperglycemic and antihyperlipidemic effects in insulin-resistant obese mice. *Mol Nutr Food Res.* 2014;58(4):903–907.

9. Miroliaei M, Khazaei S, Moshkelgosha S, Shirvani M. Inhibitory effects of Lemon balm (*Melissa officinalis*, L.) extract on the formation of advanced glycation end products. *Food Chem.* 2011;129(2):267–271.

10. Taiwo AE, Leite FB, Lucena GM, et al. Anxiolytic and antidepressant-like effects of *Melissa officinalis* (lemon balm) extract in rats: influence of administration and gender. *Indian J Pharmacol.* 2012;44(2):189–192.

11. Lina SH. A medicinal herb, *Melissa officinalis* L. ameliorates depressive-like behavior of rats in the forced swimming test via regulating the serotonergic neurotransmitter. *J Ethnopharmacol.* 2015;175:266–272.

12. Emamghoreishi M, Talebianpour MS. Antidepressant effect of *Melissa officinalis* in the forced swimming test. *DARU.* 2009;17(1):42–47.

13. Yoo DY, Choi JH, Kim W, et al. Effects of *Melissa officinalis* L. (lemon balm) extract on neurogenesis associated with serum corticosterone and GABA in the mouse dentate gyrus. *Neurochem Res.* 2011;36:250–257.

14. Kennedy DO, Scholey AB, Tildesley NTJ, Perry EK, Wesnes KA. Modulation of mood and cognitive performance following acute administration of *Melissa officinalis* (lemon balm). *Pharmacol Biochem Behav.* 2002;72(4):953–964.

15. Heydari N, Dehghani M, Emamghoreishi M, Akbarzadeh M. Effect of *Melissa officinalis* capsule on the mental health of female adolescents with premenstrual syndrome: a clinical trial study. *Int J Adolesc Med Health.* 2017;0015:2191–2278.

16. Ranjbara M, Firoozabadi A, Salehi A, et al. Effects of Herbal combination (*Melissa officinalis* L. and *Nepeta menthoides* Boiss. & Buhse) on insomnia severity, anxiety and depression in insomniacs: randomized placebo-controlled trial. *Integr Med Res.* 2018;7(4):328–332.

17. Hoffmann D. *Medical Herbalism: The Science and Practice of Herbal Medicine.* Rochester, VT: Healing Arts Press; 2003. p. 567.

18. Namjoo A, MirVakili M, Faghani M. Biochemical, liver and renal toxicities of Melissa officinals hydroalcoholic extract on balb/C mice. *J HerbMed Pharmacol.* 2013;2(2):35–40.

19. Gardner Z, McGuffin M, editors. *Botanical Safety Handbook.* 3rd ed. Boca Raton, FL: CRC Press; 2013. p. 550.

7.42 *MIMOSA PUDICA*

Mimosa pudica is a creeping flowering plant of the pea family. It is commonly known as the sensitivity plant, touch-me-not, shame plant, or shy plant. Those names arise from the remarkable folding and lowering of its leaves in response to being touched. In Ayurvedic and Unani Medicine, *Mimosa pudica* root has been used to treat fevers, jaundice, dysentery, vaginal and uterine complaints, inflammations, fatigue, and asthma. In Western medicine, mimosa root is used for treating insomnia, irritability, and premenstrual syndrome.[1]

The primary compounds isolated from *Mimosa pudica* are crocetin, crocin, 6-hydroxy flavone (2), 2'-hydroxy flavanone, p-coumaric acid, p-hydroxy benzoic acid, chlorogenic acid, jasmonic acid, and L-mimosine, caffeic acid (7), ethyl gallate (3), catechin (9), and gallic acid.[2] Some of these phytochemicals are shared by other medicinal plants, several of which have been found to have antidepressant effects in animals and humans.

7.42.1 ANTIOXIDANT

Extracts of *Mimosa pudica* were found to have substantial antioxidant effects *in vitro* in the standard DPPH radical-scavenging activity and the ferric reducing/antioxidant power assays. Several of its constituent phytochemicals, i.e., 5,7,3′,4′-tetrahydroxy-6-C-[β-D-api ose-(1→4)]-β-D-glycopyranosyl flavone, isorientin, orientin, isovitexin, and vitexin, exhibited substantial antioxidant effects. The first of them – with a formal, chemical name and not a common name due to its rarity – exerted an antioxidant power equal to that of the commercial antioxidant troxol.[3] Antioxidant effects have also been observed in the livers of mice that had been chronically administered extract of *Mimosa pudica* while also administered hepatotoxic levels of alcohol. Liver damage was reduced by the herb. Extracts also attenuated the alcohol-induced decreases in hepatic levels of protective superoxide dismutase, catalase, glutathione peroxidase, glutathione, and vitamin C. Lipid peroxidation was reduced and elimination of reactive oxygen species was enhanced.[4]

7.42.2 ANTI-INFLAMMATORY

Evaluation of various phytochemical components of *Mimosa pudica* found that several, particularly L-mimosine, crocetin, crocin and jasmonic acid, were potent inhibitors of nitric oxide, TNF-α and IL-1ß release *in vitro* from lipopolysaccharide-stimulated cultured macrophages. Others, including ethyl gallate, gallic acid and caffeic, showed less but still significant anti-inflammatory activity. Similar effects were seen in the same study in serum samples of rats injected with lipopolysaccharide.[5]

7.42.3 ANTIDIABETIC/ANTI-METABOLIC SYNDROME

Mimosa pudica has traditionally been used in Southeast Asia to treat symptoms of diabetes. Modern evidence lends support for such use. In rats made diabetic by exposure to alloxan, ethanolic extract of *Mimosa pudica* significantly lowered serum glucose levels in a manner similar to that of metformin.[6] Those results may be due in part to the finding that extract of the herb inhibits the enzymes α-amylase and α-glucosidase, and thus slows the digestion and absorption of carbohydrates.[7]

7.42.4 PRECLINICAL ANTIDEPRESSANT-LIKE EFFECTS

Rats treated for 30 days with extract of *Mimosa pudica* showed reduced immobility in the forced swim test in a manner similar to effects of desipramine and clomipramine. The herb also increased the rate of reinforcements received in the differential reinforcement of low rates of response paradigm. This procedure is uniquely sensitive to antidepressants, which uniformly increase reinforcements.[8] The hydroalcoholic extract of *Mimosa pudica* also reversed the depression-like state induced in rats by treatment with reserpine. The herb reduced immobility of rats in the forced swim test, as did imipramine and fluoxetine.[9] Along with antidepressant-like effects, extract of *Mimosa pudica* had anxiolytic-like effects in mice in the elevated-T maze.[10]

Several specific phytochemicals in *Mimosa pudica* have been found to have antidepressant-like effects. For example, crocin, thought to be one of the important contributors to the antidepressant effects of saffron in animals and humans, has been identified in extracts of *Mimosa pudica*. Chronic administration of crocin reduces immobility time in rats subjected to the forced swim test.[11] Caffeic acid, found in both *Mimosa pudica* in coffee, reduces the duration of immobility and freezing in the forced swim test in mice.[12]

7.42.5 HUMAN ANTIDEPRESSANT EFFECTS

In Mexican folk medicine, aqueous extracts from dried leaves of *M. pudica* L. are used to treat depression.[13] However, there are no published formal studies of effects of *Mimosa piduca* on major depression in human subjects.

7.42.6 DOSAGE

Dosage information is not readily available. However, in a study of *Mimosa pudica* for treatment of menorrhagia in women, doses of 500–1,500 mg a day of dried aqueous extract were effective and without adverse effects, even in daily maintenance dosing.[14] Easley and Horne give dosage recommendation for the closely related herb of the Mimosa genus, *Albizia julibrissin*, which they describe as also having antidepressant properties. They recommend tincture of dried flowers (1:2, 50% alcohol) or bark (1:5, 50% alcohol) 10 drops to 5 ml taken 1–4 times daily.[15]

7.42.7 TOXICITY

Oral administration of dried aqueous extract caused no deaths or obvious adverse signs in mice in doses up to 2,000 mg/kg.[16] Extrapolated to a human adult of 75 kg, this dose would be approximately 140 g.

7.42.8 SAFETY IN PREGNANCY

There is conflicting data on the safety of *Mimosa pudica* during pregnancy. In one report, the leaves of *Mimosa pudica* showed abortifacient effects in pregnant rats.[17] However, a case study described a woman treated chronically with *Mimosa pudica* for four years to treat menorrhagia. While being treated, this woman was able to carry a normal pregnancy to full term.[18]

7.42.9 Drug Interactions

Review of the literature revealed no studies of the possible interactions of *Mimosa pudica* with commonly used medications.

REFERENCES

1. Joseph B, George J, Mohan J. Pharmacology and traditional uses of *Mimosa pudica*. *IJPSDR*. 2013;5(2):41–44.
2. Johnson K, Narasimhan G, Krishnan C. *Mimosa pudica* Linn—a shyness princess: a review of its plant movement, active constituents, uses and pharmacological activity. *IJPSDR*. 2014;5(12):5104–5118.
3. Zhang J, Yuan K, Zhou W, Zhou J, Yang P. Studies on the active components and antioxidant activities of the extracts of *Mimosa pudica* Linn. from southern China. *Pharmacogn Mag*. 2011;7(25):35–39.
4. Nazeema TH, Brindha V. Antihepatotoxic and antioxidant defense potential of *Mimosa pudica*. *Int J Drug Discov*. 2009;1(2):1–4.
5. Patel NK, Bhutani KK. Suppressive effects of *Mimosa pudica* (L.) constituents on the production of LPS-induced pro-inflammatory mediators. *EXCLI J*. 2014;13: 1011–1021.
6. Sutar NG, Sutar UN, Behera BC. Antidiabetic activity of the leaves of *Mimosa pudica* Linn in albino rats. *J Herb Med Toxicol*. 2009;3(1):123–126.
7. Tunnaai TS, Zaidul IS, Ahmed QU, et al. Analyses and profiling of extract and fractions of neglected weed *Mimosa pudica* Linn. traditionally used in Southeast Asia to treat diabetes. *S Afr J Bot*. 2015;99:144–152.
8. Molina M, Contreras CM, Tellez-Alcantara P. *Mimosa pudica* may possess antidepressant actions in the rat. *Phytomedicine*. 1999;6(5):319–323.
9. Lawar M, Shende V, Ingle S, Sanghvi M, Hamdulay N. Effect of hydroalcoholic extract of *Mimosa Pudica* Linn. with imipramine and fluoxetine on reserpine-induced behavioural despair in rats. *PharmaTutor*. 2014;2(9):123–134.
10. Mbomo RA, Gartside S, Ngo Bum E, Njikam N, Okello E, McQuade R. Effect of *Mimosa pudica* (Linn.) extract on anxiety behaviour and GABAergic regulation of 5-HT neuronal activity in the mouse. *J Psychopharmacol*. 2012;26:575–583.
11. Hassani FV, Naser Vi, Razavi BM, Mehri S, Abnous K, Hosseinzadeh H. Antidepressant effects of crocin and its effects on transcript and protein levels of CREB, BDNF, and VGF in rat hippocampus. *DARU*. 2014;22:16.
12. Takeda H, Tsuji M, Inazu M, Egashira T, Matsumiya T. Rosmarinic acid and caffeic acid produce antidepressive-like effect in the forced swimming test in mice. *Eur J Pharmacol*. 2002;449(3):261–267.
13. Del Amo S. *Catalogo de plantas medicinales del estado de Veracruz*. 9: Xalapa: INIRB; 1979. p. 279.
14. Vaidya GH, Shetha UK. Ancient *Mimosa pudica* (Linn.) its medicinal value and pilot clinical use in patients with menorrhagia. *Sci Life*. 1986;5(3):156–160.
15. Easley T, Horne S. *The Modern Herbal Dispensary*. Berkeley, CA: North Atlantic Books; 2016. p. 267.
16. Kathikeyan M, Deepa MK. Antinociceptive activity of *Mimosa pudica* Linn. *IJPT*. 2010;9:11–14.
17. Norton SP. Antifertility activity of leaves of *Mimosa pudica* Linn in early pregnancy of albino rats. *Indian J Zool*. 1978;6:89–93.
18. Vaidya GH, Sheth UK. *Mimosa pudica* (Linn.): its medicinal value and pilot clinical use in patients with menorrhagia. *Ancient Sci Life*. 1986;5(3):156–160.

7.43   *OCIMUM BASILICUM* (SWEET BASIL)

Ocimum basilicum, often known as sweet basil, is an herb of the family *Lamiaceae*. It is an ancient herb that has been cultivated for 5,000 years or more. With origins in India and Asia, it is said to have been brought to Greece by Alexander the Great. It was used in Indian Ayurvedic Medicine, later by the ancient Greek healers, Dioscorides and Galen, and through the Renaissance by herbalists such as Culpeper and Gerard. The herb has had many medicinal uses over the centuries, including treatment of colds, cold and hay fever, indigestion, constipation, diarrhea, and vomiting. It has also been used to treat nervous conditions, such as stress, migraine, and mental fatigue.[1] *Ocimum basilicum* contains a large number of phytochemicals, with over 200 having been identified, including monoterpenes, sesquiterpenes, triterpenes, aromatics, flavonoids, polyphenols, cinnamates, and steroids.[2] Of those phytochemicals in the essential oil of *Ocimum basilicum*, linalool (about 30.0%), chavicol, and (Z)-cinnamic acid methyl ester are the main components. The essential oil of the closely related holy basil (*Ocimum sanctum*) has similar contents with a higher quantity of eugenol.[3]

7.43.1   ANTIOXIDANT

Ocimum basilicum has been shown to exert antioxidant effects. An ethanol extract of the herb exhibited *in vitro* antioxidant effects 75% of those of α-tocopherol.[4] Pretreatment with extract of *Ocimum basilicum* also prevented oxidative damage in brain tissue due to global cerebral ischemia in mice. The concentrations of thiobarbituric acid reactive substances were reduced, a sign of reduced lipid peroxidation, and antioxidant glutathione levels were spared in cerebral cortex. Treatment with *Ocimum basilicum* also attenuated the impairments in short-term memory and motor coordination.[5]

7.43.2   ANTI-INFLAMMATORY

Anti-inflammatory effects of *Ocimum basilicum* have been observed in several *in vivo* and *in vitro* models. Oral administration of extract from seed reduced the paw edema caused by histamine and prostaglandin F-2.[6] The crude alcoholic extract of *Ocimum basilicum* reduced the *in vitro* inflammatory responses of human mononuclear cells to lipopolysaccharide as shown by reduction in release of TNF- α and IL-IB. The extract also blunted the inflammatory effects of lipopolysaccharide in the murine RAW 264.7 macrophage line. In the latter cells, the inducible nitric oxide synthase (iNOS) along with subsequent production of nitric oxide were suppressed.[7]

7.43.3   ANTIDIABETIC/ANTI-METABOLIC SYNDROME

Anti-hyperglycemic and hypolipidemic effects of the aqueous extracts from *Ocimum basilicum* have also been reported in streptozotocin-induced diabetic rats. These effects occur after acute and chronic oral treatment. The herbal extract lowered serum glucose levels in both diabetic and, albeit to lesser degree, in non-diabetic rats.[8] Part of this effect could be due to the inhibition of a-glucosidase and a-amylase by *Ocimum basilicum* phytochemicals.[9] This would slow the absorption of sugars after ingestion of complex carbohydrates.

7.43.4 PRECLINICAL ANTIDEPRESSANT-LIKE EFFECTS

There is an extensive literature concerning apparent antidepressant-like effects of *Ocimum basilicum* and *Ocimum sanctum* in rodents. Many studies have shown that, in both rats and mice, treatment with *Ocimum basilicum* has an antidepressant-like effect. In some cases, the herb was evaluated simply to determine its ability to reduce stress immobility and despair in otherwise healthy animals, that is, to prevent rather than treat a depression-like syndrome. In rats, extract of seeds of *Ocimum basilicum* decreased immobility times in the forced swim and tail suspension tests.[10] Extract of *Ocimum basilicum* had the same antidepressant-like effects in mice.[11,12]

Chronic administration of *Ocimum basilicum* to rats exposed to an electro-magnetic field for eight weeks – enough to cause oxidative damage in the brain – decreased immobility time in the forced swim test.[13] Acute administration of a hydro-ethanolic extract of *Ocimum basilicum* leaves also reduced immobility in the forced swim test in rats suffering a depression-like syndrome induced by exposure to ovalbumin.[14]

In at least two studies, antidepressant like effects of *Ocimum basilicum* were observed, and additional neurochemical analyses were performed to help deter-mine the mechanisms by which the herb may have produced the effects. In one case, chronic treatment of rats with hydroethanolic extract of *Ocimum basilicum* reduced immobility in the forced swim test, and this was associated with reduction of oxida-tion products in the brains of those animals.[15] In a study of mice, chronic treatment with *Ocimum basilicum* reduced immobility in the forced swim test in mice that had been subjected to chronic unpredictable stress. These animals also had lower lev-els of serum corticosterone. Neurochemical analysis of their brains further showed decreased apoptosis in both neurons and glial cells, and increased neurogenesis in the dentate gyrus of the hippocampus. Both BDNF and glucorticoid receptor density were increased in that area of the brain.[16]

In a few studies, holy basil, rather than *Ocimum basilicum*, has been evaluated for antidepressant-like effects. A methanol extract of *Ocimum sanctum* administered to mice reduced immobility time in the forced swim test.[17] In a similar study, extract of *Ocimum sanctum* reduced immobility times in both the forced swim and tail suspension tests.[18]

Of particular note, in one study the antidepressant-like effects of *Ocimum basilicum* and *Ocimum sanctum* were compared in rats that had had intracerebral injections of beta-amyloid. That treatment can cause not only memory deficits, but also anxiety and depression-like behaviors. The two species of basil, *Ocimum sanctum L.* and *Ocimum basilicum*, were essentially equal in reducing immo-bility time in the forced swim test. Moreover, the major phytochemicals were similar in the two species, with linalool, camphor, b-elemene, α-bergamotene and bornyl-acetate, estragole, eugenol, and 1,8-cineole being the primary con-stituents of each.[19]

7.43.5   Human Antidepressant Effects

There have been at least two published studies of effects of *Ocimum sanctum* on anxiety and mood in human subjects. In one case, Ocimum sanctum was administered to patients diagnosed with generalized anxiety disorder through the Brief Psychiatric Rating Scale.

None of the patients were severely depressed, and none were being prescribed anxiolytic or antidepressant medications during the study. After two months of treatment, anxiety scores were significantly reduced. Although none had been diagnosed as suffering MDD, the depression index scores in the Brief Psychiatric Rating Scale also dropped significantly, by 30% over the two-month period. No adverse side effects were noted.[20]

In a randomized, double-blind, placebo-controlled study, *Ocimum sanctum* was evaluated for its effects on "general stress." The degree of stress was evaluated in what appeared to be an *ad hoc* list of symptoms typical of stress. Participants were not formally diagnosed with any psychiatric illness, but rather had volunteered complaints of suffering issues described in the *ad hoc* symptom list, many of which are common complaints of patients suffering anxiety or depression. After six weeks of treatment, there were significant reductions in forgetfulness, sexual problems of recent origin, frequent feeling of exhaustion, and sleep problems in comparison with both the baseline and final reports from the placebo group. There were no complaints of adverse effects during the study.[21]

7.43.6   Dosage

In the two successful human studies of *Ocimum sanctum* referred to above, one used 500 mg capsules of dried alcohol extract twice a day, whereas the other used 400 mg capsules of standardized extract taken three times a day.

7.43.7   Toxicity

The LD50 in mice for essential oil of *Ocimum basilicum* was found to be 3.64 ml per kilogram of weight.[22] Extrapolated to a 75 kg human adult, the dose would be approximately 255 ml.

7.43.8   Safety in Pregnancy

Traditional Chinese Medicine doctrine warns against use of *Ocimum basilicum* during pregnancy.[23]

7.43.9   Drug Interactions

The *Botanical Safety Handbook* notes no known interactions with drugs.[24]

REFERENCES

1. Marwat S, Khan MS, Ghulam S, Anwar N, Mustafa G, Usman K. Phytochemical constituents and pharmacological activities of sweet basil-*Ocimum basilicum* L. (*Lamiaceae*). *Asian J Chem.* 2011;23(9):3773–3782.

2. Chang X, Alderson PG, Wright CJ. Solar irradiance level alters the growth of basil (*Ocimum basilicum* L.) and its content of volatile oils. *Environ Exp Bot.* 2008;63(1–3):216–223.

3. Barbalho SM, Machado FMVF, Rodrigues JDS, Silva TH, Goulart RD. Sweet basil (*Ocimum basilicum*): much more than a condiment. *TANG.* 2012;2(1):1–5.

4. Durga RK, Karthikumar S, Jegatheesan K. Isolation of potential antibacterial and anti-oxidant compounds from *Acalyphha indica* and *Ocimum basilicum. J Med Plants Res.* 2009;3(10):703–706.

5. Boraa KS, Arorab S, Shrib R. Role of *Ocimum basilicum* L. in prevention of ischemia and reperfusion-induced cerebral damage, and motor dysfunctions in mice brain. *J Ethnopharmacol.* 2011;137(3):1360–1365.

6. Rakha P, Sharma S, Parle M. Anti-inflammatory potential of the seeds of *Ocimum basilicum* Linn. in rats. *Asian J Bio Sci.* 2010;5(1):16–18.

7. Selvakkumar C, Gayathri B, Vinaykumar KS, Lakshmi BS, Balakrishnan A. Potential antiinflammatory properties of crude alcoholic extract of *Ocimum basilicum* L. in human peripheral blood mononuclear cells. *J Health Sci.* 2007;53(4):500–505.

8. Zeggevagh AN, Sulpice T, Eddouks M. Anti-hyperglycaemic and Hypolipidemic Effects of *Ocimum basilicum* Aqueous Extract in Diabetic Rats. *Am J Pharmacol Toxicol.* 2007;2(3):123–129.

9. El-Beshbishy HA, Bahashwan SA. Hypoglycemic effect of basil (*Ocimum basilicum*) aqueous extract is mediated through inhibition of α-glucosidase and α-amylase activi-ties: an in vitro study. *Toxicol Ind Health.* 2012;28:42–50.

10. Brar B, Duhan JS, Rakha P. Antidepressant activity of various extracts from seeds of *Ocimum Basilicum* Linn. *IJSR.* 2015;4(3):41–43.

11. Sudhakar P, Gopalakrishna HN, Alva A, et al. Antidepressant activity of ethanolic extract of leaves of *Ocimum sanctum* in mice. *J Pharm Res.* 2010;3(3):624–626.

12. Suhendy H, Priatna M. Antidepressant activity of some fractions of the basil leaves [*Ocimum Basilicum* (L)] on the swiss webster male mice. *J Ilmu Kefarmasian Indonesia.* 2018;16(2):188–193.

13. Abdoly M, Farnam A, Fathiazad F, et al. Antidepressant-like activities of *Ocimum basilicum* (sweet Basil) in the forced swimming test of rats exposed to electromagnetic field (EMF). *Afr J Pharm Pharmacol.* 2012;6(3):211–215.

14. Neamati A, Talebi S, Hosseini M, Hossein Boskabady M, Beheshti F. Administration of ethanolic extract of *Ocimum Basilicum* leaves attenuates depression like behavior in the rats sensitized by Ovalbumin. *Curr Nutr Food Sci.* 2016;12(1):72–78.

15. Muneefa KI, Doss VA, Sowndarya R. Beneficial effect of hydroethanolic extract of *Ocimum basilicum* L on enzymic and non enzymic antioxidant in depression induced rats. *J Med Plant Stud.* 2017;5(3):185–188.

16. Ayoub NN, Firgany AEDL, El-Mansy AA, Ali S. Can *Ocimum basilicum* relieve chronic unpredictable mild stress-induced depression in mice? *Exp Mol Pathol.* 2017;103(2):153–161.

17. Maity TK, Mandal SC, Saha BP, Pal M. Effect of *Ocimum sanctum* roots extract on swimming performance in mice. *Phytother Res.* 2000;14(2):120–121.

18. Chatterjee M, Verma P, Maurya R, Palit G. Evaluation of ethanol leaf extract of *Ocimum sanctum* in experimental models of anxiety and depression. *J Pharm Biol.* 2011;49(5):477–483.

19. Gradinariu V, Cioanca O, Hritcu L, Trifan A, Gille E, Hancianu M. Comparative efficacy of *Ocimum sanctum* L. and *Ocimum basilicum* L. essential oils against amy-loid beta (1-42)-induced anxiety and depression in laboratory rats. *Phytochem Rev.* 2015;14(4):567–575.

20. Bhattacharyya D, Sur TK, Jana U, Debnath PK. Controlled programmed trial of *Ocimum sanctum* leaf on generalized anxiety disorders. *Nepal Med Coll J.* 2008;10:176–179.

21. Saxena RC, Singh R, Kumar P, et al. Efficacy of an extract of *Ocimum tenuiflorum* (OciBest) in the management of general stress: a double-blind, placebo-controlled study. *Evid Based Complement Alternat Med*. 2012;2012:894509.
22. Ismail M. Central properties and chemical composition of *Ocimum basilicum*. Essential oil. *Pharm Biol*. 2006;44(8):619–626.
23. Chen JK, Chen TT. *Chinese Medical Herbology and Pharmacology*. City of Industry, CA: Art of Medicine Press; 2004.
24. Gardner Z, McGuffin M, editors. *Botanical Safety Handbook*. 3rd ed. Boca Raton, FL: CRC Press; 2013. p. 591.

7.44 *ORIGANUM VULGARE* (OREGANO)

Origanum vulgare, commonly known as oregano, is a plant in the *Lamiaceae* family. It is native to temperate western and southwestern Eurasia and the Mediterranean region. This family of plants, commonly referred to as the mint family, includes many herbs used for flavoring food and treating illness. It was used by the ancient Greek physicians Theophrastus and Dioscorides, for stomach disorders and as poultices for festering wounds, and has since remained a mainstay for practitioners of herbal medicine.[1]

Among the phytochemical components of oil of *Origanum vulgare* are the thymol and carvacrol, as well as γ-terpinene, cis-terpinene, p-cymene, linalool, terpinen-4-ol, and sabinene.[2,3]

7.44.1 ANTIOXIDANT

Pre-feeding of *Origanum vulgare* leaves to rats attenuated the oxidative damage of subsequent exposure to CCl_4. Serum levels of liver transaminases were reduced, as were levels of the lipid peroxidation biomarker malondialdehyde in liver tissue. *Origanum vulgare* also improved superoxide anion and hydrogen peroxide scavenging in liver.[4]

Carvacrol, the prominent phenolic monoterpene in *Origanum vulgare*, prevents the oxidative damage to rat organ systems that result from repetitive restraint stress. It reduced levels of the lipid peroxidation biomarker, malondialdehyde, in the brain, liver, and kidney, but increased levels of glutathione in those organs. It also maintained activities of the antioxidant enzymes, superoxide dismutase, glutathione peroxidase, glutathione reductase, and catalase.[5]

7.44.2 ANTI-INFLAMMATORY

In activated human macrophages, *Origanum vulgare* essential oil reduced release of TNF-α, IL-1β, and IL-6.[6] Carvacrol purified from *Origanum vulgare* has been shown to exert potent anti-inflammatory effects in several experimental models. It dampens dextran-, histamine-, and substance P-induced paw edema in rats. Carvacrol also significantly reduced ear edema in mice induced by injection 12-O-tetradecanoylphorbol acetate or arachidonic acid. In fact, it was as effective as the standard anti-inflammatory agent, indomethacin.[7]

7.44.3 ANTIDIABETIC/ANTI-METABOLIC SYNDROME

In normal rats, aqueous extract of *Origanum vulgare* slightly decreases blood glucose. However, in streptozotocin-induced diabetic rats, the extract significantly decreased blood glucose.[8]

Administration of carvacrol to streptozotocin-induced diabetic rats slightly reduced serum glucose and total serum cholesterol, but produced no significant differences in serum insulin levels, food-water intake values and body weight changes.[9] However, carvacrol attenuated diabetes-associated cognitive deficits in rats, as shown by improved performance in the Morris water maze. This effect of carvacrol in streptozotocin-treated rats appeared to be mediated by reductions in oxidative stress, inflammation and apoptosis. The phytochemical increased depleted superoxide dismutase and glutathione, and reduced toward normal brain levels of NF-κB, TNF-α, IL-1β, and caspase-3.[10]

7.44.4 PRECLINICAL ANTIDEPRESSANT-LIKE EFFECTS

The essential oil of *Origanum vulgare* reversed depression-like effects of chronic unpredictable stress in rats, as shown by return of preference for sucrose and decreased immobility in the forced swim test. The extract used in the study was rich in thymol, gamma-terpinene, borneol, cymene, and carvacrol.[11]

Extract of *Origanum vulgare* also exhibited antidepressant-like effects in mice by reducing immobility in the forced swimming test. The *Origanum vulgare* extract inhibited the reuptake and degradation of the monoamine neurotransmitters in a dose-dependent manner. *In vivo* measurement by microdialysis showed increases in serotonin levels in the dorsal hippocampi of the experimental mice.[12]

Carvacrol, extracted from *Origanum vulgare* essential oil, also showed an antidepressant-like effect in mice in the forced swimming and tail suspension tests.[13] Antidepressant-like effects of carvacrol were also seen in rats, with improvements observed in the forced swim test. It was further determined that carvacrol increased serotonin and dopamine levels in the hippocampi and prefrontal cortices, as was seen with the full essential oil.[14] Thus, evidence from mouse and rat studies suggest that the antidepressant effects of the components of *Origanum vulgare* in rodents are due in part to enhancement of monoaminergic activity.

Thymol, a major component of essential oil of *Origanum vulgare*, also showed an antidepressant-like effect in mice that had been subjected to chronic, unpredictable stress. It restored depleted levels of norepinephrine and serotonin in the hippocampus. It also reduced levels of proinflammatory cytokines including IL-1β, IL-6, and TNF-α. This appeared to be due in part to the inhibition by thymol of the nod-like receptor protein 3 (NLRP3) inflammasome, a major player in the neuroinflammation cascade.[15]

7.44.5 HUMAN ANTIDEPRESSANT EFFECTS

Despite the long use of *Origanum vulgare* in herbal medicine, and the promising indications of antidepressant effects of *Origanum vulgare* and its phytochemical constituents, there have been no formal clinical studies of their use in the treatment of MDD in human subjects.

7.44.6 DOSAGE

Easley and Horne recommend dosage of *Origanum vulgare* in weak infusions, taking 2–4 ounces 1–4 times daily. They further recommend tincture of dried leaf (1:5, 65% alcohol and 10% glycerin) 1–2 ml taken 3–4 times daily.[16]

7.44.7 TOXICITY

Origanum vulgare has a history of safe use as an herbal food ingredient and is listed as a natural extract that is generally recognized as safe (GRAS).[17] In a study of rats, the oral LD50 of *Origanum vulgare* oil was considered to be 4,239 mg/kg in males and 5,662 mg/kg in females. Instances of death were seen at 2,500mg /kg.[18] The 2,500 mg/kg dose extrapolated to a 75 kg human adult would be approximately 175 g.

7.44.8 SAFETY IN PREGNANCY

The *Botanical Safety Handbook* notes there is no definitive guidance from the literature concerning the safety of *Origanum vulgare* during pregnancy.[19] They note no human studies, but do cite a study of pregnant pigs in which *Origanum vulgare* improves some components of health during pregnancy, including lowering mortality rates of both sows and piglets.[20]

7.44.9 DRUG INTERACTIONS

Methanolic extract of *Origanum vulgare* protects rat liver tissue from the cytotoxic effects of cyclophosphamide. This effect of the herb is due to blocking the conversion of cyclophosphamide into a toxic compound by the cyctochrome P450 system. Thus, the herb is inhibitory toward certain P450 enzymes, most likely the CYP2B family.[21] Nonetheless, the *Botanical Safety Handbook* notes no known interactions of *Origanum vulgare* with commonly used medications.

REFERENCES

1. Sakkas H, Papadopoulou C. Antimicrobial activity of basil, oregano, and thyme essential oils. *J Microbiol Biotechnol.* 2017;27(3):429–438.
2. Pirigharnaei M, Zare S, Heidary R, Khara J, EmamaliSabzi R, Kheiry F. The essential oils compositions of Iranian oregano (*Origanum vulgare* L.) populations in field and provenance from Piranshahr district, West Azarbaijan Province, Iran. *Avicenna J Phytomed.* 2011;1(2):106–114.
3. Lukas B, Schmiderer C, Novak J. Essential oil diversity of European *Origanum vulgare* L. (Lamiaceae). *Phytochemistry.* 2015;119:32–40.
4. Botsoglou NA, Taitzoglou IA, Botsoglou E. Effect of long-term dietary administration of Oregano on the alleviation of carbon tetrachloride-induced oxidative stress in rats. *J Agric Food Chem.* 2008;56(15):6287–6293.

5. Samarghandian S, Farkhondeh T, Samini F, Borji A. Protective effects of carvacrol against oxidative stress induced by chronic stress in rat's brain, liver, and kidney. *Biochem Res Int.* 2016;2016:Article ID 2645237.

6. Ocaña-Fuentes A, Arranz-Gutiérrez E, Señorans FJ, Reglero G. Supercritical fluid extraction of oregano (*Origanum vulgare*) essentials oils: anti-inflammatory properties based on cytokine response on THP-1 macrophages. *Food Chem Toxicol.* 2010;48(6):1568–1575.

7. Silva FV, Guimaraes AG, Silva ERS, et al. Anti-inflammatory and anti-ulcer activities of carvacrol, a monoterpene present in the essential oil of oregano. *J Med Food.* 2012;15(11):984–991.

8. Lemhadri A, Zeggwagh NA, Maghrani M, Jouad H, Eddouks M. Anti-hyperglycaemic activity of the aqueous extract of *Origanum vulgare* growing wild in Tafilalet region. *J Ethnopharmacol.* 2004;92(2–3):251–256.

9. Bayramoglu G, Senturk H, Bayramoglu A, et al. Carvacrol partially reverses symptoms of diabetes in STZ-induced diabetic rats. *Cytotechnology.* 2014;66(2):251–257.

10. Deng W, Lu H, Teng J. Carvacrol attenuates diabetes-associated cognitive deficits in rats. *J Mol Neurosci.* 2013;51(3):813–819.

11. Amiresmaeili A, Roohollahi S, Mostafavi A, Askari N. Effects of oregano essential oil on brain TLR4 and TLR2 gene expression and depressive-like behavior in a rat model. *Res Pharm Sci.* 2018;13(2):130–141.

12. Mechan AO, Fowler A, Seifert N, et al. Monoamine reuptake inhibition and mood-enhancing potential of a specified oregano extract. *Br J Nutr.* 2011;105:1150–1163.

13. Melo FHC, Moura BA, de Sousa DP, et al. Antidepressant-like effect of carvacrol (5-isopropyl-2-methylphenol) in mice: involvement of dopaminergic system. *Fundam Clin Pharmacol.* 2011;25(3):362–367.

14. Zotti M, Colaianna M, Morgese M, et al. Carvacrol: from ancient flavoring to neuro-modulatory agent. *Molecules.* 2013;18(6):6161–6172.

15. Deng XY, Li HY, Chen JJ, et al. Thymol produces an antidepressant-like effect in a chronic unpredictable mild stress model of depression in mice. *Behav Brain Res.* 2015;291:12–19.

16. Easley T, Horne S. *The Modern Herbal Dispensary.* Berkeley, CA: North Atlantic Books; 2016. p. 277.

17. Kwon TS, Park SC, Kim KH, Shin JY, Shin DH, Kim JC. Acute toxicity evaluation of Oregano oil in rats. *Korean J Lab Anim Res.* 2004;20(4):419–425.

18. EFSA. Scientific opinion on the use of oregano and lemon balm extracts as a food additive. *EFSA J.* 2010;8(2):1514–1533.

19. Gardner Z, McGuffin M, editors. *Botanical Safety Handbook.* 3rd ed. Boca Raton, FL: CRC Press; 2013. p. 603.

20. Allan P, Bilkei G. Oregano improves reproductive performance of sows. *Theriogenology.* 2005;63(3):716–721.

21. Habibi E, Shokrzadeh M, Chabra A, Naghshvar F, Keshavarz-Maleki R, Ahmadi A. Protective effects of Origanum vulgare ethanol extract against cyclophosphamide-induced liver toxicity in mice. *Pharm Biol.* 2015;53(1):10–15.

7.45 *PAEONIA LACTIFLORA* (PEONY)

In China, Japan, and Korea, the dried root of peony has been used for over 1,200 years for the treatment of a variety of human ailments. Uses include cleansing the blood, relaxing tense muscles and cramps, regulating women's hormonal based problems, helping treat fevers, as an antiseptic wash for wounds, and for the "falling sickness" (epilepsy). There are three main species of *Paeonia* used in herbal medicine, *Paeonia lactiflora*, *Paeonia officinalis*, and *Paeonia suffruticosa*. Among those three, *Paeonia lactiflora* is most commonly used. *Paeonia lactiflora* contains a wide variety of phytochemicals, including monoterpenoid glycosides, flavonoids, tannis, stilbenes, triterpenes, steroids, paeonols, and other phenolic compounds. These phytochemicals are thought to exhibit antioxidant, anti-inflammatory, anti-tumor, antimicrobial, immune-modulating, cardioprotective, and neuroprotective effects.[1] Paeoniflorin, from a water/alcohol extract of the root, has been shown to account for the major part of the plant's pharmacological effects. *Paeonia lactiflora* is one of the herbs Yeung et al. report to be among the most commonly used in traditional Chinese herbal combinations to treat MDD.

7.45.1 ANTIOXIDANT

Crude extract of the root of *Paeonia lactiflora* has been shown to have significant antioxidant effect in the DPPH radical scavenging activity. Further evaluation of components purified from the extract showed that the fraction containing the components catechin, tetragalloyldigalloyl B-D-glucose, and penta-galloyl-β-D-glucose were largely responsible for that antioxidant effect.[2]

In vivo sub-chronic administration of the aqueous extract of *Paeoniae lactiflora* protected rat livers from the oxidative damage of CCl_4. Serum levels of liver transaminases, nitric oxide, and hyaluronic acid (indicative of liver tissue damage) were reduced by the extract. Levels of malonyldialdehyde were reduced in liver tissue.[3]

7.45.2 ANTI-INFLAMMATORY

Paeoniflorin is a potent anti-inflammatory that blocks the synthesis of prostaglandin PGE2 and decreases cellular response to leukotriene B4, which plays a role in the inflammatory response. Furthermore, it inhibits stimulation of TNF-α and the inflammatory IL-6, and stimulates release of the anti-inflammatory IL-10.[4] Both paeoniflorin and albiflorin purified from *Paeonia lactiflora* showed ability to block the inflammatory effect of lipopolysaccharide in cultured mouse macrophages. Both attenuated production of NO, PGE2, TNF-α, and IL-6. Expression of the enzymes, iNOS and COX-2, were both reduced as well. Paeoniflorin was the more potent of the two.[5]

7.45.3 ANTIDIABETIC/ANTI-METABOLIC SYNDROME

Paeoniflorin and 8-debenzoylpaeoniflorin isolated from the dried root of *Paeonia lactiflora* produced a significant blood sugar lowering effect in streptozotocin-treated rats.[6] Paeoniflorin also restored expressions of PPARγ and insulin-sensitivity in cultured adipocytes that had been exposed to TNFα. Insulin-induced phosphorylation of AKT in adipocytes was also restored.[7]

7.45.4 PRECLINICAL ANTIDEPRESSANT-LIKE EFFECTS

A multitude of preclinical studies have demonstrated antidepressant-like effects of *Paeonia lactiflora* and its constituent phytochemicals in rats and mice.

Chronic administration of glycosides extracted from *Paeonia lactiflora* reversed the decreases in sucrose consumption and increases in immobility time in the forced swim test in rats given repeated injections of corticosterone. The glycosides also reversed the reduction of BDNF by corticosterone in rat brain.[8] Paeoniflorin also reversed the reduced consumption of sucrose and low motor activity suggestive of depression-like symptoms in rats subjected to chronic unpredictable stress.[9]

As with rats, paeoniflorin extracted from *Paeonia lactiflora* reduced immobility times of mice in both the forced swim and tail suspension tests. In the same study, paeoniflorin prevented corticosterone-induced neurotoxicity in cultured PC12 cells.[10] In mice treated with reserpine, another animal model of MDD, extract of *Paeonia lactiflora* also reduced immobilization times in the forced swim and tail suspension tests.[11] As evidence of its anti-neuroinflammatory effects, intragastric administration of paeoniflorin reversed the depression-like behaviors in mice induced by high-doses of Interferon-α. Behaviors in the sucrose preference, tail suspension, and forced swim tests were normalized. Paeoniflorin also countered the neuroinflammatory effects in the amygdala, medial prefrontal cortex and ventral hippocampus by normalizing concentrations of the cytokines IL-6, IL-1β and TNF-α, IL-9, IL-10, IL-12, and MCP-1. The density of microglia in those areas was also reduced.[12] In brain tissue, paeoniflorin also activates the akt/mTOR pathway and inhibits GSK-3β.[13]

7.45.5 HUMAN ANTIDEPRESSANT EFFECTS

Paeonia lactiflora is a common ingredient in a number of traditional Chinese herbal combinations for the treatment of depression-like syndromes. Nonetheless, there are no studies of *Paeonia lactiflora* alone as a treatment of MDD in human subjects.

7.45.6 DOSAGE

Easley and Horne recommend the dosage of *Paeonia lactiflora* to be 1–4 oz of standard decoction up to four times daily. They further suggest tincture of fresh root (1:2 95% alcohol) or dried root (1:5 60% alcohol) 10 drops to 1 ml up to four times daily.[14] Capsules of standardized peony root extract are available commercially.

7.45.7 TOXICITY

Studies show *Paeonia lactiflora* to be quite non-toxic. The LD50 of orally administered Chinese white peony in rats is 81 g/kg. Extrapolated to a 75 kg human adult, this dose would be approximately 5,650 g. The LD50 of intraperitoneally administered Chinese red peony, is reported to be 2.9g/kg of ethanol extract, and 10.8g/kg for aqueous extract.[15] Note that the distinctions between red and white peony are subtle, and perhaps even artificial, with red being the wild root. They are generally from the same species of peony, i.e., lactiflora, and rely on the same phytochemical constituents.

7.45.8 SAFETY IN PREGNANCY

Two well-regarded reference texts for Traditional Chinese Herbal Medicine give no contraindications for use of *Paeonia lactiflora* by pregnant women.[16,17] The species *Paeonia officinalis* was described as having no definitive information in the literature to confirm the safety of its use in pregnancy. However, the *Botanical Safety Handbook* stated that the tree peony, *Paeonia suffruticosa*, should be avoid in pregnancy, per traditional use.[18]

7.45.9 DRUG INTERACTIONS

At least one species of *Paeonia*, *P. emodi*, has been found to significantly reduce CYP3A and CYP2C protein expression in rat liver. This resulted in a substantial increase in serum levels of carbamazepine.[19] None of the three more commonly used species of *Paeonia* discussed in the *Botanical Safety Handbook* were known to have interactions with commonly prescribed medications.

REFERENCES

1. He CN, Peng Y, Zhang YC, Xu LJ, Gu J, Xiao PG. Phytochemical and biological studies of *Paeoniaceae*. *Chem Biodivers*. 2010;7:805–837.
2. Bang MH, Song JC. Lee SY. Isolation and structure determination of antioxidants from the root of *Paeonia lactiflora*. *Appl Biol Chem*. 1999;42(2):170–175.
3. Jiang YP, Liu YG, Chen HC. Effect of aqueous extract of *Radix Paeoniae Rubra* against carbon tetrachloride induced liver fibrosis in rats. *Her Med*. 2004-08.
4. He DY, Dai SM. Anti-inflammatory and immunomodulatory effects of *Paeonia lactiflora* Pall., a traditional chinese herbal medicine. *Front Pharmacol*. 2011;2:1–5.
5. Wang QS, Gao T, Cui TL, Gao LN, Jiang HL. Comparative studies of paeoniflorin and albiflorin from *Paeonia lactiflora* on anti-inflammatory activities. *Pharm Biol*. 2014;52(9):1189–1195.
6. Hsu FL, Lai CW, Cheng JT. Antihyperglycemic effects of paeoniflorin and 8-debenzoylpaeoniflorin, glucosides from the root of *Paeonia lactiflora*. *Planta Med*. 1997;63(4):323–325.
7. Kong P, Chi R, Zhang L, Wang N, Lu Y. Effects of paeoniflorin on tumor necrosis factor-α-induced insulin resistance and changes of adipokines in 3T3-L1 adipocytes. *Fitoterapia*. 2013;91:44–50.
8. Mao Q-Q. Peony glycosides reverse the effects of corticosterone on behavior and brain BDNF expression in rats. *Behav Brain Res*. 2012;227(1):305–309.
9. Qiua F-M, Zhong XM, Mao QQ, Huang Z. Antidepressant-like effects of paeoniflorin on the behavioural, biochemical, and neurochemical patterns of rats exposed to chronic unpredictable stress. *Neurosci Lett*. 2013;541:209–213.
10. Cui GZ. Study on the antidepressant-like effect of paeoniflorin. *Mod Pharm Clin*. 2009-04.
11. Shao JH. Anti-depressant effect of *Paeonia lactiflora* Pall on mice. *Ningxia Yixue Zazhi*. 2008;30:490–491.
12. Li J, Huang S, Huang W, et al. Paeoniflorin ameliorates interferon-alpha-induced neuroinflammation and depressive-like behaviors in mice. *Oncotarget*. 2017;8(5):8264–8282.
13. Wang D, Wong HK, Feng YB, Zhang ZJ. Paeoniflorin, a natural neuroprotective agent, modulates multiple anti-apoptotic and pro-apoptotic pathways in differentiated PC12 cells. *Cell Mol Neurobiol*. 2013;33:521–529.

14. Easley T, Horne S. *The Modern Herbal Dispensary*. Berkeley, CA: North Atlantic Books; 2016. p. 282.
15. Zhu YP. *Chinese Materia Medica: Chemistry, Pharmacology and Application*. Amsterdam: Harwood Academic Publishers; 1998.
16. Bensky D, Clavey S, Stoger E. *Chinese Herbal Medicine: Materia Medica*. 3rd ed. Seattle, WA: Eastland Press; 2004.
17. Chen JK, Chen TT. *Chinese Medical Herbology and Pharmacology*. City of Industry, CA: Art of Medicine Press; 2004.
18. Gardner Z, McGuffin M, editors. *Botanical Safety Handbook*. 3rd ed. Boca Raton, FL: CRC Press; 2013. p. 605–609.
19. Raish M, Ahmad A, Alkharfy KM, et al. Effects of *Paeonia emodi* on hepatic cytochrome P450 (CYP3A2 and CYP2C11) expression and pharmacokinetics of carbamazepine in rats. *Biomed Pharmacother*. 2017;90:694–698.

7.46 *PANAX GINSENG* (GINSENG)

The name ginseng is used in reference to several plants of the genus *Panax*, i.e., *Panax ginseng*, or Korean ginseng; *Panax notoginseng*, or Chinese ginseng; and *Panax quinquefolius*, or American ginseng. The herb called Siberian ginseng is not in the Panax family, but is rather another plant entirely, *Eleutherococcus senticosus*. Ginseng has been used for several thousand years in Asia as a tonic and in TCM has been included as an ingredient in several classical herbal combinations for the treatment of MDD. It is currently classified by many as an adaptogen, with the alleged ability to enhance physical performance, promote vitality, increase resistance to stress and aging, and strengthen immune function.

Approximately 200 substances, such as ginsenosides, polysaccharides, polyacetylenes, peptides, and amino acids, have been isolated from Korean ginseng, and more than 100 substances have been isolated from American and Chinese ginsengs. Korean ginseng, American ginseng, and Chinese ginseng differ in several ways. The most significant difference is the presence of ginsenoside Rf in Korean and Chinese ginsengs but not American ginseng. Moreover, pseudoginsenoside occurs in American ginseng but not the Korean or Chinese species. Although there are differences among the ginsengs, it has been suggested that the most important factor in determining the bioavailability and activity of the ginsengs may be differences in intestinal flora in individuals.[1] The root of *Panax ginseng* is steamed and dried to prepare red ginseng, while the peeled roots dried without steaming are designated as white ginseng. The prepared ginseng can then be further treated through fermentation. In any case, most of the research has focused on the Korean species, that is, *Panax ginseng*.

7.46.1 ANTIOXIDANT

In vitro, alcoholic extracts of *Panax ginseng* were potent in scavenging in DPPH radicals, as well as in preventing lipid peroxidation and linoleate oxidation. Interestingly, wild ginseng was found to be more potent than cultivated *Panax ginseng* in these antioxidant actions, and this appeared to be due to higher concentrations of the flavonoids quercetin and kaempferol. Although phenolic compounds are mostly responsible for the antioxidant effects of *Panax ginseng*, polysaccharides in the herb also contribute such effects. Polysaccharides purified from *Panax ginseng* exhibited potent effects in the superoxide radical, hydroxyl radical, and 1,2,3-phentriol self-oxidation assays.[2]

Pretreatment with *Panax ginseng* also protected against the oxidative liver damage produced by CCl_4 in rats. As has been found with other antioxidant herbs, extract of *Panax ginseng* prevented the large increases in levels of malondialdehyde and NO, and increased levels of glutathione and activities of antioxidant enzymes activities in liver homogenates.[3]

7.46.2 ANTI-INFLAMMATORY

Preparations of *Panax ginseng*, white, red, and fermented red, all exhibited significant anti-inflammatory effects in cultured mouse macrophages exposed to LPS. Levels of the inflammatory cytokines TNF-α and interferon gamma were reduced in these cells. Inducible NOS was reduced as were NF-kβ and activity of COX-2.[4] *Panax quinquefolius* also exhibits anti-inflammatory effects. It both prevents and treats inflammation-driven colitis in the in vivo mouse dextran sulfate model. It acts by reducing inducible nitric oxide synthase and cyclooxygenase-2.[5]

Panax ginseng also produces anti-inflammatory effects in humans. Pretreatment with the herb prevented increases in serum creatinine kinase and IL-6 in healthy adults strained by prolonged uphill treadmill walking. Treatment also improved insulin sensitivity in those subjects.[6]

7.46.3 ANTIDIABETIC/ANTI-METABOLIC SYNDROME

Panax ginseng has been reported to improve insulin resistance, reduce inflammation, and enhance resiliency to stress, which together would be expected to counter the ongoing syndrome of MDD. In one report, daily *Panax ginseng* elevated mood, improved psychophysical performance, and reduced fasting blood glucose and body weight in non-insulin-dependent diabetic patients. These changes were reflected in reductions in glycated hemoglobin levels.[7] A more recent randomized, double-blind, placebo-controlled study also showed improvement in glycemic control and enhancement of insulin sensitivity.[8]

7.46.4 PRECLINICAL ANTIDEPRESSANT-LIKE EFFECTS

Panax ginseng has shown antidepressant effects in a number of animal models of depression. Pretreatment with the herb reduced immobility time in the forced swim test in mice. It also reversed the reduction in the sucrose preference index, decrease in locomotor activity, and prolongation of latency of feeding in the novelty environment in rats subjected to chronic unpredictable stress. In neurochemical evaluation, it was found that *Panax ginseng* reversed stress-induced decreases in monoamine neurotransmitter concentration and brain-derived neurotrophic factor (BDNF) expression in the hippocampus of those animals.[9] In the corticosterone-induced mouse depression model, pretreatment with *Panax ginseng* reduced immobility time in the forced swim and tail suspension tests. It also reversed decreases in BDNF and enhanced the inhibition of GSK-3β through phosphorylation.[10]

Panax ginseng showed antidepressant effects in rats exhibiting depression-like behaviors due to exposure to lipopolysaccharide. It reduced immobility in the forced swim and rail suspension tests, and restored sucrose preference. Some of the antidepressant-like effects were likely due to anti-inflammatory effects. *Panax ginseng* dampened lipopolysaccharide-induced increases in 5-HT and tryptophan turnover in the brain, and reduced mRNA levels for IL-1β, IL-6, TNF-α, and indole-amine 2,3-dioxygenase in the hippocampi of these animals. In a rat model of trau-matic head injury, post-injury treatment with *Panax ginseng* improved neurological function, reduced neuronal loss, and increased activity of superoxide dismutase while decreasing the activity of nitric oxide synthase. It also decreased levels of interleukin-1β, interleukin-6, and tumor necrosis factor-α and increased the anti-inflammatory cytokine interleukin-10 in the cortical area surrounding the injured core.[11]

At least one specific phytochemical in *Panax ginseng*, ginsenoside Rg1,[12] and its intestinal metabolite. 20(S)-protopanaxadiol[13] has been found to have potent anti-depressant effects in animal studies. The metabolite 20(S)-protopanaxadiol was as potent as fluoxetine.

7.46.5   HUMAN ANTIDEPRESSANT EFFECTS

Like *Rhodiola*, the ginsengs have long had reputations as "adaptogens." Brief treat-ment of otherwise healthy young adults with *Panax quinquefolius* can produce sig-nificant improvement in attention, working memory capacity, and calmness.[14] In a similar group of healthy young adults, it enhanced performance of a mental arithme-tic task and ameliorated the increase in subjective feelings of mental fatigue experi-enced by participants during the later stages of a sustained, cognitively demanding task performance.[15] There are a number of studies showing the ability of ginsengs to reduce physical fatigue in cancer patients. For example, daily dosing of *Panax quin-quefolius* over eight, though not four, weeks, reduced general and physical cancer-related fatigue.[16]

Not all reports of the "adaptogenic" effects of ginseng are positive. For example, eight weeks of adding *Panax ginseng* to the diet gained no advantage for women tested in an all-out-effort, 30-second leg cycle ergometry test – the Wingate proto-col – followed by a controlled recovery under constant laboratory conditions.[17] In a study of self-rated perceptions of performance by night shift nurses, *Panax ginseng* improved the subjective sense of competence and mood. However, whereas some variables of bodily feeling were seen as improved, others were rated as worse while taking the herb.[18]

Ginsengs are time-honored in TCM as components of antidepressant treatment. There is also a large preclinical literature showing efficacy of ginseng as an antide-pressant. Thus, it is surprising that there are no published clinical studies on its use as monotherapy for MDD. Nonetheless, *Panax ginseng* significantly improved the residual symptoms of depression in women whose antidepressants provided only partial remission.[19] A similar study found that the addition of *Panax ginseng* signifi-cantly improved upon the antidepressant effect of venlafaxine.[20]

7.46.6 DOSAGE

Hoffman recommends a dose of 200 mg of 5:1 standardized extract once a day, 1–2ml tincture (1:5 in 60%) three times a day, or a standard decoction of 0.5 teaspoon powdered root in a cup of water three times a day.[21]

7.46.7 TOXICITY

The ginsengs appear quite non-toxic, with the LD50 in mice reported to be in a range between 10–30 g/kg.[22] That is far beyond typical human dosage.

7.46.8 SAFETY IN PREGNANCY

Reference texts from Traditional Chinese Medicine note no concerns about the use of ginseng during pregnancy.[23] At least two studies have described use of ginseng in pregnant women with no adverse effects on mother or fetus.[24,25]

7.46.9 DRUG INTERACTIONS

There have been case reports of bleeding in patients after adding *Panax ginseng* to warfarin treatment.[26] However, more rigorous human studies of this possible interaction have shown no such effects.[27] In healthy human volunteers, 28 days of 500 mg *Panax ginseng* twice daily significantly decreased serum levels of midazolam in comparison with controls. This was taken to be indicative of induction of the CYP3A family of cytochrome P450 enzymes.[28]

REFERENCES

1. Kom DH. Chemical diversity of *Panax ginseng*, *Panax quinquifolium*, and *Panax notoginseng*. *J Ginseng Res*. 2012;36(1):1–15.
2. Luo D, Fang B. Structural identification of ginseng polysaccharides and testing of their antioxidant activities. *Carbohyd Polym*. 2008;72(3):376–381.
3. Karadeniz A, Yıldırım A, Karakoç A, Kalkan Y, Celebi F. Protective effect of *Panax ginseng* on carbon tetrachloride induced liver, heart and kidney injury in rats. *Revue Méd Vét*. 2009;160(5):237–243.
4. Hyun M-S, Hur JM, Shin YS, Song BJ, Mun YJ, Woo WH. Comparison study of white ginseng, red ginseng, and fermented red ginseng on the protective effect of LPS-induced inflammation in RAW 264.7 cells. *J Appl Biol Chem*. 2009;52(1):21–27.
5. Jin Y, Kotakadi VS, Ying L, et al. American ginseng suppresses inflammation and DNA damage associated with mouse colitis. *Carcinogenesis*. 2008;29(12):2351–2359.
6. Jung HL, Kwak HE, Kim SS, et al. Effects of *Panax ginseng* supplementation on muscle damage and inflammation after uphill treadmill running in humans. *Am J Chin Med*. 2011;39(3)441–450.
7. Sotaniemi EA, Haapakoski E, Rautio A. Ginseng therapy in non-lnsulin-dependent diabetic patients: effects on psychophysical performance, glucose homeostasis, serum lipids, serum aminoterminalpropeptide concentration, and body weight. *Diabetes Care*. 1995;18(10):1373–1375.
8. Vuksan V, Sung MK, Sievenpiper JL, et al. Korean red ginseng (*Panax ginseng*) improves glucose and insulin regulation in well-controlled, type 2 diabetes: results of a randomized, double-blind, placebo-controlled study of efficacy and safety. *Nutr Metab Cardiovasc Dis*. 2008;18(1):46–56.

9. Dang H, Chen Y, Liu X, et al. Antidepressant effects of ginseng total saponins in the forced swimming test and chronic mild stress models of depression. *Prog Neuropsychopharmacol Biol Psychiatry.* 2009;33(8):1417–1424.

10. Chen L, Dai J, Wang Z, Zhang H, Huang Y, Zhao Y. Ginseng total saponins reverse corticosterone-induced changes in depression-like behavior and hippocampal plasticity-related proteins by interfering with GSK-3β-CREB signaling pathway. *Evid Based Complement Altern.* 2014;2014:Article ID 506735.

11. Xia L, Jiang ZL, Wang GH, Hu BY, Ke KF. Treatment with ginseng total saponins reduces the secondary brain injury in rat after cortical impact. *J Neurosci Res.* 2012;90(7)1424–1436.

12. Liu Z, Qi Y, Cheng Z, Zhu X, Fan C, Yu SY. The effects of ginsenoside Rg1 on chronic stress induced depression-like behaviors, BDNF expression and the phosphorylation of PKA and CREB in rats. *Neuroscience.* 2016;322:358–369.

13. Xu C, Teng J, Chen W, et al. 20 (S)-protopanaxadiol, an active ginseng metabolite, exhibits strong antidepressant-like effects in animal tests. *Prog Neuropsychopharmacol Biol Psychiatry.* 2010;34(8):1402–1411.

14. Scholey A, Ossoukhova A, Owen L, et al. American ginseng improves human neuro-cognitive function: a randomized, double-blind, placebo-controlled, crossover study. *Psychopharmacology.* 2010;212:345–356.

15. Reay JL, Kennedy DO, Scholey AB. Single doses of *Panax ginseng* (G115) reduce blood glucose levels and improve cognitive performance during sustained mental activity. *J Psychopharmacol.* 2005;19:357–365.

16. Barton DL, Soori GS, Bauer BA, et al. Pilot study of *Panax quinquefolius* (American ginseng) to improve cancer-related fatigue: a randomized, double-blind, dose-finding evaluation: NCCTG trial N03CA. *Support Care Cancer.* 2009;18:179–187.

17. Engels HJ. Effects of ginseng supplementation on supramaximal exercise performance and short-term recovery. *J Strength Cond Res.* 2001;15(3):290–295.

18. Hallstrom C, Fulder S, Carruthers M. Effects of ginseng on the performance of nurses on night duty. *Am J Chin Med.* 1978;6(4):277–282.

19. Jeong HG, Ko YH, Oh SY, Han C, Kim T, Joe SH. Effect of Korean Red Ginseng as an adjuvant treatment for women with residual symptoms of major depression. *Asia Pac Psychiatry.* 2015;7(3):330–336.

20. Li L, Li L, Yang L. Clinical observation on Ginseng Tiaopi powder and venlafaxine in treating 30 cases of depression. *J Fujian Univ Tradit Chin Med.* 2008-02.

21. Hoffmann D. *Medical Herbalism: The Science and Practice of Herbal Medicine.* Rochester, VT: Healing Arts Press; 2003. p. 570.

22. Brekham II, Dardymov IV. New substances of plant origin, which increase non-specific resistance. *Annu Rev Pharmacol.* 1969;9:419–430.

23. Bensky D, Clavey S, Stoger E. *Chinese Herbal Medicine: Materia Medica.* 3rd ed. Seattle, WA: Eastland Press; 2004.

24. Chin R. Ginseng and common pregnancy disorders. *Asia Ocean J Obstet Gynaecol.* 1991;17(4):379–380.

25. Chuang CH, Doyle P, Wang JD, Chang PJ, Lai JN, Chen PC. Herbal medicines used during the first trimester and major congenital malformations: an analysis of data from a pregnancy cohort study. *Drug Saf.* 2006;29(6):537–548.

26. Janetsky K, Morreale A. Probable interaction between warfarin and ginseng. *Am J Health Syst Pharm.* 1997;54:692–693.

27. Jiang X, Williams KM, Liauw WS, et al. Effect of St John's wort and ginseng on the pharmacokinetics and pharmacodynamics of warfarin in healthy subjects. *BJCP.* 2004;57(5):592–599.

28. Malati CY, Robertson SM, Hunt JD, et al. Influence of *Panax ginseng* on cytochrome P450 (CYP)3A and P-glycoprotein (P-gp) activity in healthy participants. *J Clin Pharmacol.* 2012;52(6):932–939.

7.47   *PASSIFLORACEAE INCARNATA* (PASSIONFLOWER)

The common name passionflower describes a variety of species of plants in the genus *Passifloraceae*. The German E Commission specifically considers the species, *Passifloraceae incarnata*, but many species contain similar phytochemicals and offer similar benefits. The majority of *Passifloraceae* species are found in Mexico and Central and South America, although there are additional representatives in the United States, Southeast Asia, and Oceania.[1] The herb has long been used in herbal medicine for the treatment of anxiety, dysmenorrhea, epilepsy, insomnia, neurosis, and neuralgia. M. Grieve, in her work *A Modern Herbal*, noted that "Its narcotic properties cause it to be used in diarrhea and dysentery, neuralgia, sleeplessness and dysmenorrhea."[2] There is more recent evidence of usefulness of *Passifloraceae* species in the treatment of mood disturbance.

Among the phytochemicals that have been identified in *Passifloraceae incarnata* are homoorientin, orientin, vitexin, isovitexin, chrysin, schaftoside, isoschaftoside, chlorogenic acid, hyperoside, caffeic acid, quercetin, luteolin, rutin, scutelarein, vicenin, and their various glycoside products.[3,4] The herb also contains small amounts of the harmala indole alkaloids, harman, harmin, harmalin, harmol, and harmalol.[5]

7.47.1   ANTIOXIDANT

The aqueous and ethanolic extracts of *Passifloraceae incarnata* have been found to exert potent antioxidant effects in the DPPH radical scavenging test and the ABTS radical cation decolorization assay.[6] *In vivo*, extract of *Passifloraceae incarnata* fruit showed strong antioxidant effects in the aluminum-induced Alzheimer's model in rats. Chronic oral administration of the fruit pulp, peel, and juice of the fruit of *Passifloraceae incarnata* reversed the reductions in catalase, superoxide dismutase, glutathione peroxidase, and glutathione in the hippocampus and livers of aluminum-treated rats.[7]

7.47.2   ANTI-INFLAMMATORY

Acute intraperitoneal injection of extract of *Passiflora incarnata* attenuated carrageenan-induced paw edema in rats.[8] Methanolic extract of *Passiflora foetida* showed anti-inflammatory effects in cultured mouse macrophages exposed to lipopolysaccharide. The extract prevented the production of prostaglandin E2 and the expression of inducible COX-2. It also inhibited the activation of NF-κB, but enhanced activity of ERK1/2, which has been associated with antidepressant effects.[9]

7.47.3   ANTIDIABETIC/ANTI-INFLAMMATORY

Chronic administration of dried methanolic extract of *Passifloraceae incarnata* showed significant antidiabetic effects in streptozotocin-induced diabetic mice. Treatment produced significant reductions in fasting blood glucose and urine glucose levels. Performance in oral glucose tolerance tests were improved, as were serum lipid profiles and body weights.[10] In human subjects, *Passifloraceae incarnata* extract lowered fasting glucose and hemoglobin A1c, and increased insulin sensitivity by the HOMA method.[11]

7.47.4   PRECLINICAL ANTIDEPRESSANT-LIKE EFFECTS

A hydroalcoholic extract of *Passifloraceae incarnata* decreased immobility times in the forced swim test and tail suspension test in mice.[12] Acute intraperitoneal administration of the methanolic extract of the similar species, *Passifloraceae foetida*, produced similar antidepressant-like effects in mice in the forced swim and tail suspension tests.[13]

Oral administration of alcoholic extracts of the species *Passiflora edulis* also produced antidepressant-like effects in mice by decreasing immobility in the forced swim and tail suspension tests. Further analysis showed the effects to be due to several structurally similar cyclopassifloic acids, which are triterpenoids unique to *Passifloraceae* species.[14] On the other hand, antidepressant effects of aqueous extract of *Passiflora edulis* in mice were found to be due to flavonoid glycosides.[15]

Extract of *Passifloraceae incarnata* exerted an antidepressant-like effect during the post-ictal phase after pentylenetetrazol-induced seizures in mice. It decreased immobility in the forced swim test post-ictally. The extract also offered some degree of protection from the pentylenetetrazol-induced seizure. Interestingly, while the protective effects of *Passifloraceae incarnata* against seizure were attributed to its gabaergic agonist effects, the benzodiazepine diazepam protected against seizure, but did not decrease immobility in the forced swim.[16]

7.47.5   HUMAN ANTIDEPRESSANT EFFECTS

There are no studies of effects of *Passifloraceae incarnata* on patients diagnosed specifically with MDD. However, there are several studies showing effects of the herb on mood, anxiety, and sleep in human subjects. In one such study, extract of *Passifloraceae incarnata* was administered to women suffering psychiatric symptoms of menopause that included depressed mood, insomnia, fatigue, and irritability, as per the Cooperman's Index of Menopausal Symptoms. *Passifloraceae incarnata* significantly relieved menopausal symptoms by the third and throughout the sixth week of the study. The symptoms of irritability and depressed mood were among the most rapidly improved symptoms.[17]

7.47.6   DOSAGE

Hoffman recommends 1–4 ml of tincture (1.5 in 40%) twice a day. Alternatively, he suggests up to 2 g of dried herb four times a day.[18]

7.47.7   TOXICITY

The LD50 of intraperitoneally administered extract of *Passifloraceae incarnata* has been reported to be above 900 mg/kg in mice.[19] Extrapolated to a 75 kg human adult, this dose would above 70 g. It was also reported that three weeks of oral administration of 5 g/k per day of hydroalcoholic extract of *Passifloraceae incarnata* had no deleterious effects on rats.[20]

7.47.8   SAFETY IN PREGNANCY

Pregnancy: No adverse developmental effects were observed in rats born to females administered 400mg/kg Passifloraceae incarnata extract daily on days 7 through 17 of pregnancy.[21] The *Botanical Safety Handbook* notes there is no definitive guidance from the literature concerning the safety of *Passiflora incarnata* during pregnancy.[22] They do cite an animal study in which treatment with *Passiflora incarnata* during gestation caused no adverse effects.

7.47.9   DRUG INTERACTIONS

A benzoflavone (BZF) moiety in *Passiflora incarnata* has been reported to inhibit aromatase, a member of the CYP3A4 isoenzyme family. This effect dampens the metabolic conversion of testosterone to its metabolites, thereby increasing serum levels of free testosterone and decreasing free estrogen. This may explain some reports of aphrodisiac effects of the herb.[23] However, the *Botanical Safety Handbook* notes no known interactions of *Passiflora incarnata* with commonly prescribed medications.

REFERENCES

1. Krosnick SE, Porter-Utley KE, MacDougal JM, Jørgensen PM, McDade LA. New insights into the evolution of *Passiflora* subgenus *Decaloba* (*Passifloraceae*): phylogenetic relationships and morphological synapomorphies. *Syst Bot.* 2013;38 (3):692–713.
2. Grieve M. *A Modern Herbal.* NewYork, NY: Harcourt, Brace & Company; 1931. p. 618.
3. Grundmann O, Wang J, McGregor GP, Butterweck V. Anxiolytic activity of a phytochemically characterized *Passiflora incarnata* extract is mediated via the GABAergic system. *Planta Med.* 2008;74(15):1769–1773.
4. Lee QM, van den Heuvel H, Delorenzo O, et al. Mass spectral characterization of C-glycosidic flavonoids isolated from a medicinal plant: *Passiflora incarnata. J Chromatogr B.* 1991;562(1–2):435–446.
5. Soulimani R, Younos C, Jarmouni S, Bousta D, Misslin R, Mortier F. Behavioural effects of *Passiflora incarnata* L. and its indole alkaloid and flavonoid derivatives and maltol in the mouse. *J Ethnopharmacol.* 1997;57(1):11–20.
6. Masteikova, R, Bernatoniene J, Bernatoniene R. Antiradical activities of the extract of *Passiflora incarnata. Acta Pol Pharm.* 2008;65(5):577–583.
7. Doungue HT, Kengne APN, Kuate D. Neuroprotective effect and antioxidant activity of *Passiflora edulis* fruit flavonoid fraction, aqueous extract, and juice in aluminum chloride-induced Alzheimer's disease rats. *Nutrire.* 2018;43:23.
8. Dhawan D, Kumar S, Sharma A. Evaluation of central nervous system effects of *Passiflora incarnata* in experimental animals. *Pharm Biol.* 2003;41(2):87–91.
9. Park LW, Kwon OK, Ryu HW, et al. Anti-inflammatory effects of Passiflora foetida L. in LPS-stimulated RAW264.7 macrophages. *Int J Mol Med.* 2018;41(6):3709–3716.
10. Gupta RK, Kumar D, Chaudhary AK, Maithani M, Singh R. Antidiabetic activity of *Passiflora incarnata* Linn. in streptozotocin-induced diabetes in mice. *J Ethnopharmacol.* 2012;139(3):801–806.

11. do Socoro Ramos de Queiroz M, Janebro DI, da Cunha MAL, et al. Effect of the yellow passion fruit peel flour (*Passiflora edulis f. flavicarpa deg.*) in insulin sensitivity in type 2 diabetes mellitus patients. *Nutr J.* 2012;11(89):1–7.

12. Jafarpoor N, Abbasi-Maleki S, Asadi-Samani M, Khayatnouri MH. Evaluation of antidepressant-like effect of hydroalcoholic extract of *Passiflora incarnata* in animal models of depression in male mice. *J Herb Med Pharmacol.* 2014;3(1):41–45.

13. Santosh P, Venugopl R, Nilakash AS, Kunjbihari S, Mangala L. Antidepressant activity of methanolic extract of *Passiflora foetida* leaves in mice. *Int J Pharm Pharm Sci.* 2011;3(1):112–115.

14. Wang C, Xu FQ, Shang JH. Cycloartane triterpenoid saponins from water soluble of *Passiflora edulis* Sims and their antidepressant-like effects. *J Ethnopharmacol.* 2013;148(3):812–817.

15. Ayres AS, de Araújo LLS, Soares TC, et al. Comparative central effects of the aqueous leaf extract of two populations of *Passiflora edulis*. *Rev Bras Farmacogn.* 2015;25(5):499–505.

16. Singh B, Singh D, Goel RK. Dual protective effect of *Passifloraceae incarnata* in epilepsy and associated post-ictal depression. *J Ethnopharmacol.* 2012;139(1):273–279.

17. Fahami F, Asali Z, Aslani A, Fathizadeh N. A comparative study on the effects of *Hypericum Perforatum* and passion flower on the menopausal symptoms of women referring to Isfahan city health care centers. *Iran J Nurs Midwifery Res.* 2010;15(4):202–207.

18. Hoffmann D. *Medical Herbalism: The Science and Practice of Herbal Medicine.* Rochester, VT: Healing Arts Press; 2003. p. 571.

19. Aoyagi N, Kimura R, Murata T. Studies on *Passiflora incarnata* dry extract. I. Isolation of maltol and pharmacological action of maltol and ethyl maltol. *Chem Pharm Bull.* 1974;22(5):1008–1013.

20. Sopranzi N, De Feo G, Mazzanti G, Tolu L. Biological and electroencephalographic parameters in rats in relation to *Passiflora incarnata* L. *La Clin Ter.* 1990;132(5):329–333.

21. Hirakawa T, Suzuki T, Sano Y, Kamata T, Nakamura M. Reproductive studies of *P. incarnata* extract teratological study. *Kiso To Rinsho.* 1981;15:3431–3451.

22. Gardner Z, McGuffin M, editors. *Botanical Safety Handbook.* 3rd ed. Boca Raton, FL: CRC Press; 2013. p. 621.

23. Patel SS, Verma NK, Gauthaman K. *Passiflora Incarnata* Linn: a review on morphology, phytochemistry and pharmacological aspects. *Pharmacogn Rev.* 2009;3(5):175–181.

7.48 *PIPER METHYSTICUM* (KAVA)

Kava is the product of pulverized roots and rhizomes of the *Piper methysticum* pepper plant that is indigenous to South Pacific islands. It has been used by Polynesian cultures as a ceremonial beverage for welcoming guests and honoring all degrees of social relationships. The effects of drinking kava have been described as inducing a "warm, pleasant and cheerful but lazy feeling making people sociable, though not hilarious or loquacious, and not interfering with reasoning."[1]

Among the many phytochemical constituents of kava, known collectively as kavalactones or kavapyrones, are dihydrokawain, kawain, methysticin, yangonin, dihydromethysticin, desmethoxyyangonin, flavokawin A, pinostrobinchalcone, dihydrotectochrysin, alpinetinchalcone, alpinetin, dihydrooroxylin A, and others in lesser degrees of concentration.[2] Six of these kavalactones, including kavain, dihydrokavain, methysticin, dihydromethysticin, yangonin, and desmethoxyyangonin, are responsible for nearly all of the plant's pharmacological activity.

Kava is described as having a wide range of pharmacological effects of kava including anxiolytic, anti-stress, sedative, analgesic, muscle relaxant, antithrombotic, neuroprotective, mild anesthetic, hypnotic, and anticonvulsant. Its main mechanisms of action have been thought to be modulation of GABA activity via alteration of lipid membrane structure and sodium channel function, monoamine oxidase B inhibition, and noradrenaline and dopamine reuptake inhibition.[3] However, recent research shows the constituent yangonin to be a novel CB1 receptor ligand.[4]

7.48.1 Antioxidant

Several compounds from kava have been shown to exert antioxidant effects. The flavokawains A and B, stimulated Nrf2 in cultured human hepatocytes. Nrf2, in turn, induces activity of a variety of antioxidant enzymes. Both flavokawains protected liver cells from the oxidative damage of H_2O_2. However, flavokawains B had mild hepatotoxic effects that reduced those protective effects.[5] Flavokawains A and yangonin each showed significant DPPH radical scavenging capacity *in vitro*.[6]

A common *in vivo* test of antioxidant effects of phytochemicals is to assess the ability of the substance to protect the liver from the oxidative damage of CCl_4. Unfortunately, there have been rare reports of hepatotoxicity from kava. Indeed, in a study looking at effects of kava, sesame oil, and CCl_4 on liver pathology, kava alone was found to have no adverse effect on liver tissue. However, kava did not protect the livers of rats from CCl_4, but rather made the oxidative toxicity worse. The addition of sesame lignans alleviated the hepatotoxicity of CCl_4, with and without kava.[7]

7.48.2 Anti-Inflammatory

Components of kava exhibit potent anti-inflammatory effects at a number of points in the inflammatory cascade. Several kavalactones inhibit okadaic acid-stimulated release of TNF-α in mice.[8] In addition, the kavalactone methysticin inhibits TNF-α-induced stimulation of NF-κB.[9] The kavalactone, flavokawain A, suppressed the inflammatory response of murine macrophages to lipopolysaccharide. Flavokawain A suppressed expression of iNOS and COX-2, as well as the subsequent production of NO and PGE2 in the LPS-stimulated RAW 264.7 cells. Flavokawain A inhibited lipopolysaccharide-induced activation of NF-κB and AP-1 signaling pathways, as well as the activation of JNK and p38 MAPK that are responsible for expression of the enzymes iNOS and COX-2. Finally, flavokawain A suppressed LPS-induced expression of pro-inflammatory cytokines, such as TNF-α, IL-1β, and IL-6.[10]

7.48.3 Antidiabetic/Anti-Metabolic Syndrome

There are no reports in the literature of studies of specific effects of kava on diabetes, insulin action or Metabolic Syndrome.

7.48.4 Preclinical Antidepressant-Like Effects

Curiously, there are no animal studies of effects of kava on depression-like behaviors that might reveal both efficacy and help elucidate possible mechanisms. Nonetheless, kava clearly has central effects of psychiatric significance. Chronic administration of methysticin was found to improve cognitive deficits in a mouse model of Alzheimer's disease. This effect was likely due to the ability of methysticin to decrease neuroinflammation through activation of the Nrf2 pathway in the hippocampus and cortex of mice. Indeed, methysticin treatment significantly reduced microgliosis, astrogliosis, and secretion of the pro-inflammatory cytokines TNF-α and IL-17A in those areas of the brain.[11] Stimulation of Nrf2 is currently seen as having an antidepressant effect, likely due to the dampening of neuroinflammation.

7.48.5 HUMAN ANTIDEPRESSANT EFFECTS

Kava's predominant psychiatric effect has been anxiolysis. Indeed, a 2003 Cochrane review of 12 double blind, random controlled trials concluded that "kava extract is an effective symptomatic treatment for anxiety although, at present, the size of the effect seems small."[12] However, there is evidence that it also exhibits antidepressant effects. There is at least one formal, placebo-controlled, double-blind crossover study of the effectiveness of kava in the treatment of anxiety and MDD in human subjects.[13] An aqueous extract of kava was used in this study and administered over a three-week period. Kava significantly reduced depression per the Montgomery-Asberg Depression Rating Scale. It also reduced anxiety in both the Hamilton Anxiety Scale and Beck Anxiety Inventory.

7.48.6 SAFETY/TOXICITY

It has been reported that some individuals using kava have suffered severe and even fatal hepatotoxicity. In a 2003 paper, this relationship was somewhat softened. It was noted that out of 19 cases in Germany, only one was unequivocally related to kava. Nonetheless, it was prudently suggested that patients using kava be monitored.[14] Subsequent analysis of the problem has provided little resolution. Indeed, it even became suspected that mold toxins similar to aflatoxin may have contaminated kava herb stored in hot humid conditions in the tropical Pacific.[15] However, studies over the last decade have shown that kava contains a toxic component, identified to be flavokawain B, that kills liver cells. Interestingly, exogenous glutathione rescues hepatocytes from flavokavain B-induced death.[16]

The most recent study at the time of this writing described an unequivocal instance of hepatotoxicity requiring liver transplant in a man who used standard prudent doses of the herb. Moreover, analysis of the herb he used revealed no contaminants or obvious changes in the usual mix of phytochemicals found in kava. Thus, the conclusion was that while rare, the use of unadulterated kava in standard doses can result in hepatotoxicity.[17]

In the study noted in the human antidepressant effects section, the aqueous extract was found to be safe, with no serious adverse effects and no clinical hepatotoxicity. A lack of hepatic toxicity was also seen in a study of quite large doses of aqueous extracts of kava in rats.[18] Nonetheless, other authors have noted health risks with frequent consumption of even the apparently safer, aqueous extract of kava. They reported no evidence of cognitive impairment, liver toxicity, or permanent liver damage. However, they did see evidence of scaly skin rash, weight loss, raised gamma glutamyl transpeptidase liver enzyme levels, nausea, loss of appetite, or indigestion. It was concluded that: "The health and social implications of chronic kava drinking can be significant for individuals and communities, although most effects of even heavy consumption appear to be reversible when consumption is stopped."[19]

The *Botanical Safety Handbook*[20] notes that the American Herbal Products Association has established a trade requirement (AHPA 2011) that products that containing kava bear the following statement:

> Caution: US FDA advises that a potential risk of rare, but severe liver injury may be associated with kava-containing dietary supplements. Ask a healthcare professional before use if you have or have had liver problems, frequently use alcoholic beverages, or are taking any medication. Stop use and see a doctor if you develop symptoms that may signal liver problems (e.g., unexplained fatigue, abdominal pain, loss of appetite, fever, vomiting, dark urine, pale stools, yellow eyes or skin). Not for use by persons under 18 years of age, or by pregnant or breastfeeding women. Not for use with alcoholic beverages. Excessive use, or use with products that cause drowsiness, may impair your ability to operate a vehicle or dangerous equipment.

For the above reasons, I cannot in good conscience recommend the alcoholic or even aqueous extracts of kava. If an individual has already used kava with good result and finds it an indispensable part of their herbal regimen, then it is advised that they inform their provider so that they can be monitored for hepatotoxicity.

REFERENCES

1. Pepping J. Kava: Piper methysticum. *Am J Health Syst Pharm.* 1999;56(10):957–958.
2. Shulgin AT. The narcotic pepper, the chemistry and pharmacology of *Piper methysticum* and related species. *Bull Narc.* 1973;25:59–74.
3. Sarris J, LaPorte E, Schweitzer I. Kava: a comprehensive review of efficacy, safety, and psychopharmacology. *Aust N Z J Psychiatry.* 2011;45 (1):27–35.
4. Ligressti A, Villano R, Allarà M, Ujváry I, Di Marzo V. Kavalactones and the endocannabinoid system: the plant-derived yangonin is a novel CB1 receptor ligand. *Pharmacol Res.* 2012;66(2):163–169.
5. Pinner KD, Wales CTK, Gristock RA, Vo HT, So N, Jacobs AT. Flavokawains A and B from kava (*Piper methysticum*) activate heat shock and antioxidant responses and protect against hydrogen peroxide-induced cell death in HepG2 hepatocytes. *Pharm Biol.* 2016;54(9):1503–1512.
6. Wu D, Yu L, Nair MG, De Witt DL, Ramsewak RS. Cyclooxygenase enzyme inhibitory compounds with antioxidant activities from *Piper methysticum* (kava kava) roots. *Phytomedicine.* 2002;9(1):41–47; Stuttgart.
7. Chen W-L. Sesame lignans significantly alleviate liver damage of rats caused by carbon tetrachloride in combination with kava. *J Food Drug Anal.* 2010;18(4):249–255.
8. Hashimoto T, Suganuma M, Fujiki H, Yamada M, Kohno T, Asakawa X. Isolation and synthesis of TNF-α release inhibitors from Fijian kawa (*Piper methysticum*). *Phytomedicine.* 2003;10(4):309–317.
9. Shaik AA, Hermanson DL, Xing C. Identification of methysticin as a potent and nontoxic NF-κB inhibitor from kava, potentially responsible for kava's chemopreventive activity. *Bioorg Med Chem Lett.* 2009;19(19):5732–5736.

10. Kwon DJ, Ju SM, Youn GS, Choi SY, Park J. Suppression of iNOS and COX-2 expression by flavokawain A via blockade of NF-κB and AP-1 activation in RAW 264.7 macrophages. *Food Chem Toxicol.* 2013;58:479–486.
11. Fragoulis A, Siegl S, Fendt M, et al. Oral administration of methysticin improves cognitive deficits in a mouse model of Alzheimer's disease. *Redox Biol.* 2017;12:843–853.
12. Pittler MH, Ernst E. Kava extract for treating anxiety. *Cochrane Database Syst Rev.* 2002;2:CD003383.
13. Sarris J, Kavanagh DJ, Byrne G, Bone KM, Adams J, Deed G. The Kava Anxiety Depression Spectrum Study (KADSS): a randomized, placebo-controlled crossover trial using an aqueous extract of *Piper methysticum. Psychopharmacology.* 2009;205(3):399–407.
14. Teschke R, Gaus W, Loew D. Kava extracts: safety and risks including rare hepatotoxicity. *Phytomedicine.* 2003;10(5):440–446.
15. Samuel RT, Qiu X, Lebot V. Herbal hepatotoxicity by kava: update on pipermethystine, flavokavain B, and mould hepatotoxins as primarily assumed culprits. *Digest Liver Dis.* 2011;43(9):676–681.
16. Zhou P, Gross S, Liu J-H, et al. Flavokawain B, the hepatotoxic constituent from kava root, induces GSH-sensitive oxidative stress through modulation of IKK/NF-κB and MAPK signaling pathways. *FASEB J.* 2010;24(12):4722–4732.
17. Becker MW. Liver transplantation and the use of KAVA: case report. *Phytomedicine.* 2019;56:21–26.
18. Singh YN, Devkota AK. Aqueous kava extracts do not affect liver function tests in rats. *Planta Med.* 2003;69:496–499.
19. Rychetnik L, Madronio CM. The health and social effects of drinking water-based infusions of kava: a review of the evidence. *Drug Alcohol Rev.* 2011;30;74–83.
20. Gardner Z, McGuffin M, editors. *Botanical Safety Handbook.* 3rd ed. Boca Raton, FL: CRC Press; 2013. p. 658.

7.49   *PIPER NIGRUM* (BLACK PEPPER)

Piper nigrum or, more commonly, black pepper is a member of the genus *Piper* native to India and Southeast Asia. It has been used for thousands of years as a flavoring in food, preservative, and medicine, particularly by the Ayurvedic physicians of India. It contains a wide range of phytochemicals, including phenolics, lignans, terpenes, chalcones, flavonoids, alkaloids, and steroids.[1] Piperine, pipene, piperamides, and piperamine are seen as most active with the highest potential various biological activities. The biological effects are reported to include antibacterial, analgesic, antioxidative (which has made it valuable as a spice and food preservative) antispasmodic, hepatoprotective, and antidepressant.[2]

7.49.1   ANTIOXIDANT

Both aqueous and ethanolic extracts of *Piper nigrum* exhibit potent antioxidant effects in the DPPH, free radical scavenging, superoxide anion radical scavenging, hydrogen peroxide scavenging, and metal chelating assays. Indeed, they were as potent as α-tocopherol and the standard antioxidants BHA and BHT.[3] *In vivo*, oral administration of *Piper nigrum* oil for one month significantly increased superoxide dismutase, glutathione, and glutathione reductase enzyme levels in mouse blood, and glutathione-S-transferase, glutathione peroxidase, catalase, superoxide dismutase, and glutathione enzymes in mouse liver.[4]

7.49.2 ANTI-INFLAMMATORY

Extract of *Piper nigrum* significantly inhibits COX enzymes that generate inflammatory prostaglandins.[5] Piperine, one of the main active components of *Piper nigrum*, significantly reduced the expression of IL-1β, IL-6, and TNF-α in cultured human melanoma cells. This was found to be due in part to the prevention of nuclear translocation of NF-κB.[6]

In intact mice, administration of piperine prevented lipopolysaccharide-induced endotoxic shock, the marshalling of leukocytes, and production of TNF-α.[7] In rats that had suffered induced cerebral ischemia, piperine reduced the levels of IL-1β, IL-6, and TNF-α in ischemic brain tissue. It also diminished expression of COX-2, NOS-2, and NF-κB.[8]

7.49.3 ANTIDIABETIC/ANTI-METABOLIC SYNDROME

Piper nigrum has properties that would diminish Metabolic Syndrome that is a risk factor for MDD. It has antidiabetic effects, as demonstrated by its ability to decrease serum glucose alloxan-induced diabetic mice.[9] It also exhibits hypolipidemic effects in rats fed high fat diets.[10] The addition of piperine, the primary active component of *Piper nigrum*, to rats made hyperlipidemic by high fat diets significantly reduced plasma lipids and lipoproteins with the exception of HDL, which were increased. It also reduced abnormally high serum levels of insulin back to normal.[11]

Piperine also appears to counteract effects of stress. Indeed, in one study, piperine was found to reverse corticosterone-induced depression-like behavior in mice. It also restored decreased levels of BDNF back toward normal in the hippocampi of those animals.[12]

7.49.4 PRECLINICAL ANTIDEPRESSANT-LIKE EFFECTS

In several studies, piperine has been found have antidepressant-like effects in mice. Two weeks of piperine stored preference for sucrose, increased activity in the open field, and restored normal levels of corticosterone in mice that had been subjected to chronic unpredictable stress. Treatment also restored levels of BDNF on the hippocampus and the consequent proliferation of hippocampal progenitor cells.[13]

In a very similar study, piperine was also found to show antidepressant-like effects in chronically stressed mice. Preference for sucrose was restored, and immobility was reduced in the forced swim test. BDNF was also increased in the hippocampus and frontal cortex. Interestingly, BDNF was also increased in mice that had not been subjected to chronic stress. Moreover, the antidepressant effect of piperine was prevented by co-administration of K252a that blocks the BDNF receptor TrkB. Thus, it appears that BDNF signaling is an essential mediator for the antidepressant-like effect of piperine.[14]

In yet another study in mice, pretreatment with piperine improved performance in the forced swim and tail suspension tests, suggesting antidepressant effects. The treatment also significantly increased dopamine levels in the striatum, hypothalamus, and hippocampus, and serotonin levels in the hypothalamus and hippocampus. These effects were likely explained by the finding of mild inhibition of MAO A and MAO B in those regions.[15]

Similar antidepressant-like effects of piperine were seen in studies using rats. Four weeks of administration of piperine improved performance in the forced swim test, as well as in the Morris water maze. The later suggests improvement in cognitive function.[16] In another rat study, a methanolic extract of *Piper nigrum* was used to study effects in an amyloid beta(1-42) rat model of Alzheimer's disease. Performance was improved in both the Y-maze and radial arm-maze tasks, thus indication improvement of cognitive function. Of further significance was the evaluation of antioxidant effects of the extract of *Piper nigrum* in the hippocampi of those animals. Superoxide dismutase-, catalase-, and glutathione peroxidase-specific activities and the total content of reduced glutathione were increased in the hippocampus, whereas malondialdehyde and protein carbonyl levels were reduced. Thus, it is likely that that improvement in cognitive function, as well as the antidepressants effects observed in other studies, were due in part to potent antioxidative effects of piperine in the brains of experimental animals.[17]

7.49.5   Human Antidepressant Effects

There are no trials of *Piper nigrum* for the treatment of MDD in human subjects.

7.49.6   Dosage

Easley and Horne recommend 500 mg *Piper nigrum* or 10 mg of piperine three times daily.[18] However, their recommendation is in the context of its use as a digestive aid, and in the enhancement of otherwise poor absorption of other compounds.

7.49.7   Toxicity

Piper nigrum, as a commonly used spice, is on the US government's GRAS, or Generally Regarded As Safe, list, as per the Code of Federal Regulations Title 21 Parts 172, 182, 184, and 186. In mice, the LD50 for gavaged piperine, the main active ingredient of *Piper nigrum*, is reported to be 330 mg/kg.[19]

7.49.8   Safety in Pregnancy

The *Botanical Safety handbook* gives no definitive information on the safety of using pharmacological doses of *Piper nigrum* in pregnancy. It does reference a study showing it to improve fertilization rates in domestic animals.[20]

7.49.9   Drug Interactions

The *Botanical Safety Handbook* further notes that piperine from *Piper nigrum* can enhance the absorption of various drugs. It is worth noting that piperine can also enhance the bioavailability of medicinal phytochemicals, such as resveratrol[21] and curcumin.[22]

REFERENCES

1. Parmar VS, Jain SC, Bisht KS, et al. Phytochemistry of the genus *Piper*. *Phytochemistry*. 1997;46:597–673.
2. Ahmad N, Fazal H, Abbasi BH, Farooq S, Ali M, Khan MA. Biological role of *Piper nigrum* L. (Black pepper): a review. *Asian Pac J Trop Biomed*. 2012;2(3):S1945–S1953.
3. Gülçin I. The antioxidant and radical scavenging activities of black pepper (*Piper nigrum*) seeds. *Int J Food Sci Nutr*. 2005;56(7):7491–499.
4. Jeena K, Liju VB, Umadevi NP, Kuttan R. Antioxidant, anti-inflammatory and antinociceptive properties of black pepper essential oil (*Piper nigrum* Linn). *J Essent Oil Bear Plants*. 2014;17(1):1–12.
5. Liu Y, Yadev, VR, Aggarwal BB, Nair MG. Inhibitory effects of black pepper (*Piper nigrum*) extracts and compounds on human tumor cell proliferation, cyclooxygenase enzymes, lipid peroxidation and nuclear transcription factor-kappa-B. *Nat Prod Commun*. 2010;5(8):1253–1257.
6. Pradeep CR, Kuttan G. Piperine is a potent inhibitor of nuclear factor-κB (NF-κB), c-Fos, CREB, ATF-2 and proinflammatory cytokine gene expression in B16F-10 melanoma cells. *Int Immunopharmacol*. 2004;4(14):1795–1803.
7. Bae GS, Kim MS, Jung WS, et al. Inhibition of lipopolysaccharide-induced inflammatory responses by piperine. *Eur J Pharmacol*. 2010;642(1–3):154–162.
8. Vaibhav K, Shrivastava P, Javed H, et al. Piperine suppresses cerebral ischemia-reperfusion-induced inflammation through the repression of COX-2, NOS-2, and NF-κB in middle cerebral artery occlusion rat model. *Mol Cell Biochem*. 2012;367(1–2):73–84.
9. Atal S, Agrawal RP, Vyas S, Phadnis P, Rai N. Evaluation of the effect of piperine per se on blood glucose level in alloxan-induced diabetic mice. *Acta Pol Pharm*. 2012;69(5):965–969.
10. Vijayakuma RS, Surya D, Senthilkumar R, Nalini N. Hypolipidemic effect of black pepper (*Piper nigrum* Linn.) in rats fed high fat diet. *J Clin Biochem Nutr*. 2002;32:31–42.
11. Vijayakuma RS, Nalini N. Piperine, an active principle from *Piper nigrum*, modulates hormonal and apolipoprotein profiles in hyperlipidemic rats. *J Basic Clin Physiol Pharmacol*. 2006;17:71–86.
12. Mao QQ, Huang Z, Zhong XM, Xian YF, Ip SP. Piperine reverses the effects of corticosterone on behavior and hippocampal BDNF expression in mice. *Neurochem Int*. 2014;74:36–41.
13. Li S, Wang C, Wang M, Li W, Matsumoto K, Tang Y. Antidepressant like effects of piperine in chronic mild stress treated mice and its possible mechanisms. *Life Sci*. 2007;80(15):1373–1381.
14. Mao Q-Q, Huang Z, Zhong XM, Xian YF, Ip SP. Brain-derived neurotrophic factor signaling mediates the antidepressant-like effect of piperine in chronically stressed mice. *Behav Brain Res*. 2014;261:140–145.
15. Li S, Wang C, Li W, Koike K, Nikaido T, Wang MW. Antidepressant-like effects of piperine and its derivative, antiepilepsirine. *J Asian Nat Prod Res*. 2007;9:421–430.
16. Wattanathorn J, Chonpathompikunlert P, Muchimapuraa S, Priprem A, Tankamnerdthai O. Piperine, the potential functional food for mood and cognitive disorders. *Food Chem Toxicol*. 2008;46:3106–3110.
17. Hritcu L, Noumedem JA, Cioanca O, Hancianu M, Kuete V, Mihasan M. Methanolic extract of *Piper nigrum* fruits improves memory impairment by decreasing brain oxidative stress in amyloid beta(1–42) rat model of Alzheimer's disease. *Cell Mol Neurobiol*. 2014;34(3):437–449.

18. Easley T, Horne S. *The Modern Herbal Dispensary*. Berkeley, CA: North Atlantic Books; 2016. p. 186.
19. Piyachaturawat P, Glinsukon T, Toskulkao C. Acute and subacute toxicity of piperine in mice, rats and hamsters. *Toxicol Lett*. 1983;16(3–4):351–359.
20. Gardner Z, McGuffin M, editors. *Botanical Safety Handbook*. 3rd ed. Boca Raton, FL: CRC Press; 2013. p. 663.
21. Johnson JJ, Nihal M, Siddiqui IA, et al. Enhancing the bioavailability of resveratrol by combining it with piperine. *Mol Nutr Food Res*. 2011;55(8):1169–1176.
22. Shoba G, Joy D, Joseph T, Majeed M, Rajendran R, Srinivas PS. Influence of piperine on the pharmacokinetics of curcumin in animals and human volunteers. *Planta Med*. 1998;64(4):353–356.

7.50 *POLYGALA TENUIFOLIA*

Polygala is a large genus of flowering plants belonging to the family *Polygalaceae*. The genus is distributed throughout much of the world in temperate zones and tropics. The species *Polygala tenuifolia* grows in Asia and has been an important herb in Traditional Chinese Medicine. In Chinese it is called "*yuan zhi*," and it has traditionally been used to calm the mind and restore memory.[1] It is also an important component in variety of herbal combinations used to treat syndromes analogous to MDD. The best-known combination in which it is an ingredient is *Kaixinsan*.[2] However, it is included in other well-known combinations, including *An shen yang xin cha, Gui pi wan, Tian wang bu xin dan, Ding xin wan*, and *An mian pian*. Yeung et al. did not list it among the ten herbs most often used in Traditional Chinese Medicine to treat MDD. However, it is certainly a contender. The species *Polygala senega* is native to North America. Its common name is Seneca snakeroot, which harkens back to the Seneca people that used the plant to treat snakebite.[3]

Many of the *Polygala* species are used in folk medicine as anesthetics, anti-inflammatory agents, and for the treatment of disturbances of the bowel, kidney, and central nervous system. The genus is rich in bioactive phytochemicals, such as triterpenic saponins, lignans, and, in particular, xanthones. One large group of steroid-like phenolic glycosides, the tenuifolisides, appear unique to *Polygala tenuifolia*.[4] Another group of unique and complex oligosaccharides, the polygalatenosides, have been found to inhibit noradrenergic reuptake.[5] Various, more common flavonoids, often in glycosylated form, include quercetin, kaempferol, and coumaric, caffeic, and ferulic acids.[6]

7.50.1 ANTIOXIDANT

Polygala tenuifolia has been shown to have potent antioxidant effects, both *in vitro* and *in vivo*. Aqueous extract of *Polygala tenuifolia* exerted significant antioxidant activities in the DPPH radical scavenging assay. DNA was protected from H_2O_2-induced oxidative damage. The extract also increased activity of superoxide dismutase and inhibited the production of inducible nitric oxide in cultured rat pheochromocytoma cells.[7]

In a breed of mice known as "accelerated senescence-prone, short-lived mice," extract of *Polygala tenuifolia*, as well as an oligosaccharide ester from a highly purified fraction, exhibited potent antioxidant effects. Treatment significantly increased activities of the superoxide dismutase and glutathione peroxidase enzymes in liver tissue. Levels of the oxidation marker malondialdehyde were decreased in both liver and blood.[8]

Central antioxidant effects in intact animals have also been demonstrated. Administration of extract of *Polygala tenuifolia* root attenuated lipid peroxidation and other oxidative damage in the brains of gerbils that had suffered the shock of cerebral ischemia and reperfusion.[9]

7.50.2 ANTI-INFLAMMATORY

The aqueous extract of *Polygala tenuifolia* root inhibited secretion TNF-α and IL-1 by mouse astrocytes stimulated with either Substance P or lipopolysaccharide.[10] Anti-inflammatory effects of *Polygala tenuifolia* were also seen in a mouse model of colitis. Extract of the root decreased production of INF-γ but increased levels of the anti-inflammatory cytokine, IL-4, in intraepithelial lymphocytes.[11]

Tenuigenin, an active component of *Polygala tenuifolia* root extracts, was shown to significantly decrease the release of nitric oxide from lipopolysaccharide-activated rat microglia. This was due to the scavenging of the NO radical. Additionally, TEN can significantly decrease the secretion and mRNA levels of the inflammatory protein matrix metalloproteinase-9, as well as TNF-α and IL-1β from activated microglia.[12]

Both aqueous and ethanolic extracts of *Polygala senega* dampened the response of cultured murine macrophage RAW 164.7 cells to lipopolysaccharide. The production of the cytokines IL-1, TNF-α, and IL-6 were reduced.[13] A methanolic extract of *Polygala tenuifolia* also attenuated the rat paw edema response and inhibited the production of lipopolysaccharides-induced 6-keto-PGF1α in macrophage cultures. Both of these effects were likely mediated by antagonism of the inflammatory enzyme, COX-2.[14]

7.50.3 ANTIDIABETIC/ANTI-METABOLIC SYNDROME

Polygala tenuifolia and *senega* both have antidiabetic effects, as well as other actions that help mitigate Metabolic Syndrome. Acute intraperitoneal administration of senegin-II, a primary component of *Polygala senega*, reduced blood glucose both in normal mice and in mice with genotype KK-Ay, a model of diabetes type II.[15] Treatment with extract of *Polygala tenuifolia* reduced fat accumulation in cultured 3T3-L1 adipocytes. In intact, high fat diet-induced obese mice, the extract prevented weight increase, lowered serum triglycerides, and attenuated development of liver steatosis.[16]

7.50.4 PRECLINICAL ANTIDEPRESSANT-LIKE EFFECTS

There is substantial and compelling evidence from the animal literature that *Polygala tenuifolia* and some of its phytochemical constituents produce antidepressant-like effects in various rodent models of depression. Acute administration of hydromethanolic extract of *Polygala tenuifolia* reduced immobility of mice in the forced swim and tail suspension tests.[17] In one study, the reduction of immobility in the tail suspension test was blocked by the AMPA receptor antagonist NBQX, a pattern previously demonstrated by ketamine and other ketamine-like rapid-onset antidepressants. Subacute administrations of the extract also restored the preference for saccharin in mice exposed to chronic stress.[18]

Chronic administration of corticosterone produces a depression-like state that is an animal model of MDD. Co-administration of extract of *Polygala tenuifolia* root to mice chronically treated with corticosterone reversed the exaggerated immobility in the forced swim test. It also reversed the corticosterone-induced impairment in dendritic spine maturation and the decreases in expression of glial cell line-derived neurotrophic factor in the hippocampus and nucleus accumbens.[19]

Senegenin is a major bioactive constituent of *Polygala tenuifolia* Willd that has anti-inflammatory and neuroprotection effects. In mice subjected to chronic unpredictable stress, daily treatment with senegenin ameliorated the stress-induced depression-like abnormalities. The treatment decreased immobility time in the tail suspension and forced swim test, and restored the appetite for sucrose. Subsequent analyses of the hippocampus in these mice also revealed neurochemical changes. Senegenin increased levels of both BDNF and neurotrophin-3 in this area. The cleavage of pro-IL-1β into IL-1β by the NLRP3 inflammasome pathway was also attenuated.[20] In mice subjected to chronic daily restraint stress, concomitant daily administrations of extract of *Polygala tenuifolia* also attenuated stress-induced increases in anxiety-like behavior.[21]

A component purified from *Polygala tenuifolia*, a triterpenoid referred to as Yuanzhi-1, was found to reduce immobility in the forced swim and tail suspension tests in mice. It also had quite high low nanomolar affinity for all three monoamine uptake sites. Indeed, the effective dose of Yuanzhi-1, 2.5 mg/kg, was lower than the dose of the positive control drug duloxetine, 2.5 mg/kg.[22]

Extract of *Polygala tenuifolia* also attenuated the depression-like effects of chronic unpredictable stress in rats. Chronic administration restored sucrose consumption and reduced the elevated serum levels of corticosterone in the stressed animals. Hippocampal levels of BDNF were also restored to normal.[23]

In rats subjected to chronic unpredictable stress, acute administration of a specific oligosaccharide isolated from *Polygala tenuifolia*, 3,6'-disinapoylsucrose, also reduced immobility in the forced swim and tail suspension tests. Serum levels of corticosterone, adrenocorticotropic hormone, and corticotropin-releasing hormone were substantially reduced. This treatment increased phosphorylated CREB and BDNF.[24] Those results were consistent with those seen in mice with scopolamine-induced cognitive impairments. Another specific component of extract of *Polygala tenuifolia*, polygalasaponin XXXII, improved cognitive function, and increased both the MAPk cascade and activity of BDNF in the hippocampus.[25]

Yet another specific phytochemical in *Polygala tenuifolia* was found to ameliorate the depression-like effects of chronic unpredictable stress in rats. Daily administration of 3,6'-disinapoyl sucrose purified from *Polygala tenuifolia* showed an antidepressant-like effect in restoring preference for sucrose in stressed animals. Plasma corticosterone levels were normalized. In the brains of these animals, activities of

MAO-A and MAO-B were reduced toward normal. Antioxidant effects were also noted in brain tissue. Activity of superoxide dismutase was increased, and levels of the oxidation marker malondialdehyde were reduced.[26] Treatment with 3,6'-disinapoyl sucrose has also been found to enhance the expressions of cell adhesion molecule L1, laminin, CREB, and BDNF in the hippocampi of such rats subjected to chronic stress. All of those proteins are involved in neuronal plasticity and neurite outgrowth and are associated with antidepressant effects.[27]

7.50.5   HUMAN ANTIDEPRESSANT EFFECTS

There are several reports of the use of extract of *Polygala tenuifolia* to improve memory and general cognitive function in elderly[28] and otherwise healthy young adults.[29] The results of those studies were favorable, and without obvious adverse effects.

Nonetheless, despite a long tradition of use in Traditional Chinese Medicine as a component in herbal combinations to treat depression-like syndromes, there are no published studies of the use of *Polygala tenuifolia* or other Polygala species for the treatment of MDD in human subjects.

7.50.6   DOSAGE

Polygala senega, a species closely related to *Polygala tenuifolia*, is approved by the German E Commission under the name of Senega snakeroot. However, the only application is respiratory congestion. Its only warning is the risk of gastrointestinal irritation with prolonged use. It recommends 1–3 g of root or fluid extract, or 2.5–7.5 g of tincture per day.[30]

7.50.7   TOXICITY

In a study of toxicity, doses of dried aqueous extract of up to 2,000 mg/kg to mice caused no mortalities, clinical signs, changes in body or organ weights, or histopathological changes to any of 14 different organ systems. Thus, the LD50 was deemed to be above 2,000 mg/kg.[31] Extrapolated for a 75 kg human adult, this dose is approximately 140g.

7.50.8   SAFETY IN PREGNANCY

Reference texts from Traditional Chinese medicine advise against use of *Polygala tenuifolia* during pregnancy.[32]

7.50.9   DRUG INTERACTIONS

A study showed that refined extracts compounds from *Polygala tenuifolia* Willd significantly inhibited activity of the CYP2E1 P450 isoenzyme from human liver microsomes, but had little effect on the activities of CYP1A2, CYP2A6, CYP2C8, CYP2C9, CYP2C19, CYP2D6, and CYP3A isoforms.[33] Nonetheless, the

Botanical Safety Handbook notes no known interactions with commonly prescribed medications.[34]

REFERENCES

1. Maeng S. Rapid-onset antidepressants and *Radix Polygalae. J Drug Abuse*. 2017;3(3):13.
2. Zhu KY, Xu SL, Choi RC, Yan AL, Dong TT, Tsim KW. *Kai-xin-san*, a Chinese herbal decoction containing *ginseng radix et rhizoma, polygalae radix, acori tatarinowii rhizoma*, and *poria*, stimulates the expression and secretion of neurotrophic factors in cultured astrocytes. *Evid Based Complement Altern Med*. 2013;2013.
3. Small E, Catling PM. *Polygala senega* L. (Seneca Snakeroot). *Canadian Medicinal Crops*. Agriculture and Agri-Food Canada; 2012.
4. Ikeya Y, Sugama K, Okada M, Mitsuhashi H. Four new phenolic glycosides from *Polygala tenuifolia. Chem Pharm Bull*. 1991;39(10):2600–2605.
5. Cheng MC, Li CY, Ko HC, Ko FN, Lin YL, Wu TS. Antidepressant principles of the roots of *Polygala tenuifolia. J Nat Prod*. 2006;69(9):1305–1309.
6. Klein LC, de Andrade SF, Filho VC. A pharmacognostic approach to the polygala genus: phytochemical and pharmacological aspects. *Chem Biodivers*. 2012;9:181–209.
7. Hyang N, Kim MM. Effect of *Polygala radix* hot water extract on biological activity in PC12 cells. *J Life Sci*. 2013;23(8):1041–1049.
8. Liu P, Hu Y, Guo DH, et al. Antioxidant activity of oligosaccharide ester extracted from *Polygala tenuifolia* roots in senescence-accelerated mice. *Pharm Biol*. 2010;48(7):828–833.
9. Park JH, Kim JS, Jang DS, Lee SM. Effect of *Polygala tenuifolia* root extract on cerebral ischemia and reperfusion. *Am J Chin Med*. 2006;34(1):115–123.
10. Kim HM, Lee EH, Na HJ, et al. Effect of *Polygala tenuifolia* root extract on the tumor necrosis factor-alpha secretion from mouse astrocytes. *J Ethnopharmacol*. 1998;61:201–208.
11. Hong T, Jin GB, Yoshino G, et al. Protective effects of *Polygalae* root in experimental TNBS-induced colitis in mice. *J Ethnopharmacol*. 2002;79(3):341–346.
12. Lu L, Li X, Xu P, Zheng Y, Wang X. Tenuigenin down-regulates the release of nitric oxide, matrix metalloproteinase-9 and cytokines from lipopolysaccharide-stimulated microglia. *Neurosci Lett*. 2017;650:82–88.
13. Van Q, Nayak BN, Reimer M, Jones PJ, Fulcher RG, Rempel CB. Anti-inflammatory effect of Inonotus obliquus, *Polygala senega* L. and *Viburnum trilobum* in a cell screening assay. *J Ethnopharmacol*. 2009;125:487–493.
14. Oh JJ, Kim SJ. Inhibitory effect of the root of *Polygala tenuifolia* on bradykinin and COX 2-mediated pain and inflammatory activity. *Trop J Pharm Res*. 2013;12(5):755–759.
15. Kako M, Miura T, Usami M, et al. Effect of senegin-II on blood glucose in normal and NIDDM mice. *Biol Pharm Bull*. 1995;18:1159.
16. Wang CC, Yen JH, Cheng YC, et al. *Polygala tenuifolia* extract inhibits lipid accumulation in 3T3-L1 adipocytes and high-fat diet-induced obese mouse model and affects hepatic transcriptome and gut microbiota profiles. *Food Nutr Res*. 2017;61(1):Article ID 1379861.
17. Liu P, Hu Y, Guo DH, et al. Potential antidepressant properties of Radix Polygalae (*Yuan Zhi*). *Phytomedicine*. 2010;17(10):794–799.
18. Shin IM, Son SU, Park H, et al. Preclinical evidence of rapid-onset antidepressant-like effect in radix polygalae extract. *PloS One*. 2014;9:e88617.
19. Araki R, Fujiwara H, Matsumoto K, et al. *Polygalae radix* extract ameliorates behavioral and neuromorphological abnormalities in chronic corticosterone-treated mice. *Tradit Kampo Med*. 2018;5(2):89–97.

20. Li HW, Lin S, Qin T, Li H, Ma Z, Ma S. Senegenin exerts anti-depression effect in mice induced by chronic un-predictable mild stress via inhibition of NF-κB regulating NLRP3 signal pathway. *Int Immunopharmacol.* 2017;53:24–32.

21. Lee B, Sur B, Shin S, et al. *Polygala tenuifolia* prevents anxiety-like behaviors in mice exposed to repeated restraint stress. *Anim Cells Syst.* 2015;19(1):1–7.

22. Jina ZL, Gao N, Zhang J, et al. The discovery of Yuanzhi-1, a triterpenoid saponin derived from the traditional Chinese medicine, has antidepressant-like activity. *Prog Neuropsychopharmacol Biol Psychiatry.* 2014;53(4):9–14.

23. Hu Y, Liu P, Guo DH, Rahman K, Wang DX, Xie TT. Antidepressant effects of the extract YZ-50 from *Polygala tenuifolia* in chronic mild stress treated rats and its possible mechanisms. *J Pharm Biol.* 2010;48(7):794–800.

24. Hu Y, Liao HB, Liu P, Guo DH, Rahman K. A bioactive compound from *Polygala tenuifolia* regulates efficiency of chronic stress on hypothalamic-pituitary-adrenal axis. *Die Pharm.* 2009;64(9):605–608.

25. Xue W, Hu J, Yuan Y, et al. Polygalasaponin XXXII from *Polygala tenuifolia* root improves hippocampal-dependent learning and memory. *Acta Pharmacol Sin.* 2009;30:1211–1219.

26. Ming YH, Liu M, Liu P, Guo DH, Wei RB, Rahman K. Possible mechanism of the anti-depressant effect of 3,6'-disinapoyl sucrose from *Polygala tenuifolia* Willd. *J Pharm Pharm.* 2011;63(6):869–874.

27. Hua Y, Liao HB, Dai-Hong G, Liu P, Wang YY, Rahman K. Antidepressant-like effects of 3,6'-disinapoyl sucrose on hippocampal neuronal plasticity and neurotrophic signal pathway in chronically mild stressed rats. *Neurochem Int.* 2010;56(3):461–465.

28. Shin KY, Lee JY, Won BY, et al. BT-11 is effective for enhancing cognitive functions in the elderly humans. *Neurosci Lett.* 2009;465(2):157–159.

29. Lee JY, Kim KY, Shin KY, Won BY, Jung HY, Suh YH. Effects of BT-11 on memory in healthy humans. *Neurosci Lett.* 2009;454(2)111–114.

30. Blumenthal M, Busse WR, Goldberg A, et al. *The Complete German Commission E Monographs.* Austin, TX: American Botanical Council; 1998. p. 203–204.

31. Kang BH. Single oral dose toxicity study of extracts of Polygala radix in ICR mice. *J Physiol Pathol Korean Med.* 2013;27(4):453–459.

32. Chen JK, Chen TT. *Chinese Medical Herbology and Pharmacology.* City of Industry, CA: Art of Medicine Press; 2004.

33. Li ZL, Dong XZ, Wang DX. Effect of oligosaccharide esters and polygalaxanthone III from Polygala tenuifolia willd towards cytochrome P450. *China J Chin Mater Med.* 2014;39(22):4459–4463.

34. Gardner Z, McGuffin M, editors. *Botanical Safety Handbook.* 3rd ed. Boca Raton, FL: CRC Press; 2013. p. 682.

7.51   *PORIA COCOS*

Poria cocos is a fungus belonging to the polyporaceae family of the *Basidiomycetes*. It is usually a parasite in the roots of pine trees in East Asia, Australia, America, and Africa. The mushroom is a commonly used plant in Chinese Medicine, in which it is known as *Fuling*. Traditionally, the fungus has been considered to have spleen-invigorative, stomach-tonifying, sedative, tranquilizing, diuretic, and damp-clearing effects. It is used to treat retention of phlegm and fluid, dysuria, edema, poor appetite with watery stools, palpitations, and insomnia. It is also noted by Yeung et al. to be one of the most commonly used herbs in combinations used to treat various forms of MDD. It contains a wide variety of phytochemicals. Much of its biological activity is thought to be due to the large steroid-like triterpene, pachymic acid, and similar molecules.[1] However, the plant also contains a variety of polysaccharide molecules that are biologically active and produce significant antioxidant effects.[2]

7.51.1 ANTIOXIDANT

Extract of *Poria cocos* also protected cultured PC12 neuronal cells from beta-amyloid-induced apoptosis by dampening oxidation. It reduced the concentration of malondialdehyde-modified protein adducts that are markers for oxidative damage by reactive oxygen species.[3] Interestingly, like commercial antioxidants, such as BHA or BHT, powder of *Poria cocos* has been found to be an effective antioxidant preservative when added to dough in the production of cookies.[4]

7.51.2 ANTI-INFLAMMATORY

Alcohol extracts of *Poria cocos* have anti-inflammatory effects, and inhibit the release of a variety of inflammatory cytokines, including IL-1β, IL-6, TNF-α, and granulocyte-monocyte colony stimulating factor.[5] The hydroalcoholic extract from *Poria cocos* ameliorates carrageenan-induced acute edemas and oxazolone delayed hypersensitivity in mice.[6] Ethanolic extracts of the mushroom blunt the lipolysaccharide-induced inflammatory response in a mouse macrophage cell line, in part by suppressing the NF-kappaB signaling pathway.[7]

7.51.3 ANTIDIABETIC/ANTI-METABOLIC SYNDROME

Compounds isolated from poria reduce hyperglycemia in mouse models of noninsulin-dependent diabetes mellitus, enhance insulin sensitivities, and activate PPAR-γ.[8] The ethyl acetate extract of *Poria cocos* reverses TNF-α-induced suppression of PPAR-γ in human hepatic stellate cells. The extract also dampens NF-κB activity induced by TNF-α.[9] Thus, *Poria cocos* ameliorates several components of the Metabolic Syndrome.

7.51.4 PRECLINICAL ANTIDEPRESSANT-LIKE EFFECTS

A saponin containing a triterpene similar in structure to pachymic acid reversed behavioral effects of unpredictable stress in a rat model of depression, and restored levels of BDNF.[10] Two polysaccharides purified from *Poria cocos* – quite distinct from the above noted triterpene – were also found to reduce immobility times in the tail suspension and forced swim tests in mice that had been subjected to chronic unpredictable mild stress.[11]

7.51.5 HUMAN ANTIDEPRESSANT EFFECTS

There are no reported trials of *Poria cocos* on its own in the treatment of MDD in human subjects.

7.51.6 DOSAGE

There are no reports of doses of *Poria cocos* alone for treatment of MDD, as it is traditionally used in combination for this application. I note that references in the literature for conditions such as edema show human subjects taking 3 g of dried mushroom three times daily with good results.

7.51.7 TOXICITY

The oral LD50 of dried warm water extract in mice is 10 g/kg.[12] In a human adult weighing 75 kg, this would extrapolate to approximately 700 g.

7.51.8 SAFETY IN PREGNANCY

Review of the literature revealed no studies of the safety of *Poria cocos* during human or animal pregnancy.

7.51.9 DRUG INTERACTIONS

Poria cocos has been found to inhibit the CYP3A4 enzyme.[13] A variety of neuroleptics, including aripiprazole, haloperidol, pimozide, and risperidone, are metabolized by this enzyme.[14] The enzyme is also involved in the metabolism of SSRIs and tricyclic antidepressants.[15] Thus, chronic use of *Psoralea corylifolia* could affect blood levels of a variety of important psychiatric medications.

REFERENCES

1. Shu S, Chen B, Zhou M, Zhao X, Xia H, Wang M. De novo sequencing and transcriptome analysis of *Wolfiporia cocos* to reveal genes related to biosynthesis of triterpenoids. *PLoS One.* 2013;8(8):e71350.
2. Chen XP, Tang QC, Chen Y, Wang W, Li S. Simultaneous extraction of polysaccharides from *Poria cocos* by ultrasonic technique and its inhibitory activities against oxidative injury in rats with cervical cancer. *Carbohyd Polym.* 2010;79(2):409–413.
3. Park YH, Son IH, Kim B, Lyu YS, Moon HI, Kang HW. *Poria cocos* water extract (PCW) protects PC12 neuronal cells from beta-amyloid-induced cell death through antioxidant and antiapoptotic functions. *Pharmazie.* 2009;64:(11)760–764.
4. Yu HH. Quality characteristics and antioxidant activity of cookies added with Baekbokrung (*Poria cocos* Wolf) powder. *Korean J Hum Ecol.* 2014;23(3):443–452.
5. Ríos J-L. Chemical constituents and pharmacological properties of *Poria cocos. Planta Med.* 2011;77:681–691.
6. Cuellar MJ, Giner RM, Recio MD, Just MJ, Manez S, Rios JL. Effect of the basidiomycete *Poria cocos* on experimental dermatitis and other inflammatory conditions. *Chem Pharm Bull.* 1997;45(3):492–494.
7. Jeong JW, Lee HH, Han MH, et al. Ethanol extract of *Poria cocos* reduces the production of inflammatory mediators by suppressing the NF-kappaB signaling pathway in lipopolysaccharide-stimulated RAW 264.7 macrophages. *BMC Complement Altern Med.* 2014;14:101.

8. Sato M, Tai T, Nunoura Y, Yajima Y, Kawashima S, Tanaka K. Dehydrotrametenolic acid induces preadipocyte differentiation and sensitizes animal models of noninsulin-dependent diabetes mellitus to insulin. *Biol Pharm Bull.* 2002;25:81–86.

9. Su YB, Huang YT. *Poria cocos* inhibited the activation of hepatic stellate cells. *Planta Med.* 2009;75:1034–1035.

10. Li, L-F. Antidepressant-like effects of the saponins extracted from *Chaihu-jia-longgu-muli-tang* in a rat unpredictable chronic mild stress model. *Fitoterapia.* 2012;83(1):93–103.

11. Zhang W, Chen L, Li P, Zhao J, Duan J. Antidepressant and immunosuppressive activities of two polysaccharides from *Poria cocos* (Schw.) Wolf. *Int J Biol Macromol.* 2018;120:1696–1704.

12. Yin J, Gun LG. *Modern Research and Clinical Applications of Chinese Materia Medica.* Beijing: Academic Publishing; 1993. p.489–492.

13. Dong HY, Shao JW, Chen JF, Wang T, Lin FP, Guo YH. Transcriptional regulation of cytochrome P450 3A4 by four kinds of traditional Chinese medicines. *J Chin Mater Med.* 2008;33(9):1014–1017.

14. van der Weide K, van der Weide J. The influence of the CYP3A4*22 polymorphism and CYP2D6 polymorphisms on serum concentrations of aripiprazole, haloperidol, pimozide, and risperidone in psychiatric patients. *J Clin Psychopharmacol.* 2015;35(3):228–236.

15. Haduch A, Wójcikowski J, Daniel WA. The effect of tricyclic antidepressants, selective serotonin reuptake inhibitors (SSRIs) and newer antidepressant drugs on the activity and level of rat CYP3A. *Eur Neuropsychopharmacol.* 2006;16(3):178–186.

7.52 *PSORALEA CORYLIFOLIA*

Psoralea corylifolia is a plant of the *Leguminosae*, or pea, family. The plant grows in the Indian plains, the lower Himalayas and through China. In India it is most commonly known as *babchi* or *bakuchi*, and in China as *buguzhi*. In English, it is sometimes referred to as fountain bush. Its seeds, seed oil, roots, and leaves have long been used in Ayurvedic and Chinese Medicine. Its primary use in traditional medicines has been in the treatment of skin ailments. However, it has also been used in disorders as varied as scorpion bite, indigestion, and asthma, and as an aphrodisiac.[1] It has been described as a treatment for insomnia and as a tonic nervine.[2]

Psoralea corylifolia contains a wide variety of unique flavones, coumarins, monoterpenes, chalcones, lipids, resins, stigmasteroids, and flavonoids, many of whose names reflect back to the genus and species, or Indian language terms for the plant. There are psoralens, hydroxypsoralens, corylifols, bavahins, bavichanones, bachuisoflavones, and other such namesakes. These phytochemicals exert a variety of chemical and pharmacological effects. Many are strong antioxidants, and some are estrogenic. Bakuchiol, for example, shows low micromolar affinity for the ER receptor. These various molecules also display anti-inflammatory, hypoglycemic, and, in preclinical studies, antidepressant-like effects.[3]

7.52.1 ANTIOXIDANT

Like many herbs rich in phenolic compounds, *Psoralea corylifolia* has been found to have potent antioxidant effects. Extract of the plant was an effective reactive oxygen species scavenger in the standard *in vitro* DPPH assay.[4] Bakuchiol, bavachinin, bavachin, isobavachin, and isobavachalcone isolated from the seeds of *Psoralea corylifolia* showed strong antioxidative activities in rat liver microsomes and mitochondria. They inhibited CCl_4-induced lipid peroxidation in microsomes and prevented NADH-dependent and ascorbate-induced mitochondrial lipid peroxidation.[5]

In vivo, administration of bakuchiol to rats that had received CCl_4 attenuated lipid peroxidation and intracellular glutathione depletion in hepatocytes. Reduced serum levels of aspartate transaminase and alanine transaminase demonstrated the ability of the phytochemical to spare liver tissue from the CCl_4-induced oxidative damage.[6]

7.52.2 ANTI-INFLAMMATORY

The various phytochemicals in *Psoralea corylifolia* are reported to have significant anti-inflammatory effects acting through various pathways. Bavachin A from *Psoralea corylifolia* attenuates the inflammatory effects of carrageenan in rat paw edema, an effect largely mediated by COX2.[7]

Several phytochemicals extracted from *Psoralea corylifolia* have also been found to have potent, albeit differential, anti-inflammatory effects in cultured macrophages exposed to lipopolysaccharide. TNF-α level was significantly decreased by corylifol A and bakuchiol and IL-1β secretion was significantly decreased by psoralen, corylifol A, and neobavaisoflavone, whereas IL-6 expression was decreased by corylifol A neobavaisoflavone and bakuchiol. Thus, the unfractionated, crude extract would be expected to attenuate the activity of a range of inflammatory cytokines.[8] In a similar *in vitro* study, the *Psoralea corylifolia* components, angelicin, isobavachalcone, and bakuchiol suppressed lipopolysaccharide (LPS)-stimulated nitric oxide production in microglia from mouse brain. Neobavaisoflavone and bakuchiol also spared mouse hippocampal cells from hydrogen peroxide-induced cell death.[9] Thus, brain tissue is responsive to the anti-inflammatory effects of *Psoralea corylifolia*.

In vivo, in mice, isobavachalcone attenuates MPTP-induced Parkinson's disease. It appears to do so by inhibiting over-activation of microglia and decreasing the expression of IL-6 and IL-1β. It was found, *in vitro*, that isobavachalcone produces these neuroprotective effects, in part, by blocking the NF-κB pathway in neurons.[10]

7.52.3 ANTIDIABETIC/ANTI-METABOLIC SYNDROME

The whole plant of *Psoralea corylifolia* is traditionally used in the treatment of diabetes. In mice with dexamethasone-induced insulin resistance, extract of *Psoralea corylifolia* decreased serum glucose and triglyceride levels. In liver tissue, treatment increased hepatic levels of glutathione and the antioxidant enzymes superoxide dismutase and catalase. Lipid peroxidation in liver tissue was reduced.[11]

Many of the prenylated flavonoids isolated from the seeds of *Psoralea corylifolia* have been found to act as PPAR-γ agonists.[12] PPAR-γ regulates fatty acid storage and glucose metabolism and is a target of the "glitazone" medications that treat diabetes. There is evidence that pioglitazone and other PPAR-γ agonists exert antidepressant effects in animals and humans.[13]

The neuropathology of Alzheimer's disease is exacerbated by diabetes, insulin resistance, and over-activation of GSK-3β. Those factors can also contribute to MDD.[14] In evaluating the effects of *Psoralea corylifolia* in the treatment of Alzheimer's disease it was found that the total prenylflavonoid extraction from the herb significantly improved cognitive performance of mice in a mouse model of the disease. This extract of glycogen synthase inhibited GSK-3β in brain tissue, as well as decreasing the expression of the pro-inflammatory cytokines TNFα, IL-6, and IL-1β.[15]

7.52.4 PRECLINICAL ANTIDEPRESSANT-LIKE EFFECTS

Preclinical evaluations of *Psoralea corylifolia* in rodents provides strong evidence of antidepressant effects. Orally administered crude alcoholic extract of *Psoralea corylifolia* seeds decreased immobility of mice in the forced swim and tail suspension tests. The highest dose, 300 mg/kg was effective in the forced swim test on the first day of treatment, whereas the 75 mg dose became effective after three days. Similarly, the 300 mg/kg dose became effective in the tail suspension test on the third day of treatment, whereas the 75 mg dose became effective after one week. The 300 mg/kg was as effective as fluoxetine and amitriptyline with the same latency to effect.[16]

Both acute and three days of psoralidin isolated from the seeds of *Psoralea corylifolia* decreased immobility time of mice in the forced swim test. After the three days of treatment, psoralidin increased levels of 5-hydroxytryptamine and its metabolite, 5-hydroxyindoleacetic acid, in brain tissue. Psoralidin also attenuated rises in serum corticosterone induced by swimming stress.[17]

Subchronic administrations of the total extract of furocoumarins, including psoralen, from seeds of *Psoralea corylifolia* reduced immobility of mice in the forced swim test. The 30 mg/kg dose for two weeks was found more effective than either amitriptyline or fluoxetine. The higher doses of furocoumarins significantly reduced activities of MAO-A and MAO-B in brain tissue and blocked the rise in swim-induced serum corticosterone.[18] The furocoumarin-rich extract also showed anti-depressant-like effects in mice that had been subjected to chronic unpredictable stress. Administration of the extract over several days restored preference for sucrose in stressed animals. Treatment also reduced serum levels of the stress hormone, corticosterone, and preserved activity of the antioxidant enzyme superoxide dismutase in liver. In brain tissue, the furocoumarins reversed stressed-induced increases in activities of MAO-A and MAO-B.[19] A subsequent study found that the ethanolic extract of *Psoralea corylifolia*, as well as the flavanone bavachinin purified from it, selectively inhibit MAO-B over MAO-A.[20]

As in mice, extract of *Psoralea corylifolia* seed reduced immobility time of rats in the forced swim test. Treatment also maintained monoamine levels in the brain, as well as levels of glutathione and activities of antioxidant enzymes.[21] The specific phytochemical, Δ3,2-Hydroxybakuchiol, purified from *Psoralea corylifolia* showed antidepressant-like effects in rats that had been subjected to chronic unpredictable stress. Treatment reduced immobility in the forced swim and tail suspension tests, and restored preference for sucrose. Further evaluation with microdialysis showed that Δ3,2-hydroxybakuchiol increased dopamine and norepinephrine concentration in striatum. Those findings were noted to be consistent with previous findings of Δ3,2-hydroxybakuchiol blocking dopamine and norepinephrine transporters in pheochromocytoma cells. Indeed, the IC50 values were said to show potency similar to that of bupropion.[22]

7.52.5 HUMAN ANTIDEPRESSANT EFFECTS

Despite many references to uses for depression, strong animal data, and clear mechanisms by which *Psoralea corylifolia* could act as an antidepressant, there have been no published studies of the use of the herb for the treatment of MDD in human subjects.

7.52.6 DOSAGE

The recommended dosage of powdered seed of *Psoralea corylifolia* is 3–6 g in split dose.[23]

7.52.7 TOXICITY

The LD50 of dried methanolic extract of *Psoralea corylifolia* in mice was found to be more than 10 g/kg by intraperitoneal administration, and more than 20 g/kg by oral administration.[24] Extrapolated to the weight of a 75 kg human adult, this dose would be approximately 700 g.

Psoralea corylifolia is rich in furocoumarins, which can cause photosensitivity, particularly if applied to the skin. However, ingestion of high doses of the herb can also cause photosensitivity. In one report, an individual consuming a daily dose of 30 g of seed for the self-treatment of vitiligo, developed erythema, itching, and blistering that stopped after cessation of the herb.[25] There have also been several case reports of acute cholestatic hepatitis in individuals taking very high doses of *Psoralea corylifolia* seed.[26]

7.52.8 SAFETY IN PREGNANCY

In animal studies, *Psoralea corylifolia* has been found to be toxic to both pregnant females and embryos. The herb should not be used during pregnancy.[27]

7.52.9 DRUG INTERACTIONS

Extract of *Psoralea corylifolia* has been found to inhibit CYP3A4.[28] A variety of neuroleptics, including aripiprazole, haloperidol, pimozide, and risperidone, are metabolized by this enzyme.[29] The enzyme is also involved in the metabolism of SSRIs and tricyclic antidepressants.[30] Thus, chronic use of *Psoralea corylifolia* could affect blood levels of a variety of important psychiatric medications.

Special note: Due indiscriminate and illegal collections, as well as destruction of natural habitat, *Psoralea corylifolia* has been included in the world's endangered list of plants.[31]

REFERENCES

1. Khushboo PS, Jadhav VM, Kadam VJ, Sathe NS. *Psoralea corylifolia* Linn.— "Kushtanashini". *Pharmacogn Rev.* 2010;4(7):69–76.
2. Rajput S. Brief review of bakuchi (*Psoralea corylifolia* Linn.) and its therapeutic uses. *Ayurpharm Int J Ayurveda Allied Sci.* 2014;3(11):322–330.
3. Alam F, Khan GN, Bin Asad MHH. *Psoralea corylifolia* L: ethnobotanical, biological, and chemical aspects: a review. *Phytother Res.* 2018;32:597–615.
4. Shinde AN. Determination of isoflavone content and antioxidant activity in *Psoralea corylifolia* L. callus cultures. *Food Chem.* 2010;118(1):128–132.
5. Haraguchi H, Inoue J, Tamura Y, Mizutani K. Antioxidative components of *Psoralea corylifolia* (Leguminosae). *Phtyother Res.* 2002;16(6):539–544.
6. Park E-J, Zhao Y-Z, Kim Y-C, Sohn D H. Protective effect of (S)-bakuchiol from *Psoralea corylifolia* on rat liver injury in vitro and in vivo. *Planta Med.* 2005;71(6): 508–513.
7. Kapoor LD. *Handbook of Ayurvedic Medicinal Plants.* Boca Raton, Florida: CRC Press; 2001. p. 274–275.
8. Chai L. Inhibition of inflammatory cytokines in the LPS-induced RAW264.7 cells by four components from Fructus Psoraleae. *Tradit Chin Drug Res Clin Pharmacol.* 2013-04.
9. Kim YJ, Lim HS, Lee J, Jeong SJ. Quantitative analysis of *Psoralea corylifolia* Linne and its neuroprotective and anti-neuroinflammatory effects in HT22 hippocampal cells and BV-2 microglia. *Molecules.* 2016;21(8):1076.
10. Jing H, Wang S, Wang M, Fu W, Zhang C, Xu D. Isobavachalcone attenuates MPTP-induced Parkinson's disease in mice by inhibition of microglial activation through NF-κB pathway. *PLoS One.* 2017;12:e0169560.
11. Tayade PM, Jagtap SA, Borde S, Chandrasekar N, Joshi A. Effect of *Psoralea corylifolia* on dexamethasone-induced insulin resistance in mice. *J King Saud Univ-Sci.* 2012;24:251–255.
12. Ma S, Huang Y, Zhao Y, et al. Prenylflavone derivatives from the seeds of *Psoralea corylifolia* exhibited PPAR-γ agonist activity. *Phytochem Lett.* 2016;16:213–218.
13. Colle R, De Larminat D, Rotenberg S, et al. PPAR-γ agonists for the treatment of major depression: a review. *Pharmacopsychiatry.* 2017;50(2):49–55.
14. Mendelson SD. *Metabolic Syndrome and Psychiatric Illness: Interactions, Pathophysiology, Assessment and Treatment.* London, UK: Academic Press; 2008.
15. Chen ZJ, Yang YF, Zhang YT, Yang DH. Dietary total prenylflavonoids from the fruits of *Psoralea corylifolia* L. prevents age-related cognitive deficits and down-regulates Alzheimer's markers in SAMP8 mice. *Molecules.* 2018;23(1):196.
16. Bhawya D, Anilakumar KR, Khanum F. Effect of *Psoralea corylifolia* extract on physically induced depression in mice. *Defence Life Sci J.* 2017;2(2):199–205.
17. Yi L-T, Li YC, Pan Y, et al. Antidepressant-like effects of psoralidin isolated from the seeds of *Psoralea Corylifolia* in the forced swimming test in mice. *Prog Neuropsychopharmacol Biol Psychiatry.* 2008;32(2):510–519.
18. Chen Y, Kong LD, Xia X, Kung HF, Zhang L. Behavioral and biochemical studies of total furocoumarins from seeds of *Psoralea corylifolia* in the forced swimming test in mice. *J Ethnopharmacol.* 2005;96(3):451–459.
19. Chen Y, Wang HD, Xia X, Kung HF, Pan Y, Kong LD. Behavioral and biochemical studies of total furocoumarins from seeds of *Psoralea corylifolia* in the chronic mild stress model of depression in mice. *Phytomedicine.* 2007;14(7–8):523–529.
20. Zarmouh NO, Mazzio EA, Elshami FM, Messeha SS, Eyunni SV, Soliman KF. Evaluation of the inhibitory effects of bavachinin and bavachin on human monoamine oxidases A and B. *Evid-Based Complement Altern.* 2015:Article ID 852194.

21. Selvan AT, Dwivedi K, Subramanian NS, Ramadevi M, Prasad B, Muthu SK. Pharmacological and neurobiochemical evidence for antidepressant effect of *Psoralea corylifolia* seeds extract. *Adv Pharmacol Toxicol.* 2014;15(3):31–50.

22. Zhao G, Guo L, Huang W, Hu JL. Δ3,2-Hydroxybakuchiol attenuates depression in multiple rodent models possibly by inhibition of monoamine transporters in brain. *Evid-Based Complement Altern.* 2018:Article ID 1325141.

23. Anonymous. *Pharmacopoeia of India.* 2nd ed. New Delhi: Govt. of India, Ministry of Health and Family Welfare; 1978. p. 42.

24. Kim S.-J. Composition containing an extract of *fructus psoraleae* for inhibiting anxiety and depression, improving memory and treating dementia. US Patent 20060099284A1; 2005.

25. Maurice PD, Cream JJ. The dangers of herbalism. *BMJ.* 1989 Nov 11;299(6709):1204.

26. Hwang SH, Park JA, Jang YS, et al. A case of acute cholestatic hepatitis caused by the seeds of Psoralea-corylifolia. *Korean J Hepatol.* 2001;7(3):341–344.

27. Xu M, Xiao Y, Tian XY, et al. Embryotoxicity of *Psoralea corylifolia* L.: in vivo and in vitro studies. *Birth Defects Res B Dev Reprod Toxicol.* 2012;95(6):386–394.

28. Liu Y, Flynn TJ. CYP3A4 inhibition by *Psoralea corylifolia* and its major components in human recombinant enzyme, differentiated human hepatoma HuH-7 and HepaRG cells. *Toxicol Rep.* 2015;2:530–534.

29. van der Weide K, van der Weide J. The influence of the CYP3A4*22 polymorphism and CYP2D6 polymorphisms on serum concentrations of aripiprazole, haloperidol, pimozide, and risperidone in psychiatric patients. *J Clin Psychopharmacol.* 2015;35(3):228–236.

30. Haduch A, Wójcikowski J, Daniel WA. The effect of tricyclic antidepressants, selective serotonin reuptake inhibitors (SSRIs) and newer antidepressant drugs on the activity and level of rat CYP3A. *Eur Neuropsychopharmacol.* 2006;16(3):178–186.

31. Baskaran P, Jayabalan N. Rapid micropropagation of *Psoralea corylifolia* L. using nodal explants cultured in organic additive-supplemented medium. *J Hortic Sci Biotechnol.* 2007;82:908–913.

7.53 RHODIOLA ROSEA

Rhodiola rosea is a perennial flowering plant in the family Crassulaceae. It grows naturally in wild Arctic and mountainous regions of Europe, Asia, and North America. It has been popular in traditional medical systems in Eastern Europe and Asia, with a reputation for stimulating the nervous system, decreasing depression, enhancing work performance, and eliminating fatigue. It has thus been categorized by some as an adaptogen due to its ability to increase resistance to a variety of stressors. Its benefits have been said to include antidepressant, anticancer, cardioprotective, and central nervous system enhancement. It has particularly been recommended for asthenic conditions – decline in work performance, sleep difficulties, poor appetite, irritability, hypertension, headaches, and fatigue – as consequences of intense physical or intellectual strain.[1]

Rhodiola rosea rhizomes contain essential oils, fats, waxes, sterols, glycosides, simple organic acids, and phenolics. Among the specific components thought responsible for its benefits are geraniol, rosavin, rosarian, rosiridol, rhodiolin, rosin, rosiridin, p-tyrosol, salidroside, various proanthocyanidins, and gallic acid derives.[2] Rosavin and salidroside are the constituents used for standardization of extracts.

7.53.1 ANTIOXIDANT

Aqueous extract of *Rhodiola rosea* protected cultured human keratinocytes from the damage of several different oxidative insults, including exposure to Fe(II)/ascorbate, Fe(II)/H_2O_2, and tert-butyl-hydroperoxide. Treatment with *Rhodiola rosea* also maintained levels of glutathione and activities of catalase, superoxide dismutase, glutathione peroxidase, and glutathione reductase in those cells. It decreased levels of lipid peroxidation as assessed by the TBARS method.[3]

Two specific polysaccharides extracted from *Rhodiola rosea* showed potent *in vitro* antioxidant effects in scavenging DPPH, hydroxyl, and superoxide anion radicals. They also exhibited *in vivo* hepatoprotective effects from the oxidative damage of CCl_4 in mice. Mice pretreated with the polysaccharides prior to injection of CCl_4, had reduced serum levels of transaminases. In their liver tissue, levels of glutathione, and activities of catalase and superoxide dismutase were spared.[4]

7.53.2 ANTI-INFLAMMATORY

In animal studies, *Rhodiola rosea* has also been shown to exert a potent anti-inflammatory effect. In the carrageenan-induced paw edema, formaldehyde-induced arthritis, and nystatin-induced paw edema models of inflammation rats, extracts of *Rhodiola rosea* exhibited inhibited acute and subacute inflammation. It appeared to do so by inhibiting the activities of the enzymes, cyclooxygenase-2 and Phospholipase A2.[5]

Peripherally administered *Rhodiola rosea* and its constituents do appear to enter the brain and produce central anti-inflammatory effects. The expressions of the pro-inflammatory factors iNOS, IL-1β, and TNF-α in the prefrontal cortex of mice were substantially suppressed by the oral administration of *Rhodiola rosea* to mice administered LPS by intraperitoneal injection. Similar effects were seen in microglial cell culture after exposure to LPS. Crude extracts of the herb were less effective than the isolated components, rosin and salidroside. The levels of the mediators of inflammation, phosphorylated MAPK, pJNK, and pp38 were also decreased by the treatment with isolated rosin and salidroside in cortical cell cultures exposed to high concentrations of glutamate.[6]

Rhodiola rosea has been shown to offer at least systemic ant-inflammatory effects in humans. Extracts of *Rhodiola rosea* decreased serum levels of the inflammatory C-reactive protein and creatinine kinase in otherwise healthy human volunteers after engaging in prolonged exhausting physical exercise.[7]

7.53.3 ANTIDIABETIC/ANTI-METABOLIC SYNDROME

The ability of *Rhodiola rosea* to enhance performance may due in part to its effects on insulin sensitivity and glucose metabolism. Salidroside, one of the primary biologically active components of *Rhodiola rosea,* enhances insulin-mediated Akt activation and glucose uptake in rat skeletal muscle cells *in vitro.*[8]

In the Zucker diabetic fatty rat, an animal whose metabolic anomalies simulate the condition of metabolic syndrome in humans, treatment with *Rhodiola crenulata* decreased fasting plasma insulin levels and enhanced insulin sensitivity in the homeo-stasis model assessment of insulin resistance. The crenulata species of *Rhodiola rosea* is closely related to rosea and contains significant concentrations of active sali-droside.[9] Indeed, similar to the effects of ketamine and many substances with anti-depressant effects, salidroside ameliorates insulin resistance through activation of a mitochondria-associated AMPK/PI3K/Akt/GSK-3β pathway.[10] Aqueous extracts of *Rhodiola rosea* also reduce serum glucose levels in rats made diabetic by treatment with streptozotocin. Interestingly, this effect is eliminated by adrenalectomy, and is thought to be mediated by stimulation of adrenal release of β-endorphin.[11]

7.53.4 Preclinical Antidepressant-Like Effects

Standard preclinical studies in animals have suggested antidepressant effects of *Rhodiola rosea* and its major phytochemical constituents. In female rats that had been subjected to weeks of chronic unpredictable stress, chronic administration of extract of *Rhodiola rosea* standardized to 3% rosavin and 1% salidroside restored consumption of sucrose solution, locomotor and exploratory activities, body weight, and normal estrous cycle length.[12]

In the olfactory bulbectomy-induced depression model in rats, salidroside exhib-ited antidepressant effects by alleviating OBX-induced hyperactivity in the open field test, and decreasing immobility time in tail suspension and forced swim tests. Aside from reversing behavioral deficits, salidroside reduced TNF-α and IL-1β lev-els in hippocampus, but increase glucocorticoid receptor (GR) and brain-derived neurotrophic factor (BDNF) expression in this area. It also attenuated corticotropin-releasing hormone expression in hypothalamus, significantly reducing serum levels of corticosterone.[13]

Five days of treatment with salidroside also attenuated the depression-like behav-iors of mice injected with the inflammation-inducing LPS. It did so in a manner comparable to fluoxetine. Salidroside reduced levels of NF-kβ, and attenuated the LPS-induced decreases in serotonin and norepinephrine in the frontal cortex. This major component of *Rhodiola rosea* also increased expression of BNDF and TrkB in the hippocampus.[14]

In one study, salidroside was found to be the primary active component in the antidepressant effect of *Rhodiola rosea* extract, as assayed by the forced swim test. Rosavin, rosarin, rosin, cinnamic alcohol, cinnamaldehyde, and cinnamic acid were ineffective.[15]

7.53.5 Human Antidepressant Effects

Studies on the effects of *Rhodiola rosea* on MDD in humans are few and inconsis-tent. In a head-to-head comparison with sertraline, *Rhodiola rosea* did not exhibit significant antidepressant effect, though the antidepressant did.[16] Also, the antide-pressant effects of *Rhodiola rosea* in humans with mild to moderate degrees of MDD were tested in a randomized double-blind placebo-controlled study. After six weeks of daily dosing, standardized extract SHR-5 of rhizomes of *Rhodiola rosea* L. relieved overall depression, insomnia, emotional instability, and somatization as per the Beck Depression Inventory and Hamilton Rating Scale for Depression (HAMD) questionnaires.[17]

Rhodiola rosea acquired its classification as an adaptogen through its apparent ability to relieve stress and buffer the hypothalamic-pituitary-adrenal axis. These effects of *Rhodiola rosea* were evaluated in human adults suffering "fatigue syndrome" or "other reactions to severe stress," as defined by the ICD criteria. Extract of *Rhodiola rosea* containing salidroside, tyrosol, rosavin, and triandrin, or placebo were administered over 28 days. *Rhodiola rosea* produced a significant lessening in symptoms of "burnout" as measured by the Pines Burnout scale. The herb also caused a substantial decrease in the levels of salivary cortisol upon awakening.[18] A similar study, albeit without a control group, also found *Rhodiola rosea* to provide significant relief from symptoms of burnout. Subjects received daily doses of an ethanolic extract of *Rhodiola rosea* that was high in rosalin. Substantial improvements were seen in the exhaustion and depersonalization scales of the Maslach Burnout Inventory. Treatment improved symptoms of anxiety, exhaustion, feeling of heteronomy, impaired concentration, irritability, loss of zest for life, and somatic symptoms. It also increased sexual interest, frequency, and enjoyment of sex, and improved overall sexual function and satisfaction. In general, *Rhodiola rosea* led to more alertness, calmness, and "good mood."[19]

7.53.6 DOSAGE

Marciano and Vizniak recommend tincture (1:5, 40%) 1–3 ml three times a day, or standard extract (3% rosavin 1% salidroside) 300–600 mg a day.[20]

7.53.7 TOXICITY

The oral LD50 of *Rhodiola rosea* extract in mice is reported to be 3,360 mg/kg.[21] Extrapolated to the weight of a human adult of 75 kg, this would be approximately 230 g.

7.53.8 SAFETY IN PREGNANCY

Evidence on the safety and appropriateness of *Rhodiola rosea* supplementation during pregnancy and lactation is currently unavailable.[22]

7.53.9 DRUG INTERACTIONS

In vitro studies have shown *Rhodiola rosea* to inhibit CYP3A4 and P-glycoprotein.[23] Nonetheless, a recent review notes that human studies have given no indications that such interactions have clinical significance.[24]

REFERENCES

1. Kelly GS. *Rhodiola rosea*: a possible plant adaptogen. *Altern Med Rev.* 2001;6(3):293–302.

2. Panossian A, Wikman G, Sarris J. Rosenroot (*Rhodiola rosea*): traditional use, chemical composition, pharmacology and clinical efficacy. *Phytomedicine*. 2010;17:481–493.
3. Calcabrini C, De Bellis R, Mancini U, et al. *Rhodiola rosea* ability to enrich cellular antioxidant defences of cultured human keratinocytes. *Arch Dermatol Res*. 2010;302:191–200.
4. Xu Y, Jiang H, Sun C-Y, et al. Antioxidant and hepatoprotective effects of purified *Rhodiola rosea* polysaccharides. *Int J Biol Macromol*. 2018;117:167–178.
5. Bawa PAS, Khanum F. Anti-inflammatory activity of *Rhodiola rosea* – "a second-generation adaptogen". *Phytother Res*. 2009;23(8):1099–1102.
6. Lee Y, Jung C, Jang S, et al. Anti-inflammatory and neuroprotective effects of constituents isolated from *Rhodiola rosea*. *Evid-Based Complement Altern*. 2013;2013:514049.
7. Abidov M, Grachev S, Seifulla RD, Ziegenfuss TN. Extract of *Rhodiola rosea* radix reduces the level of C-reactive protein and creatinine kinase in the blood. *Bull Exp Biol Med*. 2004;138(1):63–64.
8. Li H-B, Ge Y, Zheng XX, Zhang L. Salidroside stimulated glucose uptake in skeletal muscle cells by activating AMP-activated protein kinase. *Eur J Pharmacol*. 2008;588(2–3):165–169.
9. Wang J-W, Rong X, Li W, Yang Y, Yamahara J, Li Y. *Rhodiola crenulata* root ameliorates derangements of glucose and lipid metabolism in a rat model of the metabolic syndrome and type 2 diabetes. *J Ethnopharmacol*. 2012;142(3):782–788.
10. Zheng T, Yang X, Wu D, et al. Salidroside ameliorates insulin resistance through activation of a mitochondria-associated AMPK/PI3K/Akt/GSK-3β pathway. *BJP*. 2015;172(13):3284–3301.
11. Niu C-S, Chen LJ, Niu HS. Antihyperglycemic action of rhodiola-aqeous extract in type1-like diabetic rats. *BMC Complement Altern Med*. 2014;14:20–29.
12. Mattioli L, Funari C, Perfumi M. Effects of *Rhodiola rosea* L. extract on behavioural and physiological alterations induced by chronic mild stress in female rats. *J Psychopharmacol*. 2008;23(2):130–142.
13. Yang S-J, Yu HY, Kang DY, et al. Antidepressant-like effects of salidroside on olfactory bulbectomy-induced pro-inflammatory cytokine production and hyperactivity of HPA axis in rats. *Pharmacol Biochem Behav*. 2014;124:451–457.
14. Zhu L, Wei T, Gao J, et al. Salidroside attenuates lipopolysaccharide (LPS) induced serum cytokines and depressive-like behavior in mice. *Neurosci Lett*. 2015;606:1–6.
15. Panossian A, Nikoyan N, Ohanyan N, et al. Comparative study of *Rhodiola* preparations on behavioral despair of rats. *Phytomedicine*. 2008;15(1–2):84–91.
16. Mao JJ, Xie SX, Zee J, et al. *Rhodiola rosea* versus sertraline for major depressive disorder: a randomized placebo-controlled trial. *Phytomedicine*. 2015;22(3):394–399.
17. Darbinyan V, Aslanyan G, Amroyan E, Gabrielyan E, Malmström C, Panossian A. Clinical trial of *Rhodiola rosea* L. extract SHR-5 in the treatment of mild to moderate depression. *Nord J Psychiatry*. 2007;61(6):343–348.
18. Olsson EMG, von Schéele B, Panossian AG. A randomized, double-blind, placebo-controlled, parallel-group study of the standardized extract SHR-5 of the roots of *Rhodiola rosea* in the treatment of subjects with stress-related fatigue. *Planta Med*. 2009;75:105–112.
19. Kasper S, Dienel A. Multicenter, open-label, exploratory clinical trial with *Rhodiola rosea* extract in patients suffering from burnout symptoms. *Neuropsychiatr Dis Treat*. 2017;13:889–898.
20. Marciano M, Vizniak, N.A. *Botanical Medicine*. Toronto, Ontario: Prohealth; 2016. p. 297.
21. Kurkin V, Zapesochnaya G. Chemical composition and pharmacological characteristics of *Rhodiola rosea*. *J Med Plants*. 1985:1231–1445.

22. Kennedy DA, Lupattelli A, Koren G. Safety classification of herbal medicines used in pregnancy in a multinational study. *BMC Complement Altern Med.* 2016;16:102.
23. Hellum BH, Tosse A, Hoybakk K, Thomsen M, Rohloff J, Nilsen OG. Potent inhibition of CYP3A4 and P-glycoprotein by *Rhodiola rosea*. *Planta Med.* 2010;76(4):331–338.
24. Gerbarg PL, Brown RP. Integrating *Rhodiola rosea* in Clinical Practice. In: Gerbarg PL, Muskin PR, Brown RP, editors. *Complimentary and Intergrative Treatments in Psychiatric Practice.* Arlington, VA: American Psychiatric Association Publishing; 2017. p. 135–141.

7.54   *ROSMARINUS OFFICINALIS* (ROSEMARY)

Rosmarinus officinalis, more commonly known as rosemary, is a member of the mint family *Lamiaceae* that is native to the Mediterranean region. It is a perennial herb with fragrant, evergreen, needle-like leaves and white, pink, purple, or blue flowers. Its medicinal properties were known to the ancients. The Greeks and Romans believed it strengthened the memory. Hippocrates, Galen, and Dioscorides prescribed the herb for various ailments, and it was an essential part of the apothecary's repertoire during the Renaissance.[1] *Rosmarinus officinalis* has found wide use in traditional herbal medicine, and has been used to treat depression, anxiety, tiredness, defective memory, rheumatic complaints, circulatory disorders, headache, menstrual disorders, nervous menstrual complaints, sprains, and bruises.

Among the many phytochemicals found in *Rosmarinus officinalis* are rosmarinic acid, camphor, linalool, caffeic acid, chlorogenic acid, ursolic acid, apigenin, luteolin, pinene, borneol, betulinic acid, carnosic acid, and carnosol.[2]

7.54.1   ANTIOXIDANT

Extract of *Rosmarinus officinalis* and two of its primary phytochemicals, carnosic acid and rosmarinic acid, all showed significant antioxidant effects *in vitro* in the DPPH and ABTS radical-scavenging assays and in the ferric thiocyanate test.[3] In a similar study, the essential oil of *Rosmarinus officinalis* was found to be slightly more effective in the DPPH test than the single flavonoid components, 8-cineole, α-pinene, and β-pinene.[4]

In vivo, the crude methanolic extract of *Rosmarinus officinalis* protected rats against the oxidative damage of CCl_4. Administration of the extract prevented the elevations of serum bilirubin and transaminases. Malonylaldehyde was reduced in the liver tissue of treated rats, as were total lipid peroxides and nitric oxide levels. Pathology studies of liver tissue showed the extract of *Rosmarinus officinalis* also reduced the degree and extent of fibrosis and necrosis.[5]

7.54.2   ANTI-INFLAMMATORY

Extracts of *Rosmarinus officinalis* exhibit potent anti-inflammatory effects. Essential oil of *Rosmarinus officinalis* reduced carrageenan-induced pleurisy and paw edema in rats. It also exhibited significant antinociceptive effects.[6] In mouse skin, extracts of *Rosmarinus officinalis* leaves reduced the expression of IL-1β and TNF-α, and inhibited COX-2. Ethanolic extract, carnosic acid and carnosol also significantly inhibited the overproduction of nitric oxide (NO) in the RAW 264.7 murine macrophage cell line.[7]

Phytochemicals in *Rosmarinus officinalis* have also been shown to dampen pathological processes that underlie neuroinflammation. Carnosol from *Rosmarinus officinalis* inhibits the TNF-α-induced signaling pathways through inhibition of inhibitor of nuclear factor kappa-B (IKK-β) activity as well as the upregulation of HO-1 expression. At fairly modest concentrations, carnosol upregulates Nrf2 and HO-1 leading to downregulation of the inflammatory responses mediated by TNF-α, prostaglandin E-2, and nitrite.[8]

In vitro, phenolic compounds in *Rosmarinus officinalis* reduce the inflammatory effects of lipopolysaccharide on microglia,[9] and stimulate Nrf2 and HO-1 antioxidant protein in those same brain glial cells.[10] Carnosic acid has been reported to upregulate the expression of NGF and BDNF in human glioblastoma and dopaminergic neuronal cell lines *in vitro*.[11,12] Importantly, components of *Rosmarinus officinalis* administered to intact animals have shown ability to reverse deposition of amyloid protein[13] and mitigate drug-induced dementia.[14] Thus, these substances do cross the blood–brain barrier to produce effects in brain tissue.

7.54.3 ANTIDIABETIC/ANTI-METABOLIC SYNDROME

Rosmarinus officinalis has been shown to lower blood glucose and cholesterol levels and mitigate weight gain. In a recent study to investigate the mechanism of action, *Rosmarinus officinalis* was found to increase glucose consumption in HepG2 cells. It did so by activating AMP-activated protein kinase, an enzyme that enhances glucose and fatty acid uptake and oxidation when cellular energy is low. *Rosmarinus officinalis* also enhanced the activity of SIRT1 (which is enhanced by resveratrol), PPARγ coactivator 1α (which increases mitochondrial function), glucose-6-phosphatase, and the low-density lipoprotein receptor (which enhances uptake of cholesterol into cells.[15] Metabolic Syndrome, with insulin resistance, lipid metabolism, and inflammation, is an important risk factor for MDD. *Rosmarinus officinalis* is a source of phenolic phytochemicals that reduce the damage of metabolic syndrome through anti-oxidant, anti-inflammatory, hypoglycemic, hypolipidemic, hypotensive, anti-atherosclerotic, anti-thrombotic, hepatoprotective, and hypocholesterolemic effects.[16]

7.54.4 PRECLINICAL ANTIDEPRESSANT-LIKE EFFECTS

Animal studies have shown *Rosmarinus officinalis* to possess antidepressant-like effects. Various extractions from Rosmarinus officinalis, including essential oil and oil-free fractions, all produced antidepressant-like effects in mice in the standard tail suspension and open-field tests. Two isolated phytochemicals, carnosol and betulinic acid, were also found effective.[17]

In another study, hydroalcoholic extracts of *Rosmarinus officinalis* improved performance in the forced swimming and tail suspension tests in mice. These antidepressant-like effects were prevented by pretreatment with the serotonin synthesis inhibitor, p-chlorophenylalanine, or antagonists of various central monoaminergic receptors. The authors then argued for monoaminergic mediation.[18] Extracts of *Rosmarinus officinalis* also reversed the depression-like effects of olfactory bulbectomy in mice in a manner similar to that of fluoxetine.[19]

Rosmanol, cirsimaritin, and salvigenin extracted from *Rosmarinus officinalis* were all found to produce antidepressant-like effects in mice in the tail suspension and forced swim tests. Interestingly, all three phytochemicals also exhibited anxiolytic effects per the elevated plus maze and light/dark box paradigms. Those anxiolytic effects, were comparable to those of diazepam, were not blocked by flumazenil, but were inhibited by pentylenetetrazol, suggesting a mode of action through GABAA receptors, but at a site other than the high affinity benzodiazepine binding site.[20]

7.54.5 HUMAN ANTIDEPRESSANT EFFECTS

Despite the compelling evidence that *Rosmarinus officinalis* produces antidepressant-like effects in animals through well-known neurochemical mechanisms, there is only one study of the effects of *Rosmarinus officinalis* on mood in humans in the scientific literature. In this randomized, double-blind study, it was found that the administration of 500 mg of *Rosmarinus officinalis* twice a day for one month to normal university students improved anxiety and, to a minor degree, mood per the Hospital Anxiety and Depression Scale.[21]

7.54.6 DOSAGE

Hoffman states the standard dosing of *Rosmarinus officinalis* to be 1–2 ml three times a day of tincture (1:5, 40%), or infusion of 1–2 teaspoons per cup of water, taken three times a day.[22]

7.54.7 TOXICITY

Neither methanolic or aqueous extracts of *Rosmarinus officinalis* were found to be toxic in mice in oral doses of up to 5,000 mg of dried extract per kg of weight. Thus, the LD50 was deemed to be above 5,000 mg/kg.[23] Extrapolated for a human adult of 75 kg, this would be approximately 350 g. Oral administration of essential oil up to 2,000 mg/kg also resulted in no animal deaths.[24]

7.54.8 SAFETY IN PREGNANCY

The *Botanical Safety Handbook* notes conflicting evidence concerning the safety of *Rosmarinus officinalis* during pregnancy. Their conclusion is that while the dried herb and decoction are likely safe, the use of the essential oil should be avoided during pregnancy.[25]

7.54.9 DRUG INTERACTIONS

The essential oil of *Rosmarinus officinalis*, its dried leaves, and various extracts have been found to induced several P450 enzymes, particularly the CYP2B family.[26]

Among the major drug substrates the CYP2B family are bupropion, cyclophos-phamide, ifosfamide, pethidine, ketamine, and propofol.[27] The *Botanical Safety Handbook* notes no known interactions of *Rosmarinus officinalis* with commonly used medications.

REFERENCES

1. Begum A, Sandhya S, Vinod KR, Reddy S, Banji D. An in-depth review on the medicinal flora *Rosmarinus officinalis* (Lamiaceae). *Acta Sci Pol Technol Aliment.* 2013;12(1):61–73.
2. Vallverdú-Queralt A, Regueiro J, Martínez-Huélamo M, Alvarenga JF, Leal LN, Lamuela-Raventos RM. A comprehensive study on the phenolic profile of widely used culinary herbs and spices: rosemary, thyme, oregano, cinnamon, cumin and bay. *Food Chem.* 2014;154:299–307.
3. Erkan N, Ayranci G, Ayranci E. Antioxidant activities of rosemary (*Rosmarinus Officinalis* L.) extract, blackseed (*Nigella sativa* L.) essential oil, carnosic acid, rosmarinic acid and sesamol. *Food Chem.* 2008;110:76–82.
4. Wang W, Wu N, Zu YG, Fu YJ. Antioxidative activity of *Rosmarinus officinalis* L. essential oil compared to its main components. *Food Chem.* 2008;108(3):1019–1022.
5. Gutiérrez R, Alvarado JL, Presno M, Pérez-Veyna O, Serrano CJ, Yahuaca P. Oxidative stress modulation by *Rosmarinus officinalis* in CCl₄-induced liver cirrhosis. *Phytother Res.* 2010;24:595–601.
6. Takak, I, Bersani-Amado LE, Vendruscolo A, et al. Anti-inflammatory and antinoci-ceptive effects of *Rosmarinus officinalis* L. essential oil in experimental animal models. *J Med Food.* 2008;11(4):741–746.
7. Mengoni ES, Vichera G, Rigano LA, et al. Suppression of COX-2, IL-1β and TNF-α expression and leukocyte infiltration in inflamed skin by bioactive compounds from *Rosmarinus officinalis* L. *Fitoterapia.* 2011;82(3):414–421.
8. Foresti R, Bains SK, Pitchumony TS, et al. Small molecule activators of the Nrf2-HO-1 antioxidant axis modulate heme metabolism and inflammation in BV2 microglia cells. *Pharmacol Res.* 2013;76:132–148.
9. Kuhlmann A, Rohl C. Phenolic antioxidant compounds produced by in vitro. Cultures of Rosemary (*Rosmarinus officinalis*.) and their anti-inflammatory effect on lipopoly-saccharide-activated microglia. *Pharm Biol.* 2006;44(6):401–410.
10. de Oliveira MR. The dietary components carnosic acid and carnosol as neuroprotective agents: a mechanistic view. *Mol Neurobiol.* 2016;53(9):6155–6168.
11. Kosaka K, Yokoi T. Carnosic acid, a component of rosemary (*Rosmarinus officinalis* L.), promotes synthesis of nerve growth factor in T98G human glioblastoma cells. *Biol Pharm Bull.* 2003;26:1620–1622.
12. Park JA, Kim S, Lee SY, et al. Beneficial effects of carnosic acid on dieldrin-induced dopaminergic neuronal cell death. *Neuroreport.* 2008;19:1301–1304.
13. Rasoolijazi H, Azad N, Joghataei MT, Kerdari M, Nikbakht F, Soleimani M. The protective role of carnosic acid against beta-amyloid toxicity in rats. *Sci World J.* 2013;2013:Article ID 917082.
14. Ozarowski M, Mikolajczak PL, Bogacz A, et al. *Rosmarinus officinalis* L. leaf extract improves memory impairment and affects acetylcholinesterase and butyrylcholinester-ase activities in rat brain. *Fitoterapia.* 2013;91:261–271.
15. Tu Z, Moss-Pierce T, Ford P, Jiang TA. Rosemary (*Rosmarinus officinalis* L.) extract regulates glucose and lipid metabolism by activating AMPK and PPAR pathways in HepG2 cells. J Agric Food Chem. 2013;61(11):2803–2810.

16. Hassani FV, Shirani K, Hosseinzadeh H. Rosemary (*Rosmarinus officinalis*) as a potential therapeutic plant in metabolic syndrome: a review. *NS Arch Pharmacol.* 2016;389(9):931–949.
17. Machadoa DG, Cunha MP, Neis VB, et al. Antidepressant-like effects of fractions, essential oil, carnosol and betulinic acid isolated from *Rosmarinus officinalis* L. Food Chem. 2013;136(2):999–1005.
18. Machado DG, Bettio LEB, Cunha MP, et al. Antidepressant-like effect of the extract of *Rosmarinus officinalis* in mice: involvement of the monoaminergic system. *Prog Neuropsychopharmacol Biol Psychiatry.* 2009;33:642–650.
19. Machado DG, Cunha MP, Neis VB, et al. *Rosmarinus officinalis* L. hydroalcoholic extract, similar to fluoxetine, reverses depressive-like behavior without altering learning deficit in olfactory bulbectomized mice. *J Ethnopharmacol.* 2012;143(1):158–169.
20. Abdelhalim A, Karim N, Chebib M, et al. Antidepressant, anxiolytic and antinociceptive activities of constituents from *Rosmarinus officinalis*. *J Pharm Pharm Sci.* 2015;18(4):448–459.
21. Nematolahia P, Mehrabani M, Karami-Mohajer S, Dabaghzadeh F. Effects of *Rosmarinus officinalis* L. on memory performance, anxiety, depression, and sleep quality in university students: a randomized clinical trial. *Complement Ther Clin.* 2018;30:24–28.
22. Hoffmann D. *Medical Herbalism: The Science and Practice of Herbal Medicine.* Rochester, VT: Healing Arts Press; 2003. p. 567.
23. Alnamer R, Alaoui K, Bouidida EH, Benjouad A, Cherrah Y. Psychostimulants activity of *Rosmarinus officinalis* L. methanolic and aqueous extracts. *J Med Plants Res.* 2012;6(10):1860–1865.
24. Dipe de Faria LR, Lima CS, Perazzo FF, Carvalho JC. Anti-inflammatory and antinociceptive activities of the essential oil from *Rosmarinus officinalis* L. Int J Pharm Sci Rev Res. 2011;7(2):Article-001.
25. Gardner Z, McGuffin M, editors. *Botanical Safety Handbook.* 3rd ed. Boca Raton, FL: CRC Press; 2013. p. 740.
26. Debersac P, Heydel J-M, Amiot M-J, et al. Induction of cytochrome P450 and/or detoxication enzymes by various extracts of rosemary: description of specific patterns. *Food Chem Toxicol.* 2001;39(9):907–918.
27. Turpeinen M, Raunio H, Pelkonen O. The functional role of CYP2B6 in human drug metabolism: substrates and inhibitors in vitro, in vivo and in silico. *Curr Drug Metab.* 2006;7(7):705–714.

7.55 SALVIA DIVINORUM

Salvia divinorum is a plant native to the mountainous regions of the Mexican state of Oaxaca. It is found only in forest ravines and other moist, cloudy areas of the Sierra Mazateca between 2,000 and 5,000 feet of altitude, which likely adds to its mystique. Indeed, it is one of the more unusual plants to be evaluated for antidepressant effects, owing to its potent hallucinogenic effects. It is one of several vision-inducing plants employed by the native Mazatec Indians for curing and divination. Its properties have earned it the name "sage of the diviners," "seer's sage," or simply salvia.[1]

Salvia contains a phytochemical, salvinorin A, that produces a psychotomimetic effect through the seemingly rare ability to activate kappa opiate receptors in the brain. It appears to be the only plant-derived chemical to selectively act at kappa receptors over other subtypes of opiate receptors. Moreover, it is a non-nitrogenous diterpene with a molecular structure unrelated to all other known opioid receptor ligands.[2] Aside from "magical, divinatory uses," the plant has also been used among the native Mazatec Indians for treatment of headache, rheumatism, to regulate eliminatory functions, and as a tonic for the old and feeble.[3]

Salvia divinorum has a number of unique, possibly psychoactive phytochemicals similar to salvinorin A, named salvinorins B through J, as well as the structurally related divinatorins A through F, salvidivins A through D, and salvinicins A and B. The more common and medicinally significant phytochemicals, rosemarinic acid, stigmasterol, and oleanolic acid, have also been identified in *Salvia divinorum*.[4] Recent evidence suggests that salvinorin A may also interact with cannabinoid receptors.[5]

7.55.1 Antioxidant

The *Salvia* species, as a whole, produce a great variety of biologically active terpenoids, steroids, flavonoids, and polyphenols. Almost all species studied have been shown to exert antioxidant effects.[6] However, perhaps because of being eclipsed by interest in its psychotomimetic effects, antioxidant effect has not been a focus of study for *Salvia divinorum*.

7.55.2 Anti-Inflammatory

While the most prominent effects of salvinorin A may be psychotomimetic, it has been found to have very potent anti-inflammatory effects. *In vitro*, salvinorin A attenuated lipopolysaccharide-induced inflammatory reactions in mouse macrophages. Treatment reduced lipopolysaccharide-induced nitrite, TNF-α, IL-10, and nitric oxide synthase expression. These effects of salvinorin were blocked not only by the opiate receptor antagonist nalone, but also by the cannabinoid receptor antagonist rimonabant. In intact animals, salvinorin dampened both lipopolysaccharide- and carrageenan-induced paw edema. Those effects were also blocked by naloxone and rimonabant.[7]

7.55.3 Antidiabetic/Anti-Metabolic Syndrome

No published studies specifically concerning antidiabetic effects of *Salvia divinorum* could be found.

7.55.4 Preclinical Antidepressant-Like Effects

The results of studies of antidepressant and rewarding effects of *Salvia divinorum* in rodents are mixed. In one study, administration of *Salvia divinorum* to rats caused an aversive reaction.[8] However, in another study, administration of purified salvinorin A resulted in place preference and self-administration, which suggests the treatment was rewarding.[9] In one study, administration of salvinorin A caused an increase in immobility of rats, suggesting a depressive effect.[10]

In several studies, antidepressant-like effects were noted. Salvinorin A reduced immobility in the forced swim test with rats, and in the tail suspension test with mice. These effects could be blocked by either the κ-opioid receptor antagonist nor-binaltorphimine or the CB1 cannabinoid receptor antagonist AM251. Treatment with Salvinorin A also suggested anxiolytic effects in rats in the elevated plus maze.[11] Chronic treatment of rats with salvinorin A also reversed the reduction in preference for sucrose – considered a form of depression-like anhedonia – that was induced by chronic unpredictable stress.[12]

7.55.5 HUMAN ANTIDEPRESSANT EFFECTS

In humans, *Salvia divinorum* has been used for spiritual healing and mystical experi-ence, and not generally for treatment of what mainstream Western medicine refers to as MDD. The effects of *Salvia divinorum* can be intense and like those of other potent hallucinogens and so-called "psychedelic" drugs. Euphoric effects are rarely reported.[13,14]

Although *Salvia divinorum* has a very different pharmacological profile than ket-amine, some of the peculiar effects it produces are similar to those described in ket-amine intoxication. Such effects of *Salvia divinorum* reported by human experimental subjects included: (1) Becoming objects (yellow plaid, French fries, fresh paint, a drawer, a pant leg, a Ferris wheel, etc.); (2) Visions of various two-dimensional sur-faces, films and membranes; (3) Revisiting places from the past, especially childhood; (4) Loss of the body and/or identity; (5) Various sensations of motion, or being pulled or twisted by forces of some kind; (6) Uncontrollable hysterical laughter; (7) Overlapping realities – the perception that one is in several locations at once.[15] However, most expe-rienced users of hallucinogens report the effects of *Salvia divinorum* to be unique and different from ketamine, LSD, or marijuana. Moreover, salvinorin A does not substitute as a stimulus for either LSD or ketamine in discrimination tests using rats.[16]

There are only a few reports on persisting effects of *Salvia divinorum* on mood in humans. In one often-noted case report, a young woman with a long history of severe MDD resistant to both standard antidepressants and cognitive behavioral therapy found complete and persisting resolution of symptoms with intermittent use of leaves of *Salvia divinorum*. The number of leaves generally consumed in each treatment appears small, i.e., two or three as opposed to a dozen or more in typical shamanic ceremonies. However, she also had on occasion consumed 8–16 leaves of the herb and noted "psychospiritual" awakening, characterized by the discovery of the depth of her sense of self, greater self-confidence, increased feelings of intuitive wisdom and "connectedness to nature."[17] Generally, the reports of effects on mood are less dramatic and not entirely positive. For example, whereas some informal users describe the intensity of the hallucinogenic experience in positive ways, others find the experience so intense that they would not continue to use the substance.[18] Eight weeks after a controlled study of acute effects of *Salvia divinorum*, subjects reported a mix of positive and negative effects lasting more than 24 hours after inha-lation, including increases in positive mood, empathy, and aesthetic sensitivity as well as headache, fatigue, and difficulty concentrating.[19] Stimulation of kappa opiate receptors in the human brain have not been associated with antidepressant effects. Indeed, buprenorphine, which acts as a potent kappa receptor antagonist as well as a partial mu receptor agonist, has been well noted to exhibit antidepressant effects. In a small, open label study, buprenorphine was very effective in relieving symptoms of MDD in subjects who had been resistant to standard treatments.[20]

All in all, prudence prevents the recommendation of *Salvia divinorum* for treatment of MDD in humans. First, there is no lack of other effective and better studied herbs and medications for the treatment of this illness. Second, while the psychedelic state induced by *Salvia divinorum* may be instructive and even therapeutic, the dangers of induction of traumatic or even permanently damaging psychotic/dissociative states are too great a risk. Finally, casual, recreational use should be strongly discouraged. It is noted that in the Mazatec culture that has used the plant for healing, this is done ceremonially with guidance by experienced shaman in the context of a larger religious and mystical world view.

7.55.6 Dosage

There are no reliable doses reported for clinical use. The number of leaves generally consumed in each treatment of the young woman in the above case report was two or three as opposed to a dozen or more in typical shamanic ceremonies. However, the route of administration of *Salvia divinorum* is a critical factor in the strength of its effect. For example, an infusion prepared from ten fresh leaves, when quickly swallowed, produced no effect in any subject. However, the same amount, when held in the mouth for ten minutes and spat out, produced substantial effects in all subjects. Similarly, when vaporized and inhaled, perceptible hallucinogenic effects of pure salvinorin A are perceptible at doses of 200 μg. Inhalation of 1 mg produces full out of body experiences and marked distortion of reality. However, 10 mg of encapsulated salvinorin A produced no effects after swallowing.

7.55.7 Toxicity

There does not appear to be human toxicological data. A polyphenol-free extract of salvinorin leaf was found to be without acute effect in cats, even at doses over 500 times the typical human dose. In mice, a similar methanolic extract of whole leaf was more toxic, with an LD50 of 340 mg/kg i.p. A dosing of 1 g/kg of pure salvinorin A was found to cause no deaths in mice.[21]

In any case, the concern with *Salvia divinorum* is not so much its acute toxicity, but rather its psychotomimetic properties at higher doses.

7.55.8 Safety in Pregnancy

Review of the literature revealed no studies of the safety of *Salvia divinorum* during human or animal pregnancy. Indeed, existing studies in humans eliminated pregnant subjects from participation.

7.55.9 Drug Interactions

Review of the literature revealed no studies of the possible interactions of *Salvia divinorum* with commonly used medications.

REFERENCES

1. Valdfis LJ, Dfaz JL, Paul AG. Ethnopharmacology of Ska Maria Pastora (Salvza Dzvznor Um, Epling and Jativa-M.). *J Ethnopharmacol.* 1983;7:287–312.
2. Butelman ER, Kreek MJ. Salvinorin A, a kappa-opioid receptor agonist hallucinogen: pharmacology and potential template for novel pharmacotherapeutic agents in neuro-psychiatric disorders. *Front Pharmacol.* 2015;6:190.
3. Perry N, Howes MJ, Houghton P, et al. Why sage may be a wise remedy: Effects of Salvia on the nervous system. In: Kintzios SE, editor. *Sage: The Genus Salvia.* Netherlands: Harwood; 2000. p. 207–223.
4. Casselman I. Genetics and Phytochemistry of *Salvia divinorum* [Ph.D. dissertation]. [East Lismore, NSW]: Southern Cross University; 2016.
5. Fichna J, Schicho R, Andrews CN, et al. Salvinorin A inhibits colonic transit and neu-rogenic ion transport in mice by activating kappa-opioid and cannabinoid receptors. *Neurogastroenterol Motil.* 2009;21:S1326–Se128.
6. Bahadori MB, Mirzaei M. Cytotoxicity, antioxidant activity, total flavonoid and phe-nolic contents *of Salvia urmiensis* Bunge and *Salvia hydrangea* DC. ex Benth. *RJP.* 2015;2(2):27–32.
7. Aviello G, Borrelli F, Guida D, et al. Ultrapotent effects of salvinorin A, a hallucino-genic compound from *Salvia divinorum*, on LPS-stimulated murine macrophages and its anti-inflammatory action in vivo. *Mol Med.* 2011;89:891–902.
8. Sifka KJ, Loria MJ, Lewellyn K, et al. The effect of *Salvia divinorum* and *Mitragyna speciosa* extracts, fraction and major constituents on place aversion and place prefer-ence in rats. *J Ethnopharmacol.* 2014;151(1):361–364.
9. Braida D, Capurro V, Zani A, et al. Potential anxiolytic-and antidepressant-like effects of salvinorin A, the main active ingredient of *Salvia divinorum*, in rodents. *Br J Pharmacol.* 2009;157(5):844–853.
10. Carlezon WAC. Depressive-like effects of the κ-opioid receptor agonist salvinorin A on behavior and neurochemistry in rats. *JPET.* 2006;316(1):440–447.
11. Braida D, Capurro V, Zani A, et al. Potential anxiolytic- and antidepressant-like effects of salvinorin A, the main active ingredient of Salvia divinorum, in rodents. *BJP.* 2009;157(5):844–853.
12. Harden MT, Smith SE, Niehoff JA, McCurdy CR, Taylor GT. Antidepressive effects of the κ-opioid receptor agonist salvinorin A in a rat model of anhedonia. *Behav Pharmacol.* 2012;23(7):710–715.
13. Johnson MW, MacLean KA, Reissig CJ, Prisinzano TE, Griffiths RR. Human psycho-pharmacology and dose-effects of salvinorin A, a κ-opioid agonist hallucinogen present in the plant Salvia divinorum. *Drug Alcohol Depend.* 2011;115:150–155.
14. Ranganathan M, Schnakenberg A, Skosnik PD, et al. Dose-related behavioral, subjec-tive, endocrine, and psychophysiological effects of the κ-opioid agonist Salvinorin A in humans. *Biol Psychiatry.* 2012;72:871–879.
15. Siebert DJ. Salvia divinorum and Salvinorin A: new pharmacologic findings. *J Ethnopharmacol.* 1994;43:53–56.
16. Killinger BA, Peet MM, Baker LE. Salvinorin A fails to substitute for the discrimina-tive stimulus effects of LSD or ketamine in Sprague-Dawley rats. *Pharmacol Biochem Behav.* 2010;96(3):260–265.
17. Hanes KR. Antidepressant effects of the herb *Salvia divinorum*: a case report. *J Clin Psychopharmacol.* 2001;21(6):634–635.
18. Kelly BC. Legally tripping: a qualitative profile of *Salvia divinorum* use among young adults. *J Psychoact Drugs.* 2011;43(1):46–54.
19. Addy PH. Acute and post-acute behavioral and psychological effects of salvinorin A in humans. *Psychopharmacology.* 2012;220(1):195–204.

20. Bodkin JA, Zornberg GL, Lukas SE, Cole JO. Buprenorphine treatment of refractory depression. *J Clin Psychopharmacol.* 1995;15:49–57.
21. Munro TA. The Chemistry of *Salvia divinorum* [Master's thesis]. [Melbourne]: University of Melbourne; 2006.

7.56   *SCELETIUM TORTUOSUM*

Sceletium tortuosum, known by locals in South Africa as channa, kanna, or kougoed, is a flowering succulent plant of the family, *Mesembryanthemaceae*. For hundreds of years it has been prized by the indigenous Khoi and San tribes as a tonic and mood elevator. It is picked and prepared in ritualized fashion, likely to allow the fermentation process to rid the herb of toxic properties. Mention of its use was made in 1928 in the *South African Journal of Science*:

> The preparation is chewed and retained in the mouth for a while, when their spirits would rise, eyes brighten and faces take on a jovial air, and they would commence to dance. But if indulged in to excess it robbed them of their senses and they became intoxicated.[1]

Indeed, while it has often been used by indigenous people for toothache, abdominal pain, and hunger relief, most reports suggest the plant is primarily used for pleasure.

Plants of the genus Scelelium contain a variety of indole alkaloids, most notable of which are mesembrine and mesembranone. Serotonin is itself an indole alkaloid, and those plant alkaloids are potent 5-HT uptake inhibitors. Mesembrine is also a modestly potent inhibitor of phosphodiesterase-4, the enzyme primarily responsible for the degradation of the intracellular second messenger, cAMP.[2] PDE4 inhibitors have antidepressant effects in both animals and humans, and the enzyme is seen as a potential target of antidepressant drug action.[3] Along with being a potent 5-HT reuptake inhibitor, a high mesembrine extract of sceletium was also found to be a 5-HT releasing agent from both human astrocytes and mouse hippocampal neurons.[4]

7.56.1   ANTIOXIDANT

Along with 5-HT reuptake inhibition and phosphodiesterase inhibition, extract of *Sceletium tortuosum* also exhibits significant antioxidant effects. This is likely due to the variety of phytochemicals the plant contains aside from the psychoactive, mesembrine-type alkaloids. These phytochemicals include anthraquinones, terpenes, polyphenols, anthocyanin, tannins, alkaloids, glycosides, carbohydrates, and coumarins. Testing in the DPPH model demonstrated a significant, dose-dependent ability of the extracts to scavenge free radicals.[5]

7.56.2   ANTI-INFLAMMATORY

Extracts of *Sceletium tortuosum* also have anti-inflammatory effects. Pre-treatment of cultured human astrocytes reduced their production of IL-6 and monocyte chemoattractant protein-1 in response to exposure to lipopolysaccharide. Monocyte chemoattractant protein-1 is yet another inflammatory cytokine associated with MDD.

Curiously, higher concentrations of the extract began to enhance the inflammatory response of the astrocytes, possibly due to toxicity.[6]

7.56.3 ANTIDIABETIC/ANTIMETABOLIC SYNDROME

The role of glucocorticoids in Metabolic Syndrome is well known. This is due to effects such as enhancement of gluconeogenesis[7] and development of insulin resistance under conditions of severe and persistent stress.[8] One known effect of *Sceletium tortuosum* is the dampening of glucocorticoid production by the adrenal cortex.[9] This effect would tend to break the cycles of stress, hyperglycemia, and insulin resistance that precipitate Metabolic Syndrome. It would also prevent the development of cortisol resistance and the imbalance of mineralocorticoid and glucocorticoid receptor activity in the brain that has been seen as contributing to MDD in instances of prolonged exposure to stress.[10]

7.56.4 PRECLINICAL ANTIDEPRESSANT-LIKE EFFECTS

Sceletium tortuosum has been found to have a number of behavioral effects in rodents. Acute intraperitoneal administration of a methanolic extract of *Sceletium tortuosum* decreased immobility time of rats in the forced swim test, which is taken to be an antidepressant-like effect. Although considered to have an anxiolytic effect, this single administration had no effect on behavior in the elevated plus maze.[11] Nonetheless, daily administration of *Sceletium tortuosum* for 17 days reduced basal serum levels of corticosterone, as well as dampening the increase of corticosterone in response to restraint.[12] In mice, a low (10 mg/kg) but not high (80 mg/kg) and possibly intoxicating dose of mesembrine alkaloids also decreased immobilization time in the forced swim test.[13]

An increasingly used animal model of antidepressant action is the assay of distress vocalization in chicks. Acute administration of extract of *Sceletium tortuosum* decreased distress vocalization in chicks, which is taken as evidence of an antidepressant-like effect.[14]

7.56.5 HUMAN ANTIDEPRESSANT EFFECTS

There are few studies of effects of *Sceletium tortuosum* in human subjects. In an interesting fMRI study, acute administration of 25 mg of Zembrin, a proprietary preparation of *Sceletium tortuosum*, attenuated reactivity to fearful faces in the amygdala. Follow-up connectivity analysis on the emotion-matching task further showed that amygdala-hypothalamus coupling was also reduced. Those results were interpreted as reflecting an anxiolytic dampening of activity in threat circuitry of the human brain.[15]

There are no formal clinical studies of *Sceletium tortuosum* in the treatment of MDD in human subjects. However, several case studies have been reported.[16] In the first case, the patient was suffering from severe depression of four months duration, and was started on 50 mg sceletium daily. A sustained improvement in mood was reported, with a marked decrease in insomnia and generalized anxiety.

The sceletium was discontinued after four months of continual use with no signs or symptoms of withdrawal.

In the second case, a patient with a personality disorder and dysthymia was started on 50 mg sceletium daily. Within ten days the patient said that her mood had lifted, and she was able to feel more focused, more engaged, and not so socially "distant". She doubled her dose of sceletium to two 50 mg tablets daily just prior to her exams a month later, and described feeling less anxious and more able to cope with her usual examination anxiety.

In the third case, the patient presented with Major Depressive Disorder, with symptoms of depressed mood, increased sleep, overeating, anxiety, psychomotor agitation, and thoughts of death. The patient was started on sceletium, 50 mg in the morning and at lunchtime. Improvement in mood, sleep pattern, and energy were noted the first day. After six weeks of treatment with sceletium, the patient was fully recovered.

In what was ostensibly only a pilot study to determine safety and tolerability, subjects received either placebo, 8 mg or 25 mg of extract of *Sceletium tortuosum*. There were no apparent differences between the three treatments with regard to vital signs, 12-lead ECG, body weight, and physical examination from screening to the end of the three-month treatment period. There were no significant changes observed in hematology or biochemistry parameters between initial screening and the end of the study. Both doses of extract *Sceletium tortuosum* were well-tolerated. The only notable adverse report was a slight increase in complaints of headache in the higher dosing of *Sceletium tortuosum*.[17]

7.56.6 DOSAGE

As suggested above, the most extensive work has been done with the proprietary form of *Sceletium tortuosum*, named Zembrin. The manufacturing process was intended to mimic in some fashion the folk method used by South African natives to rid the herb of toxic properties. Dosing appears to be in the range of 50 to 100 mg of product a day, and are well tolerated.

7.56.7 TOXICITY

No LD50 data could be found in the literature.

7.56.8 SAFETY IN PREGNANCY

Review of the literature revealed no studies of the safety of *Sceletium tortuosum* during human or animal pregnancy.

7.56.9 DRUG INTERACTIONS

Studies have shown components of *Sceletium tortuosum* to inhibit the P450 enzymes, CYP11B1 and CYP21A2.[18] The blood levels of some antidepressants, for example, paroxetine[19] and citalopram,[20] are likely to be increased by treatment with *Sceletium tortuosum*.

REFERENCES

1. Laidler PW. The magic medicine of the Hottentots. *S Afr J Sci.* 1928;25:433–447.
2. Harvey AL, Young LC, Viljoen AM, Gericke NP. Pharmacological actions of the South African medicinal and functional food plant *Sceletium tortuosum* and its principal alkaloids. *J Ethnopharmacol.* 2011;137(3):1124–1129.
3. Zhang HT. Cyclic AMP-specific phosphodiesterase-4 as a target for the development of antidepressant drugs. *Curr Pharm Des.* 2009;15(14):1688–1698.
4. Coetzeea DD, López V, Smith C. High-mesembrine sceletium extract (Trimesemine™) is a monoamine releasing agent, rather than only a selective serotonin reuptake inhibitor. *J Ethnopharmacol.* 2016;177(11):111–116.
5. Kapewangoloa P. *Sceletium tortuosum* demonstrates in vitro anti-HIV and free radical scavenging activity. *S Afr J Bot.* 2016;106:140–143.
6. Bennett AC, Van Camp A, Lopez V, Smith C. *Sceletium tortuosum* may delay chronic disease progression via alkaloid-dependent antioxidant or anti-inflammatory action. *J Physiol Biochem.* 2018;74(4):539–547.
7. Khani S, Tayek JA. Cortisol increases gluconeogenesis in humans: its role in the metabolic syndrome. *Clin Sci.* 2001;101(6):739–747.
8. Andrews RC, Walker BR. Glucocorticoids and insulin resistance: old hormones, new targets. *Clin Sci.* 1999;96(5):513–523.
9. Swart AC, Smith C. Modulation of glucocorticoid, mineralocorticoid and androgen production in H295 cells by Trimesemine, a mesembrine-rich Sceletium extract. *J Ethnopharmacol.* 2016;177:35–45.
10. de Kloet ER, DeRijk RH, Meijer OC. Therapy insight: is there an imbalanced response of mineralocorticoid and glucocorticoid receptors in depression? *Nat Clin Pract Endocrinol.* 2007;3:168–179.
11. Loria MJ, Ali Z, Abe N, Sufka KJ, Khan IA. Effects of *Sceletium tortuosum* in rats. *J Ethnopharmacol.* 2014;155:731–735.
12. Smith C. The effects of *Sceletium tortuosum* in an in vivo model of psychological stress. *J Ethnopharmacol.* 2011;133:31–36.
13. Schell R. Sceletium Tortuosum and Mesembrine: A Potential Alternative Treatment for Depression [Senior thesis]. Claremont College; 2014.
14. Carpenter JM, Jourdan MK, Fountain EM, et al. The effects of *Sceletium tortuosum* (L.) N.E. Br. extract fraction in the chick anxiety-depression model. *J Ethnopharmacol.* 2016;193(4):329–332.
15. Terburg D, Syal S, Rosenberger LA, et al. Acute effects of *Sceletium tortuosum* (Zembrin), a dual 5-HT reuptake and PDE4 inhibitor, in the human amygdala and its connection to the hypothalamus. *Neuropsychopharmacology.* 2013;38:2708–2716.
16. Gericke N. Clinical application of selected South African medicinal plants. *Aust J Med Herbal.* 2001;13:3–17.
17. Nell H, Siebert M, Chellan P, Gericke N. A randomized, double-blind, parallel-group, placebo-controlled trial of extract *Sceletium tortuosum* (Zembrin) in healthy adults. *J Altern Complement Med.* 2013;19(11):898–904.
18. Johannes I. The In Vitro Inhibition of Adrenal Steroidogenic Enzymes and Modulation of Adrenal Hormone Production By *Sceletium tortuosum* [Master's thesis]. [Stellenbosch, South Africa]: Stellenbosch University; 2018.
19. Lin KM, Tsou HH, Tsai IJ, et al. CYP1A2 genetic polymorphisms are associated with treatment response to the antidepressant paroxetine. *Pharmacogenomics.* 2010;11:1535–1543.
20. Kuo H-W, Liu SC, Tsou H H, et al. CYP1A2 genetic polymorphisms are associated with early antidepressant escitalopram metabolism and adverse reactions. *Pharmacogenomics.* 2013;14:1191–1201.

7.57   SCHISANDRA CHINENSIS

Schisandra chinensis, known as magnolia-vine or schisandra, and in China as *wu wei zi*, is a deciduous woody vine native to forests of Northern China and the Russian Far East. The fruit of the plant, called magnolia berry, and known by the Chinese as five-flavor-fruit, is used the manufacture of juices, wines, sweets, and for medicinal purposes. It is one of many plants evaluated by Russians scientists for having characteristics of an adaptogen, that is, for an apparent ability to give non-specific stress resistance and resiliency to cells and organ systems. Many such plants have been used in folk medicine for treatment of weakness, depression, and the ravages of age. *Schisandra chinensis* is alleged to increase work capacity and protect against a broad spectrum of harmful factors including heat shock, skin burn, cooling, frostbite, immobilization, aseptic inflammation, irradiation, and heavy metal intoxication.[1]

A multitude of polyphenolic phytochemicals have been isolated from *Schisandra chinensis*. Lignans – some of which are so-called "phytoestrogens" – are the major and characteristic constituents of the genus *Schisandraceae*. Schizandrins are antioxidant lignans somewhat unique to these plants. Most important among them are the dibenzo[a,c]cyclooctadiene lignans. The plant also contains glycosides, for example, (+)-Isoscoparin 66 Quercetin-3-O--l-rhamnopyranosyl (1→6)--d-glucopyranoside; flavonoids, including hyperoside, isoquercitrin, rutin, and quercetin; and organic acids.[2]

7.57.1   ANTIOXIDANT

Schisandra chinensis has many properties that would be expected to both treat and reduce the risk of developing MDD. The herb exerts potent antinitrosative and antioxidative effects. It blocks the induction of NO synthesis in lipopolysaccharide-activated microglia cells.[3] A methanolic extract of dried *Schisandra chinensis* fruit containing dibenzocyclooctadiene lignans significantly attenuated the neurotoxicity induced by L-glutamate in primary cultures of rat cortical cells. These effects were evidenced by inhibition in the increase of intracellular [Ca2+], increases in the levels of glutathione, enhancement of the activity of glutathione peroxidase, and decreases in the formation of cellular peroxide.[4]

7.57.2   ANTI-INFLAMMATORY

Schisandra chinensis also exhibits strong anti-inflammatory effects. Extracts of the fruit inhibits cyclooxygenase 1 and 2.[5] Pre-treatment of murine macrophages with *Schisandra chinensis* lignans also reduced the lipopolysaccharide-induced secretion of pro-inflammatory cytokines.[6]

7.57.3   ANTIDIABETIC/ANTI-METABOLIC SYNDROME

Water-soluble polysaccharides from *Schisandra chinensis* have found to reduce hyperglycemia in alloxan-induced diabetic mice. They also increase liver glycogen content and improve serum lipid profiles.[7] Extracts of *Schisandra chinensis* inhibit protein tyrosine phosphatase 1B, which itself is a negative regulator of insulin and leptin signal transduction. Extracts also inhibit α-glucosidase in the brush border of the intestine, thus slowing digestion of complex carbohydrates and the absorption of glucose.[8]

7.57.4 PRECLINICAL ANTIDEPRESSANT-LIKE EFFECTS

There are reports of *Schisandra chinensis* exhibiting antidepressant-like effects in intact rodents. In mice, ethanol extract of the dried fruit of *S. chinensis* ameliorated the depressive-like behavior induced by repeated corticosterone injections. The extract restored normal levels of sucrose consumption and decreased the immobility time in the forced swim test. It also reversed the corticosterone-induced decreases in BDNF, tyrosine kinase receptor B, and cAMP-response element binding protein signaling in the hippocampus and the prefrontal cortex.[9]

Similarly, *Schisandra chinensis* reversed the depression-like effects and cognitive impairments due to chronic unpredictable mild stress in mice. The preference for sucrose was restored and immobility reduced in the forced swimming test. Cognitive deficits from the chronic stress were reversed, as evidence by improved performance in the Morris water maze. *Schisandra chinensis* also reversed neurochemical effects of stress in the hippocampus. The reductions in BDNF and diminished activity in the tyrosine kinase receptor B and cAMP-response element binding protein pathways were ameliorated. It was further shown that activity of the extracellular signal-regulated kinase, phosphatidylinositol 3 kinase, and AKT signaling pathways were restored with resulting inhibition of GSK-3β.[10]

Treatment with *Schisandra chinensis* has also been found to reduce serum levels of cortisol and induced nitric oxide in rabbits subjected to restraint stress.[11]

7.57.5 HUMAN ANTIDEPRESSANT EFFECTS

Despite the long history use of *Schisandra chinensis* in Traditional Chinese Medicine and the folk medicine of Russia, as well as references to reports of benefits for mood in an old Russian literature, there are no accessible published reports of effects of *Schisandra chinensis* in the treatment of MDD in human subjects.

7.57.6 DOSAGE

Therapeutic dosages are 400–450 mg powdered herb in capsules three times daily or 1–2 ml of 1:3 EtOH tincture of *Schisandra chinensis* three times daily.[12]

7.57.7 TOXICITY

LD50 value of dried ethanolic extract of *Schisandra chinensis* was estimated to be 35.63 ± 6.46 g/kg.[13] In addition, it was observed that chronic feeding of *Schisandra chinensis* fruit to mice in doses as high as 4 g/kg had no adverse effects.[14]

7.57.8 SAFETY IN PREGNANCY

The *Botanical Safety Handbook* noted the report that women who took *Schisandra chinensis* during pregnancy experienced less postpartum hemorrhaging, without adverse effects to themselves or their newborns.[15] *Schisandra chinensis* was also noted to be safe to mothers and infants when used to successfully induce labor.[16]

7.57.9 DRUG INTERACTIONS

Schisandra chinensis is metabolized primarily by the P450 enzyme, CYP3A.[17] A variety of neuroleptics, including aripiprazole, haloperidol, pimozide, and risperidone, are metabolized by this enzyme.[18] The enzyme is also involved in the metabolism of SSRIs and tricyclic antidepressants.[19] Thus, chronic use of *Schisandra chinensis* could affect blood levels of a variety of important psychiatric medications.

REFERENCES

1. Panossian AG. Pharmacology of *Schisandra chinensis* Bail.: an overview of Russian research and uses in medicine. *J Ethnopharmacol*. 2008;118(2):183–212.
2. Lu Y, Chen DF. Analysis of *Schisandra chinensis* and *Schisandra sphenanthera*. *J Chromatogr*. 2009;1216:1980–1990.
3. Hu D, Yang Z, Yao X, et al. Dibenzocyclooctadiene lignans from *Schisandra chinensis* and their inhibitory activity on NO production in lipopolysaccharide-activated microglia cells. *Phytochemistry*. 2014;104:72–78.
4. Kim SR, Lee MK, Koo KA, et al. Dibenzocyclooctadiene lignans from *Schisandra chinensis* protect primary cultures of rat cortical cells from glutamate-induced toxicity. *J Neurosci Res*. 2004;76(3):397–405.
5. Blunder M, Pferschy-Wenzig EM, Fabian WM, et al. Derivatives of schisandrin with increased inhibitory potential on prostaglandin E2 and leukotriene B4 formation in vitro. *Bioorg Med Chem*. 2010;18:2809–2815.
6. Oh SY, Kim YH, Bae DS, et al. Anti-inflammatory effects of gomisin N, gomisin J, and schisandrin C isolated from the fruit of *Schisandra chinensis*. *Biosci Biotechnol Biochem*. 2010;74:285–291.
7. Zhao T, Mao G-H, Zhang M, et al. Anti-diabetic effects of polysaccharides from ethanol-insoluble residue of *Schisandra chinensis* (Turcz.) Baill on alloxan-induced diabetic mice. *Chem Res Chin Univ*. 2013;29(1):99–102.
8. Fang,LL, Cao J-Q, Duan L-L, Tang Y, Zhao Y. Protein tyrosine phosphatase 1B (PTP1B) and α-glucosidase inhibitory activities of *Schisandra chinensis* (Turcz.) Baill. *J Funct Food*. 2014;9:264–270.
9. Yan T, Xu M, Wan S, et al. *Schisandra chinensis* produces the antidepressant-like effects in repeated corticosterone-induced mice via the BDNF/TrkB/CREB signaling pathway. *Psychiatr Res*. 2016;243:135–142.
10. Yan T, He B, Wan S, et al. Antidepressant-like effects and cognitive enhancement of *Schisandra chinensis* in chronic unpredictable mild stress mice and its related mechanism. *Sci Rep*. 2017;7:6903.
11. Panossian A, Hambardzumyan M, Hovhanissyan A, et al. The adaptogens Rhodiola and Schizandra modify the response to immobilization stress in rabbits by suppressing the increase of phosphorylated stress-activated protein kinase, nitric oxide and cortisol. *Drug Targets Insights*. 2007;1:39–54.
12. Bensky D, Gamble A. *Chinese Herbal Medicine: Materia Medica*. Revised Edition. Seattle, WA: Eastland Press; 1993.
13. Pan SY, Yu ZL, Dong H, et al. Ethanol extract of *fructus schisandrae* decreases hepatic triglyceride level in mice fed with a high fat/cholesterol diet, with attention to acute toxicity. *Evid Based Complement Alternat Med*. 2011:729412.

14. Ryu SN. Acute toxicity of fruit pigment and seed oil of *Schizandra chinensis* in mice. *J Korean Soc Int Agric.* 1998;10:37–41.

15. Gaistruk A, Taranovskij K. The treatment of arterial hypotension in pregnant women using *Schisandra chinensis*. *Urb Prob Obstet Gynecol.* 1968 1:183–186.

16. Trifinova A. Stimulation of labor activity using schizandra chinensis. *Obstet Gynecol.* 1954;4:19–22.

17. Cao Y-F, Zhang Y-Y, Li, J, et al. CYP3A catalyses schizandrin biotransformation in human, minipig and rat liver microsomes. *Xenobiotica.* 2010;40(1):38–47.

18. van der Weide K, van der Weide J. The influence of the CYP3A4*22 polymorphism and CYP2D6 polymorphisms on serum concentrations of aripiprazole, haloperidol, pimozide, and risperidone in psychiatric patients. *J Clin Psychopharmacol.* 2015;35(3):228–236.

19. Haduch A, Wójcikowski J, Daniel WA. The effect of tricyclic antidepressants, selective serotonin reuptake inhibitors (SSRIs) and newer antidepressant drugs on the activity and level of rat CYP3A. *Eur Neuropsychopharmacol.* 2006;16(3):178–186.

7.58   *SCUTELLARIA LATERIFLORA* (SKULLCAP)

Skullcap refers to two medicinal plants, American skullcap (*Scutellaria lateriflora*) and Chinese skullcap (*Scutellaria baicalensis*). *Scutellaria baicalensis* has long been a mainstay in Chinese medicine, and has been used for anxiety, depression, neurological conditions, and gastric distress. *Scutellaria lateriflora* has been used in traditional Native American medicine for nervous tension and various psychiatric and neurological problems. The main phytochemicals in this genus are the flavonoids and glycosides, scutellarin, scetellarein, baicalin, baicalein, wogonin, wogonoside, apigenin, chrysin, and oroxylin A. The plants also contain serotonin, melatonin, and various alkaloids. There are differences between the two species in the percentages of certain phytochemicals they contain.[1]

7.58.1 ANTIOXIDANT

Both *Scutellaria lateriflora* and *Scutellaria baicalensis* exhibit potent antioxidant effects, with little difference between the two. Both showed substantial oxygen radical scavenging capacity in the DPPH assay. In mouse brain homogenates, ethanolic extract of skullcap reduced tert-butyl peroxide-induced reactive oxygen species and lipid peroxides. The extract also protected plasmid DNA from hydrogen peroxide-UV induced cleavage.[2] Methanolic extract of *Scutellaria baicalensis* attenuated the oxidative damage of CCl_4 in rats. It minimized the extent of liver fibrosis and reduced lipid peroxidation in liver homogenates.[3]

7.58.2 ANTI-INFLAMMATORY

Extracts of *Scutellaria baialensis* have also been shown to possess potent anti-inflammatory effects *in vivo*, in the zymosan-induced mouse air-pouch model, and *in vitro*, by exposing mouse macrophages to LPS. In each case, extracts of the herb reduced the expression of nitric oxide (NO), inducible NOS (iNOS), Cyclooxygenase2 (COX-2), Prostaglandin E2 (PGE2), Nuclear Factor-kappaB (NF-κB), and IκBα, as well as the inflammatory cytokines, IL-1β, IL-2, IL-6, IL-12, and TNF-α.[4]

7.58.3 ANTIDIABETIC/ANTI-METABOLIC SYNDROME

Scutellaria baialensis mitigates components of Metabolic Syndrome, likely due to its constituent, baicalin. This flavonoid reduces high fat diet-induced insulin resistance and ectopic fat storage in skeletal muscle by stimulating the akt pathway with resulting inhibition of GSK-3β.[5]

7.58.4 PRECLINICAL ANTIDEPRESSANT-LIKE EFFECTS

Preclinical studies show antidepressant effects of *Scutellaria* phytochemicals. The flavonoid baicalin, isolated from the dried root of *Scutellaria baicalensis*, reduced immobility time in tail suspension test (TST) and the forced swim test (FST) in mice. It also decreased immobility time in FST in rats with effects comparable to those of fluoxetine. The effects of chronic unpredictable stress on rats were counteracted by baicalin, with rats regaining their taste for sucrose. In this study, baicalin was found to inhibit MAO type A and, at high dose, also MAO B, which suggested that part of its antidepressant effect was due to inhibition of those enzymes.[6] In a similar study of chronic unpredictable stress in mice, baicalin blocked the reduction in sucrose intake as well as increases in serum corticosterone, and increases in levels of inflammatory COX-2 and PGE2 activity.[7] Baicalin also reverse depression-like behavioral effects of repeated injections of exogenous corticosterone.[8]

Apigenin, a common flavonoid prominent in *Scutellaria* and present in many other herbs, has been shown to produce antidepressant effects in a number of animal studies. Apigenin ameliorated behavioral abnormalities, such as decreased locomotor activity and reduced sucrose consumption, induced in rats subjected to chronic unpredictable stress. It appeared to do so, at least in part, by stimulating PPARγ expression – which in turn enhances responsivity to insulin – and by blocking activation of the NLRP3 inflammasome and secretion of the inflammatory cytokine, IL-1β, in the prefrontal cortex.[9] Apigenin also blocked immobility in the tail suspension test and decrease in preference for sucrose in mice injected with lipopolysaccharide. It did so in a manner similar to fluoxetine.[10] This effect of apigenin also appeared to be mediated by blocking the inflammatory response in the brain. The flavonoid significantly inhibited the expression of iNOS and COX-2, and the activation of NF-κB in the prefrontal cortex.

Chrysin, another flavonoid in the *Scutellaria*, also exhibits antidepressant effects in rodent models. It reversed the depression-like behavioral effects of olfactory bulbectomy in mice in a manner similar to that of fluoxetine. The flavonoid also prevented the elevation of tumor necrosis factor-α, interferon-γ, interleukin-1β, interleukin-6, kynurenine (KYN) levels, and indoleamine-2,3-dioxygenase activity, and decrease in BDNF in the hippocampus.[11]

7.58.5 HUMAN ANTIDEPRESSANT EFFECTS

Scutellaria baicalensis is an ingredient in at least two of the most commonly used Traditional Chinese Medicine herbal combinations for the treatment of MDD, i.e., *tiao qi* and *chai hu jia long gu mu li*. Data do show that skullcap ameliorates many pathophysiologies that lead to depression, including systemic inflammation, Metabolic Syndrome, insulin resistance, chronic stress, and artificial elevation of corticosterone levels. These effects are also seen in the brain when animals are treated with specific phytochemical components of *Scutellaria baicalensis*, including baicalin, chrysin, and apigenin. Oroxylin A, a *scutellaria* flavonoid, stimulates neurogenesis in mouse hippocampus, which is yet another effect associated with antidepressant action.[12] There is also a report that plants of this genus contain hyperforin, the primary antidepressant component of well-studied St John's wort.[13] Interestingly, flavonoids from *Scutellaria lateriflora* displace LSD from the 5-HT7 receptors, which is believed to play a role in MDD. It should be noted that the flavonoids responsible for this effect appear to bind with rather low high micromolar affinity and it is questionable if this has clinical significance.[14]

In a placebo-controlled, double-blind, crossover study of *Scutellaria lateriflora* in healthy volunteers, participants did report significant improvements in mood, per the Profile of Mood States assessment.[15] Nonetheless, while there are bases to suspect antidepressant effects of *Scutellaria lateriflora* and *Scutellaria baicalensis*, there are no convincing reports about antidepressant effects of *Scutellaria* in patients diagnosed with MDD.

7.58.6 DOSAGE

Easley and Horne state the dose of *Scutellaria baicalensis* to be 10 drops to 5 ml of tincture of dried leaf and flower (1:60% alcohol) 2–4 times daily. Alternatively, they suggest a standard infusion 4–8 oz three times daily.[16]

7.58.7 TOXICITY

In tests of acute LD50 of aqueous extract of *Scutellaria baicalensis* root in mice, no deaths or adverse effects were seen in doses up to 2,000 mg/kg.[17] That would extrapolate to a dose of 140 g for a 75 kg human adult. I do note a report from a Traditional Chinese Medical Hospital in Beijing in which a very small percentage (0.12%) of patients suffered elevations in transaminases after treatment with *Scutellaria baicalensis radix*. In most cases, these elevations were mild, but there was at least one case in which five-fold increases were noted.[18]

7.58.8 SAFETY IN PREGNANCY

Reference texts from Traditional Chinese Medicine advise against use of *Scutellaria baicalensis* during pregnancy.[19]

7.58.9 DRUG INTERACTIONS

A number of phytochemicals isolated from *Scutellaria baicalensis* have been found to inhibit the cytochrome enzyme CYP2D6.[20] The CYP2D6 enzyme catalyzes the metabolism of a large number of clinically important drugs including antidepressants, neuroleptics, and opioids.[21]

REFERENCES

1. Cole IB, Cao J, Alan AR, Saxena PK, Murch SJ. Comparisons of *Scutellaria baialensis*, *Scutellaria laterflora*, and *Scutellaria racemosa*: genome size, antioxidant potential and phytochemistry. *Planta*. 2007;74:1–8.

2. Lohani M, Ahuja M, Buabeid MA, et al. Anti-oxidative and DNA protecting effects of flavonoids-rich *Scutellaria lateriflora*. *Nat Prod Commun*. 2013;8(10):1415–1418.

3. Nan JX, Park E-J, Kim Y-C, Ko G, Sohn DH. *Scutellaria baicalensis* inhibits liver fibrosis induced by bile duct ligation or carbon tetrachloride in rats. *JPP*. 2002;54(4): 555–563.

4. Kim EH, Shim B, Kang S, et al. Anti-inflammatory effects of *Scutellaria baicalensis* extract via suppression of immune modulators and MAP kinase signaling molecules. *J Ethnopharmacol*. 2009;126(2):320–331.

5. Xi Y-L, Hong-Xia LI, Chen C, et al. Baicalin attenuates high fat diet-induced insulin resistance and ectopic fat storage in skeletal muscle, through modulating the protein kinase B/Glycogen synthase kinase 3 beta pathway. *Chin J Nat Med*. 2016;14(1): 48–55.

6. Zhu W, Ma S, Qu R, Kang D, Liu Y. Antidepressant effect of baicalin extracted from the root of *scutellaria baicalensis* in mice and rats. *J Pharm Biol*. 2006;44:7503–7510.

7. Li Y-C. Chronic treatment with baicalin prevents the chronic mild stress-induced depressive-like behavior: involving the inhibition of cyclooxygenase-2 in rat brain. *Prog Neuropsychopharmacol Biol Psychiatry*. 2013;40(10):138–143.

8. Li Y-C, Wang LL, Pei YY, et al. Baicalin decreases SGK1 expression in the hippocampus and reverses depressive-like behaviors induced by corticosterone. *Neuroscience*. 2015;311:130–137.

9. Li R, Wang X, Qin T, et al. Apigenin ameliorates chronic mild stress-induced depressive behavior by inhibiting interleukin-1β production and NLRP3 inflammasome activation in the rat brain. *Behav Brain Res*. 2016;296:318–325.

10. Li R, Zhao D, Qu R, Qu R, Ma S. The effects of apigenin on lipopolysaccharide-induced depressive-like behavior in mice. *Neurosci Lett*. 2015;594:17–22.

11. Filho CB, Jesse CR, Donato F, et al. Chrysin promotes attenuation of depressive-like behavior and hippocampal dysfunction resulting from olfactory bulbectomy in mice. *Chem-Biol Interact*. 2016;260:154–162.

12. Lee S, Kim DH, Lee DH, et al. Oroxylin A, a flavonoid, stimulates adult neurogenesis in the hippocampal dentate gyrus region of mice. *Neurochem Res*. 2010;35(11):1725–1732.

13. Murch SJ, Rupasinghe HPV, Goodenowe D, Saxena PK. A metobolomic analysis of medicinal diversity in Huang-qin (*Scutellaria baicalnensis*) genotypes discovery of new compounds. *Plant Cell Rep*. 2004;23:419–425.

14. Gafner S, Bergeron C, Batcha LL, et al. Inhibition of [3H]-LSD binding to 5-HT7 receptors by flavonoids from *Scutellaria lateriflora*. *J Nat Prod*. 2003;66:535–537.

15. Brock C, Whitehouse J, Tewfik I, Towell T. American skullcap (*Scutellaria lateriflora*): a randomised, double-blind placebo-controlled crossover study of its effects on mood in healthy volunteers. *Phytother Res*. 2014;28(5):692–698.

16. Easley T, Horne S. *The Modern Herbal Dispensary*. Berkeley, CA: North Atlantic Books; 2016. p. 301.

17. Lee J-W, Jung Y, Jung T, Kim JD, Choi HY. Mouse single oral dose toxicity test of *Scutellariae Radix* aqueous extracts. *Korean J Orient Int Med*. 2013;34(1)46–58.

18. Melchart D, Hager S, Albrecht S, Dai J, Weidenhammer W, Teschke R. Herbal traditional chinese medicine and suspected liver injury: a prospective study. *World J Hepatol*. 2017 Oct 18;9(29):1141–1157.

19. Chen JK, Chen TT. *Chinese Medical Herbology and Pharmacology*. City of Industry, CA: Art of Medicine Press; 2004.

20. Mo SL, Liu WF, Chen Y, et al. Ligand- and protein-based modeling studies of the inhibitors of human cytochrome P450 2D6 and a virtual screening for potential inhibitors from the chinese herbal medicine, *Scutellaria baicalensis* (Huangqin, Baikal Skullcap). *Comb Chem High Throughput Screen.* 2012;15(1):36–80.
21. Bertilsson L, Dahl ML, Dalén P, Al-Shurbaji A. Molecular genetics of CYP2D6: clinical relevance with focus on psychotropic drugs. *BJCP.* 2002;53(2):111–122.

7.59 *SILYBUM MARIANUM* (MILK THISTLE)

Silybum marianum, commonly known as milk thistle, is a biennial plant of the *Asteraceae* family. The plant is native to the low mountains of Mediterranean Europe, but is now cultivated throughout the world. The herb has been used as a healing substance for the last 2,000 years. It is mentioned in the writings of the famous physicians and herbalists of the past, Dioscorides, Pliny the Elder, Hieronymus Bock, Jacobus Theodorus, and Nicholas Culpeper. It has been used primarily for treatment of liver diseases, including alcoholic liver disease, cirrhosis, steatohepatitis, and non-alcoholic toxic and drug induced-hepatitis.[1] The German Commission E recommends its use primarily for dyspeptic complaints and liver conditions, including toxin-induced liver damage and hepatic cirrhosis, and as a supportive therapy for chronic inflammatory liver conditions.[2] However, the mechanisms by which it improves liver function may be of benefit in the resolution of more general symptoms of Metabolic Syndrome, and possibly in the prevention and treatment of MDD.

The extract of *Silybum marianum*, referred to as silymarin, is the combination of the flavonolignans, silybin, isosilybin, silydianin, and silychristine.[3]

7.59.1 ANTIOXIDANT

Silymarin has potent anti-inflammatory and antioxidant effects. The *in vivo* antioxidant effects of silymarin have been shown in numerous studies of its hepatoprotective effects. Silymarin almost completely blocked the oxidative cirrhotic damage of CCl_4 in the livers of rats. Serum transaminases remained within normal limits, and lipid peroxidation in liver tissue was prevented.[4] In a similar study, silymarin prevented increases in transaminases after treatment with CCl_4, and also prevented decreases in superoxide dismutase, glutathione reductase, glutathione -S-transferase, and glutathione in liver tissue. CCl_4-induced increases in serum pro-inflammatory cytokines including TNF-α, TGF-β1, and IL-6 were also attenuated.[5]

Owing to their polyphenolic molecular structure, the flavonolignans in silymarin are highly lipophilic and have central nervous system effects. Silymarin was found to spare gerbil brain from damaging effects of oxygen radicals after 30 minutes of arterial occlusion. Superoxide dismutase levels were partially restored by silymarin treatment, whereas levels of malondialdehyde and inflammatory leukotrienes were reduced.[6]

7.59.2 ANTI-INFLAMMATORY

Silymarin attenuates inflammation in two standard rodent models, carrageenan-induced paw edema in rats and xylene-induced ear inflammation in mice. In the latter, silymarin was found to be as effective as indomethacin.[7] Silymarin also improves the survivability of lipopolysaccharide-induced shock in mice due to its potent anti-inflammatory effects. Silymarin dose-dependently suppressed the lipopolysaccharide-induced production of IL-1β and PGE2 in cultured mouse macrophages. These anti-inflammatory effects were due in part to prevention of nuclear translocation of NF-kβ.[8]

Silymarin also has central effects. Silymarin attenuates the oxidative and inflammatory damage of cerebral ischemia and reperfusion. Treatment with silymarin prior to the ligation of carotid arteries reduced the size of the subsequent brain infarction by 16–40% and improved neurological deficits. Biomarkers of lipid peroxidation, protein nitrosylation, and oxidative stress were all reduced. In addition, activities of inflammatory enzymes, including inducible nitric oxide synthase, cyclooxygenase-2, and myeloperoxidase, were reduced; the inflammation activators, NF-kβ and STAT-1 were dampened; and levels of IL-6 and TNF- were reduced.[9]

7.59.3 ANTIDIABETIC/ANTI-METABOLIC SYNDROME

NFk-B, which is reined in by silymarin, stimulates inflammation and thus contributes to insulin resistance.[10] Silymarin also reduces activity of TNF-α, which not only stimulates NFk-B, but also causes immune system cells to release reactive oxygen compounds. The resulting oxidative damage further diminishes response to insulin. In a study of patients diagnosed with both cirrhosis and diabetes, silymarin reduced abnormal oxidation of fats, increased sensitivity to insulin, and lowered abnormally high levels of insulin in the blood.[11]

7.59.4 PRECLINICAL ANTIDEPRESSANT-LIKE EFFECTS

Administration of aqueous, but not ethanolic, extract of *Silybum marianum* seeds significantly reduced the duration of immobility of mice in the forced swim test.[12] In another study, acute administration of silymarin to mice reduced immobility in both the forced swim and tail suspension tests. It was further found that co-administration of the NO synthase inhibitor, l-NAME, blocked this antidepressant-like effect, suggesting that it is mediated by production of nitric oxide.[13]

Chronic administration of silymarin reduced immobility of mice in the forced swim test, and reversed some of the depression-like effects of reserpine in this test. Silymarin also countered oxidative damage caused by reserpine as shown by decreased thiobarbituric acid reactive substances and increased glutathione levels in brain tissue.[14]

Pretreatment with silymarin also ameliorated the depression-like effects in mice caused by chronic unpredictable mild stress. Immobility was reduced in the forced swim test and preference for sucrose was restored, both of which being indications of anti-depressant effects. Treatment also reduced the high, stress-induced serum levels of corticosterone. Neurochemical analyses of brain tissue further showed that silymarin increased towards normal the levels of BDNF, and the neurotransmitters, serotonin, norepinephrine, and dopamine in the hippocampus and cerebral cortex. Levels of the inflammatory cytokines IL-6 and TNF-α were reduced in those areas of the brain; whereas signs of antioxidant effects of treatment were evident in reduced formation of malondialdehyde formation, but increases in the activities of superoxide dismutase and catalase. Chronic treatment with the antidepressant fluoxetine produced the same types of changes as silymarin.[15] Along with reserpine and chronic-stressed-induced depression, silymarin also relieved the depression-like symptoms induced by acute restraint stress and olfactory bulbectomy in mice.[16]

The flavonolignan, silibinin, isolated from extracts of *Silybum marianum*, also exhibits antidepressant-like effects in mice subjected to chronic unpredictable mild stress. Immobility times were reduced in both the forced swim and tail suspension tests. Silibinin also increased the levels of BDNF, serotonin, and norepinephrine in the prefrontal cortex and hippocampus. It is noteworthy that animals had been subjected to stress for five weeks prior to a three-week course of silibinin, suggesting that treatment can treat and not simply prevent depression-like syndromes.[17]

Oral administrations of silymarin have also been reported to provide an anxiolytic-like effect in rats demonstrated by increases in activity in the elevated plus-maze. Moreover, this effect was amplified by barely effective doses of the 5-HT1A agonist 8-OH-DPAT and blocked by the 5-HT1A antagonist, NAN190. Thus, this effect of silymarin was mediated in part by 5-HT1A receptors.[18] However, along with anxiolysis, chronic administration of 5-HT1A receptor agonists are thought to exert antidepressant effects.[19]

7.59.5 Human Antidepressant Effects

There are no formal studies of the effects of silymarin on MDD in human subjects. Moreover, the only mood related study is not rigorous in its analysis. In a group of elite female athletes, silymarin caused significant further improvements in self-rated mood and serum levels of the inflammatory serum cytokine, IL-6 beyond those obtained in a rigorous physical training program.[20]

Curiously, there are at least two reports on the use of silymarin in the treatment of OCD in human subjects. In a double-blind randomized trial, adults diagnosed with OCD through the Yale-Brown Scale for OCD, patients received either extract of silymarin or an active fluoxetine control for eight weeks. At the end of the time, both treatments significantly reduced the Yale-Brown Scale for OCD scores. There were no significant differences in the efficacy of the two treatments. There were also no differences between the two groups in terms of observed side effects.[21] In a later published series of three cases, authors have described success in treating resistant OCD with 150 mg of silymarin twice a day. In one case, only silymarin was used for the remission of symptoms after failure of standard treatments with antidepressants.[22] OCD can be very difficult to treat, and silymarin may be worth trying especially if there is comorbid MDD and/or Metabolic Syndrome.

7.59.6 Dosage

The German E Commission notes the daily dosage of silymarin to be 200–400 mg a day, standardized to silibinin content.[23]

7.59.7 Toxicity

Silymarin appears to have quite low acute toxicity. In rats, the LD50 of orally administered silymarin was 10 g/kg.[24] This would extrapolate to an approximate dose of 700 g for a 75 kg human adult.

7.59.8 Safety in Pregnancy

The *Botanical Safety Handbook* cites evidence of relative safety of the use of *Silybum marianum* during pregnancy. Limited human and animal studies of the use of silymarin during pregnancy reveal no adverse effects.[25,26]

7.59.9 Drug Interactions

The *Botanical Safety Handbook* notes no known interactions with drugs.[27] Indeed, at moderate doses, extract of *Silybum marianum* has little if any ability to either inhibit or induce P450 enzymes from human liver microsomes.[28]

REFERENCES

1. Pepping J. Milk thistle: *Silybum marianum*. *Am J Health Syst Pharm*. 1999;56(12):1195–1197.
2. Blumenthal M, Busse WR, editors. *The Complete German Commission E Monographs: Therapeutic Guide to Herbal Medicines*. Austin, TX: American Botanical Council; 1998.
3. Post-White J, Ladas EJ, Kelly KM. Advances in the use of milk thistle: (*Silybum marianum*). *Integr Cancer Ther*. 2007;6(2):104–109.
4. Mourelle M, Muriel P, Favari L, Franco T. Prevention of CCl_4-induced liver cirrhosis by silymarin. *Fundam Clin Pharmacol*. 1989;3:183–191.
5. Abdel-Moneim AM, Al-Kahtani MA, El-Kersh MA, Al-Omair MA. Free radical-scavenging, anti-inflammatory/anti-fibrotic and hepatoprotective actions of taurine and silymarin against CCl_4 induced rat liver damage. *PLoS One*. 2015;10(12):e0144509.
6. Rui YC. Effects of silybin on production of oxygen free radical, lipoperoxide and leukotrienes in brain following ischemia and reperfusion. *Zhongguo Yao Li Xue Bao*. 1990;11:418–421.
7. De La Puerta R, Martinez E, Bravo L, Ahumada MC. Effect of silymarin on different acute inflammation models and on leukocyte migration. *JPP*. 1996;48(9):968–970.
8. Kang JS, Jeon YJ, Park SK, Yang KH, Kim HM. Protection against lipopolysaccharide-induced sepsis and inhibition of interleukin-1β and prostaglandin E2 synthesis by silymarin. *Biochem Pharmacol*. 2004;67(1):175–181.
9. Hou Y-C, Liou K-T, Chern C-M. Preventive effect of silymarin in cerebral ischemia-reperfusion-induced brain injury in rats possibly through impairing NF-B and STAT-1 activation. *Phytomedicine*. 2010;17:963–973.
10. Paska A, Permana PA, Menge C, et al. Macrophage-secreted factors induce adipocyte inflammation and insulin resistance. *Biochem Biophys Res Commun*. 2006;341(2):507–514.
11. Velussi M, Cernigoi AM, Dapas F, Caffau C, Zilli M. Long-term (23 months) treatment with an anti-oxidant drug (silymarin) is effective on hyperinsulinemia, exogenous insulin needs and malondialdehyde levels in cirrhotic diabetic patients. *J Hepatol*. 1997;26(4):871–879.

12. Karimi G, Keisari MS. Evaluation of antidepressant effect of ethanolic and aqueous extracts of *Silybum marianum* L. seed in mice. *JMP*. 2007;4(24):38–43.

13. Khoshnoodi M, Fakhrael N, Dehpour AR. Possible involvement of nitric oxide in anti-depressant-like effect of silymarin in male mice. *Pharm Biol*. 2015;53(5):739–745.

14. Bansal N, Gill R, Gupta GD. Silymarin: a flavolignan with antidepressant activity. *J Pharm Innov*. 2013;3(5):93–98.

15. Thakare VN, Patil RR, Oswal RJ, Dhakane VD, Aswar MK, Patel BM. Therapeutic potential of silymarin in chronic unpredictable mild stress induced depressive-like behavior in mice. *J Psychopharmacol*. 2018;32(2):223–235.

16. Thakare V. Pharmacological Evalation of Silymarin and Protocatechuic Acid in Experimentally induced Depression in Rodents [Doctoral dissertation]. [Ahmedabad, India]: Nirma University, Ahmedabad; 2017.

17. Yan W-J, Tan YC, Xu JC, et al. Protective effects of silibinin and its possible mech-anism of action in mice exposed to chronic unpredictable mild stress. *Biomol Ther (Seoul)*. 2015;23(3):245–250.

18. Solati J, Yaghmaei P, Mohammdadi K. Role of the 5-HT1A serotonergic system in anxiolytic-like effects of silymarin. *Neurophysiology*. 2012;44(1):49–55.

19. Blier P, Ward NM. Is there a role for 5-HT1A agonists in the treatment of depression? *Biol Psychiatry*. 2003;53(3):193–203.

20. Nazarali P, Pormphamadi A, Hanachi P. Effect of six weeks of resistance training (RT) and silymarin supplement on the changes in the inflammation marker interleukin 6 and psychological profile in elite female Taekwondo players in Alborz Province. *Int J Sport Stud*. 2015;5(1):57–61.

21. Sayyah M, Boostani H, Pakseresht S, Malayeri A. Comparison of *Silybum marianum* (L.) Gaertn. with fluoxetine in the treatment of obsessive–compulsive disorder. *Prog Neuropsychopharmacol Biol Psychiatry*. 2010;34(2):362–365.

22. Grant JE, Odlaug BL. Silymarin treatment of obsessive-compulsive spectrum disor-ders. *J Clin Psychopharmacol*. 2015;35(3)340–342.

23. Blumenthal M, Busse WR, Goldberg A, et al. *The Complete German Commission E Monographs*. Austin, TX: American Botanical Council; 1998. p. 170.

24. Lecomte J. Pharmacologic properties of silybin and silymarin. *Rev Med Liege*. 1975;30:110–114.

25. Gonzalez M, Reyes H, Ribalta J, et al. Effect of sylimarin on pruritus of cholestasis. *Hepatology*. 1988;8(5):1356.

26. Hahn G, Lehmann HD, Kürten M. On the pharmacology and toxicology of sily-marin, an antihepatotoxic active principle from *Silybum marianum* (L.) Gaertn). *Arzneimittelforschung*. 1968;18(6):698–704.

27. Gardner Z, McGuffin M, editors. *Botanical Safety Handbook*. 3rd ed. Boca Raton, FL: CRC Press; 2013. p. 807.

28. Doehmer J, Weiss G, McGregor GP, Appel K. Assessment of a dry extract from milk thistle (Silybum marianum) for interference with human liver cytochrome-P450 activi-ties. *Toxicol in Vitro*. 2011;25(1):21–27.

7.60   *THEOBROMA CACAO* (CHOCOLATE)

Theobroma cacao, hereafter to be called cocoa or chocolate, was cultivated by the Olmec Indians of Central America as far back as 1500 BC, and was a drink of the Indian aristocracy. After it was brought back to Europe by the Spaniards, it became a drink of the European elite. In 1753, the great Swedish taxonomist Linnaeus quite rightfully gave the cocoa plant the scientific name *Theobroma*, which means "food of the gods." Aside from serving as a beverage and dessert, cocoa and chocolate are coming to be seen as having significant health benefits.

The cocoa bean from which chocolate is made contains nitrogenous elements, most notably the methylxanthines, theobromine and caffeine. It also contains a variety of flavonoids, the most prominent of which are epicatechin and catechin 2. Those flavonoids, in turn, can also serve as building blocks for the polyphenolic procyanidins.[1]

Along with flavonoids and methylxanthines shared with other herbs, chocolate has some unique neurochemical properties. It contains significant amounts of phenylethylamine.[2] This substance, which is the decarboxylation product of the amino acid phenylalanine, has long been suspected of having antidepressant effects.[3] Chocolate also contains small amounts of the endocannabinoid anandamide. Other substances in chocolate similar in structure to anandamide may act on the enzymes systems in the brain to increase the levels of endogenous cannabinoids in the brain.[4]

7.60.1 ANTIOXIDANT

After consumption of chocolate, particularly dark chocolate, polyphenols are measurable in the cerebral cortex, hippocampus, and cerebellum of Wistar rats. There these polyphenols exert potent antioxidant effects, as shown by increases in superoxide dismutase activity and reductions in levels of TBARS, which are indicators of lipid peroxidation.[5] Epicatechin, a primary active flavonoid in chocolate, enters the brains of intact animals after oral ingestion.[6] In a rat model of traumatic brain injury, epicatechin significantly reduced lesion volume, edema, and cell death and improved neurologic function. These changes were largely due to potent antioxidant effects, including inhibition of heme oxygenase-1 expression, and enhancement of superoxide dismutase and Nrf2 activity.[7]

7.60.2 ANTI-INFLAMMATORY

In vitro, cocoa flavanols reduce the expression of IL-2 by macrophages stimulated with phytohemagglutinin, a response similar to that of lipopolysaccharide.[8] Extracts of cocoa also prevent the synthesis of nitric oxide by cultured macrophages stimulated by lipopolysaccharide or IFN-γ.[9] Epicatechin, catechin, and procyanidins purified from cocoa have been found to block activation of NF-kβ in cultured T-lymphocytes, thus preventing initiation of an important component of the inflammatory cascade.[10] Habitual intake of chocolate can reduce cardiac risk by nearly 20%. This is thought to be due to reduction of inflammation in arterial walls, reduction in leukotrienes, increases in prostacyclin, and decreases in LDL oxidation.[11]

7.60.3 ANTIDIABETIC/ANTI-METABOLIC SYNDROME

In immortalized human liver cells, cocoa flavonoids were found to strengthen insulin signaling by activating key proteins of the insulin response pathway. These flavonoids also dampened activity of phosphoenolpyruvate carboxykinase, a key protein involved in the gluconeogenesis, which in turn led to diminished glucose production.[12]

In intact Zucker fatty rats, animals that are bred to exhibit the signs of Metabolic Syndrome, adding cocoa to the diet decreased body weight gain, glucose, and insulin levels, and improved glucose tolerance and insulin resistance.[13] Chocolate also reduces insulin resistance in healthy human subjects.[14]

7.60.4 PRECLINICAL ANTIDEPRESSANT-LIKE EFFECTS

Studies have shown antidepressant-like effects of chocolate and cocoa flavonoids in rodents. Administration of a polyphenolic extract of cocoa for 14 days significantly reduced the duration of immobility of male rats in the forced swim test.[15] Cocoa extracts also reduced immobility time in the forced swim in ovariectomized female rats.[16] In mice, a week of treatment with cocoa extract decreased immobility times in the forced swim and tail suspension tests, and these effects were similar to those produced by amitriptyline.[17] Despite formidable properties noted above, epicatechin has not been evaluated for antidepressant-like effects in rodents. However, the similar chocolate flavonoid, catechin, reduces immobility in the forced swim test in rats exhibiting depression-like behavior due to chronic administration of high doses of corticosterone.[18]

7.60.5 HUMAN ANTIDEPRESSANT EFFECTS

Human studies have found connections between depressed mood and craving for chocolate. In one case, the craving for chocolate while depressed was associated with both personality type and subtype of depression. That is, such individuals, primarily women, scored high on neuroticism scales and more often were found to suffer so-called "Atypical MDD," characterized by increases in appetite, weight gain, sensitivity to rejection, hypersomnia, and feeling limbs to be "heavy like lead." Most reported that consuming the chocolate improved their moods.[19] However, improvements in mood from eating chocolate while depressed were deemed ephemeral.[20] In evaluating the effects of chocolate on mood in healthy men and women, it was similarly found that the chocolate briefly improved mood, but only due to it being a tasty, sweet treat. Unpalatable chocolate had no such effect.[21]

In an epidemiological study, consumption of chocolate was associated with tendency for depressed mood in women as defined by higher scores on the Center for Epidemiologic Studies Depression Scale.[22] Whether the chocolate caused the depressed mood, or the feeling "blue" led to craving for chocolate was not clear. It has been reported, for example, that sweet, high fat foods are preferred by women with binge-eating disorders and that those preferences are mediated by the endogenous opioid system.[23] Nonetheless, it appears that depressed mood persisted in the above-noted women despite their intake of chocolate, which would not bode well for the notion that chocolate might serve as an antidepressant.

In a study of chronic administration of chocolate polyphenols to healthy adults, treatment was found to significantly improved calmness and contentedness, albeit through the rather imprecise method of the Bond and Lader Visual Analogue Scales.[24]

A preparation of high cocoa liquor/polyphenol rich chocolate administered over eight weeks also relieved symptoms in patients diagnosed with Chronic Fatigue Syndrome by the British Centres for Disease Control and Prevention criteria. None of the subjects were suffering comorbid psychiatric illnesses as defined by the DSM-IV, nor were any taking antidepressant or anxiolytic medications. Nonetheless, by the end of the study, the chocolate significantly lowered not only fatigue, but also anxiety and depression scores on the Hospital Anxiety and Depression Scale.[25] Unfortunately, there are no formal studies of the use of chocolate to treat patients diagnosed with MDD.

7.60.6 DOSAGE

Since chocolate is not prescribed as medication, there are no bases to determine a standard dose for augmentation of antidepressant treatment. However, in the above study in which chocolate enhanced insulin sensitivity in humans, the subjects received 100 g dark chocolate per day. In the above study of healthy adults in which treatment was found to significantly improve calmness and contentedness, subjects consumed 15 g of high polyphenol chocolate three times a day.

7.60.7 TOXICITY

In a study of toxicological study of cocoa, it was found that the extract is quite benign with an LD50 over 5000 mg of dried aqueous extract of cocoa per kg.[26] Extrapolated to a 75 kg human, this would be a single dosing of about 350 g of dried extract. Note that the pure aqueous extract is free of cocoa butter, which forms a substantial part by weight of chocolate.

The "dark" side of chocolate is the amount of sugar and saturated fat that it usually contains. The first of these problems is easily solved. There are many delicious, sugar free dark chocolates on the market now. The second problem might not be as bad as it first may seem. The cocoa butter mixed into chocolate to form bars contains about 30% oleic acid, which is the healthy, monounsaturated fatty acid in olive oil. Cocoa butter is rich in saturated fatty acids. It contains about 25% palmitate and 33% stearate. Admittedly, aside from being a natural product of our metabolism, palmitate has few redeeming features. On the other hand, stearic acid has probably got a bad rap. Stearic acid tends not to increase LDL or decrease HDL. The daily feeding of as much as 10 oz of milk chocolate bars does not adversely affect serum cholesterol profiles.[27]

7.60.8 SAFETY IN PREGNANCY

A study of chocolate consumption during pregnancy shows it to not be harmful. Indeed, it appears to bestow a number of benefits, including reduced blood pressure, improved glycemic control, and normalization of liver enzymes in the serum.[28]

7.60.9 DRUG INTERACTION

Review of the literature reveals no reports on effects of chocolate on drug action or metabolism.

REFERENCES

1. Nehlig A. The neuroprotective effects of cocoa flavanol and its influence on cognitive performance. *BJCP.* 2013;75(3):716–727.
2. Ziegleder G, Stojacic E, Stumpf B. Occurrence of beta-phenylethylamine and its derivatives in cocoa and cocoa products. *Z Lebensm Unters Forsch.* 1992;195(3):235–238.
3. Sabelli H, Fahrer R, Medina RD, Ortiz FE. Phenylethylamine relieves depression after selective MAO-B inhibition. *J Neuropsychiatry Clin Neurosci.* 1994;6(2):203.
4. James JS. Marijuana and chocolate. *AIDS Treat News.* 1996;257:3–4.
5. da Silva Medeiros N, Marder RK, Wohlenberg MF, Funchal C, Dani C. Total phenolic content and antioxidant activity of different types of chocolate, milk, semisweet, dark, and soy, in cerebral cortex, hippocampus, and cerebellum of wistar rats. *Biochem Res Int.* 2015:Article ID 294659.
6. Abd El Mohsen MM, Kuhnle G, Rechner AE, et al. Uptake and metabolism of epicatechin and its access to the brain after oral ingestion. *Free Radic Biol Med.* 2002;33:1693–1702.
7. Cheng T, Wang W, Li Q, et al. Cerebroprotection of flavanol (–)-epicatechin after traumatic brain injury via Nrf2-dependent and -independent pathways. *Free Radic Biol Med.* 2016;92:15–28.
8. Mao TK, Powell J, Van de Water J, Keen CL, Schmitz HH, Gershwin ME. The influence of cocoa procyanidins on the transcription of interleukin-2 in peripheral blood mononuclear cells. *Int J Immunother.* 1999;15:23–29.
9. Ono K, Takahashi T, Kamei M, et al. Effects of an aqueous extract of cocoa on nitric oxide production of macrophages activated by lipopolysaccharide and interferon-γ. *Nutrition.* 2003;19(7–8):681–685.
10. Mackenzie GG, Carrasquedo F, Delfino JM, Keen CL, Fraga CG, Oteiza PI. Epicatechin, catechin, and dimeric procyanidins inhibit PMA-induced NF-κB activation at multiple steps in Jurkat T cells. *FASEB J.* 2004;18:167–169.
11. Ding EL, Hutfless SM, Ding X, Girotra S. Chocolate and prevention of cardiovascular disease: a systematic review. *Nutr Metab.* 2006;3:2.
12. Cordero-Herrera I, Martín MA, Bravo L, Goya L, Ramos S. Cocoa flavonoids improve insulin signalling and modulate glucose production via AKT and AMPK in HepG2 cells. *Mol Nutr Food Res.* 2013;57(6):974–985.
13. Cordero-Herrera I, Martín MA, Escrivá F, Álvarez C, Goya L, Ramos S. Cocoa-rich diet ameliorates hepatic insulin resistance by modulating insulin signaling and glucose homeostasis in Zucker diabetic fatty rats. *J Nutr Biochem.* 2015;26(7):704–712.
14. Grassi D, Lippi C, Necozione S, Desideri G, Ferri C. Short-term administration of dark chocolate is followed by a significant increase in insulin sensitivity and a decrease in blood pressure in healthy persons. *Am J Clin Nutr.* 2005;81(3):611–614.
15. Messaoudi M, Bisson JF, Nejdi A, Rozan P, Javelot H. Antidepressant-like effects of a cocoa polyphenolic extract in Wistar-Unilever rats. *Nutr Neurosci.* 2008;11:269–276.
16. Sari ABT, Misnawi M, Pudjiastuti P, Baktir A. Depressive behavior of rats consuming cocoa powder and cocoa extract. *Pelita Perkeb.* 2018;34(2):90–94.
17. Hidayat M, Taufik AAR, Hidatul MJ, Sabariah NAW. Therapeutic role of *Theobroma cacao* in depressive disorder model in mice. *Malaysian J Med Sci.* 2007;14(Suppl):106.
18. Lee B, Sur B, Kwon S, et al. Chronic administration of catechin decreases depression and anxiety-like behaviors in a rat model using chronic corticosterone injections. *Biomol Ther (Seoul).* 2013;21(4):313–322.
19. Parker G, Crawford J. Chocolate craving when depressed: a personality marker. *Br J Psychiatry.* 2007;191:351–352.
20. Parker G, Parker I, Brotchie H. Mood state effects of chocolate. *J Affect Disord.* 2006;92:149–159.

21. Macht M, Mueller J. Immediate effects of chocolate on experimentally induced mood states. *Appetite.* 2007;49:667–674.
22. Rose N, Koperski S, Golomb BA. Mood food: chocolate and depressive symptoms in a cross-sectional analysis. *Arch Intern Med.* 2010;170(8):699–703.
23. Yanovski S. Sugar and fat: cravings and aversions. *J Nutr.* 2003;133:835S–837S.
24. Pase MP, Scholey AB, Pipingas A, et al. Cocoa polyphenols enhance positive mood states but not cognitive performance: a randomized, placebo-controlled trial. *J Psychopharmacol.* 2013;27(5):451–458.
25. Sathyapalan T, Beckett S, Rigby AS, Mellor DD, Atkin SL. High cocoa polyphenol rich chocolate may reduce the burden of the symptoms in chronic fatigue syndrome. *Nutr J.* 2010;9:55.
26. Eniola OW, Omobola OA, Joseph AS, et al. Phytochemical screening and toxicological implication of administration of aqueous cocoa powder extract obtained from Nigeria in albino rat. *Asian J Plant Sci Res.* 2015;5(5):4–8.
27. Kris-Etherton PM, Mustad VA. Chocolate feeding studies: a novel approach for evaluating the plasma lipid effects of stearic acid. *Am J Clin Nutr.* 1994;60(6):1029S–1036S.
28. Di Renzo GD, Brillo E, Romanelli M. Potential effects of chocolate on human pregnancy: a randomized controlled trial. *J Matern Fetal Neonatal Med.* 2012;25(10):1860–1867.

7.61 *TILIA* (LINDEN)

Tilia is a genus of medicinal trees commonly referred to as Linden. The genus includes American linden (*Tilia americana*), large-leaved linden (*Tilia platyphyllos*), little-leaf linden (*Tilia cordata*), and silver linden (*Tilia tomentosa*). Phytochemical studies have demonstrated that the *Tilia* species possess a wealth of flavonoids, terpenes, and glycosides that are common among medicinal plants, including quercetin, kaempferol, rutin, scopoletin, hyperoside, β-sitosterol, and tiliroside.[1] The flowers and leaves of these trees have been used in folk traditions as medicine for treatment of conditions as various as colds, cough, fever, infections, inflammation, high blood pressure, migraine, muscle spasms, nausea, and hysteria.[2] It has also been used by herbalists in the treatment of melancholia, particularly when perturbed by worry and insomnia.[3]

7.61.1 ANTIOXIDANT

In the standard *in vitro* DPPH scavenging assay, the ethanolic extract of *Tilia cordata* leaves was found to be rich in polyphenols with high antioxidant activity.[4] *In vivo*, extracts of leaves of *Tilia americana* extracts protected against CCl_4-induced oxidative damage in rats. Chronic administration of extracts of *Tilia* prevented serum elevations of liver enzymes that are indicative of CCl_4-induced hepatic damage, and preserved activity of hepatic superoxide dismutase, catalase, glutathione-S-transferase, and glutathione reductase. The extract also preserved the activity of those antioxidant enzymes and prevented lipid peroxidation in brain tissue.[5]

7.61.2 ANTI-INFLAMMATORY

The various *Tilia* trees have been shown to exhibit anti-inflammatory effects. Two of the prominent flavonoid glycosides in extract of *Tilia argentea*, kaempferol-3,7-O-α-dirhamnoside and quercetin-3,7-O-α-dirhamnoside, dampened carrageenan-induced paw edema model in mice.[6] Extract of *Tilia cordata* containing epicatechin and other procyanidins also attenuated the response of human neutrophils to lipopolysaccharide *in vitro*. It diminished production of the inflammatory cytokines IL-8 and Macrophage Inflammatory Protein-1β, and decreased release of reactive oxygen species.[7]

7.61.3 ANTIDIABETIC/ANTI-METABOLIC SYNDROME

Tiliroside, a primary a glycosidic flavonoid in the *Tilia* genus and other medicinal plants, offers many beneficial effects in managing diabetes and preventing Metabolic Syndrome. Chronic administration of tiliroside to obese, diabetic mice decreased plasma insulin, free fatty acid, and triglyceride levels, and increased plasma adiponectin levels as well as levels of adiponectin receptors in liver tissue.[8] Low adiponectin levels are seen in patients with MDD, and intracerebroventricular administration of the hormone has anti-depressant-like effects in mice.[9] Quercetin[10] and kaempferol[11] in *Tilia* species also reduce hyperglycemia and enhance insulin sensitivity.

7.61.4 PRECLINICAL ANTIDEPRESSANT-LIKE EFFECTS

Tilia is most often thought of as an anxiolytic, mildly sedating herb. There is evidence in the animal literature to support this impression. For example, a methanolic extract of *Tilia americana*, though not n-hexane, ethyl acetate, or aqueous extracts, increased the amount of time spent by mice in the open arms of the elevated plus maze. The anxiolytic-like effect as attributed to tiliroside, quercetin, quercitrin, kaempferol, and their glycosides in that particular extarction.[12] In a similar study, the anxiolytic-like effects of methanol extract of *Tilia americana* in mice were attenuated by both the benzodiazepine blocker flumazenil and the 5-HT1A receptor antagonist WAY100635.[13]

However, studies also show antidepressant-like effects of *Tilia* species in rodents. A hydroethanolic extract of *Tilia Americana*, high in tiliroside, was found to reduced immobility of mice in the forced swim test.[14] Extract of *Tilia argentea* also decreased immobility of mice in the forced swim test.[15] Also, several of the primary phytochemicals in *Tilia* trees have themselves been found to produce antidepressant-like effects in rodents. Quercetin,[16] kaempferol,[17] and rutin[18] contained in the *Tilia* species have been found to reduce immobility in forced swim tests.

7.61.5 HUMAN ANTIDEPRESSANT EFFECTS

There are no published studies on the use of *Tilia* species for the treatment of MDD in human subjects.

7.61.6 DOSAGE

The German Commission E recommends *Tilia* only for coughs and colds. It suggests a dose of 2–4 g of dried flower per day, and offers no cautions in its use.[19] *The Modern Herbal Dispensary* recommends *Tilia* for its most common application, which is to relieve tension and anxiety. They recommend 15 ml of 1:5 dried leaf tincture, 2–4 times daily.[20]

7.61.7 Toxicity

Extracts from the various *Tilia* species, appears to be relatively non-toxic. In mice, the intraperitoneal LD50 of the dried methanol extract of *Tilia americana* was 375 mg/kg, whereas the hexane extract had an LD50 > 2,900 mg/kg.[21]

7.61.8 Safety in Pregnancy

The *Botanical Safety Handbook* notes there is no definitive guidance from the literature concerning the safety of this herb during pregnancy.[22]

7.61.9 Drug Interactions

The *Botanical Safety Handbook* notes no known interactions with commonly used medications.

REFERENCES

1. Negri G, Santi D, Tabach R. Flavonol glycosides found in hydroethanolic extracts from *Tilia cordata*, a species utilized as anxiolytics. *Braz J Med Plants*. 2013;15(2):217–224.
2. Bradley P, editor. *British Herbal Compendium, I*. Dorset, Great Britain: British Herbal Medicine Association; 1992. p. 142–144.
3. Yarnell E, Abascal K. Botanical treatments for depression. *Altern Complement Ther*. 2001;7(3):138–143.
4. Wissam Z, Al Asaad Nour JB, Zein N, Saleh D. Extracting and studying the antioxidant capacity of polyphenols in dry linden leaves (*Tilia cordata*). *J Pharmacogn Phytochem*. 2017;6(3):258–262.
5. Coballase-Urrutia E, Cárdenas-Rodríguez N, Gonzales-Garcia MC, et al. Biochemical and molecular modulation of CCl_4-induced peripheral and central damage by *Tilia americana* var. *mexicana* extracts. *Saudi Pharm J*. 2017;25(3):319–331.
6. Toker G, Küpeli E, Memisoğlu M, Yesilada E. Flavonoids with antinociceptive and anti-inflammatory activities from the leaves of *Tilia argentea* (silver linden). *J Ethnopharmacol*. 2004;95(2–3):393–397.
7. Czerwińskaa ME, Dudek MK, Pawłowska KA, Pruś A, Ziaja M, Granica S. The influence of procyanidins isolated from small-leaved lime flowers (*Tilia cordata* Mill.) on human neutrophils. *Fitoterapia*. 2018;127:115–122.
8. Goto T, Teraminami A, Lee JY, et al. Tiliroside, a glycosidic flavonoid, ameliorates obesity-induced metabolic disorders via activation of adiponectin signaling followed by enhancement of fatty acid oxidation in liver and skeletal muscle in obese-diabetic mice. *J Nutr Biochem*. 2012;23(7):768–776.
9. Liu J, Guo M, Zhang D, et al. Adiponectin is critical in determining susceptibility to depressive behaviors and has antidepressant-like activity. *Proc Natl Acad Sci USA*. 2012;109(30):12248–12253.
10. Jeong S-M, Kang MJ, Choi HN, Kim JH, Kim JI. Quercetin ameliorates hyperglycemia and dyslipidemia and improves antioxidant status in type 2 diabetic db/db mice. *Nutr Res Pract*. 2012;6(3):201–207.

11. Zang Y, Zhang L, Igarashi K, Yu C. The anti-obesity and anti-diabetic effects of kaempferol glycosides from unripe soybean leaves in high-fat-diet mice. *Food Funct.* 2015;6:834–841.

12. Herrera-Ruiz M, Román-Ramos R, Zamilpa A, Tortoriello J, Jiménez-Ferrer JE. Flavonoids from *Tilia americana* with anxiolytic activity in plus-maze test. *J Ethnopharmacol.* 2008;118(2):312–317.

13. Aguirre-Hernández E, González-Trujano ME, Terrazas T, Herrera Santoyo J, Guevara-Fefer P. Anxiolytic and sedative-like effects of flavonoids from *Tilia americana* var. *mexicana*: GABAergic and serotonergic participation. *Salud Ment.* 2016;39(1):37–46.

14. Aviles Montes D. Efecto antidepresivo y caracterización química de extractos de hojas de plantas silvestres y cultivadas en condiciones de invernadero de Tilia americana var [Masters thesis]. [Yautepec, Morelos]: Instituto Politecnico Nacional; 2008.

15. Aydin S, Öztürk Y, Başer KHC, Kirimer N, Kurtar-Öztürk N. Effects of *Alcea pallida* L. (A.) and *Tilia argentea* Desf. ex DC infusions on swimming performance in mice. *Phytother Res.* 1992;6(4):219–220.

16. Anjaneyulu M, Chopra K, Kaur I. Antidepressant activity of quercetin, a bioflavonoid, in streptozotocin-induced diabetic mice. *J Med Food.* 2003;6(4):391–395.

17. Hosseinzadeh H, Motamedshariaty V, Hadizadeh F. Antidepressant effect of kaempferol, a constituent of saffron (*Crocus sativa*) petal, in mice and rats. *Pharmacologyonline.* 2007;2:367–370.

18. Nöldner M, Schötz K. Rutin is essential for the antidepressant activity of *hypericum perforatum* extracts in the forced swimming test. *Planta Med.* 2002;68(7):577–580.

19. Blumenthal M, Busse WR, editors. *The Complete German Commission E Monographs: Therapeutic Guide to Herbal Medicines.* Austin, TX: American Botanical Council; 1998.

20. Easley T, Horne S. *The Modern Herbal Dispensary.* Berkeley, CA: North Atlantic Books; 2016.

21. Aguirre-Hernández E, Martíneza AL, González-Trujanoa JME, Moreno J, Vibrans H, Soto-Hernández M. Pharmacological evaluation of the anxiolytic and sedative effects of *Tilia americana* L. var. *mexicana* in mice. *J Ethnopharmacol.* 2007;109(1):140–145.

22. Gardner Z, McGuffin M, editors. *Botanical Safety Handbook.* 3rd ed. Boca Raton, FL: CRC Press; 2013. p. 861.

7.62   *TRIGONELLA FOENUM-GRAECUM* (FENUGREEK)

Trigonella foenum-graecum, commonly known as fenugreek, is an annual plant in the family *Fabaceae*. It has a long history of human use, both as medicine and foodstuff, and is thought to have first been cultivated as long as 6,000 years ago. Consumption of seeds, sprouts, and leaves in many Near Eastern and Mediterranean cuisines speaks to its relative safety. The leaves and seeds have been used to prepare extracts or powders for medicinal use in Chinese, Ayurvedic, Greek, Roman, and European herbal traditions. It has been used as a general tonic, and for conditions as varied as used as weakness, edema, failure of lactation, indigestion, flatulence, and baldness.[1] Modern studies suggest potential for treatment of diabetes, Metabolic Syndrome, and MDD.

Trigonella foenum-graecum is rich in phytochemicals including trigonelline, cineol, camphor, thymol, eugenol, aniline, phenol, elemene, selinene, and many others.[2] One of the chemicals responsible for the characteristic odor and flavor of the herb is the furanol sotolon. It also contains a variety of sterols and saponigens, and sterol saponigens, including stigmasterol, fucosterol, solosterol, campestenol, gitogenine, smilagenin, and tigogenin.[3] *Trigonella foenum-graecum* also contains an unusual amino acid, 4-hydroxyisoleucine, that makes up 0.6% of seed weight, but up to 85% of the free amino acids in seed.[4] There is evidence that this amino acid may have unique antidiabetic and antidepressant effects.

7.62.1 ANTIOXIDANT

Extract of *Trigonella foenum-graecum* has been shown to have potent antioxidant effects in a variety of *in vitro* assay systems. The seed extract scavenged hydroxyl radicals and inhibited hydrogen peroxide-induced lipid peroxidation in rat liver mitochondria. At high concentrations, the extract was a successful radical scavenger in the standard DPPH and ABTS assays, and these effects correlated with total phenolic content.[5] Aqueous extract of germinated seeds had similar *in vitro* antioxidant effects.[6] Chronic oral administration of *Trigonella foenum-graecum* to alloxan-induced diabetic rats increased blood antioxidant levels and decreased markers of lipid peroxidation.[7]

Chronic administration of *Trigonella foenum-graecum* and sodium orthovanadate in drinking water to diabetic rats increased antioxidant activities and reduced peroxidative damage in liver, heart, kidney, and brain. Levels of the antioxidant enzymes catalase, superoxide dismutase, and glutathione peroxidase were increased in those tissues.[8]

7.62.2 ANTI-INFLAMMATORY

Studies have also shown various components of the herb to have substantial anti-inflammatory effects. The methanolic extract of *Trigonella foenum-graecum* seeds exhibited anti-inflammatory effects in the standard formalin- and carrageenan-induced paw edema tests respectively. Interestingly, alkaline chloroform fraction also had antinociceptive effects equal to those of morphine. These degree of these effects of *Trigonella foenum-graecum* extracts were correlated with flavonoid and alkaloid content.[9] Many of the steroidal saponin glycosides obtained in methanol extraction of *Trigonella foenum-graecum* also exhibit potent anti-inflammatory effects. These substances were shown to inhibit phorbol-12-myristate-13-acetate-induced release of the inflammatory cytokine TNF-α in a cultured human leukemic cell line.[10]

The anti-inflammatory benefits of *Trigonella foenum-graecum* extend to the brain. The neuroinflammatory brain damage caused by treatment with aluminum chloride was attenuated by addition of powdered herb to the daily food of rats. *Trigonella foenum-graecum* reduced brain levels of IL-1β, IL-6, TNF-α, iNOS, NF-κB, and COX-2, and increased BDNF. Treatment also partially reversed the cognitive impairments from aluminum chloride.[11] Trigonelline, a primary alkaloid of *Trigonella foenum-graecum*, reversed lipopolysaccharide-induced learning and memory impairment in the rat through its central anti-inflammatory and antioxidant effects. A week of daily administration of trigonelline to rats that began immediately after they had received lipopolysaccharide spared spatial recognition memory in the Y-maze, the discrimination ratio in novel object discrimination test, and in retention and recall in a passive avoidance paradigm. Trigonelline also dampened the inflammatory activities of hippocampal NF-κB, toll-like receptor 4, and TNF-α in those animals.[12]

7.62.3 ANTIDIABETIC/ANTI-METABOLIC SYNDROME

There are numerous studies showing reduction of Metabolic Syndrome and enhancement of insulin sensitivity by *Trigonella foenum-graecum* in human subjects. Such effects occurring in brain tissue would be expected to contribute antidepressant effects. Patients with diabetes type II that received 1 g per day of dried, ethanolic extract of the herb had improved glycemic control and increased insulin sensitivity per the HOMA model.[13] Individuals with diabetes type II that received two daily doses of 12.5 g of *Trigonella foenum-graecum* seed powder for eight weeks enjoyed lowered fasting blood glucose levels and improved glucose tolerance. Insulin levels were diminished and glycosylated hemoglobin significantly reduced.[14]

Three weeks of addition of *Trigonella foenum-graecum* to the diet of alloxan-induced diabetic rats increased serum insulin and decreased serum glucose. In brain tissue, *Trigonella foenum-graecum* reversed some of the damaging effects of diabetes by increasing levels of the antioxidant enzymes superoxide dismutase and glutathione S-transferase and reducing lipid peroxidation.[15]

Trigonelline, one of the primary active alkaloids in *Trigonella foenum-graecum*, likely plays a role in the herb's antidiabetic effects. In rodent studies, it has been found to improve glucose tolerance, enhance insulin sensitivity, decrease oxidative stress, inhibit production of TNF-α, and activate PPARγ in adipose tissue. Trigonelline also crosses the blood–brain barrier and produces similar effects in neural tissue.[16,17]

The unique amino acid 4-hydroxyisoleucine also enhances insulin sensitivity in rat studies, in part by activating PI3-kinase activity in liver and muscle. The activation of PI3k is one of the primary mechanisms of action of lithium. Chronic treatment with 4-hydroxyisoleucine significantly reduced abnormally high fasting serum insulin levels in rats with type 2 diabetic rats induced by high sucrose and high lipid diets, and slowed progression of hyperinsulinemia in genetically insulin-resistant rats. Given the behavioral and cognitive effects of 4-hydroxyisoleucine, described below, it is reasonable to assume that some of these effects occur in the brain.[18]

7.62.4 PRECLINICAL ANTIDEPRESSANT-LIKE EFFECTS

Trigonella foenum-graecum and its constituent phytochemicals have been shown to have antidepressant-like effects in rodents. A week of daily administration of methanolic extract of *Trigonella foenum-graecum* seeds reduced immobility of mice in the tail suspension and forced swim tests.[19] Ethanolic extract of *Trigonella foenum-graecum* seed reduced immobility of rats in the tail suspension and forced swim tests. Chemical analyses of the animal's brains further showed substantial reduction in MAO-A and B activities in whole brain homogenates.[20]

The 4-hydroxyisoleucine contained in *Trigonella foenum-graecum* was also shown to exert anti-depressant-like effects in mice that had been subjected to treatments known to cause depression-like patterns of behavior. The unique amino acid decreased immobility in the forced swim in reserpinized mice, as well as in un-reserpinized control animals. The 4-hydroxyisoleucine also increased numbers of 5-HTP-induced head twitches in the mice suggesting enhancement of serotonergic activity.[21] Moreover, 14 days of orally administered 4-hydroxyisoleucine reversed behavioral signs of depression in rats that had undergone olfactory bulbectomy. Rats had reduced immobility in the forced swim test, and regained their usual preference for sucrose. Serum levels of corticosterone were reduced. There were also indications of anxiolytic effects of 4-hydroxyisoleucine in the open field and novelty suppressed feeding paradigms.[22]

7.62.5 HUMAN ANTIDEPRESSANT EFFECTS

There are no studies on the use of *Trigonella foenum-graecum* or its constituent phytochemicals in the treatment of MDD in human subjects. There have been reports of *Trigonella foenum-graecum* extracts enhancing libido in men and women. This has been attributed to increases in serum testosterone, but studies have not consistently supported this explanation. Daily BID dosing of 300 mg of the proprietary extract of *Trigonella foenum-graecum*, Testofen, for six weeks was reported to improve male sex drive and quality of orgasm in healthy men between the ages of 25 and 52. The treatment also improved subjective sense of muscle strength, energy, and well-being, but had no effects on mood or sleep. Serum prolactin and testosterone levels were noted to have remained within the reference range.[23] Another study found no increases of serum testosterone after treatment with *Trigonella foenum-graecum*.[24] A study of a proprietary "protodioscin-enriched" *Trigonella foenum-graecum* seed extract reported the treatment to slightly increase free testosterone levels in otherwise healthy men ages 35–65. Sexual arousal and excitement, as well as mental alertness and "mood alleviation," were also significantly enhanced.[25]

Trigonella foenum-graecum is also reported to enhance libido in women. Healthy women between the ages of 25 and 50 years were administered 300 mg of a proprietary *Trigonella foenum-graecum* seed extract twice daily for four weeks, and were found to have significant improvement in at least one measure of libido. In a subsequent study, menstruating women between the ages of 21 and 49 took the extract of *Trigonella foenum-graecum* twice daily for two menstrual cycles. At that time, women reported significant improvements in sexual function over five domains: sexual cognition, sexual arousal, sexual experience, orgasm, and sexual drive/relationship.[26]

7.62.6 DOSAGE

The German E Commission recommends a daily dose of *Trigonella foenum-graecum* seed to 6 g.[27]

7.62.7 TOXICITY

Trigonella foenum-graecum seed powder in capsule form as low as 2.5 g BID and as high as 12.5 g BID have been used to treat diabetes type II. Toxicological evaluation of 60 diabetic patients who took powdered seeds at a dose of 25 g per day for 24 weeks disclosed no hepatic, renal, or hematological abnormalities.[28] In an animal study, the acute oral LD50 was > 5 g/kg in rats, and > 2 g/kg in rabbits.[29] Those values would extrapolate to 350 g and 140 g respectively in a 75 kg adult human. In another animal study, *Trigonella foenum-graecum* powder failed to induce any signs of toxicity or mortality in mice and rats who received acute and subchronic regimens.[30]

7.62.8 SAFETY IN PREGNANCY

The *Botanical Safety Handbook* notes conflicting evidence concerning the safety of the use of *Trigonella foenum graecum* during pregnancy. Thus, without compelling need for its use, it would be wise to avoid it.[31]

7.62.9 DRUG INTERACTIONS

It is suggested that *Trigonella foenum-graecum* not be consumed until an hour after taking medications, as the mucilage present in the herb may interfere with absorption.[32] The *Botanical Safety Handbook* otherwise notes no reports of pharmacological interactions with commonly prescribed medications. *Trigonella foenum-graecum* has little effect on the CYP3A4 and CYP2D6 isoenzymes from rat liver.[33] However, it significantly inhibits the activity of the rat CYP2C11 isoenzyme, which is the counterpart of the human CYP2C9 enzyme.[34] There is reported to be no influence of differences in CYP2C9 genotypes on plasma concentrations of patients taking second-generation antidepressants.[35]

REFERENCES

1. Basch E, Ulbricht C, Kuo G, et al. Therapeutic applications of fenugreek. *Altern Med Rev.* 2003;8(1):20–27.
2. Girardon P, Bessiere JM, Baccou JC, Sauvaire Y. Volatile constituents of fenugreek seeds. *Planta Med.* 1985;51:533–534.
3. Sauvaire Y, Petit P, Baissac Y, et al. Chemistry and Pharmacology of Fenugreek. In: Mazza G, Oomah BD, editors. *Herbs, Botanicals, & Teas.* Lancaster, PA: Technomic Publishing Company, Inc.; 2000. p. 107–129.
4. Sauvaire Y, Girardon P, Baccou JC, Risterucci AM. Changes in growth, proteins and free amino acids of developing seed and pod of fenugreek. *Phytochemistry.* 1984;23(3):479–486.
5. Kaviarasana S, Naik GH, Gangabhagirathi R, Anuradha CV, Priyadarsini KI. In vitro studies on antiradical and antioxidant activities of fenugreek (*Trigonella foenum graecum*) seeds. *Food Chem.* 2007;103(1):31–37.
6. Dixit P, Ghaskadbi S, Mohan H, Devasagayam TP. Antioxidant properties of germinated fenugreek seeds. *Phytother Res.* 2005;19:977–983.
7. Ravikumar P, Anuradha CV. Effect of fenugreek seeds on blood lipid peroxidation and antioxidants in diabetic rats. *Phytother Res.* 1999;13(3):197–201.
8. Genet S, Kale, RK, Baquer NZ. Alterations in antioxidant enzymes and oxidative damage in experimental diabetic rat tissues: effect of vanadate and fenugreek (*Trigonella foenum graecum*). *Mol Cell Biochem.* 2002;236:7–12.
9. Mandegary A, Pournamdari M, Sharififar F, Pournourmohammadi S, Fardiar R, Shooli S. Alkaloid and flavonoid rich fractions of fenugreek seeds (*Trigonella foenum-graecum* L.) with antinociceptive and anti-inflammatory effects. *Food Chem Toxicol.* 2012;50:2503–2507.

10. Kawabata T, Cui MY, Hasegawa T, Takano F, Ohta T. Anti-inflammatory and anti-melanogenic steroidal saponin glycosides from fenugreek (*Trigonella foenum-graecum* L.) seeds. *Planta Med.* 2011;77(7):705–710.

11. Prema A, Arokiasamya JT, Manivasagam T, Mohamed Essa M, Guillemin GJ. Fenugreek seed powder attenuated aluminum chloride-induced tau pathology, oxidative stress, and inflammation in a rat model of Alzheimer's disease. *J Alzheimers Dis.* 2017;60(s1):S209–S220.

12. Khalili M, Alavi M, Esmaeil-Jamaat E, Baluchnejadmojarad T, Roghani M. Trigonelline mitigates lipopolysaccharide-induced learning and memory impairment in the rat due to its anti-oxidative and anti-inflammatory effect. *Int Immunopharmacol.* 2018;61:355–362.

13. Gupta A, Gupta R, Lal B. Effect of *Trigonella foenum-graecum* (fenugreek) seeds on glycemic control and insulin resistance in type 2 diabetes mellitus: a double-blind placebo controlled study. *J Assoc Phys India.* 2001;49:1057–1061.

14. Sharma RD, Sarkar A, Hazara DK, et al. Use of fenugreek seed powder in the management of non-insulin dependent diabetes mellitus. *Nutr Res.* 1996;16(8):1331–1339.

15. Kumar P, Kale RK, McLean P, et al. Antidiabetic and neuroprotective effects of *Trigonella Foenum-graecum* seed powder in diabetic rat brain. *Prague Med Rep.* 2012;113(1):33–43.

16. Yoshinari O, Igarashi K. Antidiabetic effects of trigonelline: comparison with nicotinic acid. In: Preedy VR, editor. *Coffee in Health and Disease Prevention.* London, UK: Academic Press; 2014. p. 765–775.

17. Zhou J, Chan L, Zhou S. Trigonelline: a Plant Alkaloid with Therapeutic Potential for Diabetes and Central Nervous System Disease. Curr Med Chem. 2012;19(21): 3523–3531.

18. Broca C, Breil V, Cruciani-Guglielmacci C, et al. Insulinotropic agent ID-1101 (4-hydroxyisoleucine) activates insulin signaling in rat. *Am J Physiol Endocrinol Metab.* 2004;287(3):E463–E471.

19. Pawar VS, Hugar S, Gawade B, et al. Evaluation of antidepressant-like activity of *Trigonella foenum graecum* Linn. Seeds in mice. *Pharmacologyonline.* 2008;1:455–465.

20. Khursheed R, Rizwani GH, Sultana V, et al. Antidepressant effect and categorization of inhibitory activity of monoamine oxidase type A and B of ethanolic extract of seeds of Trigonella foenum graecum Linn. *Pak J Pharm Sci.* 2014;27(5):1419–1425.

21. Gaur V, Bodhankar SL, Mohan V, et al. Antidepressant-like effect of 4-hydroxyisoleucine from *Trigonella foenum graecum* L. seeds in mice. *Biomed Aging Pathol.* 2012;2:121–125.

22. Kalshetti PB. Effects of 4-hydroxyisoleucine from fenugreek seeds on depression-like behavior in socially isolated olfactory bulbectomized rats. *Pharmacogn Mag.* 2015;11(Suppl 3): S388–S396.

23. Steels E, Rao A, Vitetta L, et al. Physiological aspects of male libido enhanced by standardized *Trigonella foenum-graecum* extract and mineral formulation. *Phytotherapy Res.* 2011;25(9):1294–1300.

24. Bushey B, Taylor LW, Wilborn CW, et al. Fenugreek extract supplementation has no effect on the hormonal profile of resistance-trained males. *Int. J. Exercise Sci. Conf. Proc.* 2009;2(1):Article 13.

25. Bagchi D, Swaroop A, Maheshwari A, et al. A novel protodioscin-enriched fenugreek seed extract (*Trigonella foenum-graecum*, family *Fabaceae*) improves free testosterone level and sperm profile in healthy volunteers. *Funct. Food Health Dis.* 2017;7(4):235–245.

26. Bhaskaran S, Venkatesh RV, Veeravalli J, et al. Use of fenugreek extract to enhance female libido. United States Patent Application Publication #20140154346A1. June 5, 2014.

27. Blumenthal M, Busse WR, Goldberg A, et al. *The Complete German Commission E Monograph: Therapeutic Guide to Herbal Medicines.* Austin, TX: American Botanical Council; 1998. p.130.
28. Sharma RD, Sarkar A, Hazra DK, et al. Toxicological evaluation of fenugreek seeds: a long-term feeding experiment in diabetic patients. *Phytother Res.* 1996;10:519–520.
29. Opdyke DL. Fenugreek absolute. *Food Cosmet Toxicol.* 1978;16:S755–S756.
30. Narasimhamurthy K, Viswanatha S, Ramesh BS. Acute and subchronic toxicity assessment of debitterized fenugreek powder in the mouse and rat. *Food Chem Toxicol.* 1999;37:831–838.
31. Gardner Z, McGuffin M, editors. *Botanical Safety Handbook,* 3rd edition. Boca Raton, FL: CRC Press; 2013. p. 872.
32. Leung AY, Foster S. *Encyclopedia of Common Natural Ingredients Used in Food, Drugs and Cosmetics.* 2nd edition. New York: New York. 1996.
33. Ahmmed SK, Mukherjee PK, Bahadur S, et al. Interaction potential of Trigonella foenum graceum through cytochrome P450 mediated inhibition. *Indian J Pharmacol.* 2015;47(5):530–534.
34. Korashy H M, Al-Jenoobi FI, Raish M, et al. Impact of herbal medicines like *Nigella sativa, Trigonella foenum-graecum,* and *Ferula asafoetida,* on cytochrome P450 2C11 gene expression in rat liver. *Drug Res.* 2014;64:1–7.
35. Grasmader K, Verwohlt PL, Rietschel M, et al. Impact of polymorphisms of cytochrome-P450 isoenzymes 2C9, 2C19 and 2D6 on plasma concentrations and clinical effects of antidepressants in a naturalistic clinical setting. *Eur J Clin Pharmacol.* 2004;60:329–336.

7.63 *VALERIANA OFFICINALIS* (VALERIAN)

Valeriana officinalis, commonly referred to as valerian, is a perennial flowering plant native to Europe and Asia. The ancients knew the plant. The Greek physician and herbalist Dioscorides recommended the root to treat myriad disorders including heart palpitations, digestive problems, epilepsy, and urinary tract infections. During the second century Galen recommended *Valeriana officinalis* as a treatment for insomnia. The major modern uses for the herb continue to be those of a sedative and anxiolytic.[1] Indeed, during World War I, *Valeriana officinalis* was used to prevent and treat shell shock in frontline troops. During World War II, it was used to help calm civilians subjected to air raids.[2]

The chemistry of *Valeriana officinalis* is complex, and there are likely over 150 different phytochemicals, many of which are pharmacologically active. It contains a wide range of phytochemicals, including the predominant iridoid valepotriates: valtrates, isovaltrate, didrovaltrate, valerosidate, and others. The volatile essential oil contains bornyl isovalerenate and bornyl acetate; valerenic, valeric, isovaleric, and acetoxyvalerenic acids; valerenal, valeranone, and cryptofaurinol; and other monoterpenes and sesquiterpenes. The plant also contains the alkaloids valeranine, chatinine, alpha-methyl pyrrylketone, actinidine, skyanthine, and naphthyridylmethylketone, and lignans, such as hydroxypinoresinol.[3,4,5,6]

7.63.1 ANTIOXIDANT

Valeriana officinalis exhibits a potent antioxidant effect. Ethanolic extracts were found to be effective scavengers of free radicals in the DPPH assay.[7] In brain homogenates exposed to quinolinic acid, 3-nitropropionic acid, sodium nitroprusside, or iron sulfate – all of which are known to cause oxidative damage – ethanolic extract of *Valeriana officinalis* decreased reactive oxygen species and products of lipid oxidation.[8] Pretreatment with extract was also found to protect against ischemic injury in cultured hippocampal pyramidal neurons by decreasing microglial activation and lipid peroxidation.[9]

7.63.2 ANTI-INFLAMMATORY

It has been noted that *Valeriana officinalis* has been used in folk medicines of Europe for inflammatory disorders. This led to the finding that in human HeLa cells, ethanolic extract of *Valeriana officinalis* antagonized the activity of NF-κB. Bioassay-guided fractionation of that extract resulted in the isolation of three sesquiterpenes responsible for that anti-inflammatory action: acetylvalerenolic acid, valerenal, and valerenic acid. Of the three, acetylvalerenolic acid was most potent in reducing the activity of NF-κB by 96%. In the same study, the ethanolic extract also reduced the NMDA and kainate-induced excitotoxicity in cultures of neurons that contained microglia and astrocytes. Interestingly, in neuron cultures that did not contain microglia and astrocytes, the excitotoxicity was not prevented, suggesting specific action on those subtypes of cells that mediate inflammation.[10]

In vivo, orally administered aqueous extract of *Valeriana officinalis* attenuated carrageenan-induced rat paw edema with efficacy similar to that of acetylsalicylic acid.[11]

7.63.3 ANTIDIABETIC/ANTI-METABOLIC SYNDROME

There is no evidence seen in the literature to suggest that *Valeriana officinalis* has any usefulness in the treatment of diabetes or Metabolic Syndrome. In fact, one study found no effect of essential oil of *Valeriana officinalis* on blood glucose or serum insulin levels in rats with a condition similar to diabetes type II.[12]

7.63.4 PRECLINICAL ANTIDEPRESSANT-LIKE EFFECTS

Sensitization to ovalbumin protein causes an inflammatory reaction in rats that includes neuroinflammation. This, in turn, causes a depression-like syndrome. When such rats were administered *Valeriana officinalis*, their performance in the forced swim test improved. That is, their immobility, a sign of behavioral despair, was reduced. Moreover, despite *Valeriana officinalis* having a mildly sedating effect, the animals treated with the herb also had increased activity in the open field.[13]

In another study of rats, 16 days of twice daily administration of *Valeriana officinalis* extract again reduced immobility time in the forced swim test.[14] A month of treatment with *Valeriana officinalis* extract was also found to increase preference for sucrose in rats that had been subjected to chronic unpredictable stress, and this antidepressant-like effect was similar to that observed in stressed rats treated with fluoxetine for that month.[15]

Similar antidepressant-like effects of *Valeriana officinalis* have been observed in mice. Three weeks of extract helped reverse the depression-like effects of electric shock in mice, as well as the stress mice sustained by simply observing and hearing the responses of other mice receiving the shock. Stressed animals that received the extract exhibited significantly less immobility in the forced swim test. Corticosterone levels in those animals were also reduced compared to stressed animals that did not receive *Valeriana officinalis*.[16]

Among the mechanisms by which *Valeriana officinalis* could produce antidepressants effects is through induction of BDNF. Both *Valeriana officinalis* root extract and purified valerenic acid stimulate expression of BDNF *in vitro* in human derived neuroblasts in a dose-dependent manner.[17] However, it was reported that valepotriates isolated from the Brazilian plant of the Valeriana genus, *Valeriana glechomifolia*, produced antidepressant effects in mice through anti-inflammatory effects without affecting levels of BDNF. Oral pretreatment with valepotriates reduced immobility in the tail suspension test in animals injected with lipopolysaccharide. That treatment also reduced levels of proinflammatory cytokines IL-1β and TNF-α in the cerebral cortices of those animals. Curiously, in animals given the valepotriates after injection of lipopolysaccharide, immobility was also reduced, but the levels of inflammatory cytokines were not decreased. In neither case were levels of BDNF affected.[18]

7.63.5 HUMAN ANTIDEPRESSANT EFFECTS

There are no studies of the use of *Valeriana officinalis* in human subjects suffering depression. There are a number of studies of the effects of the herb on sleep and anxiety in human subjects. Sleep is disturbed in many individuals that suffer MDD, and restoring sleep plays an important role in recovery. Unfortunately, despite its reputation as a sleep agent, the reviews of the effect of *Valeriana officinalis* on sleep have been surprisingly inconsistent, if not outright disappointing. Whereas some reviews of the many existing studies have concluded that it is useful for the treatment of insomnia,[19] others have found no basis for such use.[20] The consensus is that more research is needed.

A similar state of uncertainty exists in regard to the use of *Valeriana officinalis* as a treatment for anxiety. In a 2006 Cochrane review, it was decided that only one study met the rigorous criteria to include in the determination of the efficacy of *Valeriana officinalis* in the treatment of anxiety. In that particular study, it was found to be not significantly better than placebo in relieving symptoms of anxiety. However, it was also found to be not significantly different from diazepam in affecting anxiety.[21] In a more recent 2013 review of treatments of anxiety, it was similarly concluded, "Although well-tolerated, the limitation on studies performed with humans and, to some degree the contradictory results, current information is insufficient to warrant an indication of V. officinalis for clinical treatment of anxiety."[22]

7.63.6 DOSAGE

The German E Commission recommends the dose of *Valeriana officinalis* to be 2–3 g of dried root per cup in infusion, one to several times day; 1–3ml of tincture, one or more times daily; or extract equivalent to 1–3 g of drug daily.[23]

7.63.7 TOXICITY

In a test of acute toxicity of *Valerian officinalis*, doses of extract up to 2,000 mg/kg caused no deaths in mice.[24] This dose would extrapolate to 140 g in a 75 kg human adult.

7.63.8 SAFETY IN PREGNANCY

The *Botanical Safety Handbook* notes human and animal studies that found no adverse effects of even high doses of *Valeriana officinalis* root during pregnancy.[25]

7.63.9 DRUG INTERACTIONS

Review of the literature revealed no studies of the possible interactions of *Valeriana officinalis* with commonly used medications. Extract of *Valeriana officinalis* has no significant effect on the activities of the human CYP1A2, CYP2D6, CYP2E1, or CYP3A4 isoenzymes.[26]

REFERENCES

1. Murti K, Kaushik M, Sangwan Y, et al. Pharmacological properties of *Valeriana offi-cinalis* – A review. *Pharmacology*. 2011;3:641–646.
2. Mowrey DB. *The Scientific Validation of Herbal Medicine*. New Canaan, Conn: Keats Pub. 1986;2:316.
3. Becker H, Chavadej S. Valepotriate production of normal and colchicine-treated cell suspension cultures of *Valeriana wallichii*. *J Nat Prod*. 1985; 48:17–21.
4. Morazzoni P, Bombardelli E. *Valeriana officinalis*: traditional use and recent evalua-tions of activity. *Fitoterapia*. 1995; 66:99–112.
5. Franck B, Petersen U, Hüper F, et al. Valerianie, a tertiary monoterpene alkaloid from valerian. *Angew Chem Int Ed Engl*. 1970; 9:891.
6. Janot MM, Guilhem J, Contz O, et al. Contribution to the study of valerian alkaloids (*Valeriana officinalis*, L.): actinidine and naphthyridylmethylketone, a new alkaloid (author's transl). *Ann Pharm Fr*. 1979; 37:413–20.
7. Pilerood SA, Prakash J. Evaluation of nutritional composition and antioxidant activity of Borage (*Echium amoenum*) and Valerian (*Valerian officinalis*), *J Food Sci Technol*. 2014;51(5):845–854.
8. Haigert J, Fachinetto R, Pereira RP, et al. In vitro antioxidant activity of *Valeriana officinalis* against different neurotoxic Agents. *Neurochem Res*. 2009;34:1372.
9. Yoo DY, Jung HY, Nam SM, et al. *Valeriana officinalis* extracts ameliorate neuronal damage by suppressing lipid peroxidation in the gerbil hippocampus following tran-sient cerebral ischemia. *J Med Food*. 2015;18:642–647.
10. Jacobo-Herrera NJ, Vartiainen N, Bremner P, et al. NF-κB modulators from *Valeriana officinalis*. *Phytother Res*. 2006;20:917–919.
11. Dyayiya NA, Oyemitan IA, Matewu R, et al. Chemical analysis and biological potential of valerian root as used by herbal practitioners in the Eastern Cape province of South Africa. *Afr J Tradit Complement Altern Med*. 2016; 13(1):114–122.
12. Ling C, Ruhan J, Guohua D, et al. Effect of valerian oil on oxidative stress in type II diabetic rats. *China J Mod Med*. 2003-17.
13. Ali Neamati A, Chaman F, Hosseini M, et al. The effects of *Valeriana officinalis* L. hydro-alcoholic extract on depression like behavior in ovalbumin sensitized rats. *J Pharm Bioallied Sci*. 2014;6(2):97–103.

14. Hattesohla M, Feistel B, Sievers H, et al. Extracts of *Valeriana officinalis* L. s.l. show anxiolytic and antidepressant effects but neither sedative nor myorelaxant properties. *Phytomedicine.* 2008;15(1–2):2–15.

15. Tang JY, Zeng Y, Chen Q, et al. Effects of valerian on weight and behavior of depressive rats induced by chronic mild stress. *J Sun Yat-Sen Univ (Med Sci).* 2008;29(5):541–545.

16. Jung HY, Yoo DY, Kim W, et al. *Valeriana officinalis* root extract suppresses physical stress by electric shock and psychological stress by nociceptive stimulation-evoked responses by decreasing the ratio of monoamine neurotransmitters to their metabolites. *BMC Compl Alt Med.* 2014;14:476.

17. Gonulalan EM, Bayazeid O, Yalcin FN, et al. The roles of valerenic acid on BDNF expression in the SH-SY5Y cell. *Saudi Pharm J.* 2018;26(7):960–964.

18. Müller LG, Borsoi M, Stolz ED, et al. Diene valepotriates from *Valeriana glechomifolia* prevent lipopolysaccharide-induced sickness and depressive-like behavior in mice. *Evid-Based Compl Alt.* 2015; Article ID 145914.

19. Bent S, Padula A, Moore D, et al. Valerian for sleep: a systematic review and meta-analysis. *Am J Med.* 2006;119(12):1005–1012.

20. Taibi DM, Landis CA, Petry H, et al. A systematic review of valerian as a sleep aid: safe but not effective. *Sleep Med Rev.* 2007;11(3):209–230.

21. Miyasaka L, Atallah A, Soares B. Valerian for anxiety disorders. *Cochrane Database Syst Rev.* 2006:CD004515.

22. Gelfuso EA, Santos Rosa D, Lúcia Fachin A, et al. Anxiety: a systematic review of neurobiology, traditional pharmaceuticals and novel alternatives from medicinal plants. *CNS Neurol Disord Drug Targets.* 2014;543:150–65.

23. Blumenthal M, Busse WR, Goldberg A, et al. *The Complete German Commission E Monographs.* Austin, TX: American Botanical Council; 1998. p. 227.

24. Atmojo DD, Wijayahadi N. Acute Toxicity Test LD50 Value of Valerian (*Valeriana officinalis*) in Balb/c Mice [Doctoral Dissertation]. [Semarang, Indonesia]: Fakultas Kedokteran Universitas Diponegroro; 2009.

25. Gardner Z, McGuffin M, editors. *Botanical Safety Handbook*, 3rd edition. Boca Raton, FL: CRC Press; 2013. p. 903.

26. Gurley BJ, Gardner SF, Hubbard MA, et al. In vivo effects of goldenseal, kava kava, black cohosh, and valerian on human cytochrome P450 1A2, 2D6, 2E1, and 3A4 phenotypes. *Clin Pharmacol Ther.* 2005;77(5):415–426.

7.64   *VERBENA OFFICINALIS* (VERVAIN)

Verbena officinalis, commonly known as vervain, is a plant that grows wild throughout most of Europe, North Africa, China, and Japan. It has been used for medicinal purposes for thousands of years in both Asian and European medical traditions. In Traditional Chinese Medicine, it has been used for clearing away heat, detoxification, promoting blood circulation, and removing blood stasis.[1] As far back as the first century AD, the Greek physician Dioscorides called *Verbena officinalis* the "sacred herb," and for many centuries, it was taken as a cure-all.[2] In the traditional Ayurvedic medicine of India, *Verbena officinalis* has been used for treatment of a variety of psychiatric conditions, such as anxiety, depression, hysteria, and insomnia.[3] Modern study of *Verbena officinalis* has revealed antioxidant, anti-inflammatory, neuroprotective, and possibly antidepressant effects. Among the prominent components of *Verbena officinalis* are beta-sitosterol, ursolic acid, oleanolic acid, 3-epiursolic acid, 3-epioleanolic acid, verbenone, verbenalin, hastatoside, verbascoside, and beta-sitosterol-D-glucoside. Many of those, when isolated and purified, have been found to be biologically active.[4]

7.64.1 ANTIOXIDANT

Several specific phytochemicals in *Verbena officinalis* have also been found to have antioxidant effects, including oleanolic acid and ursolic acid. Oleanolic acid is a triterpene in *Verbena officinalis* also found in olive leaves, pokeweed, Wood's rose, mesquite, mistletoe, and other medicinal plants. Ursolic acid, another triterpene in *Verbena officinalis*, is also found in basil, elderflower, peppermint, rosemary, lavender, oregano, thyme, hawthorn, and other plants. Oleanolic acid protected brain tissue from the damaging oxidative effects of ischemia. In mice that had undergone bilateral common carotid artery ligation, pre-treatment with oleanolic acid increased the activities of the antioxidant enzymes superoxide dismutase and glutathione peroxidase. Levels of malondialdehyde, an indicator of oxidative damage, were reduced. Together, those effects of oleanolic acid spared neurological function, both reduced cerebral edema and minimized the size of the infarct. Similarly, pretreatment of PC12 cells with oleanolic acid prior to exposure to hydrogen peroxide improved cell survival, enhanced superoxide dismutase activity, increased glutathione content, and lowered malondialdehyde levels.[5]

Ursolic acid offered similar antioxidative protection to PC12 cells exposed to hydrogen peroxide. Pretreatment with ursolic acid maintained levels of glutathione, decreased malondialdehyde formation, and reversed hydrogen peroxide induced impairment in catalase and superoxide dismutase activities. In the same study, both oleanolic acid and ursolic acid attenuated hydrogen peroxide-induced release of the cytokines IL-6 and TNF-α from the PC12 cells, thus demonstrating anti-inflammatory effects.[6]

7.64.2 ANTI-INFLAMMATORY

Verbena officinalis has long been used for arthritic conditions due to anti-inflammatory effects. Modern studies have borne out the rationale for this use. In a standard test of anti-inflammatory effect, oral administration of extract of *triterpene* greatly reduced 12-O-tetradecanoylphorbol-13-acetate-induced ear edema in rats.[7] Oral administration of extract also ameliorated carrageenan-induced paw edema in rats, in a manner similar to that of the anti-inflammatory agent indomethacin and helped prevent the gastric ulceration induced by exposure to alcohol, almost to the extent of protection given by misoprostol.[8]

Verbena officinalis has been shown to have anti-inflammatory neuroprotective effects in cultured neurons. Aqueous extract of the herb dampened the cytotoxic effects of beta-amyloid on cortical neurons. It appeared to do so by blocking the inflammatory interferon-induced protein kinase, which activates NFkB, and c-Jun N-terminal kinase, which contributes to inflammatory responses in mammals.[9] Standard *in vitro* chemical tests have also shown that aqueous and hydroalcoholic extracts of *Verbena officinalis* have powerful antioxidant effects. *In vitro* assays of DPPH/radical scavenging activity, ABTS/radical scavenging activity, and superoxide anion radical inhibition activity, all show *Verbena officinalis* to protect against reactive oxygen species.[10]

7.64.3 Antidiabetic/Anti-Metabolic Syndrome

Ursolic and luteolin-7-glucoside, a flavonoid in *Verbena officinalis*, produce a hypoglycemic effect by inhibiting liver GSK-3β.[11] Similarly, oleanolic acid attenuated hepatic damage following ischemia and reperfusion, due in part to stimulation of PI3K and Akt, and inhibition of GSK-3β in hepatocytes. Oleanolic acid also decreased serum levels of the inflammatory cytokine, IL-1β.[12] Oleanolic acid enhances the effects of insulin through a variety of mechanisms, including: activation of PI3K/Akt; inhibition of GSK-3β; inhibition of the protein-tyrosine phosphatases PTP1B and TCPTP that dampen insulin signaling; antagonism of inflammatory factors that downregulate the insulin response, such as NF-kβ, IL-6, and TNF; alleviation of the oxidative stress-induced insulin resistance; and stimulation of peroxisome proliferator-activated receptors.[13] All such effects, if occurring in brain tissue, would likely help ameliorate MDD.

7.64.4 Preclinical Antidepressant-Like Effects

There is only one study showing antidepressant-like effects of extract of *Verbena officinalis* in intact animals. In that study, treatment with extract of the herb decreased immobility times of mice in the forced swim and tail suspension tests.[14] However, two of its major triterpenes, ursolic acid and oleanolic acid, have been found to produce antidepressant-like effects in rodents. Pretreatment with a single dose of ursolic acid decreased immobility time of mice subjected to the tail suspension and forced swimming tests. It was further found that sub-effective doses of ursolic acid in addition to sub-effective doses of dopamine agonists summated to give the antidepressant-like effect, suggesting that it was mediated by dopaminergic systems. The effects were also seen to be similar to those of fluoxetine, imipramine, and bupropion.[15] In a similar study, acute pretreatment of mice with ursolic acid had an antidepressant-like effect in the tail suspension test, and this effect was blocked by naloxone and sub-chronic treatments with ρ-chlorophenylalanine or α-methyl-ρ-tyrosine that deplete the brain of serotonin and catecholamines, respectively.[16]

Several studies have also found antidepressant-like effects of oleanic acid. Chronic, but not acute, treatment with oleanolic acid decreased immobility time of mice in the forced swim test. It was found that oleanolic acid also increased levels of serotonin, norepinephrine, and BDNF in the frontal cortex and hippocampus of those animals.[17] Further study showed that treatment with oleanolic acid led to the BDNF-mediated activation of ERK and CREB in the hippocampus, that in turn upregulated miR-132 responsible for neuronal proliferation.[18] Interestingly, in the latter study, oleanolic acid ameliorated the depression-like effects of chronic unpredictable stress in mice as evidenced by decreasing immobility times in the forced swim test. That finding suggests usefulness of oleanolic acid and, perhaps, *Verbena officinalis* in the treatment as well as prevention of MDD.

7.64.5 Human Antidepressant Effects

Herbal tea made from *Verbena officinalis* has also traditionally been used for the treatment of insomnia. Two of the herb's prominent iridoid glycosides, hastatoside and verbenalin, were evaluated for effects on sleep in rats through electroencephalographic analyses. Both phytochemicals substantially increased the total time of non-rapid eye movement sleep as well as delta activity during non-rapid eye movement sleep.[19] Thus, *Verbena officinalis* may be useful in the common situation in which depressed mood is accompanied by insomnia. Nonetheless, there are no published studies of the use of *Verbena officinalis* or its constituent phytochemicals in the treatment of MDD in human subjects.

7.64.6 Dosage

Hoffman gives the dosage of *Verbena officinalis* to be infusion of 1–3 teaspoons of dried herb per cup of water, three times a day; tincture (1.5:40%) 2.5–5 ml, three times a day; or 2–4 grams of dried herb daily.[20]

7.64.7 Toxicity

The LD50 of aqueous and hydroethanolic extracts of *Verbena officinallis* in mice has been reported to be above 5,000 mg/kg.[21] This dose would extrapolate to approximately 350 g for an adult human weighing 75 kg.

7.64.8 Safety in Pregnancy

Reference texts from Traditional Chinese Medicine advise against unsupervised use of *Verbena officinallis* during pregnancy.[22,23]

7.64.9 Drug Interactions

Verbenone, a major component of *Verbena officinalis*, is metabolized largely by the CYP2A6 isoenzyme and, to a lesser extent, the CYP2B6 isoenzyme in human liver.[24] However, the *Botanical Safety Handbook* notes no known interactions with commonly prescribed medications.[25]

REFERENCES

1. Liu Z, Xu Z, Zhou H, et al. Simultaneous determination of four bioactive compounds in *Verbena officinalis* L. by using high-performance liquid chromatography. *Pharmacogn Mag.* 2012; 8(30):162–165.
2. UcarTurker A, Yucesan B, Gurel E. Adventitious shoot regeneration from stem internode explants of *Verbena officinalis* L., a medicinal plant. *Turkish J Biol.* 2010;34:297–304.
3. Khare CP. *Indian Medicinal Plants: An Illustrated Dictionary.* Heidelberg: Springer; 2007. p. 698.
4. Deepak M, Handa SS. Antiinflammatory activity and chemical composition of extracts of *Verbena officinalis*. *Phytother Res.* 2000;14(6):463–465.

5. Rong ZT, Gong XJ, Sun HB, et al. Protective effects of oleanolic acid on cerebral ischemic damage in vivo and H_2O_2-induced injury in vitro. *J Pharmaceut Biol.* 2011;49(1):78–85.
6. Tsai S-J, Yin M-C. Antioxidative and anti-inflammatory protection of oleanolic acid and ursolic acid in PC12 cells. *J Food Sci.* 2008;73(7):H174–H178.
7. Calvo MI, Vilalta N, San Julián A, et al. Anti-inflammatory activity of leaf extract of *Verbena officinalis* L. *Phytomedicine.* 1998;5(6):465–467.
8. Speroni E, Cervellati R, Costa S, et al. Effects of differential extraction of *Verbena officinalis* on rat models of inflammation, cicatrization and gastric damage. *Planta Med.* 2007;73:227–235.
9. Lai SW, Yu MS, Yuen WH, Chang RC. Novel neuroprotective effects of the aqueous extracts from *Verbena officinalis* Linn. *Neuropharmacol.* 2006;50:641–650.
10. Rehechoa S, Hidalgo O, de Cirano MGI, et al. Chemical composition, mineral content and antioxidant activity of *Verbena officinalis* L. *Food Sci Technol.* 2011;44:875–882.
11. Azevedo MF, Camsari C, Sá CM, et al. Ursolic acid and luteolin-7-glucoside improve lipid profiles and increase liver glycogen content through glycogen synthase kinase-3. *Phytotherapy.* 2010;24(S2):S220–S224.
12. Gui B, Hua F, Chen F, et al. Protective effects of pretreatment with oleanolic acid in rats in the acute phase of hepatic ischemia-reperfusion injury: role of the PI3K/Akt pathway. *Mediat Inflamm.* 2014; ID 451826.
13. Castellano JM, Guinda A, Delgado T, et al. Biochemical basis of the antidiabetic activity of oleanolic acid and related pentacyclic triterpenes. *Diabetes,* 2013;62:1791–1799.
14. Jawaid T, Imam S.A, Kamal M. Antidepressant activity of methanolic extract of *Verbena officinalis* linn. plant in mice. *Asian J Pharm Clin Res.* 2015;8:308–310.
15. Machadoa DG, Neis VB, Balen GO, et al. Antidepressant-like effect of ursolic acid isolated from *Rosmarinus officinalis* L. in mice: evidence for the involvement of the dopaminergic system. *Pharmacol Biochem Behav.* 2012;103(2):04–211.
16. Colla ARS, Oliveira A, Pazini FL, et al. Serotonergic and noradrenergic systems are implicated in the antidepressant-like effect of ursolic acid in mice. *Pharmacol Biochem Behav.* 2014;124:108–116.
17. Yi L-T, Li J, Liu Q, et al. Antidepressant-like effect of oleanolic acid in mice exposed to the repeated forced swimming test. *J Psychopharmacol.* 2013;27(5):459–468.
18. Yi L-T, Li J, Liu BB, et al. BDNF-ERK-CREB signalling mediates the role of miR-132 in the regulation of the effects of oleanolic acid in male mice. *J Psychiatry Neurosci.* 2014;39(5):348–359.
19. Makino Y, Kondo S, Nishimura Y, et al. Hastatoside and verbenalin are sleep-promoting components in *Verbena officinalis*, *Sleep Biol Rhythms.* 2009;7(3):211–217.
20. Hoffmann D. *Medical Herbalism: The Science and Practice of Herbal Medicine.* Rochester, VT: Healing Arts Press; 2003. p. 593.
21. Hrytsyk AR, Posatska NM, Klymenko AO. Obtaining and studying properties of extracts of *Verbana officinallis*. *Pharmaceutical Rev.* 2016;3:39–44.
22. Bensky D, Clavey S, Stoger E. *Chinese herbal medicine: Materia Medica.* 3rd edition. Seattle, WA: Eastland Press; 2004.
23. Chen JK, Chen TT. *Chinese Medical Herbology and Pharmacology.* City of Industry, CA: Art of Medicine Press; 2004.
24. Miyazawa M, Sugie A, Shimada T. Roles of human CYP2A6 and 2B6 and rat CYP2C11 and 2B1 in the 10-hydroxylation of (–)-verbenone by liver microsomes. *Drug Metab Dis.* 2003;31(8):1049–1053.
25. Gardner Z, McGuffin M, editors. *Botanical Safety Handbook, 3rd edition.* Boca Raton, FL: CRC Press; 2013. p. 911.

7.65 *VITEX AGNUS-CASTUS* (CHASTE TREE)

Vitex agnus-castus, the chaste tree, is a deciduous shrub that is native to Mediterranean Europe and Central Asia. The berries have been used since ancient times for treatment of many women's health problems. Hippocrates and Dioscorides recommended the herb for disorders of the uterus, as did Gerard, the great Renaissance herbalist.[1] In recent times, extract of *Vitex agnus-castus* berries have been used to treat menstrual disorders, premenstrual syndrome, infertility, acne, discomforts of menopause, and disrupted lactation.[2,3]

The fruit and leaves of the tree contain a complicated mix of phytochemicals, including flavonoids, iridoid glycosides, alkaloids, essential oils, fatty oils, diterpenoids, and steroids.[4] Among those phytochemicals – some that are shared with other medicinal species and others that are unique to *Vitex agnus-castus* – are: vitexin, casticin, agnuside, aucubin, apigenin, thujene, pinene, sabinene, myrcene, terpinene, limonene, cineole, linalool, cryptone, citronellol, cumin aldehyde, carvacrol, b-caryophyllene, farnesene, myristicin, and scores of others.[5]

7.65.1 ANTIOXIDANT

Extract of *Vitex agnus-castus* exerts anti-inflammatory and antioxidant effects, as do many medicinal plants with antidepressant effects. The essential oil components of *Vitex agnus-castus* were deemed to have "modest" antioxidant effects per the DPPH free radical-scavenging method.[6] However, an ethanolic extract of the tree's fruit produced phenolic compounds with substantial antioxidant capacity. They were successful in the DPPH free radical-scavenging method and effective in preventing auto-oxidation of rat brain homogenates.[7]

7.65.2 ANTI-INFLAMMATORY

Some of the anti-inflammatory effect of *Vitex agnus-castus* is due to inhibition of lipoxygenase.[8] The very similar species, *Vitex negundo*, also known as the Chinese chaste tree, is used for similar medicinal purposes and has a similar phytochemical profile. Extracts of *Vitex negundo* have been found to attenuate inflammation in the carrageenan-induced rat paw edema test. The extract also exhibited a strong free radical scavenging activity by the DPPH method and significantly reduced formation of thiobarbituric acid reacting substances when evaluated for its lipid peroxidation inhibitory activity.[9]

7.65.3 ANTIDIABETIC/ANTI-METABOLIC SYNDROME

There are at least two studies showing hypoglycemic and/or insulin-enhancing effects of hydroalcoholic extracts of the fruit of *Vitex agnus-castus*. In one such study, the extract reduced serum glucose in rats with streptozotocin-induced diabetes.[10] In mice subjected to the so-called galactose-induced aging procedure, the extract lowered serum glucose and improved insulin sensitivity in normally aged and galactose-aged rats. Curiously, in rats that were young or did not receive galactose, the extract increased serum glucose and insulin resistance.[11]

7.65.4 PRECLINICAL ANTIDEPRESSANT-LIKE EFFECTS

There are no preclinical studies of antidepressant effects of *Vitex agnus-castus*.

7.65.5 HUMAN ANTIDEPRESSANT EFFECTS

There are no formal studies of the effects of *Vitex agnus-castus* extracts on major depression in human subjects. However, there is a large literature concerning its use in premenstrual and menopausal mood disorders.

Many of the complaints of perimenopause are physical in nature, such as hot flashes, night sweats, breast changes, and vaginal dryness. However, mood disorders are common. Despite the herb's reputation for relieving the discomforts of menopause, there is only one study that has evaluated the effects of essential oil of *Vitex agnus-castus* on menopausal symptoms. It was found to reduce both physical and "psychological" symptoms.[12]

Far more work has been done in evaluating its use in the treatment of premenstrual dysphoric disorder (PMDD). Through several editions, DSM criteria for PMDD have included: unstable mood; persistent irritability or anger; anxiety, nervous tension, or feeling of being overwrought; depressive mood, feeling of hopelessness, or reduction in self-esteem; reduced interest in activities such as work, family, friends, hobbies; rapid tiring and clear lack of energy; feeling of not being able to concentrate; changes in eating behavior, such as increased appetite or craving for sweet foods; and sleep disorders such as too much sleep or sleeplessness. In addition to those emotional and neurovegetative symptoms are physical discomforts, such as tender breasts, feeling of bloating, headache, joint, or muscle pain, and weight gain.

There have been at least seven formal evaluations of chronic, three-month courses of administration of extract of the fruit of *Vitex agnus-castus* on the symptoms of PMDD. Various instruments were employed to evaluate changes in severity of symptoms, including the Premenstrual Tension Syndrome Scale, Clinical Global Impression scale, Penn Daily Symptom Report, Hamilton Depression Rating scale, and Moos' Menstrual Distress Questionnaire. Uniformly, these studies found that extracts of *Vitex agnus-castus* resulted in significant improvement of emotional and neurovegetative symptoms of PMDD. When comparisons were made, the treatment compared favorably with other treatments, such as fluoxetine and pyridoxine.[13–19]

The evidence clearly shows that chronic treatment with *Vitex agnus-castus* can significantly reduce the emotional and neurovegetative complaints of PMDD, many of which closely resemble the symptoms of MDD. The interesting question is whether these benefits of *Vitex agnus-castus* are limited to specific, neuroendocrine syndromes in women or if they have more general application to MDD in both women and men. The most obvious basis for efficacy in treating PMDD would be endocrine effects. Extracts of *Vitex agnus-castus* do contain phytoestrogens, e.g., apigenin, casticin, quercetagetin, and isovitexin.[20] However, it has been noted that these phytochemicals are only weakly estrogenic and present in relatively low levels. Thus, it has been suggested that other mechanisms, such as dopamine agonism to reduce hyperprolactinemia, stimulation of opioid receptors, or enhancement of melatonin release may be at least partially responsible for the effects of *Vitex agnus-castus*.[21] Interestingly, chronic administration of *Vitex agnus-castus* extract has been shown to decrease serum testosterone in male rodents.[22] It might thus be a poor choice to treat men with MDD.

7.65.6 DOSAGE

Hoffman recommends 2.5 ml of tincture (1:5 in 60% ethanol) three times a day. He also notes the German E commission recommendation of 175 mg per day of 20:1 fruit extract standardized to 0.5% aguniside. He goes on to say that benefits may be seen within 2–3 weeks, but more often require 3–6 months of treatment.[23]

7.65.7 TOXICITY

The oral LD50 value of the essential oil of *Vitex agnus-castus* in mice was estimated to be more than 5 g/kg.[24]

In humans, the most frequent adverse events associated with the use of *Vitex agnus castus* are nausea, headache, gastrointestinal disturbances, menstrual disorders, acne, pruritus, and erythematous rash. The adverse events following *Vitex agnus-castus* treatment are mild and reversible.[25] The German Commission E has approved its use for treatment of mastalgia.[26] No drug interactions have been reported.

7.65.8 SAFETY IN PREGNANCY

The *Botanical Safety handbook* notes that *Vitex agnus-castus* has traditionally been used to prevent miscarriage in pregnant women. Thus, one might suspect that it is safe in pregnancy. They also note, however, that no follow-up studies have been performed to assess rates of adverse outcomes.[27]

7.65.9 DRUG INTERACTIONS

The *Botanical Safety Handbook* notes no known interactions with drugs.

REFERENCES

1. Hobbs C. Vitex: *The Women's Herb*. 2nd ed. Santa Cruz, CA: Botanica Press; 1990.
2. Daniele C, Coon JT, Pittler MH, et al. *V. agnus castus*: a systematic review of adverse events. *Drug Safety*. 2005;28:319–332.
3. Prilepskaya VN. *V. agnus castus*: Successful treatment of moderate to severe premenstrual syndrome. *Maturitas*. 2006;(Suppl. 1):S55–S63.
4. Hoberg E, Meier B, Sticher O. An analytical high performance liquid chromatographic method for the determination of agnuside and p-hydroxybenzoic acid contents in Agnicasti fructus. *Phytochem Anal*. 2000;11 (5):327–329.
5. Stojković D, Soković M, Glamočlija J, et al. Chemical composition and antimicrobial activity of *Vitex agnus-castus* L. fruits and leaves essential oils. *Food Chem*. 2011;128 (4):1017–1022.

6. Asdadi A, Hamdouch A, Oukacha A, et al. Study on chemical analysis, antioxidant and in vitro antifungal activities of essential oil from wild *Vitex agnus-castus* L. seeds growing in area of Argan Tree of Morocco against clinical strains of Candida responsible for nosocomial infections. *Journal de Mycologie Médicale.* 2015;25(4):e118–e127.

7. Hajdú Z, Hohmann J, Forgo P, et al. Diterpenoids and flavonoids from the fruits of *Vitex agnus-castus* and antioxidant activity of the fruit extracts and their constituents. *Phytother Res.* 2007;21(4):391–394.

8. Choudhary MI, Jalil S, Nawazet SA, et al. Antiinflammatory and lipoxygenase inhibitory compounds from *vitex agnus-castus*. *Phytother Res.* 2009;23(9):1336–1339.

9. Kulkarni RR, Virkar AD, D'mello P. Antioxidant and antiinflammatory activity of *Vitex negundoIndian. J Pharm Sci.* 2008;70(6):838–840.

10. Stella J, Krishnamoorthy P, Mohamed AJ. Hypoglycemic effect of *Vitex agnus castus* in streptozotocin induced diabetic rats. *Asian J Biochem Pharm Res.* 2011;2:206–212.

11. Ahangarpour A, Oroojan AA, Khorsandi L, et al. Pancreatic protective and hypoglycemic effects of *Vitex agnus-castus* L. fruit hydroalcoholic extract in D-galactose-induced aging mouse model. *Res Pharm Sci.* 2017; 12(2):137–143.

12. Lucks BC, Sørensen J, Veal L. *Vitex agnus-castus* essential oil and menopausal balance: A self-care survey. *Complement Ther Nurs Midwifery.* 2002;8:148–154.

13. Propping D, Burger HG, Teede HJ, et al. *Vitex agnus castus.* The treatment of gynaecological syndromes. *Therapeutikon.* 1991;5:581–585.

14. Lauritzen CH, Reuter HD, Repges R, et al. Treatment of premenstrual tension syndrome with *Vitex agnus castus*: Controlled, double-blind study versus pyridoxine. *Phytomedicine.* 1997;4:183–189.

15. Loch EG, Selle H, Boblitz N. Treatment of premenstrual syndrome with a phytopharmaceutical formulation containing *Vitex agnus castus. J Womens Health Gend Based Med.* 2000;9:315–320.

16. Berger D, Schaffner W, Schrader E, et al. Efficacy of *Vitex agnus castus* L. extract Ze 440 in patients with pre-menstrual syndrome (PMS). *Arch Gynecol Obstet.* 2000;264:150–153.

17. Schellenberg R. Treatment for the premenstrual syndrome with *agnus castus* fruit extract: prospective, randomised, placebo-controlled study. *BMJ.* 2001;322:134–137.

18. Atmaca M, Kumru S, Tezcan E. Fluoxetine versus *Vitex agnus castus* extract in the treatment of premenstrual dysphoric disorder. *Hum Psychopharmacol.* 2003;18:191–195.

19. Momoeda M, Sasaki H, Tagashira E, et al. Efficacy and safety of *Vitex agnus-castus* extract for treatment of premenstrual syndrome in Japanese patients: a prospective, open-label study. *Adv Ther.* 2014;31(3):362–373.

20. Jarry H, Spengler B, Porzel A, et al. Evidence for estrogen receptor beta-selective activity of *Vitex agnus-castus* and isolated flavones. *Planta Med.* 2003;69:945–947.

21. van Die MD, Burger HG, Teede HJ et al. *Vitex agnus-castus* (Chaste-Tree Berry) in the treatment of menopause-related complaints. *J Alt Compl Med.* 2009;15(8):853–862.

22. Nasri S, Oryan S, Haeri Rohani A, et al. The effects of *Vitex agnus castus* L. extract on gonadotrophines and testosterone in male mice. *Iranian Int J Sci.* 2004;5(1):25–30.

23. Hoffmann D. *Medical Herbalism: The Science and Practice of Herbal Medicine.* Rochester, VT: Healing Arts Press; 2003. p.596

24. Khalilzadeh E, Saiah GV, Hasannejad H, et al. Antinociceptive effects, acute toxicity and chemical composition of *Vitex agnus-castus* essential oil. *Avicenna J Phytomed.* 2015; 5(3):218–230.

25. Daniele C, Coon JT, Pittler MH, et al. *Vitex agnus castus*: a systematic review of adverse events. *Drug Safety.* 2005;28:319–332.

26. Blumenthal M, Busse WR, Goldberg A, et al. *The Complete German Commission E Monographs.* Austin, TX: American Botanical Council; 1998. p. 108.

27. Gardner Z, McGuffin M, editors. *Botanical Safety Handbook*, 3rd edition. Boca Raton, FL: CRC Press; 2013. p. 923.

7.66   *WITHANIA SOMNIFERA* (ASHWAGANDHA)

Withania somnifera, also known as Ashwagandha, is a medicinal herb from the nightshade family. It is native to India, China, and Nepal, and commonly used in the Ayurvedic tradition of India to treat chronic fatigue, nervous exhaustion, memory loss, and neurodegenerative disorders. *Withania somnifera* is said to exhibit sedative, aphrodisiac, immunomodulatory, and anti-inflammatory actions. Many of the underlying causes of MDD – for example, stress, insulin resistance, and inflammation – and the resulting neurophysiological deficits are addressed by the herb.

The plant is rich in various phytochemicals, and more than 13 alkaloids, 138 withanolides, and several sitoindosides (a withanolide containing a glucose molecule at carbon 27) have been isolated and reported from aerial parts, roots and berries of *Withania* species. The withanolides, a group of steroidal lactones, are of primary medical interest.[1]

7.66.1   ANTIOXIDANT

In various *in vitro* assays of antioxidant potential, extracts of *Withania somnifera* were shown to exert strong scavenging capacity for free radicals, hydroxyl radicals, superoxide radicals, and hydrogen peroxide.[2] Administration of powdered *Withania somnifera* root to healthy mice over 30 days, was found to decrease lipid peroxidation in liver homogenates as per the TBARS assay. Treatment also increased the activities of the antioxidant enzymes superoxide dismutase and catalase activities in liver tissue.[3] Similarly, the aqueous suspension of extract of *Withania somnifera* root prevented the rise in levels of lipid peroxidation that occurred in rabbits and mice administered lipopolysaccharide.[4]

Four weeks of orally administered extract of *Withania somnifera* attenuated oxidative damage in the livers of rats administered CCl_4. Levels of the oxidation product malondialdehyde were reduced in liver homogenates, whereas levels of the antioxidant enzymes glutathione peroxidase, glutathione reductase, and glutathione-S-transferase, were increased. Levels of glutathione also rose significantly. The sparing of liver tissue was further evident in the herb's ability to significantly reduce serum transaminase levels.[5]

7.66.2   ANTI-INFLAMMATORY

Withania somnifera exhibits anti-inflammatory effects that are attributed to the steroid-like withanolides. These effects are reported to be comparable to those of hydrocortisone.[6] Methanol extracts inhibit the COX-2 enzyme and prevent TNF-induction of inhibition of NF-κB in human central nervous system cancer cell lines.[7]

Aqueous and chloroform extractions of *Withania somnifera* also exerts potent anti-inflammatory effects in microglial cell lines. Both the extracts attenuated the

TNF-α, IL-1β, IL-6, RNS, and ROS production via downregulating the expression of inflammatory proteins like NFkB and AP1.[8]

In intact rats, *Withania somnifera* prevented lipopolysaccharide-induced neuroinflammation. Prior daily feeding of aqueous extract of leaves of *Withania somnifera* over an eight-week period suppressed reactive gliosis production of TNF-α, IL-1β, and IL-6, and expression of nitro-oxidative stress enzymes in the brains of rats that were later peripherally administered lipopolysaccharide. The inflammatory NFkB, P38, and JNK/MAPKs pathways were also suppressed.[9] *Withania somnifera* also showed neuroprotective effects in reducing by 80% cell death in the hippocampus of rats subjected to 14 hours of immobilization stress.[10]

7.66.3 Antidiabetic/Anti-Metabolic Syndrome

Chronic feeding of fructose to rats results in a syndrome similar to metabolic syndrome in humans. Metabolic Syndrome, in turn, predisposes individuals to MDD. Addition of *Withania somnifera* to the high fructose diets fed to rats significantly reduced fructose-induced increases in glucose, fasting insulin, insulin resistance, IL-6, and TNF-α.[11] The oral administration of extract of *Withania somnifera* root stabilized mitochondrial functions and prevented oxidative damage in the hypothalamus of streptozotocin-induced diabetic rat. The increases in hypothalamic lipid peroxidation and protein carbonyl content, and decreases in glutathione and glutathione peroxidase activity seen in these diabetic animals were also reversed by treatment with the herb.[12]

7.66.4 Preclinical Antidepressant-Like Effects

Withania somnifera showed significant antidepressant effects in the forced swim, tail suspension, and reserpine mouse models of MDD. The herb was not quite as effective as imipramine, and the effects did not appear to be additive.[13] In a similar study of mice, *Withania somnifera* extract produced an antidepressant effect similar to that of fluoxetine.[14] In a study using rats, it produced an antidepressant effect comparable with that of imipramine in the forced swim-induced "behavioral despair" and "learned helplessness" tests.[15]

Interestingly, *Withania somnifera* has been investigated as a component in the treatment of various cancers. This is due to its ability to enhance apoptosis, a form of programmed cell death, through inhibition of the PI3/akt/mTOR cascade.[16] This effect would seem to be at odds with antidepressant effects. However, it is possible that the effects of *Withania somnifera* on the PI3/akt/mTOR cascade are tissue-specific. In any case, the active component of the herb, Withanolide A, enhances the activity of this cascade in hippocampal neurons.[17] Moreover, *Withania somnifera* potently inhibits GSK-3β through enhancement of its phosphorylation in mouse brain. Activation of mTOR is one way in which GSK-3β is inhibited.[18] Treatment with *Withania somnifera* also spares BDNF in the hippocampus and reduces memory impairment of rats subjected to hypoxia.[19] A similar sparing of BDNF by *Withania somnifera* was observed in mice treated with scopolamine.[20]

It is of note that a methanolic extract of *Withania somnifera* inhibited the specific binding of [3H]GABA and [35S]TBPS, and enhanced the binding of [3H]flunitraz-epam to their putative receptor sites. Those findings are consistent with its traditional use to treat agitation and nervous exhaustion.[21]

7.66.5   HUMAN ANTIDEPRESSANT EFFECTS

Human studies of effects of *Withania somnifera* on MDD are few. An Indian study evaluated use of *Withania somnifera* plus shirodhara therapy in patients diagnosed with MDD by DSM-IV criteria. Shirodhara is the Ayurvedic technique of allowing oil to flow continuously down onto the forehead. After 42 days of treatment, patients responded successfully to the *Withania somnifera* plus shirodhara therapy, and in a manner similar to the active control group receiving fluoxetine 40 mg a day.[22]

7.66.6   DOSAGE

The recommended dosage is 1–10 ml of tincture of dried root or 1–3 g dried root powder three times daily. Easley and Horne suggest caution when using during pregnancy.[23]

7.66.7   TOXICITY

Withania somnifera appears to have quite low toxicity. The LD50 intraperitoneal injection of alcohol extract of the herb is reported to be 1,260 mg/kg.[24] Extrapolated to a 75 kg human adult, this dose would be approximately 88 g.

7.66.8   SAFETY IN PREGNANCY

The *Botanical Safety Handbook* advises that information of the safety of using *Withania somnifera* during pregnancy is conflicting. Some traditional sources have described the herb as an abortifacient, while others have noted its use in preventing miscarriage. It does appear safe in lactation, with traditional use in stimulating lactation.[25]

7.66.9   DRUG INTERACTIONS

The *Botanical Safety Handbook* notes no known interactions with drugs. At least one study showed no interactions of extracts of *Withania somnifera* with CYP3A4 or CYP2D6 enzyme activity in human liver microsomes.[26]

REFERENCES

1. Jain R, Kachhwaha S, Kothari SL. Phytochemistry, pharmacology, and biotechnol-ogy of *Withania somnifera* and *Withania coagulans*: a review. *J Med Plants Res.* 2012;6(41):5388–5399.

2. Pal A, Naika M, Khanum F, et al. In-vitro studies on the antioxidant assay profiling of *Withania somnifera* L. (Ashwagandha) Dunal root: Part 1. *Pharmacog J.* 2011;3(20):47–55.
3. Panda S, Kar A. Evidence for free radical scavenging activity of Ashwagandha root powder in mice. *Indian J Physiol Pharmacol.* 1997; 41(4):424–426.
4. Dhuley JN. Effect of ashwagandha on lipid peroxidation in stress-induced animals. *J Ethnopharmacol.* 1998;60(2):173–178.
5. Elberry AA, Harraz FM, Ghareib SA, et al. Antihepatotoxic effect of *Marrubium vulgare* and *Withania somnifera* extracts on carbon tetrachloride-induced hepatotoxicity in rats. *J Basic Clin Pharm.* 2010;1:247–254.
6. Anbalagan K, Sadique J. Influence of an Indian medicine (Ashwagandha) on acute-phase reactants in inflammation. *Indian J Exp. Biol.* 1981;19:245–249.
7. Mulabagal V, Subbaraju GV, Rao CV, et al. Withanolide sulfoxide from Aswagandha roots inhibits nuclear transcription factor-kappa-B, cyclooxygenase and tumor cell proliferation. *Phytother Res.* 2009;23(7):987–992.
8. Gupt M, Kaur G. Aqueous extract from the *Withania somnifera* leaves as a potential anti-neuroinflammatory agent: a mechanistic study. *J Neuroinflamm.* 2016; 13:193–210.
9. Gupta M, Kaur G. *Withania somnifera* as a potential anxiolytic and anti-inflammatory candidate against systemic lipopolysaccharide-induced neuroinflammation. *NeuroMol Med.* 2018;20(3):343–362.
10. Jain S, Shukla SD, Sharma K, et al. Neuroprotective effects of *Withania somnifera* Dunn. in hippocampal sub-regions of female albino rat. *Phytother Res.* 2001;15:544–548.
11. Noshahr ZS, Shahraki MR, Ahmadvand H, et al. Protective effects of *Withania somnifera* root on inflammatory markers and insulin resistance in fructose-fed rats. *Rep Biochem Mol Biol.* 2015;3(2):62–67.
12. Parihar P, Shetty R, Ghafourifar P, et al. Increase in oxidative stress and mitochondrial impairment in hypothalamus of streptozotocin treated diabetic rat: antioxidative effect of *Withania somnifera*. *Cell Mol Biol.* 2016;62(1):77–83.
13. Jayanthi MK. Anti-depressant effects of *withania somnifera* fat (ashwagandha ghrutha) extract in experimental mice. *Int J Pharm Bio Sci.* 2012;3(1):33–42.
14. Attari M, Jamaloo F, Shadvar S, et al. Effect of *Withania somnifera* dunal root extract on behavioral despair model in mice: a possible role for nitric oxide. *Acta Medica Iranica.* 2016;54(3)165–172.
15. Bhattacharya SK, Bhattacharya A, Sairam K, et al. Anxiolytic-antidepressant activity of Withania somnifera glycowithanolides: an experimental study. *Phytomedicine.* 2000;7(6):463–469.
16. Grogan PT, Sleder KD, Samadi AK, et al. Cytotoxicity of withaferin A in glioblastomas involves induction of an oxidative stress-mediated heat shock response while altering Akt/mTOR and MAPK signaling pathways. *Invest New Drugs.* 2013;31(3):545–557.
17. Hwang D, Vasquez I, Galvez L, et al. Ashwagandha and its active ingredient, withanolide A, increase activation of the phosphatidylinositol 3′ kinase/akt cascade in hippocampal neurons. *Eur J Med Plant.* 2017;20(2):1–19.
18. Raghavan A. Neuroprotective Potential of *Withania Somnifera* in Cerebral Ischemia [Doctoral Dissertation]. [Toledo, Ohio]: University of Toledo; 2014.
19. Baitharu I, Jain V, Deep SN, et al. *Withania somnifera* root extract ameliorates hypobaric hypoxia induced memory impairment in rats. *J Ethnopharmacol.* 2013;145:431–441.
20. Konar A, Shah N, Singh R, et al. Protective role of Ashwagandha leaf extract and its component withanone on scopolamine-induced changes in the brain and brain-derived cells. *PLoS One,* 2011;6(11): Article: e27265.
21. Mehta AK, Binkley P, Gandhi SS, et al. Pharmacological effects of *Withania somnifera* root extract on GABAA receptor complex. *Indian J Med Res.* 1991;94:312–315.

22. Fulzele A, Hudda N. Selective ayurvedic therapy for the management of major depressive disorder: a randomised control trial. *Ancient Sci Life.* 2012;32(Suppl 1): S41.
23. Easley T, Horne S. *The Modern Herbal Dispensary.* Berkeley, CA: North Atlantic Books; 2016. p. 178.
24. Sharada AC, Solomon FE, Devi PU. Toxicity of *Withania somnifera* root extract in rats and mice. *Int J Pharmacog.* 1993;31(3):205–212.
25. Gardner Z, McGuffin M, editors. *Botanical Safety Handbook*, 3rd edition. Boca Raton, FL: CRC Press; 2013. p. 274.
26. Savai J, Varghese A, Pandita N, et al. Investigation of CYP3A4 and CYP2D6 interactions of *Withania somnifera* and *Centella asiatica* in human liver microsomes. *Phytotherapy Res.* 2015;29(5):785–790.

8 The Antidepressant Effects of *Yueue* and the Herbs of Traditional Chinese Medicine

8.1 FUNDAMENTAL CONSIDERATIONS

In Traditional Chinese Medicine (TCM), MDD has never been seen as an illness solely of the mind and brain. Rather, various depression-like syndromes have been considered to be manifestations of more fundamental body-wide imbalances of *Chi*, or life energy. Discussions of syndromes resembling MDD are found in the 2,000-year-old medical treatise, *The Esoteric Scripture of the Yellow Emperor.* The general term used for depression is "*yu,*" which also means stagnation. Stagnation is reflected in many of the conditions that are seen as leading to states of emotional depression. In severe depression, stagnation or deficiency of liver *Chi* is suspected. It must be emphasized that the Western notion of the liver as a specific organ with metabolic, synthetic, and detoxifying roles is quite different from the Chinese notion of the liver being a vehicle for the movement of specific forms of *Chi* through the body. Indeed, in TCM the liver houses *Hun*, the Ethereal Soul. The Ethereal Soul is thought to control the "comings and goings" of *Shen*, the mind. Deficiency or stagnation of *Hun* thus deflates and limits the activity of the mind, leading to withdrawal, lack of motivation, and lack of enjoyment of life all around. On the other hand, sadness is attributed to stagnation or deficiency of lung and heart *Chi*, worrying and overthinking to deficiency of spleen, and irritability to rising liver *yin* and *yang* deficiency. *Yin* and *yang* themselves are each constellations of complimentary characteristics that include female and male, but also many other characteristics such as cold and hot, soft and hard, dark and light, and many other pairs of opposites. Because each system derives *Chi* from other systems and delivers *Chi* to yet others in turn, the causes of depression are ultimately seen as imbalances throughout the entire body.[1] Among the syndromes associated with psychiatric symptoms are phlegm-*Chi* stagnation, liver-*Chi* stagnation, heart *yin* deficiency, heart-kidney disharmony, spleen deficiency, *yang Chi* stagnation, liver-spleen disharmony, heart-spleen deficiency, *yin* deficiency with fire effulgence, and "Syndrome of six stagnations." Clearly, unlike mainstream Western psychiatry, TCM has never considered MDD to be a single specific illness to be treated with a single class of medications.

Another characteristic of TCM is seen in the fact that most of the ancient Chinese treatments of depression are combinations of herbs. Indeed, in the traditional Chinese medical theory, every herb in a Chinese medicine formula is essential and plays its own necessary role. This is defined according to the principle of "*Jun-Chen-Zuo-Shi*," or "monarch, minister, assistant, and envoy."[2] Thus, Chinese herbalists long recognized that one herb in a combination, the "monarch," may provide the primary benefits. However, others in the combination are necessary to prepare the body, block adverse effects, improve general health, or to regulate activity in pathways that supply or receive *Chi* from the system primarily targeted. This philosophy is quite at odds with mainstream Western medicine in which the ideal treatment is a single, potent, and incisive medication that solves the problem with as few collateral effects as possible. Indeed, the use of combinations of herbs might be seen as a form of the much maligned, "polypharmacy." Yet, in many respects the *Jun-Chen-Zuo-Shi* approach is quite insightful and comprehensive. For example, a single antidepressant medication may correct deficits in synthesis of BDNF, but not address the underlying inflammation, insulin resistance, or hypercortisolemia that led to the depression. Certainly, every prescriber has experienced poor compliance of patients prescribed antidepressants due to weight gain, anorgasmia, nausea, or disturbed sleep. Alleviating those side effects with co-administered herbs can maintain successful treatment.

8.2 *YUEJU*

One of the most fascinating and significant of recent events has been the use of modern neuroscience technology by Chinese scientists to evaluate the efficacy and determine the neurochemical mechanisms of action of the classical herbal remedies of TCM. Of particular interest is the combination known as *Yueju*. This time-honored compound is said to have been developed by the Chinese master herbalist Zhu Danxi over 800 years ago for treatment of depressed mood, anxiety, and irritability. It is a combination of equal amounts of the herbs *Gardenia jasminoides*; *Cyperus rotundus*; *Chuan Xiong*, a Chinese herb of the carrot family with the Latin name of *Ligusticum chuanxiong*, but no common English name; *Chang Zu*, another Chinese herb with the Latin name of *Atractylodes lancea*, but no common English name; and *Shen Qu*, itself a combination of fermented wheat and herbs.

A 2008 study in China found *Yueju* and venlafaxine to be quite similar in antidepressant effect in depressed patients, but with *Yueju* having fewer side effects.[3] The Japanese traditional herbal medicine mixture, *kamishoyosan*, which contains *Gardenia jasminoides*, was soon after found to ease anxiety and depression in menopausal women in a manner similar to that of paroxetine. The herbal treatment also reduced serum levels of the inflammatory cytokine IL-6.[4]

A number of studies have further explored the efficacy of *Yueju* in animal models of depression. One study examined the time course of *Yueju* in mouse models of depression, as well as the neurochemical mechanisms by which the drug acts. The preparation had antidepressant-like effects in the learned helplessness, tail suspension, and open field tests.[5] Of interest was the rapidity with which the *Yueju* acted. Whereas repeated doses of standard antidepressants were necessary to produce such results, a single dosing of *Yueju* was effective. Neurochemical evaluations further showed that BDNF was increased in the hippocampi the mice as early as 30 minutes after injection. *Yueju* also rapidly increased activity of eEF2 by decreasing phosphorylation. In addition, *Yueju* dampened NMDA receptor activity and stimulated mTOR in a manner similar to that of ketamine.[6] It was subsequently shown that in human patients with MDD, addition of *Yueju* to fluoxetine resulted in a more rapid antidepressant response.[7]

Because *Yueju* is a mixture of herbs, it had not been obvious what component was most responsible for the antidepressant effect. However, it has since been found that extracts of *Gardenia jasminoides* replicate both the antidepressant effects in mice and the rapid upregulation of BDNF in the hippocampi produced by the *Yueju* combination.[8] Most recently, geniposide, a terpenoid glycoside component extracted from *Gardenia jasminoides*, was found on its own to increase BDNF in mouse brain.[9]

In recent reviews, Chinese neuroscientists discussed a number of other herbal mixtures that traditionally have been used for conditions with prominent components of MDD.[10] These papers exemplify the current efforts by Chinese scientists to evaluate and interpret the effects of traditional herbal remedies from the perspective of modern molecular neuroscience. Yeung et al. have noted the ten most commonly used TCM herbal treatments for depression to be: *xiao yao san, chai shu gan, gan mai dazo, gui pi, shi wei wen dan tang, ban xia hou pu, chai hu jia long gu mu li, tiao qi, yi pi,* and *tang shen kang.*[11] Curiously, this top ten list did not include the above-noted *Yueju.* Feng et al. have also reviewed the apparently lesser used but effective herbal combinations *kaix san, shu gan jie yu, sini san,* and *wu ling.*

As an informative note: In the names of TCM herbal medications, common suffixes often refer to the form of the combination. Thus, *pian* refers to flat tablets, *wan* is the form of a round pill, *san* is a powdered form, and *tang* refers to its use as a tea.

8.3 XIAO YAO SAN

Xiao yao san, known by the lyrical name of "Free and easy wanderer," consists of *Bupleuri falcatum radix, Angelica sinensis, Paeonia lactiflora radix, Poria incrassate, Atractylodes macrocephala, Mentha haplocalyx, Zinjiber officinallis,* and *Glycyrrhiza glabara.* Among the phytochemicals identified in *xiao yao san* are: Gallic acid, catechin, albiflorin, paeoniflorin, liquiritin apioside, liquiritin, ferulic acid, Z-ligustilide, atractylenolide, and isorhamnetin. Note that several of the constituent herbs are discussed in the previous chapter with more detailed phytochemical identification.

Animal studies have provided indications as to the mechanism by which *xiao yao san* acts as an antidepressant. The combination decreased immobility in the forced swim test. This antidepressantlike effect was accompanied by reduced serum levels of the inflammatory cytokine, IL-1, and corticosterone, the rat stress hormone. Treatment also reduced IL-1 in the paraventricular nucleus and hippocampus of the stressed rats.[12]

Rats subjected to chronic restraint stress show deficits in time spent in the open field test and in the open arm of the elevated plus maze, which are taken as anxiety-like behavior. Treatment with *xiao yao san* reversed these effects. Examination of the rats' brains revealed that stress-induced changes in α-amino-3-hydroxy-5-methyl-4-isoxazolepropionic acid (AMPA) receptors are also reversed by *xiao yao san.*[13] The antidepressant effects of ketamine are thought to be due at least in part to increased activity at AMPA receptors and subsequent increases in mTOR and BDNF in the hippocampus and prefrontal cortex.[14]

Xiao yao san has also been tested in human clinical studies. In a meta-analysis of 26 randomized trials (involving 1,837 patients), *xiao yao san* combined with antidepressants was more effective than the antidepressants alone in reducing HAMD and SDS scores. Head to head, *xiao yao san* was superior to antidepressants in reducing the HAMD score. No adverse effects of *xiao yao san* were reported.[15]

In a study based in traditional Chinese perspective,[16] 50 patients with affective disorders were typed into the categories of depressed liver resulting in fire, mild *Yang* deficiency, and mild *Yin* deficiency and were treated with *xiao yao san*. The results were 26 patients with marked improvement, 17 patients with improvement, and 7 patients with no improvement. It is possible that the marked differences in responses among subjects reflect genuine subtypes of depression defined by TCM.

8.4   CHAI HU SHU GAN

Chai hu shu gan (translated approximately as "soothe the liver with *Bupleuri falcatum*") is a mixture of *Bupleuri falcatum radix*, *Citrus reticulata*, *ligusticum chuanxiong*, *Cyperus rotundus*, *Citrus aurantium*, *Paeonia lactiflora radix*, and *Glycyrrhiza glabara*.

Among the many phytochemicals in the decoction of *Chai hu shu gan* are: paeoniflorin, naringin, hesperidin, saikosaponin A, glycyrrhizic acid, nobiletin, tangeretin, ferulic acid, gallic acid, oxypaeoniflorin, albiflorin, liquiritin, narirutin, meranzin hydrate, liquiritigenin, quercetin, benzoylpaeoniflorin, isoliquiritigenin, and formononetin.

Chai hu shu gan blocks the effects of stress on the reduction of ERK5 in rat hippocampus. ERK5 mediates and stimulates the cascade initiated by various nerve growth factors stimulating trk receptors, including BDNF. It also mitigates the effects of stress in stimulating the JNK pathway that mediates inflammatory cascades.[17]

The effects of *Chai hu shu gan* were studied in a large meta-analysis of ten human studies including 835 subjects. Comparisons were made against fluoxetine, paroxetine, and doxepin monotherapies versus those treatments augmented by *Chai hu shu gan*. *Chai hu shu gan* enhanced the antidepressant effects of those standard Western antidepressants. However, the efficacy of *Chai hu shu gan* as monotherapy was also significantly better than antidepressants in improving depressive symptoms.[18] Also, of interest is the finding that *Chai hu shu gan* is helpful in treating fatty liver disease,[19] suggesting that some of its benefits for depression may derive from its ability to improve insulin sensitivity and diminish the signs of Metabolic Syndrome.

8.5   GAN MAI DA ZAO

Gan mai da zao is a relatively simple combination of *Glycyrrhiza glabara*, *Shen Qu* (dried, shriveled wheat berries), and *Zizyphus jujube*. The combination was first noted around AD 1000 by the later Han Dynasty physician, Zhang Zhong Jing. It was described in his treatise, *Treatment of Various Female Diseases* and was said to have the capability to "treat hysteric psychoneurosis-sad and prone to crying, frequently yawning, as if haunted by spirits."[20]

Among the phytochemicals in *Gan mai da zao* are glycyrrhizic acid, isoliquiritin, and isoliquiritigenin, as well as rutin, cathechin, apigenin-7-glucoside, quercetin, over 150 cyclopeptide alkaloids and complex terpenoids from jujube fruit, and various poorly characterized saponins, flavonoids, alkaloids, anthraquinones, tannins, and terpenoids from wheat berries.

In a study on rats subjected to unpredictable mild stress, *Gan mai da zao* restored appetite for sucrose. Further evaluation showed that the herbal mixture reversed the increases in glutamate activity and compensatory decreases in NMDA receptor proteins in the frontal cortex and hippocampus.[21] Ketamine acts, at least in part, by inhibiting NMDA receptors.

In a meta-analysis of ten Chinese clinical studies, *Gan mai da zao* alone compared well with antidepressants with fewer side effects. The mixture also improved response when added to standard antidepressant treatment. The studies were randomized and controlled, but still deemed to be of low quality by the standards of the Cochrane reviews. The authors recommended further studies with stronger methodology.[22] In another study, in which subjects suffered what was described as "*yang* deficiency climacteric depression," a term unknown in Western psychiatry, a fortified *Gan mai da zao* treatment gave significant relief of symptoms of depression and poor sleep.[23]

8.6   *GUI PI*

Gui pi contains *Panax ginseng*, *Astragalus membranaceus*, *Atractylodes macrocephala*, *Poria cocos*, *Angelica sinensis*, *Ziziphus jujube*, *Dimocarpus longan*, *Polygala tenuifolia*, *Silybum marianum*, *Glycyrrhiza glabara*, and *Zinjiber officinallis*. The name *gui pi* can be translated as "restore the spleen." The combination is attributed to the Chinese physician Yan Yonghe, from the thirteenth century, and it has traditionally been used to address various manifestations of spleen-*Chi* deficiency.

Descriptions of studies in English are wanting. However, at least one English abstract of a study states that *Gui pi* in combination with *Gan mai da zo* (described above) was as effective in treating depression and insomnia as zolpidem plus deanxit, the combination of flupentixol and the tricyclic antidepressant, melitracen.[24]

8.7   *SHI WEI WEN DAN TANG*

Shi wei wen dan tang (translated as Ten Flavors Warm the Gallbladder Decoction) contains *Pinellia ternata*, *Citrus reticulata*, *Poria cocos*, *Schisandra chinensis*, *Panax ginseng*, *Polygala tenuifolia*, *Citru aurantium*, *Ziziphus jujube*, *Glycyrrhiza glabara*, and *Zinjiber officinallis*.

There are no references in the literature to *Shi wei wen dan tang* as a treatment for depression. There are studies reporting effectiveness in treating Parkinson's Disease,[25] and in treating schizophrenia in human patients.[26] Indeed, it was able to enhance the effects of the potent antipsychotic medication clozapine. It also improved memory and learning in a rat model of schizophrenia. In electrophysiological tests, the Chinese herbal preparation was shown to enhance long-term potentiation in the hippocampus in a manner similar to clozapine.[27]

8.8   BAN XIA HOU PU

Ban xia hou pu is the combination of Pinellia ternata, Poria cocos, Magnolia officinalis, Zinjiber officinallis, and Perilla frutescens. The name refers to the primary ingredients, Pinellia ternate and Magnolia officinalis, and it is used to address liver Chi stagnation. Ban xia hou pu prevented effects of stress in the cortex and hippocampus of rats subjected to persistent restraint in a manner similar to fluoxetine.[28]

A human study showed that Ban xia hou pu was more effective than an active control treatment in treating symptoms of anxiety and depression in a group of subjects with globus hystericus.[29]

Studies suggested Magnolia officinalis, Poria cocos, and Perilla frutescens to be the critical components in these effects of Ban xia hou pu. The phytochemicals that have been identified in decoctions of the mixture were zingiberol, rosemarinic acid, magnolol, honokiol, linalool, citral, nerolidol, caryophyllene, bisaboline, apiole, eudesmol, and farnesene.

8.9   CHAI HU JIA LONG GU MU LI

Chai hu jia long gu mu li contains some typical ingredients, including Bupleuri falcatum radix, Scutellaria lateriflora, Zinjiber officinallis, Panax ginseng, Cinnamomi cassia, Poria cocos, Pinellia ternata, Rheum palmatum, and Ziziphus jujube. It sometimes also includes the rather unusual oyster shell, lead tetroxide, and ground fossilized mastodon bone. Its name refers to "dragon bone" –ostensibly the fossilized mastodon bone – and it is used to treat "Liver and spleen disharmony with depressive heat, phlegm dampness, and disquieted heart spirit manifesting as irritability, heart palpitations, insomnia, vexation and agitation."

In at least one published study in Chinese, Chai hu jia long gu mu li was found to be as effective as fluoxetine in patients suffering post-stroke depression.[30] The very similar combination Chai hu gui zhi gan jiang tang was found to improve mood and improve sleep in in a study of women with climacteric depression.[31] It was equal in effect with a variety of antidepressants. Interestingly, the herbal combination was found to reduce serum levels of the inflammatory cytokine IL-6, which was not seen with standard antidepressant treatment.

8.10   TIAO QI

Tiao qi, which literally means, "adjust Chi," contains Bupleuri falcatum radix, Scutellaria lateriflora, Curcuma longa, Zinjiber officinallis, Lilium lancifolium, Liriope spicata, and Panax quinquefolius.

Access to English translations of clinical data on Tiao qi is limited. There is at least one reference in the Chinese literature showing that variants of the Tiao qi combination are effective in patients with co-morbid digestive disturbances and MDD.[32]

8.11   *YI PI*

Yi pi, or in English, "benefit the spleen," consists of *Astragalus membranaceus, Dioscorea polystachya, Poria cocos,* chicken egg, *Amomum globosum, Eriobotrya japonica,* and *Areca catechu.* There is a least one reference in the Chinese literature showing it to be marginally effective for treatment of climacteric depression.[33]

8.12   *TANG SHEN KANG*

Tang shen kang contains *Panax ginseng,* deer antler velvet, *Rehmannia glutinosa, Cornus kousa, Ligustrum lucidum,* and *Polygonum aviculare.* In Chinese medicine, the combination is used not only for symptoms indicative of MDD, but also for apparently unrelated conditions, such as diabetic nephropathy. In the one study with results available in English, *Tang shen kang* was at least as effective as fluoxetine in the treatment of depression in patients with comorbid depression and diabetes.[34]

8.13   *KAI XIN SAN*

Kai xin san is a decoction first described by Sun Simiao in *Beiji Qianjin Yaofang* about 1,400 years ago. It is used in China to treat stress-related psychiatric diseases with the symptoms of depression and forgetfulness. It is a combination of *Panax ginseng, Polygala tenuifolia, Acorus calamus,* and *Poria cocos.*

The *Kai xin san* mixture contains a number of unique complex terpenoids from ginseng known as ginsenosides; the Polygala saponins (arillatanoside A, tenuifolin), terpenoids (tumulosic acid, pachymic acid, dehydropachymic acid), oligosaccharide esters (18, including sibiricoses A1 and A6), and tenuifolisides A; as well as alpha, beta, and gamma asarone, cis- and trans-methylisoeugenol, methyl eugenol, polygalaxanthone, and pachymic acid from *Acorus tatarinowii* and Poria.

In a clinical study, *Kai xin san* was more effective than fluoxetine at two, four, and eight weeks in moderately depressed patients.[35] Several studies have shown the likely molecular basis for the antidepressant effect. The *Kai xin san* mixture stimulated cultured astrocytes to release BDNF and other nerve growth factors.[36,37] Along with stimulation of neurotrophic factors, *Kai xin san* may also produce antidepressant effects in rats by enhancing various aspects of serotonergic activity.[38]

8.14   *SHU GAN JIE YU*

Shu gan jie yu, roughly translated as "soothe the liver," is the combination of *Hypericum perforatum* and *Eleutherococcus senticosus.* Twenty-two compounds were identified in *Shu gan jie yu* by ultraperformance liquid chromatography tandem mass spectrometry. These included: neochlorogenic acid, syringin, chlorogenic acid, 1'-O-caffeoylquinic acid, 1',5'-, 3',5'-, and 4',5'Odicaffeoylquinic acid, eleutheroside E, rutin, hyperoside, isoquercitrin, isofraxidin, quercetin3-β-d-arabinose, quercitrin, acetyl-hyperoside, quercetin, II3,II8-biapigenin, protopseudohypericin, hyperforin, adhyperforin, hypericin, and pseudohypericin.[39]

Shu gan jie yu exhibited antidepressant-like effects in several rat models of depression. It restored preference for sucrose and reduced immobility in the forced swim test.[40] Of particular note, the combination increased levels of p-CREB and BDNF expression in the medial prefrontal cortex and hippocampal CA3 region.[41]

In view of the many positive reports on the component herbs St John's wort and Siberian ginseng in Western medicine, it is not surprising that *Shu gan jie yu* has been found effective in clinical trials. A meta-analysis of studies on the use of *Shu gan jie yu* for treatment of mild-to-moderate depression included 15 RCTs with 1,240 patients. *Shu gan jie yu* was more effective than the active controls that included paroxetine, citalopram, sertraline, fluoxetine, or melitracen-flupentixol. *Shu gan jie yu* also caused fewer side effects.[42] Another meta-analysis found *Shu gan jie yu* as effective as venlafaxine and better tolerated. Benefits of *Shu gan jie yu* and venlafaxine were also additive.[43] Effects of *Shu gan jie yu* in geriatric depression were significant, but less substantial than in younger populations.[44]

8.15   *SI NI SAN*

Si ni, meaning "frigid extremities," consists of four of the main ingredients from the larger combination *Chai hu shu gan san* described above, i.e., *Bupleuri falcatum radix*, *Citrus aurantium*, *Paeonia lactiflora radix*, and *Glycyrrhiza glabara*. Feng et al. presented compelling behavioral and neurochemical data from rodent studies, but no clinical studies of *Si ni san*. *Si ni san* reverses deficits in spatial learning and memory, aggressive behavior, and reduced growth rate in rats subjected to chronic restraint stress. Its effects were comparable to those of fluoxetine, although slower in onset and more persistent over several months. Its effects were also additive to those of fluoxetine. Although many phytochemicals have been identified in *Si ni san*, the active components were said to be saikosaponins, paeoniflorin, naringin, and glycyrrhizic acid.

It was not made clear how *Si ni san* acts to produce its antidepressant effect. A somewhat imprecise explanation was modulation of cerebral monoamine-neurotransmitter expression, hippocampal neuroprotection, inhibition of HPA-axis hyperfunction, and immune and second-messenger regulation.[45]

8.16   *WU LING*

Wu ling capsules are unusual as a Chinese herbal treatment in that they contain the single ingredient, fermented mycelia of the *Xylaria nigripes* mushroom. *Xylaria* is a genus of ascomycetous fungi commonly found growing on dead wood. *Wu ling* has no common English name, and the mushroom is likely indigenous to China. (I have also seen the name *Wu ling* referring to a combination of Chinese herbs that includes poria cocos mushroom. Indeed, the name may be translated as Poria five powder.) Among the chemical constituents are 5-methylmellein, 5hydroxymellein, 5-carboxylmellein, and genistein.[46]

Animals studies have established the efficacy of *Wu ling* in models of depression, as well as providing insight into the mushroom's mechanism of antidepressant action. *Wu ling* restored the preference for sucrose in rats that had been chronically subjected to unpredictable mild stress.[47] It also reversed the learned helplessness syndrome in mice.[48]

Wu ling stimulated neurogenesis in the hippocampi of rats subjected to unpredictable mild stress. However, this was not due to increases in the activity of BDNF, but rather to enhancement of the astrocyte protein connexin. In fact, BDNF-positive cells remained low in the cortical neurons of these animals.[49] Astrocytes maintain normal neuronal function, and a number of abnormalities of astrocyte function have been found in post-mortem studies of brain tissue from suffers of MDD. Prominent among these abnormalities is depletion of connexins 30 and 43.[50] These proteins located on astrocytic end-feet are gap junction-forming membrane proteins that allow communication between astrocytes and, in turn, proper husbandry of neurons. Feng also noted data suggesting enhancement of the translocator protein (TSPO) signaling pathway as a mechanism of *Wu ling* action. TSPO is a mitochondrial membrane protein that plays important roles in inflammation and steroid production in the brain.[51]

Clinical studies of *Wu ling* have established its efficacy in humans. A meta-analysis showed that in patients suffering post-stroke depression, *Wu ling* as monotherapy relieved symptoms of depression, while in combination therapy it augmented the benefits of SSRIs.[52] It relieved symptoms in women suffering menopausal depression[53] and in patients being treated for epilepsy.[54]

8.17   OTHER TCM HERBS USED IN THE TREATMENT OF MDD

Yeung et al. noted that the most commonly used single ingredients in TCM herbal treatments of depression, in descending order of use, are *Bupleuri falcatum radix*, *Angelica sinensis*, *Paeonia lactiflora radix*, *Poria cocos*, *Curcuma longa*, *Atractylodes macrocephala*, *Glycyrrhiza glabara*, *Ligusticum chuanxiong*, and *Cyperus rotundus*. These were discussed in the previous chapter exploring individual herbs. One of the herbs mentioned by Yeung et al., *Acorus calamus*, was not discussed. It has been used for centuries in Chinese and Western herbalism for a variety of purposes, including the treatment of depression-like syndromes. It has been held in very high regard. Unfortunately, both α- and β-asarone in *Acorus calamus* are known to have carcinogenic, mutagenic, genotoxic, and teratogenic effects in mice.[55]

Presented above are the most common combinations of herbs used in TCM for the treatment of MDD. The efficacy of these traditional herbal combinations is supported in some cases by clinical data, in more cases with preclinical data, and in most cases by elucidation of reasonable mechanisms by which they might work. These herbal combinations are effective on their own, but also appear to enhance the effects of more standard Western pharmacotherapy. Many of the phytochemicals in these herbs act similarly to ketamine in stimulating the PI3/akt/mTor intracellular pathway, inhibiting GSK-3β, and enhancing the synthesis of BNDF. Others act to reduce inflammation and stress, and enhance insulin sensitivity. These latter effects indirectly activate PI3/akt/mTor, inhibit GSK-3β, and increase BNDF. In some cases, the herbs act in other non-classical ways, such as enhancing Nrf2 and Peroxisome proliferator-activated receptor-gamma coactivator-1alpha, which plays a central role in the regulation of cellular energy metabolism.

It is not clear if all of phytochemicals brought together in the combinations of herbs used in TCM are necessary for synergistic, and thus uniquely effective, action. When presented with a combination of herbs with psychotherapeutic benefits, the Western-trained neuroscientist would likely determine which herb in the combination is most effective, and then extract and separate the phytochemicals to discover exactly which molecule in the herbal combination is specifically responsible for the antidepressant effect. The fact that so many patients with treatment-resistant MDD require a second augmenting medication should argue for combinations of herbs being useful and necessary. Similarly, the fact many therapeutically active molecules – the atypical antipsychotics come to mind – are pleomorphic and thus simultaneously active at many different receptor sites, further suggests that wide-ranging action of medication may be useful if not necessary for optimal effect.

8.18   A MEDICAL "THEORY OF EVERYTHING"

Western psychiatrists must also be willing to set aside their myopic skepticism and admit the possibility that the ancient Chinese perspective and inexplicable methods of diagnosis allow the discrimination of subtypes of depression – liver-*Chi* stagnation, heart *yin* deficiency, heart-kidney disharmony, spleen deficiency, liver-spleen disharmony – that respond to specific combinations of herbs (or polypharmacy) and not a single pharmacological "magic bullet." At the same time, there is danger in romanticizing TCM, as research has revealed many of its own weaknesses and incongruencies. For example, the traditional methods of diagnosis in Chinese Medicine, such as pulse diagnosis, suffer poor inter-rater reliability.[56] The ancient method of tongue diagnosis also has poor inter-and intra-rater reliability.[57] There is also a lack of rigorous and comprehensive studies of the efficacies of the entire range of TCM treatments that would lend support to its basic tenets of diagnosis and treatment.

Unfortunately, the marriage of ancient, mystical TCM with modern, analytical Western medicine may seem as impossible as the reconciliation of quantum mechanics and general relativity. Yet it may be possible to discover a medical "Theory of Everything." Concepts such as *Chi*, meridians, *yin* and *yang*, heat and cold, systems related to organs that stretch far beyond their anatomical confines, may seem utterly foreign and irreconcilable with "modern" Western medicine. Yet, some Chinese scientists have begun to integrate Chinese and Western medical concepts. For example, the '*Yin-Yang*' theory, the ancient philosophical position that underlies much of traditional Chinese medicine, has been interpreted in terms of mitochondrial activity. *Yang* strengthening herbs were seen as boosting body function through enhancing the mitochondrial oxidative processes, with protective *Yin* herbs maintaining antioxidant potential.[58] Another group of researchers, including both Western and Eastern physicians, note similarities between the concepts of *Yin* and *Yang*, and those of the sympathetic versus the parasympathetic nervous system, the effects of cAMP versus cGMP, and the shifting of circadian rhythms from daytime to night.[59]

Certainly, the stress response, inflammation, innate and acquired immune function, the sympathetic and parasympathetic nervous systems, endocrine interaction, Metabolic Syndrome, gut-brain axis, epigenetics, allostatic load – all phenomenon well known and accepted by Western science – exist at the level of complex systems and not within specific organs. In fact, as was previously noted, the etiology of treatment of depression is probably best understood in a body-wide, systems-oriented perspective. Metabolic Syndrome perhaps best exemplifies pathological interactions across systems that cannot be properly seen by focusing on any one organ or physiological process. When Western medicine takes a larger view – holistic, if you prefer – the Chinese notions couched in ancient language and philosophy may become more amenable to Western approaches and sensibilities.

REFERENCES

1. Maciocia G. *The Foundations of Chinese Medicine.* London, UK: Elsevier; 2005. p. 342–345.
2. Zhang E, Shen J, So KF. Chinese traditional medicine and adult neurogenesis in the hippocampus. *J Tradit Complement Med.* 2014;4(2):77–81.
3. Zhou-ke G, Qi Lu, Zhang CL, et al. The effect of Yu-Le decoction to major depression of stagnated heat type of heart-liver and level change of 5-HT and NE in the blood plasma. *Chin Arch Tradit Chin Med.* 2009;8.
4. Yasui T, Yamada M, Uemura H, et al. Changes in circulating cytokine levels in midlife women with psychological symptoms with selective serotonin reuptake inhibitor and Japanese traditional medicine. *Maturitas.* 2009;62(2):146–152.
5. Xue W, Zhou X, Yi N, et al. Yueju pill rapidly induces antidepressant-like effects and acutely enhances BDNF expression in mouse brain. *Evid Based Complement Alternat Med.* 2013;2013:1– 9.
6. Tang J-J, Xue W, Xia B, et al. Involvement of normalized NMDA receptor and mTOR-related signaling in rapid antidepressant effects of Yueju and ketamine on chronically stressed mice. *Sci Rep.* 2015;5:Article number:13573.
7. Wu R, Zhu D, Xia Y, et al. A role of Yueju in fast-onset antidepressant action on major depressive disorder and serum BDNF expression: a randomly double-blind, fluoxetine-adjunct, placebo-controlled, pilot clinical study. *Neuropsychiatr Dis Treat.* 2015;11:2013–2021.
8. Ren L, Tao W, Zhang H, et al. Two standardized fractions of Gardenia jasminoides Ellis with rapid antidepressant effects are differentially associated with BDNF up-regulation in the hippocampus. *J Ethnopharmcol.* 2016;187(1):66–73.
9. Wang J, Duan P, Cui Y, et al. Geniposide alleviates depression-like behavior via enhancing BDNF expression in hippocampus of streptozotocin-evoked mice. *Metab Brain Dis.* 2016;31(5):1113–1122.
10. Feng DD, Tang T, Lin XP, et al. Nine traditional Chinese herbal formulas for the treatment of depression: an ethnopharmacology, phytochemistry, and pharmacology review. *Neuropsychiatr Dis Treat.* 2016;12:2387–2402.
11. Yeung WF, Chung KF, Ng KY, et al. A systematic review on the efficacy, safety and types of Chinese herbal medicine for depression. *J Psychiatr Res.* 2014;57:165–175.
12. Park HJ, Shim HS, Chung SY, et al. Soyo-san reduces depressive-like behavior and proinflammatory cytokines in ovariectomized female rats. *BMC Complement Altern Med.* 2014;14:34.
13. Liang Y, Guo XL, Chen JX, et al. Effects of the Chinese traditional prescription xiaoyaosan decoction on chronic immobilization stress-induced changes in behavior and ultrastructure in rat hippocampus. *Evid Based Complement Alternat Med.* 2013;2013:984797.
14. Zhou W, Wang N, Yang C, et al. Ketamine-induced antidepressant effects are associated with AMPA receptors-mediated upregulation of mTOR and BDNF in rat hippocampus and prefrontal cortex. *Eur Psychiatry.* 2014;29(7):419–423.

15. Zhang Y, Han M, Liu Z, et al. Chinese herbal formula xiao yao san for treatment of depression: a systematic review of randomized controlled trials. *Evid Based Complement Alternat Med.* 2012;2012:931636.

16. Zhang LD, Zhang YL, Xu SH, et al. Traditional Chinese medicine typing of affective disorders and treatment. *Am J Chin Med.* 1994;22(3–4):321–327.

17. Li YH, Zhang CH, Qiu J, et al. Antidepressant-like effects of Chaihu-Shugan-San via SAPK/JNK signal transduction in rat models of depression. *Pharmacogn Mag.* 2014;10(39):271–277.

18. Wang Y, Fan R, Huang X. Meta-analysis of the clinical effectiveness of traditional Chinese medicine formula Chaihu-Shugan-San in depression. *J Ethnopharmacol.* 2012;141(2):571–577.

19. Liu Z, Xie LZ, Zhu J, et al. Herbal medicines for fatty liver diseases. *Cochrane Database Syst Rev.* 2013;8:1–249.

20. Zhang ZJ. *Treatise on Febrile Diseases Caused by Cold (Shanghan lun).* New York, NY: New World Press; 1986. p. 442.

21. Lou JS, Li CY, Yang XC, et al. Protective effect of gan mai da zao decoction in unpredictable chronic mild stress-induced behavioral and biochemical alterations. *Pharm Biol.* 2010;48(12):1328–1336.

22. Yeung WF, Chung KF, Ng KY, et al. A meta-analysis of the efficacy and safety of traditional Chinese medicine formula Ganmai Dazao decoction for depression. *J Ethnopharmacol.* 2014;153(2):309–317.

23. Ma XJ, Zhao J, Feng ZY, et al. Effects of modified ganmai dazao decoction on neuroendocrine system in patients with climacteric depression. *Zhongguo Zhong Yao Za Zhi.* 2014;39(23):4680–4684.

24. Chen H. Guipi and Ganmai Chinese date decoction treat depressed disordered concerned insomnia. *J Zhejiang Univ Tradit Chin Med.* 2010;3.

25. Zheng GQ. Therapeutic history of Parkinson's disease in Chinese medical treatises. *J Altern Complement Med.* 2009;15(1):1223–1230.

26. Tan B. Clinical study on Wendan decoction combined with clozapine for treatment of 30 cases of schizophrenia. *J Tradit Chin Med.* 1993;(5).

27. Yang C, Cai C, Yang X, et al. Wendan decoction improves learning and memory deficits in a rat model of schizophrenia. *Neural Regen Res.* 2012;7(15):1132–1137.

28. Zhang W, Li J, Zhu J, Shi Z, et al. Chinese medicine Banxia-houpu decoction regulates c-fos expression in the brain regions in chronic mild stress model in rats. *Phytother Res.* 2004;18(3):200–203.

29. Guo Y, Kong L, Wang Y, et al. Antidepressant evaluation of polysaccharides from a Chinese herbal medicine Banxia-houpu decoction. *Phytother Res.* 2004;18(3):204–207.

30. Ta G, Wang J. Clinical observation on treating post-stroke depression with chaihuijialonggu muli decoction in 24 patients. *Jilin J Tradit Chin Med.* 2008;28:179–180.

31. Ushiroyama T, Ikeda A, Sakuma K, et al. Chai-Hu-Gui-Zhi-Gan-Jiang-Tang regulates plasma interleukin-6 and soluble interleukin-6 receptor concentrations and improves depressed mood in climacteric women with insomnia. *Am J Chin Med.* 2005;33(5):703–711.

32. Hongbo G. Digestive system diseases complicated with depression treated with Tiaoqi Yangwei Anshen decoction. *J Shanxi Coll Tradit Chin Med.* 2004;1.

33. Jingyan X, Qingxiang H, Liyu Z. Therapeutic observation on sixty cases of climacteric syndromes treated by Yipi Ninggeng decoction. *Henan Tradit Chin Med.* 2004;8.

34. Chao Z, Li X, Zhou Y. Effect of time-sharing differentiation treatment of traditional Chinese medicine on diabetes mellitus combined with depressive disorder. *Hebei J Tradit Chin Med.* 2011;5.

35. Bao ZX, Zhao GP, Sun W, et al. Clinical curative effects of Kaixin powder on depression with mild or moderate degree. *Chin Arch Tradit Chin Med.* 2011;28(5):987–988.

36. Zhu KY, Xu SL, Choi RC, et al. Kai-xin-san, a Chinese herbal decoction containing ginseng radix et rhizoma, polygalae radix, acori tatarinowii rhizoma, and poria, stimulates the expression and secretion of neurotrophic factors in cultured astrocytes. *Evid Based Complement Alternat Med*. 2013;2013:731385.

37. Zhu KY, Mao QQ, Ip SP, et al. A standardized Chinese herbal decoction, kai-xin-san, restores decreased levels of neurotransmitters and neurotrophic factors in the brain of chronic stress-induced depressive rats. *Evid Based Complement Alternat Med*. 2012;2012:149256.

38. Dong XZ, Li ZL, Zheng XL, et al. A representative prescription for emotional disease, Ding-Zhi-Xiao-Wan restores 5-HT system deficit through interfering the synthesis and transshipment in chronic mild stress-induced depressive rats. *J Ethnopharmacol*. 2013;150(3):1053–1061.

39. Qiao HY, Luo R, Wu J, et al. UPLC-MS fingerprint of Shugan Jieyu capsules. *Zhong Cao Yao*. 2014;45(2):208–213.

40. Fu JH, Liu Y, Wang QY, et al. Effect of Shuganjieyu capsules on neuronal apoptosis in hippocampal CA3 area and the expression of caspase-3 in the brain of rat depression model. *Zhong Nan Da Xue Xue Bao Yi Xue Ban*. 2012;37(12):1198–1204.

41. Fu J, Zhang Y, Wu R, et al. Shuganjieyu capsule increases neurotrophic factor expression in a rat model of depression. *Neural Regen Res*. 2014;9(5):489–497.

42. Huang L, Chen LJ, Liu LL, et al. A systematic review of therapeutic efficacy and safety of shugan jieyu capsules in the treatment of mild to moderate depression. *China Pharm*. 2013;24(32):3043–3046.

43. Zhang X, Kang D, Zhang L, et al. Shuganjieyu capsule for major depressive disorder (MDD) in adults: a systematic review. *Aging Ment Health*. 2014;18(8):941–953.

44. Xie M, Jiang W, Yang H. Efficacy and safety of the Chinese herbal medicine shuganjieyu with and without adjunctive repetitive transcranial magnetic stimulation (rTMS) for geriatric depression: a randomized controlled trial. *Shanghai Arch Psychiatry*. 2015;27(2):103–110.

45. Qin L. Study on the antidepressant mechanism of Sini San. *Dang Dai Yi Xue*. 2010;16(14):29–30.

46. Lu JX. Chemical constituents of wuling fermentative powder. *Zhongguo Xian Dai Ying Yong Yao Xue*. 2014;31(5):541–543.

47. Li DQ, Li XJ, Duan JF, et al. Wuling capsule promotes hippocampal neurogenesis by improving expression of connexin 43 in rats exposed to chronic unpredictable mild stress. *Zhong Xi Yi Jie He Xue Bao*. 2010;8(7):662–669.

48. Li D, Zheng J, Wang M, et al. Wuling powder prevents the depression-like behavior in learned helplessness mice model through improving the TSPO mediated-mitophagy. *J Ethnopharmacol*. 2016;186:181–188.

49. Li DQ, Li XJ, Duan JF, et al. Wuling capsule promotes hippocampal neurogenesis by improving expression of connexin 43 in rats exposed to chronic unpredictable mild stress. *Zhong Xi Yi Jie He Xue Bao*. 2010;8(7):662–669.

50. Rajkowska G, Stockmeier CA. Astrocyte pathology in major depressive disorder: insights from human postmortem brain tissue. *Curr Drug Targets*. 2013;14(11):1225–1236.

51. Chen MK, Guilarte TR. Translocator protein 18 kDA (TSPO): molecular sensor of brain injury and repair. *Pharmacol Ther*. 2008;118(1):1–17.

52. Peng L, Zhang X, Kang DY, et al. Effectiveness and safety of Wuling capsule for post stroke depression: a systematic review. *Complement Ther Med*. 2014;22(3):549–566.

53. Wang XJ, Li J, Zou QD, et al. Wuling capsule for climacteric patients with depression and anxiety state: a randomized, positive parallel controlled trial. *Zhong Xi Yi Jie He Xue Bao*. 2009;7(11):1042–1046.

54. Peng WF, Wang X, Hong Z, et al. The anti-depression effect of Xylaria nigripes in patients with epilepsy: a multicenter randomized double-blind study. *Seizure*. 2015;29:26–33.

55. Chellian R, Pandy V, Mohamed Z. Pharmacology and toxicology of α- and β-asarone: a review of preclinical evidence. *Phytomedicine.* 2017;32:41–58.

56. Kass R. *Traditional Chinese Medicine in San Francisco: Reliability of Traditional Chinese Pulse Diagnosis* [Doctoral dissertation]. [Berkeley, CA]: University of California, Berkeley; 1987.

57. Kim M, Cobbin D, Zaslawski C. Traditional Chinese medicine tongue inspection: an examination of the inter- and intrapractitioner reliability for specific tongue characteristics. *J Altern Complement Med.* 2008;14(5):527–536.

58. Ko KM, Mak DHF, Chiu PY, et al. Pharmacological basis of 'Yang-invigoration' in Chinese medicine. *Trends Pharmacol Sci.* 2004;25(1):3–6.

59. Seki K, Chisaka M, Eriguchi M, et al. An attempt to integrate Western and Chinese medicine: rationale for applying Chinese medicine as chronotherapy against cancer. *Biomed Pharmacother.* 2005;59(Suppl 1):S132–S140.

9 Flavonoids with Preclinical Antidepressant-Like Effects

In the previous chapters, scores of different herbs and several traditional combinations of herbs were shown to exhibit antidepressant effects in preclinical and/or clinical studies. Those herbs also showed antioxidant, anti-inflammatory, and antidiabetic effects. It is doubtful that each exerts a unique mechanism of action. Rather, it is likely that the various phytochemical constituents of the herbs share mechanisms of action. Indeed, because of similarities in the biochemistries and physiologies of plants, it is likely that many of the same bioactive phytochemicals are shared among herbs with antidepressant effects.[1-3]

Flavonoids almost certainly contribute to the antidepressant effects of herbs. Human studies have shown that high intake of dietary flavonoids has a protective effect against MDD. A large study revealed that midlife women who consumed two or more servings a day of foods rich in flavonoids, flavan-3-ol polymers, and proanthocyanidins over a ten-year period had a roughly 10% lower risk of developing MDD than those that consumed less than one serving a day of such food.[4] In a similar large study of women in mid- and late life, consumption of flavonoid-rich foods over a 15-year period improved mental health and lowered the incidence of MDD per the Mental Health Index. This type of diet also decreased incidence of chronic illness and cognitive loss, ostensibly due to the antioxidant and anti-inflammatory effects of the flavonoids.[5] Many specific flavonoids – representing all of the subgroups of this class of phytochemicals – have been found to have antidepressant effects in rodents.

9.1 AMENTOFLAVONE

Amentoflavone is unique biflavonoid that occurs in a number of medicinal plants, including *Ginkgo biloba* and *Hypericum perforatum*.[6] It is also found in *Cnestis ferruginea*, which is used in traditional African medicine in the management of psychiatric disorders. This biflavonoid has antidepressant and anxiolytic-like effects in mice. The anxiolytic effects appear to have been due to action at benzodiazepine receptors, as they were reversed by the benzodiazepine antagonist flumazenil.[7] In fact, amentoflavone displaced [3H]flumazenil from rat brain benzodiazepine binding sites with a potent IC50 of 14.9 nM.[8]

9.2 APIGENIN

Apigenin is a common flavone found in a variety of herbs and spices, including avena sativa, chamomile, celery, cecropia, chaste tree, *Chrysanctinia mexicana*, garlic, lemon balm, marjoram, oregano, rosemary, and skullcap. This flavonoid has been found to have antidepressant-like effects in several preclinical studies. Acute administration reduced immobility in the forced swim test, and, in subsequent evaluation, it was found to attenuate the depression-like effects of chronic unpredictable stress.[9] In a related paradigm, apigenin reversed the depression-like effects of chronic treatment with corticosterone and restored levels of BDNF.[10]

Apigenin also attenuated the depression-like effects of lipopolysaccharide in mice due to dampening of the inflammatory response. It suppressed activation of NF-κB in the prefrontal cortex and, in turn, the synthesis of inducible nitric oxide synthase and COX-2. IL-1β and TNF-α levels were also reduced.[11] This attenuation of the inflammatory response was found to be deep and wide-ranging. In its attenuation of stress-induced depression-like behavior, apigenin reduced levels of stressinduced synthesis of IL-1β, and the stimulation of the NLRP3 inflammasome by upregulation of PPARγ that is involved in insulin response.[12]

9.3 ASTILBIN

Astilbin is a flavanonol found in St John's Wort, grapes, wine, and *Smilacis glabrae* used in Chinese Medicine.[13] Chronic administration of astilbin attenuated depression-like behaviors in mice that had been subjected to chronic unpredictable stress. This effect was due in part to upregulation of serotonergic and dopaminergic activities, and activation of the BDNF signaling secondary to restoration of stress-inhibited Akt function.[14] Astilbin was also shown to offer neuroprotective effects in a mouse model of Alzheimer's Disease. This effect was due to increases in CREB and BDNF activity in the hippocampus, as well as enhancement of Akt with subsequent dampening of GSK-3β in that area.[15]

9.4 BAICALEIN AND BAICALIN

Baicalein, a flavone, is found in *Scutellaria* species and *Oroxylum indicum* that is used in Ayurvedic Medicine.[16] Baicalin, a glycosylated form of baicalein, can also be isolated from skullcap, other *Scutellaria* species, and *Oroxylum indicum*. Both baicalein and baicalin were found to decrease immobility of mice in the forced swim and tail suspension tests. Curiously, the glycoside baicalin was deemed more potent.[17] Baicalein also dampened the depression-like effect of chronic unpredictable stress in rats, and this effect was due in part to enhancement of phosphorylation of extracellular signal-regulated kinase that, in turn, stimulated synthesis of BDNF.[18]

9.5 CHRYSIN

Chrysin is a flavone that can be isolated from skullcap and chamomile. This flavone attenuated the depression-like effect of chronic unpredictable stress in mice. Part of this effect was due to restoration of levels of BDNF and NGF in the hippocampus and prefrontal cortex.[19] Chrysin also reversed the depression-like effects of olfactory bulbectomy in mice. The treatment reduced the elevated levels of TNF-α, INF-γ, IL-1β, and IL-6 in the hippocampus. Also noted was that olfactory bulbectomy increased indoleamine-2,3-dioxygenase activity, with resulting increases in kynurenine, both of which were reduced by chrysin. At the same time, levels of serotonin and BDNF were elevated back towards normal.[20]

9.6 7,8,DIHYDROXYFLAVONE

7,8,dihydroxyflavone is a flavonoid found in a number of herbs, including *Godmania aesculifolia*, *Tridax procumbens*, and *primula* tree leaves. It has been the subject of intense research after reports that it acts as a centrally active, small molecule, selective TrkB agonist.[21] Activation of the TrkB receptors appears necessary for antidepressant action.[22] Thus, the significance of such a molecule in the treatment of MDD is readily apparent. So many authors have reported that 7,8,dihydroxyflavone activates the TrkB receptor that it appeared to have become an established fact. However, at least one group found discrepancies in this conclusion.[23] In any case, whether due to activation of TrkB receptor or not, 7,8,dihydroxyflavone has been reported to have antidepressant-like effects in rodents.

Chronic oral administration of 7,8-dihydroxyflavone decreases immobility time of mice in the forced swim test. Treated mice also had increased activation of TrkB receptors as well as increases in neurogenesis in the hippocampus.[24]

7,8-Dihydroxyflavone also reversed the depression-like symptoms of mice subjected to chronic stress. Appetite for sucrose in those animals was restored, and immobility times in the forced swimming and tail suspension tests were reduced by chronic, but not acute, treatment with the flavonoid. Levels of BDNF were increased, as was the degree of TrkB receptor activation, as demonstrated by increases in synaptic proteins downstream from that receptor.[25]

Four weeks of unpredictable stress caused decreases in sucrose consumption and increased immobility in the forced swim test in rats in a manner similar to the depression-like behaviors seen in mice. Intraperitoneal administration of 7,8-dihydroxyflavone throughout the course of stress reversed those depression-like effects. This treatment also reduced serum levels of corticosterone and increased levels of BDNF in the hippocampus.[26]

Chronic social defeat stress is yet another animal model of MDD. In this model, mice are placed with larger, more aggressive mice and thus experience persistent threat of harm. 7,8-Dihydroxyflavone decreased immobility times in the forced swim and tail suspension tests, and increased sucrose consumption in the stressed mice. This effect of 7,8-dihydroxyflavone was similar to that of ketamine in this study, though not as persistent. 7,8-dihydroxyflavone was also less effective than ketamine in restoring levels of BDNF and density of postsynaptic proteins in prefrontal cortex and hippocampus.[27]

7,8-Dihydroxyflavone is also effective in the inflammatory model of depression. Administration of lipopolysaccharide to mice increased immobility in the forced swim and tail suspension tests, and these effects were reversed by 7,8-dihydroxyflavone.

Lipopolysaccharide-induced decreases in BDNF levels and dendritic spine densities in the hippocampus and prefrontal cortex were also reversed. The complexity of changes in the depressed brain were highlighted in this study with the findings that lipopolysaccharide increased BDNF and dendritic spine density in the nucleus accumbens. Those specific changes were reversed by the TrkB antagonist, ANA-12. Surprisingly, that treatment also exerted antidepressant effects.[28]

9.7   FISETIN

Fisetin is an anthocyanidin in many fruits and vegetables, such as strawberries, apples, persimmons, onions, and cucumbers.[29] Fisetin reduced immobility in mice in the forced swim and tail suspension tests. The flavonoid increased serotonin and noradrenaline levels in the frontal cortex and hippocampus of those animals, apparently due in part to inhibition of MAO-A.[30] Pretreatment with fisetin also attenuated the depression-like effects of lipopolysaccharide injection in mice. Fisetin reversed the lipopolysaccharide-induced overexpression of IL-1β, IL-6, and TNF-α in the hippocampus and the prefrontal cortex (PFC). It also antagonized increases in iNOS mRNA expression and nitrite levels through modulation of NF-κB in those brain areas.[31] Another component of the antidepressant-like effect of fisetin in mice is the activation of the trkB receptor, which is the target of BDNF.[32]

9.8   HEPTOMETHOXYFLAVONE

Heptomethoxyflavone is a flavone found in the Japanese herbal medicine *Rikkunshito*. The general formulation of *Rikkunshito* is the combination of *Atractylodis lanceae rhizoma*, *Ginseng radix*, *Pinelliae tuber*, *Hoelen*, *Zizyphi fructus*, *Aurantii nobilis pericarpium*, *Glycyrrhizae radix*, and *Zingiberis rhizoma*.[33] The exact source of the heptomethoxyflavone isn't clear. Heptamethoxyflavone isolated from strawberry and citrus ameliorates corticosterone-induced depression-like behavior in mice. It does so in part by increasing BDNF and neurogenesis in the hippocampus. A similar activating effect of heptamethoxyflavone on BDNF was seen in the hippocampus in the transient global ischemia mouse model.[34]

9.9   HESPERIDIN AND HESPERITIN

The flavanone glycoside, hesperidin, is found in citrus fruit, onion, and peppermint. The aglycone form is hesperitin. Both have been shown to have antidepressant-like effects, with hesperidin being particularly well-studied. A number of studies have demonstrated antidepressant effects of hesperidin and shown plausible mechanisms by which this might occur. Intraperitoneal administration of hesperidin decreased immobility time of mice in the forced swim and tail suspension tests, and these antidepressant-like effects were blocked by the 5-HT1A receptor antagonist, WAY100635. Moreover, combining sub-effective doses of hesperidin and the SSRI fluoxetine produced an antidepressant-like effect. Together, those data suggest that stimulation of the 5-HT1A receptor played a role.[35] In a similar study of mice, hesperidin reduced immobility in the forced swim test, and this effect was blocked by naloxone and the selective κ-opioid receptor antagonist, DIPPA. Thus, kappa opiate receptors may also be involved in this antidepressant effect.[36]

Chronic treatment with hesperidin exerted an antidepressant-like effect in mice. This antidepressant effect was accompanied by increases in BDNF and decreases of nitric oxide in the hippocampus of these animals. Moreover, the addition of inhibitors of nitric oxide synthase enhanced the antidepressant effect. Thus, it was concluded that the antidepressant-like effect of hesperidin was mediated by inhibition of l-arginine-NO-cGMP pathway and by increases of the BDNF levels in the hippocampus. The aglycone form of hesperidin, hesperitin, produced a similar antidepressant-like effect.[37] Antidepressant-like effects of hesperidin were also seen in mice subjected to chronic mild stress (CMS). The increases in serum corticosterone levels, and the decreases in extracellular signalregulated kinase phosphorylation and BDNF levels in the hippocampus were reversed by hesperidin.[38]

Chronic treatment with hesperidin was also able to reverse the depression-like effects of olfactory bulbectomy in mice. This effect was accompanied by reductions in IL-1β, IL-6, and acetylcholinesterase activity, and increases in BDNF and NGF in the hippocampus.[39] The antidepressant and anti-inflammatory effects of hesperidin seen in the bulbectomized mice were also seen in the brains of mice that had been treated with lipopolysaccharide. Preference for sucrose was restored in those animals. Treatment decreased serum corticosterone levels, as well as levels of IL-1β, IL-6, and TNF-α in the prefrontal cortex.[40]

9.10 HYPEROSIDE

Hyperoside is an anthocyanin glycoside in St John's wort, *Drosera rotundifolia*, *Lamiaceae Stachys*, *Prunella vulgaris*, *Rumex acetosella*, *Cuscuta chinensis*, *Camptotheca acuminate*, birch leaves, various fruits, marigold, and tea. Hyperoside reduces immobility of both rats and mice in the forced swim test.[41,42] In cultured mouse neurons that had been exposed to toxic levels of corticosterone, hyperoside protects against cytotoxicity by increasing levels of BDNF and CREB.[43] Some of the antidepressant-like effect of hyperoside could be due to such increases.

9.11 ICARIIN

Icariin is a flavonol glycoside from horny goat weed and other species of *Epimedium*. Some of the antidepressant-like effects of icariin are described in Chapter 7 in the discussion of horny goat weed. For example, icariin reversed the depression-like behavior of rats that had experienced chronic unpredictable stress. This effect of icariin was due in part to its anti-inflammatory effects. It dampened activation of the hippocampal NLRP3 inflammasome/caspase-1/IL-1β axis, and thus decreased levels of TNF-α, IL-1β, and NF-κB.[44] Icariin also attenuated glucocorticoid-induced depression in rats. It restored sucrose intake, decreased immobility time in the forced swim test, and increased levels of BDNF in the hippocampus.[45]

9.12 ISOSAKURENTIN-5-O-RUTINOSIDE

Isosakurentin-5-O-rutinoside is a flavanone glycoside from Mexican oregano and related *Salvia* species. It exhibited antidepressant activity in mice in the forced swim test.[46]

9.13 KAEMPFEROL

Kaempferol is a common flavanol from caraway, clove, cumin, daylily, garlic, ginkgo biloba, ginseng, hibiscus, hops, linden, *Polygala tenuifolia*, yerba mate, and many other herbs and vegetables. Intraperitoneal administration of kaempferol decreased immobility of both mice and rats in the forced swim test.[47] Chronic feeding of kaempferol also produced antidepressant effects in mice subjected to chronic restraint stress.[48]

9.14 LIQUIRITIN AND ISOLIQUIRTIN

Liquiritin is a flavanone glycoside from the *Glycyrrhiza* species. The flavonoid showed antidepressant effects in mice in the forced swim and tail suspension tests. The compound also increased the concentrations of serotonin and norepinephrine in the hippocampus, hypothalamus, and cortex. Chalcones are phytochemicals similar to flavonoids, but with open middle or "C" rings. Isoliquirtin, a chalcone with a structure analogous to liquiritin, also produced antidepressant effects.[49] Chronic oral administration of liquiritin reduced forced swim immobility and restored sucrose appetite in rats that had been subjected to variable stresses over five weeks. Treatment also increased superoxide disumutase, but decreased malondialdehyde in erthyrocytes.[50]

9.15 LUTEOLIN

Luteolin is a flavone in *Avena sativa*, *cecropia* species, chamomile, celery, fennel, garlic, ginkgo biloba, lemon balm, oregano, rosemary, sage, St John's wort, thyme, vervain, and many other herbs and vegetables. Luteolin decreased immobility in the forced swim test in mice and, in the same report, increased activity at GABA-A receptors in cultured human neuroblastoma cells. It was suggested that GABA-A receptors may have played a role in its antidepressant effects. However, it is more likely that it adds an anxiolytic effect that is shared by many flavonoids.[51] Luteolin treatment also showed antidepressant-like effects in mice that had been chronically treated with corticosterone, which is a common model of depression. It appeared to do so by attenuating expression of endoplasmic reticulum stress-related proteins, such as caspace-3, in the hippocampus.[52] In addition, luteolin protected rat brains from the neurotoxic effects of kainic acid. It attenuated kainic acid-induced neuronal cell death and inflammation in the hippocampus, and restored activity of Akt in that area. Akt is an inhibitor of GSK-3β.[53]

9.16 MIQUELIANIN

Miquelianin is a glucuronide of quercetin present in St John's wort, *Nelumbo nucifera*, and green beans. Both acute and repeated doses of miquelianin decreased immobility of rats in the forced swim test.[42]

9.17 MYRICETIN

Myricetin is a flavonol from birch leaves, skullcap, garlic, ginkgo biloba, hibiscus, marigold, onion, tea, and wine. Sub-chronic treatment with myricitrin had an antidepressant-like effect in mice, and this effect was accompanied by increases in neurogenesis in the hippocampus.[54] Myricetin also attenuated depression-like behavior in mice subjected to a prolonged course of restraint stress. The flavonol also decreased serum levels of corticosterone in stressed animals. In the hippocampus, myricetin attenuated decreases in glutathione peroxidase but enhanced levels of BDNF.[55]

9.18 NARINGENIN AND NARINGIN

Naringenin is a flavanone found in chamomile, citrus, hops, rosemary, and other herbs. Naringin is a glycoside of naringenin. The flavanone decreased immobility of mice in tail suspension test, though not in the forced swimming test.[56]

When repeated tail suspension tests served as a source of stress, the antidepressant effects of naringenin in mice became more apparent in comparison with untreated mice. Naringenin treatment markedly decreased the immobility time. It also increased levels of serotonin, norepinephrine, and glucocorticoid receptors in the hippocampus, as well as reducing serum levels of corticosterone.[57] In another study of mice subjected to chronic stress, chronically administered naringenin restored appetite for sucrose and increased levels of BDNF in the hippocampus. The antidepressant-like effect of naringenin was blocked by K252a, an inhibitor of the TrkB receptor.[58]

Naringin, a glycoside of naringenin, has also been found to have antidepressant-like effects. Acute intraperitoneal administration reduced immobility of mice in the forced swim and tail suspension tests, and increased social behavior. It also exhibited anxiolytic and memory enhancing effects.[59]

Treatment with naringin also reversed depression-like behaviors induced by physiological and chemical trauma to the brain. In mice subjected to brain ischemia and reperfusion, chronic pretreatment with naringin reduced immobility in the tail suspension test. This was accompanied by decreases in lipid peroxidation, and increases in glutathione, glutathione-S-transferase, superoxide dismutase, and catalase levels in brain tissue. Treatment also reversed pathophysiological changes in mitochondrial activity.[60]

In mice, the neuroinflammation, mitochondrial dysfunction, and depression-like effects caused by the neurotoxin doxorubicin were attenuated by acute treatment with naringin.[61] In a study of toxicity, the "no-observed-adverse-effect-level" of naringin in rats was greater than 1250 mg/kg/day when administered orally for six consecutive months.[62] Thus, treatment with naringin appears to be quite safe.

9.19 NOBILETIN

Nobiletin is an O-methylated flavone from citrus. Acute oral administration of nobiletin to mice decreased the immobility time in both the forced swim and tail suspension tests.[63] Chronically administered nobiletin restored appetite for sucrose in rats that had been subjected to chronic unpredictable stress. This antidepressant-like effect was accompanied by decreases in serum corticosterone. This treatment also attenuated the decreases in BDNF and enhanced activation of the TrkB and synapsin I in the hippo-campi of these animals.[64] Nobiletin also spared cultured neurons from the cytotoxic effects of prolonged exposure to corticosterone, thus showing an antistress effect.[65]

Other studies have suggested additional mechanisms by which these antidepressant effects of nobiletin might occur. For example, the flavonoid enhanced bioenergetics of rat hippocampal mitochondria.[66] Nobiletin also prevents amyloid-induced decreases in activation of CREB in cultured hippocampal neurons. CREB is a major regulator of BDNF, thus, the protective effect of nobiletin on CREB may offer antidepressant-like effects under stressful conditions.[67]

9.20   ORIENTIN

Orientin is a flavone glycoside from *cecropia, mimosa pudica,* passionflower, acai berries, olives, buckwheat, and millet. Chronic oral administration of orientin decreased immobility and restored appetite for sucrose in mice that had been subjected to chronic unpredictable stress. Orientin treatment also reduced levels of oxidative stress markers, and increased the concentrations of serotonin and norepinephrine in the hippocampus and prefrontal cortex. Levels of BDNF and synapse-associated proteins in those areas of the brain were increased.[68]

9.21   QUERCETIN

Quercetin is one of the most common flavonoids. It is a flavonol found in chamomile, *Chrysanctinia Mexicana,* daylily, fennel, garlic, ginkgo biloba, ginseng, hibiscus, hops, linden, mint, *Polygala tenuifolia,* St John's wort, Schisandra, tea, turmeric, yerba mate, and many other herbs, fruits, and vegetables.

Studies show antidepressant-like effects of quercetin in a variety of animal models of MDD. Two weeks of oral treatment with quercetin reduced immobility of mice in the forced swim and tail suspension tests. The treatment also increased hypothalamic proopiomelanocortin mRNA and plasma β-endorphin levels.[69]

In mice that had been subjected to chronic unpredictable stress, chronic administration of quercetin restored appetite for sucrose. Quercetin also reduced the stress-induced elevations of the cytokines, IL-6, TNF-α, and IL-1β, and the activity of the COX-2 enzyme in the hippocampus. In addition, treatment prevented the loss of hippocampal neurons due to oxidative and inflammatory damage.[70] In a paradigm related to chronic stress, quercetin also reversed the depression-like behavior caused by treatment with corticotrophin releasing factor. Treatment with the flavonoid reduced immobility in the forced swim test and normalized social behavior.[71] The ability of quercetin to protect against stress is directly shown in its protection of cultured mouse neurons from the damaging effects of high concentrations of corticosterone. It increased cell viability in a manner similar to that of the antidepressant, clomipamramine.[72]

Chronic administration of quercetin also reversed the depression-like effects of olfactory bulbectomy in mice. Immobility in the forced swim and tail suspensions tests was reduced. This effect was attributed, in part, to antioxidant and anti-nitrosative effects, as levels of glutathione and activity of superoxide dismutase in the hippocampus were increased.[73] Others found antidepressant-like effects of quercetin in olfactory bulbectomized rats that were associated with reduced hippocampal levels of TNF-α and IL-6, as well as reduced activity of the apoptotic factor caspase-3.[74]

In one study, quercetin exhibited antidepressant-like effects in streptozotocin-induced diabetic mice using the forced swim test. Curiously, this effect was not seen in non-diabetic mice.[75] In fact, the same group later saw quercetin-induced behavioral despair in non-diabetic mice, and this effect was reversed by the alpha2 adrenoceptor antagonist, yohimbine.[76]

Sickness behavior in response to lipopolysaccharide treatment has often been compared to some of the components of MDD in humans and to the depression-like behaviors that arise from the activity of inflammatory cytokines in animals. Thus, it is of interest that quercetin has been found to attenuate some of the behavioral effects of lipopolysaccharide induced-sickness behavior and oxidative stress in rats. Treatment with quercetin reduced lipopolysaccharide-induced increases in brain TNF-α, IL-1β, and IL-6; normalized social behavior; and restored consumption of food and water.[77] Similarly, quercetin reversed the depression-like behaviors induced by the cytotoxic antibiotic Adriamycin in rats. The flavonoid reduced immobility in the forced swim test. It also decreased serum corticosterone levels, and reduced the degree of oxidative damage in brain tissue.[78]

One mechanism by which quercetin may produce an antidepressant effect is through stimulation of neurogenesis. It has been found that chronic administration of quercetin significantly increased cell proliferation in the hippocampal neurons of mice. This was likely due to the observed enhanced phosphorylation of CREB in these cells, and subsequent enhanced synthesis of BDNF. Indeed, immunofluorescence staining of synaptic markers showed substantial growth of dendritic processes in hippocampal neurons treated with quercetin.[79]

9.22   VITEXIN

Vitexin is the glucoside of the flavone apigenin. It is found in buckwheat, cecropia, chaste tree, *mimosa pudica*, passion flower, hawthorn, and other herbs. Vitexin reduced immobility time of mice in both the tail suspension and modified forced swimming tests.[80] Pretreatment with vitexin significantly reduced cell death in cultured mouse cortical neurons that were exposed to NMDA. This appeared to be related to effects on the NR2B sub-unit of the NMDA receptor.[81] NR2B antagonists are thought to have antidepressant effects.[82] In intact mice, peripherally administered vitexin also protected cerebral neurons from the damage of ischemia and reperfusion. This effect was due to upregulation of ERK1/2, and downregulation of apoptosis signaling pathways.[83] Enhancement of ERK1/2 inhibits GSK-3β activity[84] and decreased activity of ERK1/2 in the prefrontal cortex[85] is associated with depression and suicide in humans.

9.23   WOGONIN AND WOGONOSIDE

Wogonin is an O-methylated flavone found in skullcap and *Oroxylum indicum*. Wogonoside is the glycoside form of the flavone. Both wogonin and wogonoside were found to decrease immobility of mice in the forced swim and tail suspension tests.[86] Wogonin has also been found to be a relatively potent inhibitor of MAO-A and MAO-B, with respective IC50s of 6.35 and 20.8 μM.[87]

9.24   SYNTHETIC FLAVONOIDS

It is noteworthy that structure/activity studies of the relationship between flavonoids and activation of the TrkB receptor led researchers to the synthesis of a flavonoid with even stronger TrkB activation and antidepressant effects than the naturally occurring 7,8-dihydroxyflavone. That group had shown that the 7,8-dihydroxy groups are essential for those effects. They later identified 4′dimethylamino-7,8-dihydroxyflavone as having greater TrkB agonistic activity than 7,8 dihydroxyflavone. Chronic oral administration of 4′-dimethylamino-7,8-dihydroxyflavone also strongly promoted neurogenesis in dentate gyrus and exerted strong antidepressant effects.[88] Another group synthesized a variety of 5,7, dihydroxy flavonones and ran structure activity studies on their abilities to produce antidepressant effects. Several different halogen additions to the "B" ring resulted in potent antidepressant effects, as did some additions of methoxy groups to that ring. That group also synthesized a series of substituted chalcones, which are similar to flavonoids except in having an open middle or "C" ring.[89]

9.25   MECHANISMS OF FLAVONOID ANTIDEPRESSANT ACTION

Flavonoids of differing classes produce antidepressant effects by mechanisms that include reducing oxidation and dampening neuroinflammation. These phytochemicals also improve components of the Metabolic Syndrome including reducing hyperglycemia and insulin resistance. These effects help normalize the activities of the PIK3/Akt/mTOR pathway, minimize activity of GSK-3β, and allow expression of BDNF and other proteins that support cell maintenance and survival.

Flavonoids are potent antioxidants. Indeed, in *in vitro* assays of oxidation, such as the DPPH test, the flavonoid content of an herb has a nearly perfect correlation with the herb's antioxidant capacity. Flavonoids not only exert antioxidant effects, but also stimulate the activity of antioxidant enzymes, such as superoxide dismutase and glutathione peroxidase in brain tissue. Oxidative stress can cause gross damage to lipids, proteins, and other neuro-cellular structures, and thus compromise survival, function, and efficiency of cells. However, oxidation can also serve more precise roles in activating systems that increase risk of MDD. For example, TLR receptors initiate a specific inflammatory cascade in response to general, oxidative damage. Such a response of the TLR can be stimulated by so-called "damage associated molecular pattern molecules," or DAMPs, that are created during conditions of oxidative stress such as in experimental ischemia and reperfusion.[90]

Whereas DAMPs are a rather non-specific stimulus of the inflammatory cascade, certain changes in redox can be more specific as signaling mechanisms. For example, ROS can oxidize cysteine residues in protein to form disulfide bridges with glutathione. The resulting molecular structures can then act more directly upon various intracellular signaling pathways. Some of these redox sensitive signaling mechanisms are local and rapidly terminated, not unlike classical transmitter systems.[91] The NADPH oxidase enzyme, which is intimately linked with oxidation processes in the cell, is also capable of upregulating TLR4 receptors in accordance with its redox status.[92]

Whereas some antidepressant-like effects of flavonoids are likely mediated by preventing oxidative damage, flavonoids tend to be stronger antioxidants *in vitro* than *in vivo*.[93] The process of digestion, exposure to intestinal microflora, and interactions with enzymes in the intestinal wall tend to reduce the antioxidant capacity of flavonoids. Thus, other mechanisms of action may be more significant for some flavonoids and the herbs that carry them.

Neuroinflammation furthers the pathophysiology of MDD in many ways. Inflammatory cytokines released by activated microglia activate neuronal GSK-3β.[94] GSK-3β, in turn, stimulates synthesis of those same inflammatory actors,[95] TNF-α[96] and IL-1β,[97] that then act to reduce the activity of BDNF in neurons. Whereas oxidative damage can stimulate inflammatory responses, the inflammatory response itself includes activation of iNOS and NADPH oxidase.[98] Another pathological manifestation of NO-induced nitrosative damage in the brain is disruption of mitochondrial function. It has been reported that NO can inhibit mitochondrial cytochrome c oxidase. This interferes with synthesis of ATP and at the same time increases generation of ROS, which in turn causes further oxidative damage.[99] Thus, the cycle can be perpetuated.

Flavonoids exert potent anti-inflammatory effects in the brain that can produce neuroprotective and antidepressant effects. Extract of blueberry, rich in proanthocyanidin flavonoids, inhibits iNOS, IL-1β and TNF-α production by lipopolysaccharide-activated microglia cells.[100] Flavonoids of other subtypes, including quercetin, wogonin, baicalein, catechin, and genistein have also been shown to attenuate microglia- and astrocyte-mediated neuroinflammation. Among those anti-inflammatory mechanisms are inhibition of cytokine release, and dampening of the activities of iNOS, COX-2, and NADPH oxidase in astrocytes and microglia.[101]

Flavonoids have also been found to inhibit, directly or indirectly, activation of the NLRP3 inflammasome. The inflammasome is a multi-protein complex that regulates major components of the inflammatory response, including caspase-1 activation and IL-1β secretion. Inflammasome activation is mediated by NLR proteins that respond to a variety of stimuli, including infection, oxidative damage, and physical injury, but also psychological stress. Among the NLRs, NLRP3 senses the widest array of stimuli. The flavonoid apigenin has been found to inhibit lipopolysaccharide-induced synthesis of the inflammatory cytokine IL-1β by inhibiting caspase-1 activation through the disruption of the NLRP3 inflammasome assembly.[102] Luteoloside also decreases expression of NLRP3 inflammasome resulting in decreases in proteolytic activation of caspase-1. Inactivation of caspase-1 in turn results in inhibition of IL-1β production.[103]

The flavonoids isorhamnetin and hyperoside, extracted from the water dropwort plant, also inhibit the inflammasome in human and mouse macrophages.[104] However, the ability of flavonoids to inhibit the NLRP3 inflammasome is not necessarily a class-wide trait. For example, in one study, quercetin effectively inhibited the inflammasome, but the flavonoids naringenin and silymarin did not.[105]

It has further been suspected that some of the biological effects of flavonoids may be mediated by direct interaction with intracellular signaling systems. One suggested mechanism is the action of flavonoids at various levels of the intracellular mitogen-activated protein kinase (MAPK) cascades. Flavonoids interact with both kinases that phosphorylate substrates and with phosphorylases that remove phosphate.[106] MAPK enzymes are involved in a variety of intracellular activities, including transduction of stimuli that induce synthesis of BDNF. For example, rats deprived of docosahexaenoic acid are found to have reduced frontal cortex BDNF expression, CREB transcription factor, and p38 MAPK activity. These deficits, in turn, increase aggression and depression-like behaviors. The addition of docosahexaenoic acid to rat primary cortical astrocytes previously deprived of that nutrient restores BDNF protein expression. However, that restoration of BDNF is blocked by a p38 MAPK inhibitor.[107] Many flavonoids are found to increase brain levels of BDNF, which is an effect associated with antidepressant effects in animals and humans. Indeed, inhibition of MAPK also prevents the antidepressant-like effect of ketamine in rats.[108] In fact, many flavonoids with antidepressant-like effects increase brain levels of BDNF. These flavonoids include apigenin, astilbin, baicalin, chrysin, 7,8 dihydroxyflavone, hesperidin, icariin, myricetin, naringenin, and orientin.[109]

The anti-inflammatory effects of flavonoids may also be mediated by actions upon the MAPK pathways. One of the common pathways of inflammation is the activation of NO synthase with subsequent increases in NO. NO synthesis is induced by the inflammatory cytokines, IL-1b and TNF-a. NO synthesis is also stimulated by NADPH oxidase, which is activated in microglia in inflammatory states.[110] The induction of iNOS and synthesis of inflammatory cytokines in activated glial cells are dependent on MAPK signaling, including activation of ERK1/2.[111]

It has been noted that the MEK inhibitor PD98059 that can prevent ERK1/2 stimulation of INOS is a synthetic flavonoid very similar in structure to the common flavonoid quercetin. That in turn suggests that some of the anti-inflammatory effects of some flavonoids may be due to interference with the MAPK-mediated induction of NO synthase. It was further noted that many flavonoids of differing classes inhibit the release of NO by activated microglia through the downregulation of iNOS gene expression. However, it is not known if such effects are mediated by changes in signaling the MAPK system.[112]

The PI3K/Akt pathways can also be activated by flavonoids. This leads to activation of mTOR, CREB, eEF2, BDNF, and other cell survival and maintenance mechanisms that are necessary for antidepressant action, including the antidepressant effects of ketamine. In one study, the flavonoid, myricetin, activated IP3K, as well as increased the level of the anti-apoptotic factor, Bcl-2, and decreased levels of Bax, active caspase-9 and -3, which are pro-apoptotic factors.[113] That constellation of effects would be helpful for the treatment of MDD as well as management of other neurodegenerative illnesses, such as Alzheimer's disease.

MAPKs also modulate the PI3K/AKT pathway in the response to insulin. Thus, they are involved in the activities of mTOR and GSK-3β, as well as playing roles in glucose homeostasis and the development of insulin resistance.[114] Many flavonoids enhance insulin activity and combat hyperglycemia. Indeed, increased dietary intake of flavonoids increases insulin sensitivity and improves signs of Metabolic Syndrome in human subjects with diabetes type II.[115]

Data has shown that increases in monoaminergic activity are neither necessary nor sufficient for antidepressant action. Moreover, evidence shows that antidepressants that act primarily through monoaminergic mechanisms often additionally exert antioxidant, anti-inflammatory, and other effects that have little to do with serotonin, norepinephrine, or dopamine. Nonetheless, there is a basis to conclude that enhancement of monoaminergic activity can at least contribute to other antidepressant mechanisms. Thus, it is worth noting that many flavonoids have been found to increase the levels and activities of serotonin, norepinephrine, or dopamine.

In a study of antidepressant effects of quercetin in mice, it was found that the flavonoid reduced turnover of serotonin by attenuating mitochondrial MAO-A activity in the brain.[116] Chrysin was also found to inhibit MAO-A.[117] The flavonoids jaceosidine, eupafolin, leuteolin, and apigenin, have also been found to have significant inhibitory effects on MAO-A. Such effects appear to be quite common in this class of phytochemical.[118] Indeed, it is common to see reports of increases in concentrations of monoamines along with antidepressant effects produced by flavonoids. Antidepressant effects of naringenin in mice were accompanied by increases in hippocampal levels of serotonin, norepinephrine, and dopamine.[57] Liquiritin increased serotonin and norepinephrine concentrations in the hippocampus, hypothalamus, and cortex of mice.[119] Antidepressant-like effects of orientin were similarly accompanied by increases in levels of serotonin and norepinephrine in the hippocampus and prefrontal cortex of mice.[120] Astilbin increased dopamine concentrations in the hippocampus in rats.[121] Thus, some of the antidepressant effects of flavonoids may be due in part to increases in monoaminergic activity, as is alleged to be the case with standard antidepressants.

REFERENCES

1. Haytowitz DB, Eldridge AL, Bhagwat S, et al. Flavonoid content of vegetables. International Research Conference on Food, Nutrition and Cancer. 17–18 July 2003, Washington, DC.
2. Miean KH, Mohamed S. Flavonoid (myricetin, quercetin, kaempferol, luteolin, and apigenin) content of edible tropical plants. *J Agric Food Chem.* 2001;49(6):3106–3112.
3. Heigl D, Franz G. Stability testing on typical flavonoid containing herbal drugs. *Pharmazie.* 2003;58(12):881–885.
4. Chang SC, Cassidy A, Willett WC, et al. Dietary flavonoid intake and risk of incident depression in midlife and older women. *Am J Clin Nutr.* 2016;104(3):704–714.
5. Samieri C, Sun Q, Townsend MK, et al. Dietary flavonoid intake at midlife and healthy aging in women. *Am J Clin Nutr.* 2014;100(6):1489–1497.
6. Pan X, Tan N, Zeng G, et al. Amentoflavone and its derivatives as novel natural inhibitors of human cathepsin B. *Bioorg Med Chem.* 2005;13(20):5819–5825.
7. Ishola IO, Chatterjee M, Tota S, et al. Antidepressant and anxiolytic effects of amentoflavone isolated from Cnestis ferruginea in mice. *Pharmacol Biochem Behav.* 2012;103(2):322–331.
8. Baureithel KH, Büter KB, Engesser A, et al. Inhibition of benzodiazepine binding in vitro by amentoflavone, a constituent of various species of Hypericum. *Pharm Acta Helv.* 1997;72(3):153–157.

9. Yi LT, Li JM, Li YC, et al. Antidepressant-like behavioral and neurochemical effects of the citrusassociated chemical apigenin. *Life Sci.* 2008;82(13–14):741–751.

10. Weng L, Guo X, Li Y, et al. Apigenin reverses depression-like behavior induced by chronic corticosterone treatment in mice. *Eur J Pharmacol.* 2016;774:50–54.

11. Li R, Zhao D, Qu R, et al. The effects of apigenin on lipopolysaccharide-induced depressive-like behavior in mice. *Neurosci Lett.* 2015;594:17–22.

12. Li R, Wang X, Qin T, et al. Apigenin ameliorates chronic mild stress-induced depressive behavior by inhibiting interleukin-1β production and NLRP3 inflammasome activation in the rat brain. *Behav Brain Res.* 2016;296:318–325.

13. Zhang Q-F, Zhang Z-R, Cheung HY. Antioxidant activity of Rhizoma Smilacis Glabrae extracts and its key constituent-astilbin. *Food Chem.* 2009;115(1):297–303.

14. Lv QQ, Wu WJ, Guo XL, et al. Antidepressant activity of astilbin: involvement of monoaminergic neurotransmitters and BDNF signal pathway. *Biol Pharm Bull.* 2014;37(6):987–995.

15. Wang D, Li S, Chen J, et al. The effects of astilbin on cognitive impairments in a transgenic mouse model of Alzheimer's disease. *Cell Mol Neurobiol.* 2017;37(4):695–706.

16. Harminder AK, Singh V, Chaudhary AK. A review on the taxonomy, ethnobotany, chemistry and pharmacology of Oroxylum indicum Vent. *Indian J Pharm Sci.* 2011;73(5):483–490.

17. Li Y-C, Shen J-D, Liu Y-M, et al. Screening of antidepressant effects of four main flavonoids compounds from *Scutellaria baicalensis. Chin J Exp Tradit Formulae.* 2012;11.

18. Xiong Z, Jiang B, Wu PF, et al. Antidepressant effects of a plant-derived flavonoid baicalein involving extracellular signal-regulated kinases cascade. *Biol Pharm Bull.* 2011;34(2):253–259.

19. Filho CB, Jesse CR, Donato F, et al. Chronic unpredictable mild stress decreases BDNF and NGF levels and Na+, K+-ATPase activity in the hippocampus and prefrontal cortex of mice: antidepressant effect of chrysin. *Neuroscience.* 2015;289:367–380.

20. Filho CB, Jesse CR, Donato F, et al. Chrysin promotes attenuation of depressive-like behavior and hippocampal dysfunction resulting from olfactory bulbectomy in mice. *Chem Biol Interact.* 2016;260:154–162.

21. Jang SW, Liu X, Yepes M, et al. A selective TrkB agonist with potent neurotrophic activities by 7,8dihydroxyflavone. *Proc Natl Acad Sci USA.* 2010;107(6):2687–2692.

22. Li Y, Luikart BW, Birnbaum S, et al. TrkB regulates hippocampal neurogenesis and governs sensitivity to antidepressive treatment. *Neuron.* 2008 Aug 14;59(3):399–412.

23. Boltaev U, Meyer Y, Tolibzoda F, et al. Multiplex quantitative assays indicate a need for reevaluating reported small-molecule TrkB agonists. *Sci Signal.* 2017;10(493).

24. Liu X, Chan CB, Jang SW, et al. A synthetic 7,8-dihydroxyflavone derivative promotes neurogenesis and exhibits potent antidepressant effect. *J Med Chem.* 2010;53(23):8274–8286.

25. Zhang MW, Zhang SF, Li ZH, et al. 7,8-Dihydroxyflavone reverses the depressive symptoms in mouse chronic mild stress. *Neurosci Lett.* 2016;635:33–38.

26. Chang HA, Wang YH, Tung CS, et al. 7,8-Dihydroxyflavone, a tropomyosin-kinase related receptor B agonist, produces fast-onset antidepressant-like effects in rats exposed to chronic mild stress. *Psychiatry Investig.* 2016;13(5):531–540.

27. Zhang JC, Yao W, Dong C, et al. Comparison of ketamine, 7,8-dihydroxyflavone, and ANA-12 antidepressant effects in the social defeat stress model of depression. *Psychopharmacology.* 2015;232(23):4325–4335.

28. Zhang JC, Wu J, Fujita Y, et al. Antidepressant effects of TrkB ligands on depression-like behavior and dendritic changes in mice after inflammation. *Int J Neuropsychopharmacol.* 2014;18(4):77.

29. Sahu BD, Kalvala AK, Koneru M, et al. Ameliorative effect of fisetin on cisplatin-induced nephrotoxicity in rats via modulation of NF-κB activation and antioxidant defence. *PLoS One.* 2014;9(9):e105070.

30. Zhen L-L, Zhu J-J, Zhao X, et al. The antidepressant-like effect of fisetin involves the serotonergic and noradrenergic system. *Behav Brain Res.* 2012;228(2):359–366.

31. Yu X-F, Jiang X, Zhang X-M, et al. The effects of fisetin on lipopolysaccharide-induced depressive-like behavior in mice. *Metab Brain Dis.* 2016;31(5):1011–1021.

32. Wang Y, Wang B, Lu J, et al. Fisetin provides antidepressant effects by activating the tropomyosin receptor kinase B signal pathway in mice. *J Neurochem.* 2017;143(5):561–568.

33. Araki Y, Mukaisho KI, Fujiyama Y, et al. The herbal medicine rikkunshito exhibits strong and differential adsorption properties for bile salts. *Exp Ther Med.* 2012;3(4):645–649.

34. Sawamoto A, Okuyama S, Yamamoto K, et al. 3,5,6,7,8,3′,4′-Heptamethoxyflavone, a citrus flavonoid, ameliorates corticosterone-induced depression-like behavior and restores brain-derived neurotrophic factor expression, neurogenesis, and neuroplasticity in the hippocampus. *Molecules.* 2016;21:1–13.

35. Souza LC, de Gomes MG, Goes ATR, et al. Evidence for the involvement of the serotonergic 5-HT1A receptors in the antidepressant-like effect caused by hesperidin in mice. *Prog Neuropsychopharmacol Biol Psychiatry.* 2013;40:103–109.

36. Filho CB, del Fabbro L, de Gomes MG, et al. Kappa-opioid receptors mediate the antidepressant-like activity of hesperidin in the mouse forced swimming test. *Eur J Pharmacol.* 2013;698(1–3):286–291.

37. Donato F, de Gomes MG, Goes ATR, et al. Hesperidin exerts antidepressant-like effects in acute and chronic treatments in mice: possible role of l-arginine-NO-cGMP pathway and BDNF levels. *Brain Res Bull.* 2014;104:19–26.

38. Li CF, Chen SM, Chen XM, et al. ERK-dependent brain-derived neurotrophic factor regulation by hesperidin in mice exposed to chronic mild stress. *Brain Res Bull.* 2016;124:40–47.

39. Antunes MS, Jesse CR, Ruff JR, et al. Hesperidin reverses cognitive and depressive disturbances induced by olfactory bulbectomy in mice by modulating hippocampal neurotrophins and cytokine levels and acetylcholinesterase activity. *Eur J Pharmacol.* 2016;789:411–420.

40. Li M, Shao H, Zhang X, Qin B. Hesperidin alleviates lipopolysaccharide-induced neuroinflammation in mice by promoting the miRNA-132 pathway. *Inflammation.* 2016;39(5):1681–1689.

41. Haas JS, Stolz ED, Betti AH, et al. The anti-immobility effect of hyperoside on the forced swimming test in rats is mediated by the D2-like receptors activation. *Planta Med.* 2011;77(4):334–339.

42. Butterweck V, Jürgenliemk G, Nahrstedt A, et al. Flavonoids from Hypericum perforatum show antidepressant activity in the forced swimming test. *Planta Med.* 2000;66(1):3–6.

43. Zheng M, Liu C, Pan F, et al. Antidepressant-like effect of hyperoside isolated from Apocynum venetum leaves: possible cellular mechanisms. *Phytomedicine.* 2012;19(2):145–149.

44. Liu B, Xu C, Wu X, et al. Icariin exerts an antidepressant effect in an unpredictable chronic mild stress model of depression in rats and is associated with the regulation of hippocampal neuroinflammation. *Neuroscience.* 2015;294:193–205.

45. Gong M-J, Han B, Wang S, et al. Icariin reverses corticosterone-induced depression-like behavior, decrease in hippocampal brain-derived neurotrophic factor (BDNF) and metabolic network disturbances revealed by NMR-based metabonomics in rats. *J Pharm Biomed Anal.* 2016;123:63–73.

46. González-Cortazar M, Maldonado-Abarca AM, Jiménez-Ferrer E, et al. Isosakuranetin-5-O-rutinoside: a new flavanone with antidepressant activity isolated from Salvia elegans Vahl. *Molecules*. 2013;18(11):13260–13270.

47. Hosseinzadeh H, Motamedshariaty V, Hadizadeh F. Antidepressant effects of kaempferol, a constituent of saffron (Crocus sativus) petal in mice and rats. *Pharmacologyonline*. 2007;2:367–370.

48. Park SH, Sim YB, Han PL, et al. Antidepressant-like effect of kaempferol and Quercitirin, isolated from Opuntia ficus-indica var. saboten. *Exp Neurobiol*. 2010;19(1):30–38.

49. Wang W, Hu X, Zhao Z, et al. Antidepressant-like effects of liquiritin and isoliquiritin from Glycyrrhiza uralensis in the forced swimming test and tail suspension test in mice. *Prog Neuropsychopharmacol Biol Psychiatry*. 2008;32(5):1179–1184.

50. Zhao Z, Wang W, Guo H, et al. Antidepressant-like effect of liquiritin from Glycyrrhiza uralensis in chronic variable stress induced depression model rats. *Behav Brain Res*. 2008;194(1):108–113.

51. de la Peña JB, Kim CA, Lee HL, et al. Luteolin mediates the antidepressant-like effects of Cirsium japonicum in mice, possibly through modulation of the GABAA receptor. *Arch Pharm Res*. 2014;37(2):263–269.

52. Ishisaka M, Kakefuda K, Yamauchi M, et al. Luteolin shows an antidepressant-like effect via suppressing endoplasmic reticulum stress. *Biol Pharm Bull*. 2011;34(9): 1481–1486.

53. Lin TY, Lu CW, Wang SJ. Luteolin protects the hippocampus against neuron impairments induced by kainic acid in rats. *Neurotoxicology*. 2016;55:48–57.

54. Meyer E, Mori MA, Campos AC, et al. Myricitrin induces antidepressant-like effects and facilitates adult neurogenesis in mice. *Behav Brain Res*. 2017;316:59–65.

55. Ma Z, Wang G, Cui L, et al. Myricetin attenuates depressant-like behavior in mice subjected to repeated restraint stress. *Int J Mol Sci*. 2015;16(12):28377–28385.

56. Yi LT, Li CF, Zhan X, et al. Involvement of monoaminergic system in the antidepressant-like effect of the flavonoid naringenin in mice. *Prog Neuropsychopharmacol Biol Psychiatry*. 2010;34(7):1223–1228.

57. Yi LT, Li J, Li HC, et al. Antidepressant-like behavioral, neurochemical and neuroendocrine effects of naringenin in the mouse repeated tail suspension test. *Prog Neuropsychopharmacol Biol Psychiatry*. 2012;39(1):175–181.

58. Yi LT, Liu BB, Li J, et al. BDNF signaling is necessary for the antidepressant-like effect of naringenin. *Prog Neuropsychopharmacol Biol Psychiatry*. 2014;48:135–141.

59. Ben-Azu B, Nwoke EE, Umukoro S, et al. Evaluation of the neurobehavioral properties of naringin in Swiss mice. *Drug Res (Stuttg)*. 2018;68(8):465–474.

60. Aggarwal A, Gaur V, Kumar A. Nitric oxide mechanism in the protective effect of naringin against post-stroke depression (PSD) in mice. *Life Sci*. 2010;86(25–26):928–935.

61. Kwatra M, Jangra A, Mishra M, et al. Naringin and sertraline ameliorate doxorubicin-induced behavioral deficits through modulation of serotonin level and mitochondrial complexes protection pathway in rat hippocampus. *Neurochem Res*. 2016;41(9):2352–2366.

62. Li P, Wang S, Guan X, et al. Six months chronic toxicological evaluation of naringin in Sprague-Dawley rats. *Food Chem Toxicol*. 2014;66:65–75.

63. Yi LT, Xu HL, Feng J, et al. Involvement of monoaminergic systems in the antidepressant-like effect of nobiletin. *Physiol Behav*. 2011;102(1):1–6.

64. Li J, Zhou Y, Liu BB, et al. Nobiletin ameliorates the deficits in hippocampal BDNF, TrkB, and synapsin I induced by chronic unpredictable mild stress. *Evid Based Complement Alternat Med*. 2013;2013:Article ID 359682.

65. Wu M, Zhang H, Zhou C, et al. Identification of the chemical constituents in aqueous extract of Zhi-Qiao and evaluation of its antidepressant effect. *Molecules*. 2015;20(4):6925–6940.

66. Jojua N, Sharikadze N, Zhuravliova E, et al. Nobiletin restores impaired hippocampal mitochondrial bioenergetics in hypothyroidism through activation of matrix substrate-level phosphorylation. *Nutr Neurosci.* 2015;18(5):225–231.
67. Matsuzaki K, Yamakuni T, Hashimoto M, et al. Nobiletin restoring β-amyloid-impaired CREB phosphorylation rescues memory deterioration in Alzheimer's disease model rats. *Neurosci Lett.* 2006;400(3):230–234.
68. Liu Y, Lan N, Ren J, et al. Orientin improves depression-like behavior and BDNF in chronic stressed mice. *Mol Nutr Food Res.* 2015;59(6):1130–1142.
69. Park SH, Sim YB, Han PL, et al. Antidepressant-like effect of Kaempferol and Quercitirin, isolated from Opuntia ficus-indica var. saboten. *Exp Neurobiol.* 2010;19(1):30–38.
70. Mehta V, Parashar A, Udayabanu M. Quercetin prevents chronic unpredictable stress induced behavioral dysfunction in mice by alleviating hippocampal oxidative and inflammatory stress. *Physiol Behav.* 2017;171:69–78.
71. Bhutada P, Mundhada Y, Bansod K, et al. Reversal by quercetin of corticotrophin releasing factor induced anxiety- and depression-like effect in mice. *Prog Neuropsychopharmacol Biol Psychiatry.* 2010;34(6):955–960.
72. Chen QY, Gan X. The protective effect of quercetin on the cultured PC12 cells lesioned by corticosterone. *J Chem Bioeng.* 2009;26:47–49.
73. Holzmann I, da Silva LM, Corrêa da Silva JA, et al. Antidepressant-like effect of quercetin in bulbectomized mice and involvement of the antioxidant defenses, and the glutamatergic and oxidonitrergic pathways. *Pharmacol Biochem Behav.* 2015;136:55–63.
74. Rinwa P, Kumar A. Quercetin suppress microglial neuroinflammatory response and induce antidepressent-like effect in olfactory bulbectomized rats. *Neuroscience.* 2013;255:86–98.
75. Anjaneyulu M, Chopra K, Kaur I. Antidepressant activity of quercetin, a bioflavonoid, in streptozotocin-induced diabetic mice. *J Med Food.* 2003;6(4):391–395.
76. Kaur R, Chopra K, Singh D. Role of alpha2 receptors in quercetin-induced behavioral despair in mice. *J Med Food.* 2007;10(1):165–168.
77. Sah SP, Tirkey N, Kuhad A, et al. Effect of quercetin on lipopolysaccharide induced-sickness behavior and oxidative stress in rats. *Indian J Pharmacol.* 2011;43(2):192–196.
78. Merzoug S, Toumi ML, Tahraoui A. Quercetin mitigates Adriamycin-induced anxiety- and depression-like behaviors, immune dysfunction, and brain oxidative stress in rats. *Naunyn Schmiedebergs Arch Pharmacol.* 2014;387(10):921–933.
79. Tchantchou F, Lacor PN, Cao Z, et al. Stimulation of neurogenesis and synaptogenesis by bilobalide and quercetin via common final pathway in hippocampal neurons. *J Alzheimers Dis.* 2009;18(4):787–798.
80. Can ÖD, Demir Özkay Ü, Üçel Uİ. Anti-depressant-like effect of vitexin in BALB/c mice and evidence for the involvement of monoaminergic mechanisms. *Eur J Pharmacol.* 2013;699(1–3):250–257.
81. Yang L, Yang ZM, Zhang N, et al. Neuroprotective effects of vitexin by inhibition of NMDA receptors in primary cultures of mouse cerebral cortical neurons. *Mol Cell Biochem.* 2014;386(1–2):251–258.
82. Preskorn SH, Baker B, Kolluri S, et al. An innovative design to establish proof of concept of the antidepressant effects of the NR2B subunit selective N-methyl-D-aspartate antagonist, CP-101,606, in patients with treatment-refractory major depressive disorder. *J Clin Psychopharmacol.* 2008;28(6):631–637.
83. Wang Y, Zhen Y, Wu X, et al. Vitexin protects brain against ischemia/reperfusion injury via modulating mitogen-activated protein kinase and apoptosis signaling in mice. *Phytomedicine.* 2015;22(3):379–384.
84. Pal R, Bondar VV, Adamski CJ, et al. Inhibition of ERK1/2 restores GSK-3β activity and protein synthesis levels in a model of tuberous sclerosis. *Sci Rep.* 2017;7(1):4174.

85. Dwivedi Y, Rizavi HS, Roberts RC, et al. Reduced activation and expression of ERK1/2 MAP kinase in the post-mortem brain of depressed suicide subjects. *J Neurochem.* 2001;77(3):916–928.

86. Li Y-C, Shen J-D, Liu Y-M, et al. Screening of antidepressant effects of four main flavonoids compounds from *Scutellaria baicalensis. Chin J Exp Tradit Med Formulae.* 2012;11.

87. Lee HW, Ryu HW, Kang MG, et al. Potent inhibition of monoamine oxidase A by decursin from Angelica gigas Nakai and by wogonin from Scutellaria baicalensis Georgi. *Int J Biol Macromol.* 2017;97:598–605.

88. Liu X, Chan CB, Jang SW, et al. A synthetic 7,8-dihydroxyflavone derivative promotes neurogenesis and exhibits potent antidepressant effect. *J Med Chem.* 2010;53(23):8274–8286.

89. Guan LP, Liu BY. Antidepressant-like effects and mechanisms of flavonoids and related analogues. *Eur J Med Chem.* 2016;121:47–57.

90. Gill R, Tsung A, Billiar T. Linking oxidative stress to inflammation: toll-like receptors. *Free Radic Biol Med.* 2010;48(9):1121–1132.

91. Singh DK, Kumar D, Siddiqui Z, et al. The strength of receptor signaling is centrally controlled through a cooperative loop between Ca2+ and an oxidant signal. *Cell.* 2005 Apr 22;121(2):281–293.

92. Gustot T, Lemmers A, Moreno C, et al. Differential liver sensitization to toll-like receptor pathways in mice with alcoholic fatty liver. *Hepatology.* 2006;43(5):989–1000.

93. Spencer JPE, Schroeter H, Rechner AR, et al. Bioavailability of Flavan-3-ols and procyanidins: gastrointestinal tract influences and their relevance to bioactive forms in vivo. *Antioxid Redox Signal.* 2001;3(6):1023–1039.

94. Hu X, Paik PK, Chen J, et al. IFN-γ suppresses IL-10 production and synergizes with TLR2 by regulating GSK3 and CREB/AP-1 proteins. *Immunity.* 2006;24(5):563–574.

95. Green HF, Nolan YM. GSK-3 mediates the release of IL-1β, TNF-α and IL-10 from cortical glia. *Neurochem Int.* 2012;61(5):666–671.

96. Li C, Li M, Yu H, et al. Neuropeptide VGF C-terminal peptide TLQP-62 alleviates lipopolysaccharide-induced memory deficits and anxiety-like and depression-like behaviors in mice: the role of BDNF/TrkB signaling. *ACS Chem Neurosci.* 2017;8(9):2005–2018.

97. Barrientos RM, Sprunger DB, Campeau S, et al. Brain-derived neurotrophic factor mRNA downregulation produced by social isolation is blocked by intrahippocampal interleukin-1 receptor antagonist. *Neuroscience.* 2003;121(4):847–853.

98. Kozuka N, Itofusa R, Kudo Y, et al. Lipopolysaccharide and proinflammatory cytokines require different astrocyte states to induce nitric oxide production. *J Neurosci Res.* 2005;82(5):717–728.

99. Moncada S, Bolaños JP. Nitric oxide, cell bioenergetics and neurodegeneration. *J Neurochem.* 2006;97(6):1676–1689.

100. Lau FC, Bielinski DF, Joseph JA. Inhibitory effects of blueberry extract on the production of inflammatory mediators in lipopolysaccharide-activated BV2 microglia. *J Neurosci Res.* 2007;85(5):1010–1017.

101. Spencer JPE. Flavonoids and brain health: multiple effects underpinned by common mechanisms. *Genes Nutr.* 2009;4(4):243–250.

102. Zhang X, Wang G, Gurley EC, et al. Flavonoid apigenin inhibits lipopolysaccharide-induced inflammatory response through multiple mechanisms in macrophages. *PLoS One.* 2014;9(9):e107072.

103. Fan SH, Wang YY, Lu J, et al. Luteoloside suppresses proliferation and metastasis of hepatocellular carcinoma cells by inhibition of NLRP3 inflammasome. *PLoS One.* 2014;9(2):e89961.

104. Ahn H, Lee GS. Isorhamnetin and hyperoside derived from water dropwort inhibits inflammasome activation. *Phytomedicine.* 2017;24:77–86.

105. Domiciano TP, Wakita D, Jones HD, et al. Quercetin inhibits inflammasome activation by interfering with ASC oligomerization and prevents interleukin-1 mediated mouse vasculitis. *Sci Rep.* 2017;7:41539.
106. Spencer JPE. The interactions of flavonoids within neuronal signalling pathways. *Genes Nutr.* 2007;2(3):257–273.
107. Rao JS, Ertley RN, Lee HJ, et al. n-3 Polyunsaturated fatty acid deprivation in rats decreases frontal cortex BDNF via a p38 MAPK-dependent mechanism. *Mol Psychiatry.* 2007;12(1):36–46.
108. Réus GZ, Vieira FG, Abelaira HM, et al. MAPK signaling correlates with the antidepressant effects of ketamine. *J Psychiatr Res.* 2014;55:15–21.
109. German-Ponciano LJ, Rosas-Sánchez GU, Rivadeneyra-Domínguez E, et al. Advances in the preclinical study of some flavonoids as potential antidepressant agents. *Scientifica (Cairo).* 2018;2018:2963565.
110. Brown GC. Mechanisms of inflammatory neurodegeneration: iNOS and NADPH oxidase. *Biochem Soc Trans.* 2007;35(5):1119–1121.
111. Bhat NR, Zhang P, Lee JC, et al. Extracellular signal-regulated kinase and p38 subgroups of mitogen-activated protein kinases regulate inducible nitric oxide synthase and tumor necrosis factor-alpha gene expression in endotoxin-stimulated primary glial cultures. *J Neurosci.* 1998;18(5):1633–1641.
112. Spencer JPE. Flavonoids: modulators of brain function? *Br J Nutr.* 2008;99 (E-Suppl 1):ES60–ES77.
113. Kang KA, Wang ZH, Zhang R, et al. Myricetin protects cells against oxidative stress-induced apoptosis via regulation of PI3K/Akt and MAPK signaling pathways. *Int J Mol Sci.* 2010;11(11):4348–4360.
114. Schultze SM, Hemmings BA, Niessen M, et al. PI3K/AKT, MAPK and AMPK signaling: protein kinases in glucose homeostasis. *Expert Rev Mol Med.* 2012;14:e1.
115. Curtis PJ, Sampson M, Potter J, et al. Chronic ingestion of flavan-3-ols and isoflavones improves insulin sensitivity and lipoprotein status and attenuates estimated 10-year CVD risk in medicated postmenopausal women with type 2 diabetes: a 1-year, double-blind, randomized, controlled trial. *Diabetes Care.* 2012;35(2):226–232.
116. Yoshino S, Hara A, Sakakibara H, et al. Effect of quercetin and glucuronide metabolites on the monoamine oxidase-A reaction in mouse brain mitochondria. *Nutrition.* 2011;27(7–8):847–852.
117. Filho CB, Jesse CR, Donato F, et al. Chrysin promotes attenuation of depressive-like behavior and hippocampal dysfunction resulting from olfactory bulbectomy in mice. *Chem Biol Interact.* 2016;260:154–162.
118. Lee S-J, Chung H-Y, Lee I-K, et al. Phenolics with inhibitory activity on mouse brain monoamine oxidase (MAO) from whole parts of Artemisia vulgaris L (Mugwort). *Food Sci Biotechnol.* 2000;9(3):179–182.
119. Wang W, Hu X, Zhao Z, et al. Antidepressant-like effects of liquiritin and isoliquiritin from Glycyrrhiza uralensis in the forced swimming test and tail suspension test in mice. *Prog Neuropsychopharmacol Biol Psychiatry.* 2008;32(5):1179–1184.
120. Liu Y, Lan N, Ren J, et al. Orientin improves depression-like behavior and BDNF in chronic stressed mice. *Mol Nutr Food Res.* 2015;59(6):1130–1142.
121. Lv QQ, Wu WJ, Guo XL, et al. Antidepressant activity of astilbin: involvement of monoaminergic neurotransmitters and BDNF signal pathway. *Biol Pharm Bull.* 2014;37(6):987–995.

10 Preclinical Antidepressant-Like Effects of Terpenes, Polyphenolics, and Other Non-Flavonoid Phytochemicals

Many flavonoids have been shown to produce antidepressant effects, and they often appear to be the primary active components of antidepressant herbs. However, medicinal herbs contain a plethora of phytochemicals other than flavonoids, such as terpenes, stilbenes, lignans, alkaloids, phenolic acids, and others, that are also pharmacologically active. The effects of these non-flavonoid phytochemicals can also contribute to the antidepressant action of herbs, and in some cases may be the predominant antidepressant mechanism.

As with flavonoids, phenolic acids, polyphenolics, and other non-flavonoid phytochemicals can act as potent antioxidant agents. The aromatic features and highly conjugated systems with multiple hydroxyl groups make many of these substances good electron or hydrogen atom donors. This allows them to neutralize free radicals and other reactive oxygen species.[1] Structural differences among phenolic compounds, such as positioning of hydroxyl groups, can be a source of variance in the potency of antioxidant effects.[2] Bioavailability, alterations of molecular structures due to interactions with gut bacteria, and metabolism in the liver may also affect potency.

Aside from simple reduction of reactive oxygen specifies, some non-flavonoid phytochemicals – like some flavonoids – can act as ligands within the intracellular cascades controlling responses to oxidative stress and inflammation.[3] Examples are resveratrol,[4] a stilbene, and ellagic acid,[5] a dilactone of the biphenolic compound, hexahydroxydiphenic acid. Although structurally distinct from flavonoids and each other, both can directly inhibit the NLRP3 inflammasome.

Many flavonoids diminish inflammatory effects by stimulating counteractive antioxidant cascades, in part by activation of Nrf2. However, non-flavonoids can also produce this effect. For example, the phenolic acid 3,4,5-trihydroxycinnamic acid blocked the inflammatory effect of lipopolysaccharide in cultured microglial cells, at least in part by activating Nrf2. This was accompanied by dampening of release of NO, TNF-α, and IL-1β. It also suppressed expression of MCP-1, which plays a key role in the migration of activated microglia. Similar to what has been found with some effects of flavonoids, these effects of 3,4,5-trihydroxycinnamic acid were blocked by the p38 MAPK inhibitor, SB203580.[6]

Some flavonoids, such as quercetin, reduce inflammation and enhance insulin sensitivity by stimulating SIRT1.[7] Such actions would be expected to promote antidepressant effects. The stimulation of SIRT1 is also a well-known effect of the stilbene, resveratrol.[8] Other stilbenes, including piceatannol and its metabolite, Isorhapontigenin, have also been shown to stimulate SIRT1.[9] The hydroxycinnamic acid derivative, ferulic acid, is yet another non-flavonoid that stimulates SIRT1.[10]

Tetra- and pentacyclic triterpenes are groups that have been found to be particularly capable of affecting inflammatory and cell maintenance systems through interactions with NF-κB, Akt, mTOR, iNOS, COX-2, TNF-α, IL-1, IL-6, caspases-3, -8, and -9, ICAM-1, and VEGF. These actions are consistent with antidepressant effects. However, most of these neuropharmacologically active triterpenes have not been evaluated for antidepressant effects, and include the substances ganoderic acid, asiatic acid, avicin, betulinic acid, boswellic acid, celastrol, escin, lupeol, madecassic acid, momordin, pristimerin, and many others.[11]

The notion of the adaptogen is recognized in many areas of the world, particularly Eastern Europe. The characteristics of adaptogens are normalization of hypothalamic sensitivity to regulatory signals, restoration of adrenocortical function, and rescuing the organism from the stage of exhaustion as defined in the general adaptation syndrome.[12] Many plants identified as adaptogens are rich in tetracyclic and pentacyclic triterpenes that structurally resemble human steroids. Interactions of these substances with the immune system and the hypothalamic-pituitary-adrenal axis are thought to be responsible for the stress reduction and resiliency the adaptogens are said to offer. Some lignans, phenylproganes, monoterpenes, and flavonoids are also found to have adaptogenic qualities.[13] The quality of being an adaptogen contributes to antidepressant effects, especially with chronic treatment. Thus, the constellation of antidepressant neurochemical actions of the flavonoids appears to be shared by other, structurally dissimilar classes of phytochemicals. Below are nonflavonoid phytochemicals that have been reported to have antidepressant-like effects.

10.1 AMYRINS

Alpha- and beta-amyrins are pentacyclic triterpenes found in a variety of medicinal plants, including *Protium heptaphyllum*, *Lobelia inflata* and *Couroupita guianensis*. The amyrins have a variety of neuropharmacological effects, including antinociceptive, anti-inflammatory, and cannabinoid agonist effects.[14]

Both oral and intraperitoneal administrations of amyrin decreased immobility of mice in the forced swim test. Amyrin also had anxiolytic-like effects, as demonstrated by the open-field and elevatedplus-maze tests. Those effects were reversed with flumazenil, suggesting mediation by benzodiazepine receptors.[15] Others have observed such anxiolytic effects.[16] A derivative molecule, beta-amyrin palmitate, isolated from leaves of *Lobelia inflata*, produced similar antidepressant effects in mice. It decreased immobility in the forced swim test. It reversed the immobility induced by storage depletion agent, tetrabenazine, but not that induced by the norepinephrine synthesis inhibitor, alpha-methyl-para-tyrosine. Those results suggested that amyrin acted in part by increasing norepinephrine activity by releasing the neurotransmitter from newly synthesized pools.[17] Interestingly, extracts of the herb *Couroupita guianensis*, which is rich in alpha and beta-amyrin and other structurally similar triterpenoids, has been found to have antidepressant-like effects in mice.[18]

10.2   BACOPASIDES

Bacopasides are steroid-like triterpene saponins isolated from the medicinal herb, *Bacopa monnieri*. There are at least eleven variants of bacopaside found in the plant. *Bacopa monnieri* is revered in Ayurvedic medicine and has been used for thousands of years for enhancement of memory and general cognitive function. It has also found use as a tonic, tranquilizer, and sedative to treat "mental strain, insanity, epilepsy, hysteria, esthenia, and nervous breakdown."[19]

Chronic oral administration of Bacopaside I decreased the immobility time of mice in the forced swim and tail suspension tests. It had a similar antidepressant-like effect in reserpinized mice. Further neurochemical assays showed that bacopaside I increased levels of superoxide dismutase and glutathione, and decreased malondialdehyde levels, indicating antioxidant effects. However, it had no effects on either MAO-A or MAO-B activity in the brain.[20]

Bacopaside I also showed antidepressant-like effects in mice that had been subjected to chronic unpredictable stress. It decreased immobility times in the forced swim and tail suspension tests, as well as restored appetite for sucrose. It also reduced serum levels of corticosterone. Subsequent neurochemical analysis of brain tissue showed that it elevated levels of BDNF, and phosphorylated ERK and CREB.[21] Bacopaside II and closely related bacopasaponin C also showed antidepressant-like effects in mice.[22] An extract of *Bacopa monnieri* rich in bacopasides was also shown to attenuate the depression-like behavior seen in mice precipitously withdrawn from morphine. Further testing suggested this effect was due primarily to effects of Bacoside A3.[23]

10.3   BERBERINE

Berberine is a benzylisoquinoline alkaloid found in a variety of plants including European barberry, goldenseal, goldthread, Oregon grape, phellodendron, and tree turmeric. It is also contained in *Berberis aristata*, a major herb widely used in Indian and Chinese systems of medicine. Berberine exhibits a wide range of biological activities including antidiarrheal, antimicrobial, and antiinflammatory effects, as well as central nervous system activity. There is a large literature describing antidepressant-like effects of berberine in rodents. However, there are also studies of berberine for various human ailments, including diabetes. Indeed, in human subjects, berberine lowers serum glucose and fasting insulin, and increases insulin sensitivity.[24]

Berberine has an antidepressant-like effect in mice as demonstrated by decreases in immobility in the forced swim and tail suspension tests. Those effects were accompanied by increases in norepinephrine and serotonin levels in the hippocampus and frontal cortex.[25]

In a similar study, berberine produced the same antidepressant effects in mice, including some with reserpine-induced depression-like behavior. In that case, the antidepressant-like effect was prevented by pretreatment with nitric oxide stimulators arginine or sildenafil, but enhanced nitric oxide simulators, by 7-nitroindazole or methylene blue.[26] The latter findings suggest a role for inhibition of nitric oxide synthesis. In yet another mouse study, antidepressant-like effects of berberine were due in part to blockade of organic cation transporters 2 and 3, which would have resulted in inhibition of reuptake of serotonin and norepinephrine.[27]

Berberine was also found to be protective in two related rodent models of MDD. In mice, chronic oral administration of berberine prevented the depression-like effects of chronic unpredictable stress. This antidepressant effect of berberine was accompanied by attenuation of signs of neuroinflammation. The activation of microglia, stimulation of NF-κB signaling, and elevations of hippocampal IL-1β, IL-6, and TNF-α that occurred in the chronically stressed animals were attenuated by treatment. Inducible nitric oxide synthase, a downstream target in NF-κB signaling pathway, was also inhibited by berberine.[28] Berberine similarly prevented the depression-like effects of chronic administration of corticosterone in mice. It restored sucrose preference and decreased immobility in the forced swim, and these changes were accompanied by increases of BDNF levels in the hippocampus.[29]

Finally, the effects of berberine were investigated in ovariectomized female mice, as it has been shown that ovariectomy, *per se*, increases the immobility in the forced swim test. Again, berberine showed antidepressant-like effects in reducing that immobility. Along with the antidepressant effect, treatment with berberine also increased hippocampal levels of BDNF, peEF2 and phosphorylated CREB, in a manner similar to that seen with ketamine.[30]

10.4   3-N-BUTYLPHTHALIDE

3-n-Butylphthalide is one of the chemical constituents of celery oil. It had an anti-depressant-like effect in rats that had been subjected to two weeks of injections of the inflammatory agent lipopolysaccharide. 3-n-Butylphthalide inhibited expression of pro-inflammatory cytokines, including IL-1β and IL-6, and downregulated the NF-κB signal pathway. It also dampened reduced lipopolysaccharide-induced oxidative reactions in the hippocampus and enhanced Nrf2-targeted signals.[31] In the mouse model of Alzheimer's disease, 3-n-butylphthalide activated the PI3K/AKT pathway, increased BDNF levels, and stimulated TrkB. All of these effects are strongly associated with antidepressant effects.[32]

There are no studies of the use of 3-n-Butylphthalide to treat depression humans. However, it has been safe and effective in treating cognitive impairment in patients with vascular dementia.[33]

10.5   CAFFEIC ACID

Caffeic acid is a phenylpropanoid, classified as one of many hydroxycinnamic acid derivatives. It is found in virtually all plants because it is a key intermediate in the biosynthesis of lignin, one of the principal components of woody plant biomass and its residues.[34]

Intraperitoneal administration of caffeic acid reduced the duration of immobility in the forced swimming test in mice.[35] In a similar study of mice, caffeic acid reduced the duration of immobility of mice in the forced swimming test, and also maintained levels of BDNF mRNA in the frontal cortex as well as TrkB mRNA in the amygdala.[36] Caffeic acid reversed lipopolysaccharide-induced deficits in the forced swim test in mice, and this was accompanied by reductions in levels of inflammatory markers in serum and whole brain, as well as levels of markers of oxidative stress in whole brain homogenates.[37]

Caffeic acid counters some effects of Metabolic Syndrome. It increased glucose uptake capacity in insulin-resistant mouse neuroblastoma cells. In intact rats exhibiting Metabolic Syndrome from chronic high fat diets, it significantly reduced plasma glucose and insulin levels, alleviated insulin resistance, and ameliorated memory impairment. Finally, Western blot analysis demonstrated that caffeic acid increased the expression of the insulin receptor, phosphatidylinositol-3-kinase, Akt/protein kinase B, as well as the expression of leptin signaling-related proteins in the cortex of the high fat diet rats.[38]

10.6 β-CAROTENE

β-Carotene is a terpenoid that is used by the human body to produce vitamin A. The substance is common in green leafy vegetables, carrots, squash, and other vegetables with red, yellow, and orange color. The intake of such plant sources of beta carotene, as well as blood levels of vitamin A and other antioxidants, tend to be low in patients with MDD.[39]

Chronic oral administration of β-carotene decreased immobility in tail suspension test and restored sucrose preference in mice that had been subjected to chronic unpredictable mild stress. The βcarotene reversed stress-induced increases in brain catalase and MAO-A activities. It also reduced oxidative and nitrosative effects, as demonstrated by reduced levels of the oxidation markers thiobarbituric acid-reactive substances and plasma nitrite, and by increased levels of glutathione. The treatment reduced stress-induced levels of corticosterone.[40]

Baked *Cucurbita moschata* squash has long been a folk remedy to treat depression in Korea, and its active ingredient has been thought to be β-carotene. Chronic oral feeding of β-carotene or baked squash reduced immobility of mice in the forced swim test. Both treatments produced similar increases in levels of serotonin and norepinephrine in the brain, as well as increased levels of BDNF, pERK, and ER-β. The two treatments also similarly reduced levels of TNF-α and IL-6.[41]

10.7 CARVACROL

Carvacrol is a monoterpenic phenol present in the essential oil of many plants. Oregano is particularly rich in carvacrol. It decreases the immobility of mice in the forced swimming and tail suspension tests.[42] Antidepressant-like effects of carvacrol were also seen in rats, with improvements observed in the forced swim test. It was further determined that carvacrol increased serotonin and dopamine levels in the hippocampi and prefrontal cortices, as was seen with the full essential oil.[43] Carvacrol also produced anxiolytic effects in mice that were reversed by flumazenil.[44]

10.8   β-CARYOPHYLLENE

β-Caryophyllene, a CB2 receptor agonist, is a bicyclic sesquiterpene found in clove oil, the oil from the stems and flowers of *Syzygium aromaticum*, cannabis, rosemary, and hops.[45] In several studies, it has been found to have antidepressant-like effects in mice through decreases in immobility times in the forced swim and tail suspension tests. It also exhibits anxiolytic like effects in mice in the elevated plus maze, open field test, and marble burying tests.[46] In one study, the antidepressantlike effects of chronically administered β-caryophyllene in mice were accompanied by increases in hippocampal levels of BDNF.[47]

10.9   CHLOROGENIC ACID

Chlorogenic acid is the ester of caffeic acid and quinic acid. It is ubiquitous in plants, being yet another intermediate in lignin biosynthesis. Orally administered chlorogenic acid decreased immobilization in the tail suspension and forced swim tests in rats that had been chronically stressed by restraint. It also attenuated restraint-induced increases in serum pro-opiomelanocortin mRNA and plasma β-endorphin.[48] It also reduced immobility in the forced swim and tail suspension tests of mice that had been chronically administered corticosterone. In those animals, the antidepressantlike effect was accompanied by reductions in steroid-induced activation of MAO-B and production of reactive oxidative species.[49]

Chlorogenic acid was also found to have anxiolytic effects in mouse models of anxiety, including the light/dark test, the elevated plus maze, and the free exploratory test. Those anxiolytic effects were blocked by the benzodiazepine antagonist flumazenil.[50]

10.10   CROCIN

Crocin is a carotenoid found in the flowers crocus and gardenia.[51] Acute and sub-chronic administration of crocin decreased immobility of mice in the forced swimming test.[52,53] Crocetin, a simpler component molecule of crocin, also showed antidepressant effects in mice. Chronic administration of crocin similarly showed anti-depressant-like effects in rats by reducing immobility in the forced swim test.[54]

Of particular interest is that crocin has been found to be an effective augmentation strategy in the treatment of resistant MDD. Crocin was significantly more effective than placebo in improving the effects of SSRIs in human subjects.[55]

10.11   CURCUMIN

Curcumin, found primarily in the herb turmeric, is a diarylheptanoid. It is a symmetrical molecule consisting of two phenolic rings connected by two α,β-unsaturated carbonyl groups. It is the primary active constituent of turmeric. Curcumin reverses the depression-like syndrome produced in rats by chronic corticosterone treatment, and appears to do so by increasing BDNF expression in the brain.[56] It also has an antidepressant-like effect in mice that may be due in part to MAO inhibition.[57]

Chronic treatment with curcumin has also been found to have significant antidepressant effects in humans.[58] Curcumin also enhances the effects of standard antidepressants, perhaps by adding anti-inflammatory effects and enhancing brain levels of BDNF.[59]

10.12   3,6'-DISINAPOYL SUCROSE

3,6'-Disinapoyl sucrose is a complex oligosaccharide ester that can be extracted from the roots of *Polygala tenuifolia* Willd, a plant widely used in Traditional Chinese Medicine. It is an ingredient of several traditional combinations used for the treatment of symptoms of MDD, including the popular treatment *Kai Xin San*. At least three studies have shown 3,6'-disinapoyl sucrose to produce antidepressant-like effects in rats.

Daily oral administration of 3,6'-Disinapoyl sucrose restored appetite for sucrose in rats that had been chronically subjected to unpredictable stress. The oligosaccharide was also found to reduce serum corticosterone, and to inhibit the activity of brain MAO-A and MAO-B enzymes. Further antioxidant effects were seen in increases in brain superoxide dismutase activity and decreases in levels of the oxidative damage marker, malondialdehyde.[60] In a similar study, chronically administered 3,6'-disinapoyl sucrose again restored sucrose intake in stressed rats. However, it was further shown that 3,6'-disinapoyl sucrose increased phosphorylated CREB and BDNF levels in the hippocampus, which in turn increased neuronal plasticity and neurite outgrowth.[61] Yet another study showed antidepressant effects of 3,6'-disinapoyl sucrose, per the forced swim and tail suspension tests. In the same report, 3,6'-disinapoyl sucrose stimulated proliferation of cultured neural progenitor cells.[62]

Further studies confirmed the ability of 3,6'-disinapoyl sucrose to stimulate BDNF expression and CREB phosphorylation. It also showed a neuroprotective effect against glutamate and H_2O_2-induced toxicity in neurons. Pharmacological inhibition of MAPK 1, CaMKII, and Trk blocked those effects of 3,6'-disinapoyl sucrose. Interestingly, blockade of PI3K or Akt had no such antagonistic effect.[63]

10.13   ELLAGIC ACID

Ellagic acid is a polyphenolic compound found in a variety of herbs, including *Quercus rubus, Euphorbia antisyphilitica, Quercus alba, Rubus idaeus*, and *Rubus occidentalis*.[64] It is also common in a variety of nuts, fruits, and vegetables. It possesses potent neuroprotective effects through its free radical scavenging properties, iron chelation, activation of different cell signaling pathways, and mitigation of mitochondrial dysfunction.[65] It also exerts potent anti-inflammatory effects.[66]

Both acute and chronic oral administration of ellagic acid to mice reduced immobility in the forced swim and tail suspension tests.[67] In a similar study, ellagic acid also produced this antidepressantlike effect in mice that had been subjected to restraint stress. In those animals, the antidepressant-like effect was accompanied by reductions in activity of nitric oxide synthase.[68]

Ellagic acid also showed anxiolytic-like effects in mice. It increased time spent in the open arms of the elevated plus maze in a manner similar to diazepam. Moreover, that effect was blocked by pretreatment with the GABA-A receptor antagonist, pic-rotoxin, and the benzodiazepine antagonist, flumazenil.[69]

10.14 EUGENOL

Eugenol is a highly aromatic member of the allylbenzene class of chemical com-pounds. It is found in the essential oils of a variety of herbs and spices, most notably clove, nutmeg, cinnamon, basil, and bay leaf. It is known to have a variety of phar-macological properties, including anti-inflammatory and antioxidant effects.[70] It also produces the remarkable effect of dental anesthesia.[71]

Eugenol has antidepressant effects in mice, as demonstrated by the forced swim and tail suspension tests. These effects may have been due in part to increases in hippocampal levels of BDNF and metallothionein-III.[72] Metallothionein-III is a brain-specific member of the metallothionein family of metal-binding proteins. It is abundant in glutamatergic neurons that release zinc from their synaptic termi-nals, such as hippocampal pyramidal neurons and dentate granule cells. MT-III may be an important regulator of zinc in the nervous system, and its absence has been implicated in the development of Alzheimer's disease.[73] The protein has come under investigation as playing a role in MDD as a possible contributor to disrupted zinc metabolism that itself has been associated with MDD.[74]

Acute administration of methyl-eugenol, a derivative of eugenol found in *Croton zehntneri*, had antidepressant-like effects in mice. It decreased the immobility time during the forced swim test. It was not found to have anxiolytic effects.[75]

Another natural derivative of eugenol, Methyl-isoeugenol from *Pimenta pseu-docaryophyllus*, produced antidepressant- and anxiolytic-like effects in mice. It decreased immobility time in the forced swim test and increased time in the open arm of the elevated plus maze. These effects were blocked by the 5-HT1A receptor antagonist, WAY100635, which suggested a serotonergic mediation of this effect.[76] Bis-eugenol, the dimerization product of eugenol, also has antidepressant-like effects in mice.[77] The antidepressant-like effects of eugenol and various derivatives in rodents have been attributed in part to their ability to inhibit MAO-A and, to a lesser extent, MAO-B.[78]

10.15 FERULIC ACID

Ferulic acid is a phenolic compound, a derivative of cinnamic acid, that is present in a variety of food and medicinal plants, including bamboo shoots, popcorn, chocolate, coffee, red cabbage, grapefruit, eggplant, spinach, and whole grains.[79] It is also a com-ponent of many medicinal plants used in Traditional Chinese Medicine for the treat-ment of MDD and other neurological conditions. Those include *Angelica sinensis*, *Avena sativa*, *Curcuma longa*, *Hericium erinaceus*, *Ligusticum chuanxiong*, and oth-ers. It has been identified as an important phytochemical constituent of the Chinese herbal combinations, *chaihu shugan* and *xiaoyaosan* that are used for MDD.

Ferulic acid protects against the damage of cerebral ischemia and reperfusion through its antioxidant and anti-inflammatory effects.[80] There is also evidence that ferulic acid, like ketamine, may act as an NMDA receptor antagonist.[81]

Acute treatment with ferulic acid decreased immobility in mice in the tail suspension test, and that effect appeared to be mediated in part by interaction with MAPK and the PI3K/Akt pathways.[82] In a similar study, ferulic acid decreased immobility in the tail suspension and forced swim tests in mice, and these effects were thought to be due to increases in serotonergic and noradrenergic activities in the hippocampus and frontal cortex.[83] In yet another report, the antidepressant-like effects of ferulic acid were reversed by co-administration of 5-HT1A and 5-HT2 receptor antagonists, further suggesting serotonergic mediation of the effect.[84]

Ferulic acid ameliorated the stress-induced depression-like behavior of mice treated with high doses of corticosterone, and increased levels of BDNF and proliferation of neural progenitor cells in the hippocampi of those animals.[85] It also showed antidepressant effects in reserpinized mice. Ferulic acid normalized monoamine levels in the hippocampus and frontal cortex, and mitigated oxidative and nitrosative stress and inflammation.[86]

10.16 GALLIC ACID

Gallic acid, or trihydroxybenzoic acid, is a type of phenolic acid. It is found in a wide variety of medical plants including *Allan blackia floribunda*, *Garcinia densivenia*, *Bridelia micrantha*, *Caesalpinia sappan*, *Dillenia indica*, *Diospyros cinnabarina*, *Paratecoma peroba*, *Psidium guajava*, *Syzygium cordatum*, *Rhus typhina*, *Tamarix nilotica*, *Vitis vinifera*, *Hamamelis virginiana*, *Toona sinensis*, *Oenothera bienni*, and *Rubus suavissimus*. It has been reported to possess antiinflammatory, antioxidant, anti-diabetic, and neuroprotective effects.[87] In animal studies, it has also shown antidepressant-like effects.

Subchronic, intraperitoneal administration of gallic acid decreased immobility times of mice in the forced swim and tail suspension tests. It was further found to have decreased MAO-A activity, as well as malondialdehyde levels and catalase activity in brain homogenates.[88] Others observed similar antidepressant-like effects, as well as evidence of serotonergic, noradrenergic, and dopaminergic mediation.[89]

Chronically administered gallic acid prevented the depression-like effects of chronic stress in mice. In both stressed and unstressed mice, it decreased immobility in the forced swim test and restored appetite for sucrose. Again, it was found to decrease MAO-A and catalase activity in brain homogenates, as well as to decrease levels of malondialdehyde. Plasma levels of corticosterone and nitrite were also reduced.[90]

In yet another animal model of depression, gallic acid attenuated post-stroke depression-like behavior in mice. This effect of gallic acid was attributed to antioxidant effects, as it increased super oxide dismutase and glutathione in brain homogenates, but decreased thiobarbituric acid-reactive substances, which are markers of lipid peroxidation.[91]

Along with antidepressant-like effects, chronic administration of gallic acid also produced anxiolytic-like effects in rodents. Rats given the phytochemical spent greater amounts of time in the open arm of the elevated plus maze.[92] Another group that observed such anxiolytic effects found the effects were only partially blocked by flumazenil, but fully blocked by the 5-HT1A antagonist, WAY-100635, suggesting serotonergic mediation of the effect.[93]

10.17   GASTRODIN

Gastrodin is the glucoside of the simple phenolic compound, 4-(Hydroxymethyl) phenol that is found in the rhizome of *Gastrodia elata*.[94] In Traditional Chinese Medicine, *Gastrodia elata* is referred to as *Tian ma*, and it has long been used in in the treatment of various mood disorders.[95]

A number of studies have shown antidepressant-like effects of gastrodin in rats. Oral administration of gastrodin reduced immobility time in in the forced swim test. Neurochemically, it enhanced activity of serotonin and dopamine in the frontal cortex, amygdala, and hippocampus of those animals. Further analysis revealed that gastrodin positively affected proteins that modulate maintenance of neuronal cytoskeleton. It inhibited activity of the negative modulators, Slit1 and RhoA, but stimulated the positive growth modulators, CRMP2 and PFN1.[96]

Two other studies have shown that gastrodin attenuates the depression-like effects of chronic unpredictable stress in rats. In one of those studies, the restoration of sucrose preference by daily administration of gastrodin appeared to be due in part to proliferation of hippocampal stem cells and dampening of neuroinflammatory effects. The stress-induced increases in expression of p-iκB, NF-κB, and IL-1β in the hippocampus were reversed by gastrodin.[97]

In a similar study, daily administration of gastrodin again restored sucrose consumption in stressed rats. In addition, gastrodin increased expressions of both BDNF and glial fibrillary acidic protein in the hippocampus.[98] Levels of glial fibrillary acidic protein have been found to be reduced in the prefrontal cortices of individuals with MDD.[99]

In yet another model of stress-induced depression, the so-called single prolonged stress procedure, post-stress daily treatment with gastrodin decreased immobility in the forced swim test. This antidepressant-like effect was accompanied by restoration of normal levels of norepinephrine, BDNF, and Neuropeptide Y in the hippocampus.[100] After the same single prolonged stress procedure, another group saw anxiolytic-like effects of gastrodin, as it increased activity in the open field, as well as time spent in the open arm of the elevated plus maze. Along with the behavioral effects, gastrodin also decreased levels of IL-6 and IL-1β, and expression of iNOS and p38MAPK phosphorylation.[101]

10.18   GENIPIN

Genipin is a monoterpenoid extracted from *Gardenia jasminoides,* an herb commonly used in Traditional Chinese Medicine. It is a component of *Yueju,* an herbal combination used for the treatment of symptoms of depression. Genipin has numerous neuroprotective effects, including prevention of lipid peroxidation and production of nitric oxide, anti-inflammatory effects, and protection of hippocampal neurons from the toxicity of amyloid β protein.[102] It also has shown antidepressant-like effects in rodent models.

Sub-chronic oral administration of genipin decreased immobility of mice in the forced swim and tail suspension tests. It also antagonized reserpine-induced depletions of serotonin and norepinephrine in the hippocampus, which suggested these antidepressant-like effects may have been due in part to enhancement of monoaminergic activity.[103]

Genipin also reversed the decrease in sucrose consumption in rats that had suffered chronic unpredictable stress. It attenuated the rise in serum levels of corticosterone in those animals and prevented the stress-induced decreases in hippocampal levels of serotonin and norepinephrine. Reductions in 5-HT1A receptor RNA, a consequence of high corticosterone levels,[104] were also prevented, as were increases in 5-HT2A receptors. Genipin also maintained levels of hippocampal CREB and BDNF.[105]

A great deal of work has been performed in examining the mechanisms by which early or even prenatal experience of stress manifests as depression-like syndromes. One such mechanism is epigenetic change due to methylation of DNA. In a study of the antidepressant-like effects of genipin in pre-natally stressed mice, it was found that the substance normalized the expression of BDNF in the hippocampus, in part by preventing methylation of DNA through inhibiting DNA methyltransferase 1.[106]

10.19   GINSENOSIDE RG1

Ginsenoside Rg1 is a steroid glycoside that has been shown to exert a number of neuroprotective effects in *in vivo* and *in vitro* studies. Studies have shown that it exerts antidepressant-like effects in both mice and rats. In each case, these antidepressant-like effects were seen in animals that had experienced chronic unpredictable stress. In mice, Ginsenoside Rg1 exhibited its antidepressant-like activity in the forced swim and tail suspension tests. Those effects were accompanied by decreases in serum corticosterone levels and up-regulation of the BDNF signaling pathway in the hippocampus.[107] In rats, Ginsenoside Rg1 increased sucrose consumption and decreased immobility in the forced swim test. Those ameliorations of depression-like behaviors were mediated, at least in part, by CREB-regulated increases of BDNF expression in the amygdala of those rats.[108] The intestinal metabolite of ginseng, 20(S)-protopanaxadiol, has also been found to have potent antidepressant effects in animal studies. It was as potent as fluoxetine.[109]

10.20   GLYCYRRHIZIN

Glycyrrhizin, also known as glycyrrhizic acid, is a saponin found in licorice root. It is a glycated form of the pentacyclic triterpenoid, enoxolone. In at least three studies it has been shown to produce antidepressant-like effects. Chronic intraperitoneal administration of glycyrrhizin reduced immobility of mice in the forced swim and tail suspension tests.[110] Glycyrrhizin also reduced immobility in the tail suspension test and restored sucrose consumption in mice that had been chronically subjected to chronic unpredictable stress. Interestingly, in that study, glycyrrhizin also attenuated activity in the kynurenine synthetic pathway. Activation the rate-limiting enzyme, indoleamine2,3dioxygenase, leads to diversion of tryptophan from serotonin synthesis and into the pathway that generates the NMDA agonist and neurotoxin quinolinic acid.[111] The depression-like effects of lipopolysaccharide in mice were also attenuated by glycyrrhizin.[112]

10.21   4-HYDROXYISOLEUCINE

4-Hydroxyisoleucine is a unique amino acid found in fenugreek. Several studies have demonstrated its antidepressant-like effects. It decreased immobility in the forced swim in reserpinized mice, as well as in un-reserpinized control animals. The 4-hydroxyisoleucine also increased numbers of 5HTP-induced head twitches in the mice suggesting enhancement of serotonergic activity.[113] Chronic orally administered 4-hydroxyisoleucine also reversed signs of depression in olfactory bulbectomized rats. It reduced immobility in the forced swim test, and restored their usual preference for sucrose. Serum levels of corticosterone were reduced. There were also indications of anxiolytic effects of 4-hydroxyisoleucine in the open field and novelty suppressed feeding paradigms.[114]

4-Hydroxyisoleucine enhances insulin sensitivity in rats by activating PI3-kinase activity in liver and muscle. The activation of PI3k is one of the primary mechanisms of action of lithium. Chronic treatment with 4-hydroxyisoleucine significantly reduced abnormally high fasting serum insulin levels in rats with type II diabetic rats, and slowed progression of hyperinsulinemia in genetically insulin-resistant rats.[115]

10.22   HYPERFOLIATIN

Hyperfoliatin is a prenylated phloroglucinol derivative found in St John's wort. It decreased immobility of mice in forced swim test. It also inhibited synaptosomal reuptake of dopamine, serotonin, and noradrenaline, but did not compete for binding at monoamine transporters.[116] The similar acylphloroglucinol derivatives andinin A,[117] from *Hypericum andinum*, and crassipin A,[118] from *Elaphoglossum crassipes*, also produce antidepressant-like effects. Uliginosin B, discussed below, has similar structure and antidepressant properties.

10.23   LINALOOL

Linalool, a monoterpene, is found in a wide variety of plants, including lavender, cinnamon, marijuana, birch trees, mint, rosewood, laurel, citrus, and perhaps as many as 200 others.[119]

Two weeks of pretreatment with linalool protected rats from the depression-like effects of chronic restraint stress. When subjected to the forced swim test, the treated rats exhibited significantly less immobility.[120] Linalool also showed antidepressant-like effects in mice by reducing immobility in the forced swim test. This effect was blocked by the 5-HT1A antagonist, WAY 100635, which suggested a serotonergic contribution.[121]

Orally administered,[122] and even inhaled,[123] linalool has been reported to produce anxiolytic-like effects in mice, as demonstrated in the open field and elevated plus maze tests.

Linalool also attenuated amyloid-beta-induced oxidative neurotoxicity in mice. Oxidative enzymes were preserved. Linalool stimulated the Nrf2/HO-1 signaling pathway that has been associated with antidepressant effects.[124]

10.24 MACRANTHOL

Macranthol (2,4-bis(2-hydroxy-5-prop-2-enylphenyl)-6-prop-2-enylphenol) is a chain molecule consisting of three linked propylphenols. It is found in the medicinal plant, *Illicium dunnianum* tutch. Indeed, this plant is of the family, *Schisandraceae*, which contains a variety of medicinal plants used in Traditional Chinese Medicine.

Chronic administration of macranthol restored preference for sucrose and reduced immobility time in the forced swim and tail suspension tests in mice that had been subjected to chronic unpredictable stress. It also reduced serum levels of corticosterone. Neurochemically, it maintained levels of serotonin in the frontal cortex and hippocampus. Chronic, though not acute administration of macranthol, also increased levels of BDNF in stressed mice.[125] Indeed, in a follow-up study, the antidepressant-like effects of macranthol were blocked by the TrkB antagonist K252a. It was concluded that the antidepressant effect was due to activation of BDNF and TrkB and downstream stimulation of the PI3K/Akt-Bcl-2/caspase-3 signaling pathway.[126]

Sub-chronically administered macranthol attenuated the depression-like effects of lipopolysaccharide injection in mice. It reduced concentrations of IL-1β, IL-6, and TNF-α in the frontal cortex, as well as the number of cells expressing iba-1, a marker of activated microglia. Thus, macranthol could alleviate depressive-like behaviors in mice by suppressing microglia-related neuroinflammation in the prefrontal cortex and other brain areas.[127]

10.25 METHYL JASMONATE

Methyl jasmonate is a single-ringed, cyclopentane-containing organic acid released by certain plants, including *Jasminum sambac*, in response to external stress, injury, or pathogenic invasions. Interestingly, this is also a role played by resveratrol in plants. Both acute[128] and sub-chronic[129] intraperitoneal administration of methyl jasmonate decreased immobility of mice in the forced swim and tail suspension tests. Sub-chronic intraperitoneal administration of methyl jasmonate also produced antidepressant-like effects in mice injected with lipopolysaccharide. It lowered serum corticosterone levels, and reduced concentrations of malondialdehyde and TNF-α in brain homogenate. Brain levels of glutathione were increased.[130] In mice subjected to chronic unpredictable stress, methyl jasmine exhibited a so-called "adaptogenic" effect. It reduced serum corticosterone levels, attenuated oxidative stress and malondialdehyde levels in brain tissue, maintained glutathione levels, and decreased neuronal cell loss.[131]

10.26 MITRAGYNINE

Mitragynine is a complex indole alkaloid extracted from *Mitragyna speciosa* Korth. This plant from Southeast Asia is more commonly known as kratom. Mitragynine acts, at least in part, as a μ-opioid receptor agonist.[132] However, there is also evidence of monoaminergic effects.[133] The leaves of *Mitragyna speciose* are commonly used to treat pain and as a minor stimulant among natives of Thailand and surrounding areas. More recently it has been used in treating opiate addiction.

Acute, intraperitoneal administration of mitragynine reduced the immobility time of mice in both the forced swim and tail suspensions tests. The alkaloid also dampened serum levels of corticosterone.[134] The alkaloid also produced anxiolytic-like effects in rats as evidenced by behaviors in the open field and elevated plus maze. Those anxiolytic-like effects of mitragynine were antagonized by intra-peritoneal administration of naloxone, flumazenil, or sulpiride, suggesting inter-actions with opioidergic, GABAergic, and dopaminergic systems in the brain.[135] Mitragynine can inhibit 5-methoxy-N,N-dimethyltryptamine-induced head-twitch in mice. That effect is characteristic of 5-HT2A receptor antagonists, and is an effect shared by tricyclic antidepressants and atypical antipsychotics that augment antidepressant effects.[136]

10.27   OLEANOLIC ACID

Oleanolic acid is a pentacyclic triterpene found in olives, various species of *Silphium*, *Calendula officinalis*, and *Panax quinquefolium*.[137] Oleanolic acid decreased immo-bility times of mice in the forced swim and tail suspension tests. Chronic adminis-tration of oleanolic acid increased hippocampal BDNF level, but MAO activity was unaffected.[138] In a similar study, oleanolic acid reduced immobility in the forced swim test. However, in this study, the forced swim test was repeated over several weeks and thus became a significant source of stress. Indeed, over these periods of retesting, the immobility time of control animals increased, whereas latency to immobility tended to decrease. However, chronic, though not acute, treatment with oleanolic acid reduced immobility and prevented the cumulative effect of stress. That treatment also increased serotonin levels in the frontal cortex and hippocam-pus, levels of norepinephrine in the hippocampus. Chronic treatment with oleanolic acid also increased levels of BDNF in the frontal cortex and hippocampus.[139] In a subsequent study, oleanolic acid restored sucrose preference in mice subjected to chronic unpredictable stress. These antidepressant-like effects were again accompa-nied by increases in BDNF. However, it was further determined that oleanolic acid had activated ERK and CREB, as part of its mechanism of action. These actions were further associated with upregulation of miR-132 RNA and hippocampal neu-ronal proliferation.[140]

10.28   ORCINOL

Orcinol, with the chemical name, 5-methylbenzene-1,3-diol, is a phenolic compound found in various species of lichen.[141] Orcinol glucoside can be extracted from the rhizomes of *Curculigo orchioides*, a common traditional Chinese medicinal herb with diverse pharmacological activities, including immunomodulatory activity and antioxidant effects. That herb is also considered an adaptogen with neuroprotective effects.[142] Of note, *Curculigo orchioides* is considered an endangered plant.

Orcinol restored sucrose preference in rats subjected to chronic unpredictable mild stress, and reduced immobility in the forced swim and tail suspension tests. Orcinol also reduced serum corticosterone levels and upregulated BDNF and phos-phorylated-ERK1/2 in the hippocampus.[143]

10.29 PAEONIFLORIN

Paeoniflorin is a monoterpene glycoside that is a major active component of *Paeonia lactiflora*. This herb has long been used in Traditional Chinese Medicine, and is one of the most commonly used ingredients in herbal combinations used to treat MDD, including *xiao yao san*, *chai hu shugan*, *sini san*, and others. It has potent anti-inflammatory and neuroprotective effects, as well as anti-allergic, antihyperglycemic, analgesic, and nootropic properties.[144] There is an extensive literature showing antidepressant-like effects in rodents.

Intraperitoneally injected paeoniflorin significantly reduced the duration of immobility in the forced swim and tail suspension tests in mice. Treatment also increased levels of serotonin and its metabolite, 5-hydroxyindoleacetic acid, in the hippocampus.[145] Sub-chronic administration of paeoniflorin attenuated reserpine-induced depression in mice, which further suggested a monoaminergic component to its effect.[146]

Chronic intragastrically administered paeoniflorin attenuated interferon-α-induced depression-like behavior in mice, which is thought to be mediated by neuroinflammatory processes. Indeed, it reduced levels of IL-6, IL-1β, and TNF-α in serum, prefrontal cortex, hippocampus and amygdala. Treatment also prevented interferon-induced increases in microglial densities in the brain.[147]

Paeoniflorin showed similar antidepressant-like effects in rats, with several studies finding such effects in rats subjected to chronic stress. In one such study, paeoniflorin markedly increased sucrose consumption but decreased serum corticosterone and adrenocorticotropic hormone levels in the stressed rats. Treatment also attenuated stress-induced reductions in hippocampal levels of norepinephrine, serotonin, and its metabolite 5-hydroxyindoleacetic acid.[148] Similar effects of paeoniflorin on behavior, hypothalamic-pituitary-adrenal activity, and levels of brain monoamines were seen in rats stressed by the method of serial bouts of restraint stress.[149] In that study, paeoniflorin enhanced soluble guanylyl cyclase but inhibited phosphodiesterase, each of which would have enhanced cyclic-GMP activity in cells and thus added to antidepressant effects.[150]

The depression-like effects of ovariectomy in rats are also improved with paeoniflorin. In ovariectomized rats that were also subjected to chronic stress, paeoniflorin increased sucrose solution consumption. It reduced serum ACTH and corticosterone concentrations and density of brain 5-HT2 receptors, but increased density of brain 5-HT1A receptors.[151]

Paeoniflorin has neuroprotective effects that might at least partially explain its antidepressant-like effect. For example, it protected cultured rat pheochromocytoma cells from the toxic effects of corticosterone. Paeoniflorin decreased levels of intracellular reactive oxygen species and malondialdehyde, and reversed the reductions in nerve growth factor in the corticosterone-treated cells. Overall, it increased cell viability.[152] It was further shown to protect rat pheochromocytoma cells from the toxic effects of NMDA. This was seen to be due to attenuation of NMDA-induced increases in intracellular calcium activation of Calbindin-D28K.[153]

10.30 PAEONOL

Paeonol is one of many phenolic components of the root bark of *Paeonia suffruticosa*. Unlike paeoniflorin, it is an aglycone. It has been found to reduce immobility of both mice and rats in the tail suspension and forced swim tests. It also exhibits neu-roprotective effects, as it spares cultured mouse pheochromocytoma cells from the neurotoxic effects of prolonged exposure to corticosterone.[154] Chronic administration of paeonol produced a similar antidepressant-like effect in chronically stressed rats. In the brain, paeonol increased activity of BDNF, but downregulated the activity of the Rac1/RhoA pathway that inhibits the growth and repair of neural pathways and axons. Those neurochemical effects manifested as increases in the dendritic density, length, and complexity in the CA1 and dentate gyrus areas of the hippocampus.[155]

Paeonol attenuated lipopolysaccharide-induced depression-like behavior in mice. In doing so, it reduced serum levels of IL-1β, IL-6, and TNF-α, and maintained hip-pocampal levels of serotonin and norepinephrine. It also diminished NF-κB activa-tion and normalized BDNF and TrkB activity in the hippocampus.[156]

10.31 PALMATINE

Palmatine is a protoberberine alkaloid found in several plants including *Phellodendron amurense, Rhizoma coptidis, Coptis Chinensis* and *Corydalis yan-husuo*.[157] Palmatine has been found to exert MAO oxidase inhibition,[158] as well as to enhance memory.[159] It is an object of research for a variety of neuropsychiatric conditions.

Chronic intraperitoneal administration of palmatine exerted antidepressant-like effects in mice that had been subjected to chronic unpredictable stress, as well as in non-stressed mice. In both groups, palmatine decreased duration of immobility in the forced swim and tail suspension tests. The highest doses of palmatine restored sucrose preference in stressed mice. Palmatine was further found to reverse stress-induced increases in brain catalase and MAO-A activities, brain lipid peroxidation, and plasma levels of nitrite and corticosterone.[160] Palmatine also showed antidepres-sant-like effects in diabetic rats. Palmatine decreased immobility in the forced swim test and restored appetite for sucrose. Palmatine decreased levels of TNF-α and IL-1β in the hippocampus of the rats. Another effect was a decrease in the number of P2X7 receptors in the hippocampus. Increases in the density of those receptors in the hippocampus is associated with both pain and depression-like behaviors.[161]

10.32 PLUMBAGIN

Plumbagin (5-hydroxy-2-methyl-1,4-naphthoquinone) (PL) is a naturally occur-ring yellow pigment found in the plants of the *Plumbaginaceae, Droseraceae, Ancistrocladaceae,* and *Dioncophyllaceae* families. It has been reported to exhibit analgesic and anti-inflammatory activities. For example, it suppresses paw edema of rats induced by carrageenan and various pro-inflammatory mediators, including histamine, serotonin, bradykinin, and prostaglandin E2. Those anti-inflammatory effects include dampening of production of IL-1β, IL-6, and TNF-α. It also inhibits expression of inducible nitric-oxide synthase and COX-2. Some of its anti-inflamma-tory effects are due to inhibition of NF-κB activation.[162] Plumbagin also appears to reverse many of the pathologies of the Metabolic Syndrome. In rats made diabetic by high fructose diets, plumbagin reduced weight gain; improved insulin resistance and dyslipidemia; and reduced liver weight and adiposity, oxidative stress, inflammation, and liver fibrosis.[163]

Chronic oral administration of plumbagin decreased immobility in the tail suspension test and restored sucrose appetite in mice that had been subjected to chronic unpredictable stress. Its effects were comparable to those of imipramine. Plumbagin inhibited brain MAO-A activity, and reduced oxidative and nitrosative damage. It decreased plasma nitrite, brain malondialdehyde, and catalase levels, and increased glutathione levels. Stress-induced increases in plasma corticosterone levels were also attenuated by plumbagin.[164]

10.33 PODOANDIN

Intraperitoneal administration of podoandin, a sesquiterpene lactone extracted from the leaves of the medicinal plant, *Hedyosmum Brasiliense*, decreased immobility in tail suspension and forced swimming tests in mice. Pharmacological challenges suggested these effects to be mediated by serotonergic, noradrenergic, and dopaminergic systems, but not by GABAergic or opioid systems.[165] The sesquiterpene lactones are common among plants[166] and are found in a wide variety of medicinal herbs. However, even some sesquiterpene lactones found in *Hedyosmum Brasiliense* with podoandin are without antidepressant effect. Nonetheless, many of these molecules exert anti-inflammatory and antioxidant effects that offer neuroprotection. For example, aromadendrane-4β,10α-diol from that plant protects against neurodegeneration in the mouse model of Alzheimer's Disease.[167] As a cautionary note, some sesquiterpene lactones are thought to produce genotoxic and embryotoxic effects.[168]

10.34 PUNARNAVINE

Punarnavine, also known as lunamarine, is an alkaloid from the medicinal plant, *Boerhaavia diffusa*. The root of the plant has been thought to possess potent antioxidant, anti-stress, and anticonvulsant activities, and has long been used in Ayurvedic Medicine. At least two studies have shown antidepressant-like effects of punarnavine in mice. Chronic administration of both the ethanolic extract of *Boerhaavia diffusa* and purified punarnavine decreased immobility in the forced swim and tail suspension tests. Both the extract and punarnavine also significantly reduced brain MAO-A levels.[169]

Chronically administered punarnavine also reversed the depression-like effects of chronic unpredictable mild stress in mice. It decreased immobility in the forced swim test, in both stressed and unstressed mice, and restored sucrose preference. The alkaloid decreased MAO-A activity, malondialdehyde levels, and reversed the stress-induced decrease in reduced glutathione and catalase activity. It also significantly attenuated the stress-induced increase in plasma nitrite and corticosterone levels.[170]

10.35 RESVERATROL

Resveratrol (3,5,4'-trihydroxystilbene, both cis and trans forms) is a polyphenol found in grapes, various berries, peanuts, and some medicinal plants, including the

Japanese knotweed (*Polygonum cuspidatum*).[171] Knotweed is a component of the Traditional Chinese Medicine herbal combination *tang shen kang* that is used in the treatment of symptoms of MDD. Studies using purified enzymes, cultured cells, and laboratory animals have suggested that resveratrol has anti-aging, anticarcinogenic, anti-inflammatory, and anti-oxidant properties. There is also an extensive preclinical literature describing antidepressant-like effects of resveratrol.

Intraperitoneal administration of resveratrol reduced immobility of mice in the forced swim and tail suspension tests. The treatment also reduced serum corticosterone levels, and increased levels of BDNF and ERK phosphorylation in the prefrontal cortex and hippocampus.[172] In mice, chronic intraperitoneal administration of resveratrol attenuated lipopolysaccharide-induced depressive-like behavior, as it decreased immobility in the forced swim and tail suspension tests. This was likely due to anti-inflammatory effects of resveratrol. The antidepressant-like effects were accompanied by increased levels of BDNF and phosphorylated CREB, but also by attenuation of NF-κB activation in the prefrontal cortex and hippocampus and reductions in levels of inflammatory cytokines.[173]

Resveratrol displayed similar antidepressant-like effects in rats exposed to chronic stress in restoring preference for sucrose. In that study, the treatment increased levels of serotonin in the frontal cortex, hippocampus, and hypothalamus, and levels of noradrenaline and dopamine in the frontal cortex and striatum. It also inhibited MAO-A activity in those brain regions.[174] Resveratrol showed antidepressant-like effects in the WKY strain of rat that is genetically prone to depression-like behaviors.[175]

There are no formal studies of antidepressant effects of resveratrol in humans. However, in a study of menopausal women not diagnosed with MDD, 14 weeks of supplementation with resveratrol 75 mg twice daily improved cognitive function and mood per the Center for Epidemiological Studies Depression Scale.[176] In a similar study, menopausal women with symptoms of menopause, but not diagnosed with MDD, showed a variety of improvements after 12 weeks treatment with resveratrol at 25 mg a day along with equol, a metabolite produced by gut microbiota from the soy isoflavone daidzein. Scores on the HAM-D were significantly improved in comparison with control subjects. Subjects treated with resveratrol also had significantly better sleep per the Nottingham Health Profile.[177]

It must be noted that resveratrol has very poor bioavailability and the usefulness of oral supplementation can be compromised. However, there is evidence that addition of piperine can improve the bioavailability and potential mood-enhancing effects of resveratrol.[178]

10.36 RIPARIN

Various analogs of riparin, a double ringed alkaloid (O-methyl-N-2-hydroxi-benzoyl tyramine) from the green fruit of *Aniba riparia*, have been found to exhibit antidepressant and anxiolytic effects in mice, and these were attributed to effects on central noradrenergic, dopaminergic, and serotonergic systems.[179] Both oral and intraperitoneal administration of riparin I decreased immobility time in FST and TST.[180] The riparin analogues also exhibit significant anti-inflammatory effects.[181]

10.37   ROSMARINIC ACID

Rosmarinic acid is a polyphenolic phytochemical characterized as a caffeic acid ester of 3-(3,4dihydroxyphenyl)lactic acid. It is particularly abundant in the *Salvia* species of plants, including *S. officinalis*, *S. glutinosa*, *S. aethiopis*, *S. sclarea*, and *Borago officinalis*, and other medical plants, such as *Artemisia capillaris*, *Calendulla officinalis*, and *Melissa officinalis*. Among its pharmacological properties are potent antioxidant[182] and anti-inflammatory[183] effects.

Acute administration of rosmarinic acid decreased immobility time in mice. It was found that this effect was due to either monoamine reuptake inhibition or to inhibition of monoamine oxidase.[35] Pretreatment with rosmarinic acid also reduced the length of immobility in the tail suspension test in mice subjected to chronic stress. In doing so, it downregulated Mkp-1 and increased BDNF in cortical tissue to levels of those of unstressed mice. Treatment also reduced serum corticosterone and maintained dopamine levels in the levels in the limbic system.[184]

In a study with rats, rosmarinic acid similarly decreased immobility time in the forced swim test.[185] These antidepressant-like effects of rosmarinic acid in rats were accompanied by increases in hippocampal BDNF.[186]

Rosmarinic acid has been shown to possess significant anxiolytic-like effects. It increased the number of entries by mice in the open arms of the elevated plus maze, as well as increasing locomotion and motivation of the animals. These effects occurred without affecting the short-term or long-term memory.[187]

10.38   SAFRANAL

Safranal, a single-ringed compound with the IUPAC designation, 2,6,6-trimethyl-1,3-cyclohexadiene1-carboxaldehyde, can extracted from saffron. It in large part gives the spice its characteristic aroma. Saffron itself is used in Iran and surrounding areas as an antidepressant herb. Safranal appears to be one of several active components.

Acute intraperitoneal administration of safranal reduced immobility of mice in the forced swim test.[188] Safranal also showed anxiolytic-like effects in mice and increased sleep time.[189] Some of the antidepressant-like effect of safranal may be mediated by antioxidant and anti-inflammatory effects in the brain. The substance was found to attenuate hippocampal damage due to ischemia and reperfusion.[190]

10.39   SALIDROSIDE

Salidroside is a glucoside of the simple phenolic compound tyrosol. Salidorside and rosavin are likely the active components of *Rhodiola rosea*.[191] Sub-chronic treatment with salidroside attenuated the depression-like behaviors of mice injected with the inflammation-inducing lipopolysaccharide. Salidroside reduced levels of NF-kβ, and attenuated the lipopolysaccharide-induced decreases in serotonin and norepinephrine in the frontal cortex. Salidroside also increased expression of BNDF and TrkB in the hippocampus.[192]

In the olfactory bulbectomy-induced depression model in rats, salidroside decreased immobility time in tail suspension and forced swim tests. These behavioral effects of salidroside were accompanied by reduced levels of TNF-α and IL-1β

in the hippocampus, but increased glucocorticoid receptor and BDNF expression in this area. It also reduced serum levels of corticosterone.[193]

Five days of treatment with salidroside also attenuated the depression-like behaviors of mice injected with lipopolysaccharide. It did so in a manner comparable to fluoxetine. Salidroside reduced levels of NF-kβ, and attenuated the LPS-induced decreases in serotonin and norepinephrine in the frontal cortex. This major component of *Rhodiola* also increased expression of BNDF and TrkB in the hippocampus.[192]

10.40   SARSASAPOGENIN

Sarsasapogenin is a steroidal triterpene that was initially identified in plants of the genus *Smilax*. That genus includes various sarsaparilla plants, which lends the triterpene its name.[194] Sarsasapogenin is also found, even more abundantly, in the Asian plant, *Anemarrhena asphodeloides*. *Anemarrhena* is a plant genus in family *Asparagaceae*. It has only one species, *Anemarrhena asphodeloides*, which is native to China, Korea, and Mongolia.[195] Sarsasapogenin extracted from *Anemarrhena asphodeloides* is used as a starting point for synthesis of steroids. *Anemarrhena asphodeloides* itself has long been used in Traditional Chinese Medicine. In Chinese, the name of the herb is *zhi mu*, and it is commonly used for inflammatory conditions and to treat symptoms of menopause.[196] It is also used in the herbal combination *zhi bai di huang wan* to treat depression-like symptoms in human patients, and *suan zao ren tang* to treat restlessness and insomnia.[197]

Sarsasapogenin was shown to decrease immobility of mice in the forced swim test. Neurochemical assays further showed it to increase central noradrenergic and serotonergic levels in hippocampus and hypothalamus. This was found to be due to inhibition of MAO-A.[198] Sarsasapogenin also showed antidepressant-like effects in olfactory bulbectomized rats.[199]

Timosaponins, the natural glycosides of sarsasapogenin, also have antidepressant-like effects. The total timosaponins fraction extracted from *Anemarrhenae asphodeloides* reduced immobility of mice in the forced swim test, attenuated abnormal behaviors in the learned helplessness paradigm, and increased sucrose consumption in mice that had suffered chronic stress. Further neurochemical challenges showed enhancement of the lethality of yohimbine and the inducement of head-twitch by 5-HTP.[200] Administration of purified timosaponin B-II decreased immobility of mice in the forced swim and tail suspension tests, as well as enhancing the effects of co-administered dopamine and 5-HTP, suggesting monoaminergic mediation.[201] Timosaponin B-III, as well as several derivatives generated by activity of β-glucosidase, showed similar antidepressant-like effects in the tail suspension and forced swim tests.[202] Several metabolites of timosaponin B-III have shown rapid antidepressant-like effects in naive mice, as well as in chronically stressed animals. One such metabolite, YY-21, enhanced glutamatergic transmission in the prefrontal cortex. Treatment also rapidly reversed stress-induced decreases in p-mTOR, BDNF, and synaptic-related proteins, such as PSD-95 and GluA1. Pre-treatment with the mTOR-selective inhibitor rapamycin blocked the YY-21-induction of that synaptic enhancement.[203] The timosaponin B-III derivative YY-23 showed a similar rapid antidepressant-like effect and was found to act as non-competitive NMDA receptor antagonist, not unlike ketamine.[204]

The aqueous extract of *Anemarrhena asphodeloides*, which contains sarsasapo-genin and the timosaponins also produces anxiolytic like effects in mice, as demonstrated by behavior in the elevated plus maze and Vogel conflict drinking test.[205]

10.41 SCOPOLETIN

Scopoletin is a coumarin, a common class of phytochemicals. Coumarins are essentially flavonoids without the intact, third "B" ring. In the study of interest,[206] the substance was extracted from the Brazilian plant, *Polygala sabulosa*. However, scopoletin is also contained in the root of plants in the genus *Scopolia*, chicory, *Artemisia scoparia*, the roots and leaves of stinging nettle, passion flower, *Brunfelsia*, *Viburnum prunifolium*, and fenugreek.[207–209] Scopoletin reduced the immobility time of chronically stressed mice in the tail suspension test. However, it did not show an antidepressant effect in the forced swim test.

Scolpoletin also has hypoglycemic and hypolipidemic effects in streptozotocin-induced diabetic rats,[210] and has been found to exert MAO-A inhibitory effects.[211]

10.42 SULPHORAPHANE

Sulforaphane is an isothiocyanate that is common in cruciferous vegetables, including cabbage, kale, broccoli, cauliflower, Brussels sprouts, collard greens, bok choy, radish, and wasabi. It has been found to have antioxidative, anti-inflammatory, and neuroprotective activities.

Chronic intraperitoneal administration of sulforaphane decreased the immobility times of mice in the forced swim and tail suspension tests, as well as reduced the latency time for feeding in the novelty suppressed feeding test. Similar antidepressant-like effects were seen in mice that had been subjected to chronic unpredictable stress. Treatment also prevented the rise in serum corticosterone in the stressed mice, as well as serum levels of IL-6 and TNF-α.[212]

Sulphoraphane exerts anti-inflammatory effects, and pretreatment with the substance has been shown to prevent the depression-like behaviors induced by lipopolysaccharide. The decreases in immobility in the forced swim and tail suspensions tests were accompanied by reduction of serum levels of TNF-α, but increases in levels of the anti-inflammatory cytokine IL-10. In brain tissue, sulphoraphane blocked microglial activation, but increased concentrations of BDNF, postsynaptic density protein 95, and AMPA receptor 1. It also increased dendritic spine density.[213] Part of this antidepressant-like effect appears to be due to activation of Nrf2 by sulphoraphane.[214] Others similarly showed that in mice injected with lipopolysaccharide, sulphoraphane reduced spatial learning and memory impairment by activating mTOR, normalizing levels of BDNF, and increasing activity at TrkB receptors in the hippocampus.[215] Sulphoraphane also exerts an antidepressant-like effect in olfactory bulbectomized-mice.[216]

10.43 TETRANDRINE

Tetrandrine is a bisbenzylisoquinoline alkaloid found in the Chinese medicinal plant, *Stephania tetrandra*.[217] Administration of tetrandrine decreased immobility in the forced swim and tail suspension tests in mice. The substance also restored sucrose preference in mice that had been subjected to chronic unpredictable stress. Tetrandrine further restored hippocampal levels of serotonin and norepinephrine in stressed mice and in mice that been treated with reserpine, and maintained hippocampal levels of BDNF in stressed mice.[218] Other structurally similar bisbenzylisoquinolines have shown antidepressant-like effects. Those include liensinine and neferine from *Nelumbo nucifera*[219] and warifteine,[220] a bisbenzylisoquinoline from *Cissampelos sympodialis*.

10.44   L-THEANINE

L-theanine (γ-glutamylethylamide), is an amino acid component of green tea. It has been shown to reduce mental and physical stress, and to improve memory function in human subjects. Chronic oral administration of L-theanine reduced immobility of mice in the forced swim and tail suspension tests.[221] Others have found similar anti-depressant-like effects in mice. In one such study, the antidepressant-like effect of L-theanine in mice was accompanied by increases in hippocampal level of BDNF. Interestingly, unlike ketamine, L-theanine appeared to have NMDA agonist effects.[222]

An open label study has also shown the addition of L-theanine (250 mg/day) to enhance the effects of standard antidepressants in otherwise treatment-resistant human subjects diagnosed with MDD. No adverse effects were noted.[223]

10.45   ULIGINOSIN B

Uliginosin B is a compound extracted from various plants of the genus *Hypericum*, which includes St John's wort. It is a dimeric structure consisting of filicinic acid and phloroglucinol.[224] It is structurally similar to hyperfoliatin discussed above. It has neuroactive properties, including nociception, and appears to act through monoaminergic and glutamatergic systems.[225] In both rats and mice, oral administration of uliginosin B reduced immobility in the forced swim test. In mice, it was further found that sub-effective doses of uliginosin B combined with sub-effective doses of imipramine, bupropion, or fluoxetine to produce a full antidepressant-like effect. Curiously, although it had little affinity for monoamine reuptake sites, it was found to inhibit synaptosomal uptake of dopamine, serotonin, and noradrenaline at high nanomolar concentrations.[226]

10.46   URSOLIC ACID

Ursolic acid (UA) is a pentacyclic triterpenoid carboxylic acid found in a variety of herbs and fruits, including apples, basil, bilberries, cranberries, elder flower, peppermint, rosemary, lavender, oregano, thyme, and hawthorn.[227] At least two studies have shown that acute oral administration of ursolic acid decreases immobility of mice in the tail suspension and forced swimming test.[228,229] Anxiolytic-like effects were also noted in mice in the open field, elevated plus maze, and marble burying tests.[230]

10.47 VANILLIN

Vanillin is a phenolic aldehyde that is a primary component of the extract of the vanilla bean. Both chronic and acute oral administration of vanillin decrease immobilization of mice in the forced swim and tail suspension tests.[231]

Chronic administration of vanillin also reversed depression-like effects in rats subjected to chronic unpredictable stress. The treatment restored preference for sucrose and decreased immobilization in the forced swim test. Anxiolytic-like effects were also apparent in the reduction of time spent in the covered section of the maze. Neurochemical analyses showed increases in brain serotonin levels, as well as evidence of antioxidant effects in increases of glutathione, but decreases in NO and the oxidation product, malonyldialdehyde.[232] Interestingly, the use of vanillin in aromatherapy, a method that has been shown to introduce substances into brain tissue, was also effective in attenuating the depression-like effects of chronic unpredictable stress in mice, though not in olfactory bulbectomized-mice. In the former group, vanillin increased levels of serotonin and dopamine in the brain.[233]

REFERENCES

1. Zhang H, Tsao R. Dietary polyphenols, oxidative stress and antioxidant and anti-inflammatory effects. *Curr Opin Food Sci.* 2016;8:33–42.
2. Göçer H, Gülçin I. Caffeic acid phenethyl ester (CAPE): correlation of structure and antioxidant properties. *Int J Food Sci Nutr.* 2011;62(8):821–825.
3. Köhle C, Bock KW. Activation of coupled Ah receptor and Nrf2 gene batteries by dietary phytochemicals in relation to chemoprevention. *Biochem Pharmacol.* 2006;72(7):795–805.
4. Yang SJ, Lim Y. Resveratrol ameliorates hepatic metaflammation and inhibits NLRP3 inflammasome activation. *Metabolism.* 2014;63(5):693–701.
5. Tang B, Chen GX, Liang MY, et al. Ellagic acid prevents monocrotaline-induced pulmonary artery hypertension via inhibiting NLRP3 inflammasome activation in rats. *Int J Cardiol.* 2015;180:134–141.
6. Lee JW, Choi YJ, Park JH, et al. 3,4,5-Trihydroxycinnamic acid inhibits lipopolysaccharide-induced inflammatory response through the activation of Nrf2 pathway in BV2 microglial cells. *Biomol Ther (Seoul).* 2013;21(1):60–65.
7. Dong J, Zhang X, Zhang L, et al. Quercetin reduces obesity-associated ATM infiltration and inflammation in mice: a mechanism including AMPKα1/SIRT1. *J Lipid Res.* 2014;55(3):363–374.
8. Borra MT, Smith BC, Denu JM. Mechanism of human SIRT1 activation by resveratrol. *J Biol Chem.* 2005;280(17):17187–17195.
9. Kawakami S, Kinoshita Y, Maruki-Uchida H, et al. Piceatannol and its metabolite, isorhapontigenin, induce SIRT1 expression in THP-1 human monocytic cell line. *Nutrients.* 2014;6(11):4794–4804.
10. Chen X, Guo Y, Jia G, et al. Ferulic acid regulates muscle fiber type formation through the Sirt1/AMPK signaling pathway. *Food Funct.* 2019;10(1):259–265.
11. Parmar SK, Sharma TP, Airao VB. Neuropharmacological effects of triterpenoids. *Phytopharmacology.* 2013;4(2):354–372.
12. Udintsev SN, Krylova SG, Konovalova ON. Correction by natural adaptogens of hormonal-metabolic status disorders in rats during the development of adaptation syndrome using functional tests with dexamethasone and ACTH. *Biull Eksp Biol Med.* 1991;112(12):599–601.

13. Wagner H, Nörr H, Winterhoff H. Plant adaptogens. *Phytomedicine.* 1994;1(1): 63–76.
14. da Silva KAB, Paszcuk AF, Passos GF, et al. Activation of cannabinoid receptors by the pentacyclic triterpene α,β-amyrin inhibits inflammatory and neuropathic persistent pain in mice. *Pain.* 2011;152(8):1872–1887.
15. Aragão GF, Carneiro LMV, Junior APF, et al. A possible mechanism for anxiolytic and antidepressant effects of alpha- and beta-amyrin from Protium heptaphyllum (Aubl.) March. *Pharmacol Biochem Behav.* 2006;85(4):827–834.
16. Aragão GF, Carneiro LMV, Junior APF, et al. Evidence for excitatory and inhibitory amino acids participation in the neuropharmacological activity of alpha- and beta-amyrin acetate. *Open Pharmacol J.* 2009;3:9–16.
17. Subarnas A, Tadano T, Nakahata N, et al. A possible mechanism of antidepressant activity of beta-amyrin palmitate isolated from Lobelia inflata leaves in the forced swimming test. *Life Sci.* 1993;52(3):289–296.
18. Wankhede SS, Gambhire M, Juvekar A. Couroupita guianensis Abul: evaluation of its antidepressant activity in mice. *Pharmacologyonline.* 2009;2:999–1013.
19. Jain PK, Das D, Jain P, et al. Pharmacognostic and pharmacological aspect of Bacopa monnieri: a review. *Innov J Ayurvedic Sci.* 2016;4(3):7–11.
20. Liu X, Liu F, Yue R, et al. The antidepressant-like effect of bacopaside I: possible involvement of the oxidative stress system and the noradrenergic system. *Pharmacol Biochem Behav.* 2013;110:224–230.
21. Zu X, Zhang M, Li W, et al. Antidepressant-like effect of Bacopaside I in mice exposed to chronic unpredictable mild stress by modulating the hypothalamic-pituitary-adrenal axis function and activating BDNF signaling pathway. *Neurochem Res.* 2017;42(11):3233–3244.
22. Zhou Y, Shen YH, Zhang C, et al. Triterpene saponins from Bacopa monnieri and their antidepressant effects in two mice models. *J Nat Prod.* 2007;70(4):652–655.
23. Rauf K, Subhan F, Abbas M, et al. Inhibitory effect of Bacopasides on spontaneous morphine withdrawal induced depression in mice. *Phytother Res.* 2014;28(6): 937–939.
24. Yin J, Xing H, Ye J. Efficacy of berberine in patients with type 2 diabetes mellitus. *Metabolism.* 2008;57(5):712–717.
25. Peng WH, Lo KL, Lee YH, et al. Berberine produces antidepressant-like effects in the forced swim test and in the tail suspension test in mice. *Life Sci.* 2007;81(11):933–938.
26. Kulkarni SK, Dhir A. On the mechanism of antidepressant-like action of berberine chloride. *Eur J Pharmacol.* 2008;589(1–3):163–172.
27. Sun S, Wang K, Lei H, et al. Inhibition of organic cation transporter 2 and 3 may be involved in the mechanism of the antidepressant-like action of berberine. *Prog Neuropsychopharmacol Biol Psychiatry.* 2014;49:1–6.
28. Liu YM, Niu L, Wang LL, et al. Berberine attenuates depressive-like behaviors by suppressing neuro-inflammation in stressed mice. *Brain Res Bull.* 2017;134:220–227.
29. Shen JD, Ma LG, Hu CY, et al. Berberine up-regulates the BDNF expression in hippocampus and attenuates corticosterone-induced depressive-like behavior in mice. *Neurosci Lett.* 2016;614:77–82.
30. Fan J, Li B, Ge T, et al. Berberine produces antidepressant-like effects in ovariectomized mice. *Sci Rep.* 2017;7(1):1310.
31. Yang M, Dang R, Xu P, et al. Dl-3-n-Butylphthalide improves lipopolysaccharide-induced depressive-like behavior in rats: involvement of Nrf2 and NF-κB pathways. *Psychopharmacology.* 2018;235(9):2573–2585.
32. Xiang J, Pan J, Chen F, et al. L-3-n-butylphthalide improves cognitive impairment of APP/PS1 mice by BDNF/TrkB/PI3K/AKT pathway. *Int J Clin Exp Med.* 2014;7(7):1706–1713.

33. Jia J-P, Wei C, Liang J, et al. The effects of DL-3-n-butylphthalide in patients with vascular cognitive impairment without dementia caused by subcortical ischemic small vessel disease: a multicentre, randomized, double-blind, placebo-controlled trial. *Alzheimer's Dement.* 2016;12(2):89–99.

34. Boerjan W, Ralph J, Baucher M. Lignin biosynthesis. *Annu Rev Plant Biol.* 2003;54:519–546.

35. Takeda H, Tsuji M, Inazu M, et al. Rosmarinic acid and caffeic acid produce anti-depressive-like effect in the forced swimming test in mice. *Eur J Pharmacol.* 2002;449(3):261–267.

36. Takeda H, Tsuji M, Yamada T, et al. Caffeic acid attenuates the decrease in cortical BDNF mRNA expression induced by exposure to forced swimming stress in mice. *Eur J Pharmacol.* 2006;534(1–3):115–121.

37. Basu Mallik S, Mudgal J, Nampoothiri M, et al. Caffeic acid attenuates lipopolysac-charide-induced sickness behaviour and neuroinflammation in mice. *Neurosci Lett.* 2016;632(6):218–223.

38. Chang WC, Kuo PL, Chen CW, et al. Caffeic acid improves memory impairment and brain glucose metabolism via ameliorating cerebral insulin and leptin signaling pathways in high-fat diet-induced hyperinsulinemic rats. *Food Res Int.* 2015;77:24–33.

39. Payne ME, Steck SE, George RR, et al. Fruit, vegetable, and antioxidant intakes are lower in older adults with depression. *J Acad Nutr Diet.* 2012;112(12):2022–2027.

40. Dhingra D, Bansal Y. Antidepressant-like activity of β-carotene in unstressed and chronic unpredictable mild stressed mice. *J Funct Foods.* 2014;7:425–434.

41. Kim NR, Kim HY, Kim MH, et al. Improvement of depressive behavior by Sweetme Sweet Pumpkin™ and its active compound, β-carotene. *Life Sci.* 2016;147:39–45.

42. Melo FHC, Moura BA, de Sousa DP, et al. Antidepressant-like effect of carvacrol (5-isopropyl-2-methylphenol) in mice: involvement of dopaminergic system. *Fundam Clin Pharmacol.* 2011;25(3):362–367.

43. Zotti M, Colaianna M, Morgese MG, et al. Carvacrol: from ancient flavoring to neuro-modulatory agent. *Molecules.* 2013;18(6):6161–6172.

44. Melo FHC, Venâncio ET, de Sousa DP, et al. Anxiolytic-like effect of Carvacrol (5-iso-propyl-2-methylphenol) in mice: involvement with GABAergic transmission. *Fundam Clin Pharmacol.* 2010;24(4):437–443.

45. Gertsch J, Leonti M, Raduner S, et al. Beta-caryophyllene is a dietary cannabinoid. *Proc Natl Acad Sci USA.* 2008;105(26):9099–9104.

46. Bahi A, Al Mansouri S, Al Memari E, et al. β-Caryophyllene, a CB2 receptor agonist produces multiple behavioral changes relevant to anxiety and depression in mice. *Physiol Behav.* 2014;135:119–124.

47. de Oliveira DR, da Silva DM, Florentino IF, et al. Monoamine involvement in the antidepressant-like effect of β-caryophyllene. *CNS Neurol Disord Drug Targets.* 2018;17(4):309–320.

48. Park S-H, Sim Y-B, Han P-Y, et al. Antidepressant-like effect of chlorogenic acid isolated from *Artemisia capillaris* Thunb. *Anim Cells Syst.* 2010;14(4):253–259.

49. Lim DW, Han T, Jung J, et al. Chlorogenic acid from hawthorn berry (Crataegus pinnatifida Fruit) prevents stress hormone-induced depressive behavior, through mono-amine oxidase B-reactive oxygen species signaling in hippocampal astrocytes of mice. *Mol Nutr Food Res.* 2018;62(15):e1800029.

50. Bouayed J, Rammal H, Dicko A et al. Chlorogenic acid, a polyphenol from Prunus domestica (Mirabelle), with coupled anxiolytic and antioxidant effects. *J Neurol Sci.* 2007;262(1–2):77–84.

51. Escribano J, Alonso GL, Coca-Prados M, et al. Crocin, safranal and picrocrocin from saffron (Crocus sativus L.) inhibit the growth of human cancer cells in vitro. *Cancer Lett.* 1996;100(1–2):23–30.

52. Hosseinzadeh H, Karimi G, Niapoor M. Antidepressant effects of Crocus sativus stigma extracts and its constituents, crocin and safranal, in mice. *Acta Hort (ISHS)*. 2004;650:435–445.

53. Amin B, Nakhsaz A, Hosseinzadeh H. Evaluation of the antidepressant-like effects of acute and sub-acute administration of crocin and crocetin in mice. *Avicenna J Phytomed*. 2015;5(5):458–468.

54. Vahdati Hassani F, Naseri V, Razavi BM, et al. Antidepressant effects of crocin and its effects on transcript and protein levels of CREB, BDNF, and VGF in rat hippocampus. *DARU J Pharm Sci*. 2014;22(1):16.

55. Talaei A, Hassanpour Moghadam M, Sajadi Tabassi SA, et al. Crocin, the main active saffron constituent, as an adjunctive treatment in major depressive disorder: a randomized, double-blind, placebo-controlled, pilot clinical trial. *J Affect Disord*. 2015;174:51–56.

56. Huang Z, Zhong XM, Li ZY, et al. Curcumin reverses corticosterone-induced depressive-like behavior and decrease in brain BDNF levels in rats. *Neurosci Lett*. 2011;493(3):145–148.

57. Kulkarni SK, Bhutani MK, Bishnoi M. Antidepressant activity of curcumin: involvement of serotonin and dopamine system. *Psychopharmacology*. 2008;201(3):435–442.

58. Lopresti AL, Maes M, Maker GL, et al. Curcumin for the treatment of major depression: a randomized, double-blind, placebo-controlled study. *J Affect Dis*. 2014;167:368–375.

59. Yu JJ, Pei LB, Zhang Y, et al. Chronic supplementation of curcumin enhances the efficacy of antidepressants in major depressive disorder: a randomized, double-blind, placebo-controlled pilot study. *J Clin Psychopharmacol*. 2015;35(4):406–410.

60. Hu Y, Liu M, Liu P, et al. Possible mechanism of the antidepressant effect of 3,6′-disinapoyl sucrose from Polygala tenuifolia Willd. *J Pharm Pharmacol*. 2011;63(6):869–874.

61. Hu Y, Liao HB, Dai-Hong G, et al. Antidepressant-like effects of 3,6′-disinapoyl sucrose on hippocampal neuronal plasticity and neurotrophic signal pathway in chronically mild stressed rats. *Neurochem Int*. 2010;56(3):461–465.

62. bookZhenguo S, Hong Y, Yuan H. Effect of 3,6′-disinapoyl sucrose on proliferation of neural progenitor cells from neonatal rat hippocampus. *China Pharm*. 2009;21.

63. Hu Y, Liu MY, Liu P, et al. Neuroprotective effects of 3,6′-disinapoyl sucrose through increased BDNF levels and CREB phosphorylation via the CaMKII and ERK1/2 pathway. *J Mol Neurosci*. 2014;53(4):600–607.

64. Ascacio-Valdés JA, Aguilera-Carbó A, Martínez-Hernández JL, et al. *Euphorbia antisyphilitica* residues as a new source of ellagic acid. *Chem Pap*. 2010;64(4):528–532.

65. Ahmed T, Setzer WN, Nabavi SF, et al. Insights into effects of ellagic acid on the nervous system: a mini review. *Curr Pharm Des*. 2016;22(10):1350–1360.

66. Mansouri MT, Hemmati AA, Naghizadeh B, et al. A study of the mechanisms underlying the anti-inflammatory effect of ellagic acid in carrageenan-induced paw edema in rats. *Indian J Pharmacol*. 2015;47(3):292–298.

67. Girish C, Raj V, Arya J, Balakrishnan S. Evidence for the involvement of the monoaminergic system, but not the opioid system in the antidepressant-like activity of ellagic acid in mice. *Eur J Pharmacol*. 2012;682(1–3):118–125.

68. Dhingra D, Chhillar R. Antidepressant-like activity of ellagic acid in unstressed and acute immobilization-induced stressed mice. *Pharmacol Rep*. 2012;64(4):796–807.

69. Girish C, Raj V, Arya J, et al. Involvement of the GABAergic system in the anxiolytic-like effect of the flavonoid ellagic acid in mice. *Eur J Pharmacol*. 2013;710(1–3):49–58.

70. Christaki E, Bonos E, Giannenas I, et al. Aromatic plants as a source of bioactive compounds. *Agriculture*. 2012;2(3):228–243.

71. Park CK, Kim K, Jung SJ, et al. Molecular mechanism for local anesthetic action of eugenol in the rat trigeminal system. *Pain*. 2009;144(1–2):84–94.

72. Irie Y, Itokazu N, Anjiki N, et al. Eugenol exhibits antidepressant-like activity in mice and induces expression of metallothionein-III in the hippocampus. *Brain Res.* 2004;1011(2):243–246.
73. Erickson JC, Hollopeter G, Thomas SA, et al. Disruption of the metallothionein-III gene in mice: analysis of brain zinc, behavior, and neuron vulnerability to metals, aging, and seizures. *J Neurosci.* 1997;17(4):1271–1281.
74. Nowak G, Kubera M, Maes M. Neuroimmunological aspects of the alterations in zinc homeostasis in the pathophysiology and treatment of depression. *Acta Neuropsychiatr.* 2000;12(2):49–53.
75. Norte MCB, Cosentino RM, Lazarini CA. Effects of methyl-eugenol administration on behavioral models related to depression and anxiety in rats. *Phytomedicine.* 2005;12(4):294–298.
76. Fajemiroye JO, Galdino PM, De Paula JAM, et al. Anxiolytic and antidepressant like effects of natural food flavour (E)-methyl isoeugenol. *Food Funct.* 2014;5(8):1819–1828.
77. Do Amaral JF, Silva MI, de Aquino Neto MR, et al. Antidepressant-like effect of bis-eugenol in the mice forced swimming test: evidence for the involvement of the monoaminergic system. *Fundam Clin Pharmacol.* 2013;27(5):471–482.
78. Tao G, Irie Y, Li DJ, et al. Eugenol and its structural analogs inhibit monoamine oxidase A and exhibit antidepressant-like activity. *Bioorg Med Chem.* 2005;13(15):4777–4788.
79. Kumar N, Pruthi V. Potential applications of ferulic acid from natural sources. *Biotechnol Rep (Amst).* 2014;4:86–93.
80. Cheng CY, Ho TY, Lee EJ, et al. Ferulic acid reduces cerebral infarct through its antioxidative and anti-inflammatory effects following transient focal cerebral ischemia in rats. *Am J Chin Med.* 2008;36(6):1105–1119.
81. Yu L, Zhang Y, Liao M, et al. Neurogenesis-enhancing effect of sodium ferulate and its role in repair following stress-induced neuronal damage. *World J Neurosci.* 2011;1:9–18.
82. Zeni ALB, Zomkowski ADE, Maraschin M, et al. Involvement of PKA, CaMKII, PKC, MAPK/ERK and PI3K in the acute antidepressant-like effect of ferulic acid in the tail suspension test. *Pharmacol Biochem Behav.* 2012;103(2):181–186.
83. Chen J, Lin D, Zhang C, et al. Antidepressant-like effects of ferulic acid: involvement of serotonergic and noradrenergic systems. *Metab Brain Dis.* 2015;30(1):129–136.
84. Zeni ALB, Zomkowski ADE, Maraschin M, et al. Ferulic acid exerts antidepressant-like effect in the tail suspension test in mice: evidence for the involvement of the serotonergic system. *Eur J Pharmacol.* 2012;679(1–3):68–74.
85. Yabe T, Hirahara H, Harada N, et al. Ferulic acid induces neural progenitor cell proliferation in vitro and in vivo. *Neuroscience.* 2010;165(2):515–524.
86. Xu Y, Zhang L, Shao T, et al. Ferulic acid increases pain threshold and ameliorates depression-like behaviors in reserpine-treated mice: behavioral and neurobiological analyses. *Metab Brain Dis.* 2013;28(4):571–583.
87. Nayeem N, Asdaq SMB, Salem H, et al. Gallic acid: a promising lead molecule for drug development. *J Appl Pharm.* 2016;8:213–216.
88. Nagpal K, Singh SK, Mishra DN. Nanoparticle mediated brain targeted delivery of gallic acid: in vivo behavioral and biochemical studies for improved antioxidant and antidepressant-like activity. *Drug Deliv.* 2012;19(8):378–391.
89. Can ÖD, Turan N, Demir Özkay Ü, et al. Antidepressant-like effect of gallic acid in mice: dual involvement of serotonergic and catecholaminergic systems. *Life Sci.* 2017;190:110–117.
90. Chhillar R, Dhingra D. Antidepressant-like activity of gallic acid in mice subjected to unpredictable chronic mild stress. *Fundam Clin Pharmacol.* 2013;27(4):409–418.
91. Nabavi SF, Habtemariam S, Di Lorenzo A, et al. Post-stroke depression modulation and in vivo antioxidant activity of gallic acid and its synthetic derivatives in a murine model system. *Nutrients.* 2016;8(5):248.

92. Singh P, Rahul MK, Thawani V, et al. Anxiolytic effect of chronic administration of gallic acid in rats. *J Appl Pharm Sci.* 2013;3(7):101–104.

93. Mansouri MT, Soltani M, Naghizadeh B, et al. A possible mechanism for the anxiolytic-like effect of gallic acid in the rat elevated plus maze. *Pharmacol Biochem Behav.* 2014;117:40–46.

94. Hayashi J, Sekine T, Deguchi S, et al. Phenolic compounds from Gastrodia rhizome and relaxant effects of related compounds on isolated smooth muscle preparation. *Phytochemistry.* 2002;59(5):513–519.

95. Bensky D, Barolet R. *Chinese Herbal Medicine Formulas and Strategies.* Seattle: Eastland Press; 1990. p. 424.

96. Chen WC, Lai YS, Lin SH, et al. Anti-depressant effects of Gastrodia elata Blume and its compounds gastrodin and 4-hydroxybenzyl alcohol, via the monoaminergic system and neuronal cytoskeletal remodeling. *J Ethnopharmacol.* 2016;182:190–199.

97. Wang H, Zhang R, Qiao Y, et al. Gastrodin ameliorates depression-like behaviors and up-regulates proliferation of hippocampal-derived neural stem cells in rats: involvement of its anti-inflammatory action. *Behav Brain Res.* 2014;266:153–160.

98. Zhang R, Peng Z, Wang H, et al. Gastrodin ameliorates depressive-like behaviors and up-regulates the expression of BDNF in the hippocampus and hippocampal-derived astrocyte of rats. *Neurochem Res.* 2014;39(1):172–179.

99. Fatemi SH, Laurence JA, Araghi-Niknam M, et al. Glial fibrillary acidic protein is reduced in cerebellum of subjects with major depression, but not schizophrenia. *Schizophr Res.* 2004;69(2–3):317–323.

100. Lee B, Sur B, Yeom M, et al. Gastrodin reversed the traumatic stress-induced depressed-like symptoms in rats. *J Nat Med.* 2016;70(4):749–759.

101. Peng Z, Wang H, Zhang R, et al. Gastrodin ameliorates anxiety-like behaviors and inhibits IL-1β level and p38 MAPK phosphorylation of hippocampus in the rat model of posttraumatic stress disorder. *Physiol Res.* 2013;62(5):537–545.

102. Muzzarelli RAA. Genipin-crosslinked chitosan hydrogels as biomedical and pharmaceutical aids. *Carbohydr Polym.* 2009;77(1):1–9.

103. Tian JS, Cui YL, Hu LM, et al. Antidepressant-like effect of genipin in mice. *Neurosci Lett.* 2010;479(3):236–239.

104. Mendelson SD, McEwen BS. Autoradiographic analyses of the effects of adrenalectomy and corticosterone on 5-HT1A and 5-HT1B receptors in the dorsal hippocampus and cortex of the rat. *Neuroendocrinology.* 1992;55(4):444–450.

105. Wang QS, Tian JS, Cui YL, et al. Genipin is active via modulating monoaminergic transmission and levels of brain-derived neurotrophic factor (BDNF) in rat model of depression. *Neuroscience.* 2014;275:365–373.

106. Ye D, Zhang L, Fan W, et al. Genipin normalizes depression-like behavior induced by prenatal stress through inhibiting DNMT1. *Epigenetics.* 2018;13(3):310–317.

107. Jiang B, Xiong Z, Yang J, et al. Antidepressant-like effects of ginsenoside Rg1 are due to activation of the BDNF signaling pathway and neurogenesis in the hippocampus. *Br J Pharmacol.* 2012;166(6):1872–1887.

108. Liu Z, Qi Y, Cheng Z, Zhu X, et al. The effects of ginsenoside Rg1 on chronic stress induced depression-like behaviors, BDNF expression and the phosphorylation of PKA and CREB in rats. *Neuroscience.* 2016;322:358–369.

109. Xu C, Teng J, Chen W, et al. 20(S)-protopanaxadiol, an active ginseng metabolite, exhibits strong antidepressant-like effects in animal tests. *Prog Neuropsychopharmacol Biol Psychiatry.* 2010;34(8):1402–1411.

110. Dhingra D, Sharma A. Evaluation of antidepressant-like activity of glycyrrhizin in mice. *Indian J Pharmacol.* 2005;37(6):390–394.

111. Wang B, Lian YJ, Dong X, et al. Glycyrrhizic acid ameliorates the kynurenine pathway in association with its antidepressant effect. *Behav Brain Res.* 2018;353(1):250–257.

112. Wu T-Y, Liu L, Zhang W, et al. High-mobility group box-1 was released actively and involved in LPS induced depressive-like behavior. *J Psychiatr Res.* 2015;64: 99–106.

113. Gaur V, Bodhankar SL, Mohan V, et al. Antidepressant-like effect of 4-hydroxyisoleucine from Trigonella foenum graecum L. Seeds in mice. *Biomed Aging Pathol.* 2012;2(3):121–125.

114. Kalshetti PB, Alluri R, Mohan V, et al. Effects of 4-hydroxyisoleucine from fenugreek seeds on depression-like behavior in socially isolated olfactory bulbectomized rats. *Pharmacogn Mag.* 2015;11(Suppl 3):S388–S396.

115. Broca C, Breil V, Cruciani-Guglielmacci C, et al. Insulinotropic agent ID-1101 (4hydroxyisoleucine) activates insulin signaling in rat. *Am J Physiol Endocrinol Metab.* 2004;287(3):E463–E471.

116. Do Rego JC, Benkiki N, Chosson E, et al. Antidepressant-like effect of hyperfoliatin, a polyisoprenylated phloroglucinol derivative from Hypericum perfoliatum (Clusiaceae) is associated with an inhibition of neuronal monoamines uptake. *Eur J Pharmacol.* 2007;569(3):197–203.

117. Ccana-Ccapatinta GV, Stolz ED, da Costa PF, et al. Acylphloroglucinol derivatives from Hypericum andinum: antidepressant-like activity of andinin A. *J Nat Prod.* 2014;77(10):2321–2325.

118. Socolsky C, Rates SMK, Stein AC, et al. Acylphloroglucinols from Elaphoglossum crassipes: antidepressant-like activity of crassipin A. *J Nat Prod.* 2012;75(6):1007–1017.

119. Peana AT, Moretti MDL. Linalool in Essential Plant Oils: Pharmacological Effects. In: Watson RR, Preedy VR, editors. *Botanical Medicine in Clinical Practice.* Wallingford, UK: CABI; 2008. p. 716–724.

120. Saiyudthong S, Srijittapong D, Mekseepralard C. Subchronic administration of linalool decreases depressive-like behaviour in restrained rats. *J Pharm Pharmacol.* 2017;7:401–407.

121. Guzmán-Gutiérrez SL, Bonilla-Jaime H, Gómez-Cansino R, et al. Linalool and β-pinene exert their antidepressant-like activity through the monoaminergic pathway. *Life Sci.* 2015;128:24–29.

122. Cheng BH, Sheen LY, Chang ST. Evaluation of anxiolytic potency of essential oil and S-(+)-linalool from Cinnamomum osmophloeum ct. linalool leaves in mice. *J Tradit Complement Med.* 2015;5(1):27–34.

123. Linck VM, da Silva AL, Figueiró M, et al. Effects of inhaled linalool in anxiety, social interaction and aggressive behavior in mice. *Phytomedicine.* 2010;17(8–9):679–683.

124. Xu P, Wang K, Lu C, et al. Protective effects of linalool against amyloid beta-induced cognitive deficits and damages in mice. *Life Sci.* 2017;174:21–27.

125. Li J, Geng D, Xu J, et al. Antidepressant-like effect of macranthol isolated from Illicium dunnianum tutch in mice. *Eur J Pharmacol.* 2013;707(1–3):112–119.

126. Luo L, Liu XL, Li J, et al. Macranthol promotes hippocampal neuronal proliferation in mice via BDNF-TrkB-PI3K/Akt signaling pathway. *Eur J Pharmacol.* 2015;762:357–363.

127. Weng L, Dong S, Wang S, et al. Macranthol attenuates lipopolysaccharide-induced depressive-like behaviors by inhibiting neuroinflammation in prefrontal cortex. *Physiol Behav.* 2019;204:33–40.

128. Aluko OM, Umukoro S, Annafi OS, et al. Effects of methyl jasmonate on acute stress responses in mice subjected to forced swim and anoxic tests. *Sci Pharm.* 2015;83(4):635–644.

129. Umukoro S, Akinyinka AO, Aladeokin AC. Antidepressant activity of methyl jasmonate, a plant stress hormone in mice. *Pharmacol Biochem Behav.* 2011;98(1):8–11.

130. Adebesin A, Adeoluwa OA, Eduviere AT, et al. Methyl jasmonate attenuated lipopolysaccharide-induced depressive-like behaviour in mice. *J Psychiatr Res.* 2017;94:29–35.

131. Umukoro S, Aluko OM, Eduviere AT, et al. Evaluation of adaptogenic-like property of methyl jasmonate in mice exposed to unpredictable chronic mild stress. *Brain Res Bull.* 2016;121:105–114.

132. Takayama H, Ishikawa H, Kurihara M, et al. Studies on the synthesis and opioid agonistic activities of mitragynine-related indole alkaloids: discovery of opioid agonists structurally different from other opioid ligands. *J Med Chem.* 2002;45(9): 1949–1956.

133. Warner ML, Kaufman NC, Grundmann O. The pharmacology and toxicology of kratom: from traditional herb to drug of abuse. *Int J Legal Med.* 2016;130(1):127–138.

134. Idayu NF, Hidayat MT, Moklas MA, et al. Antidepressant-like effect of mitragynine isolated from Mitragyna speciosa Korth in mice model of depression. *Phytomedicine.* 2011;18(5):402–407.

135. Hazim AI, Ramanathan S, Parthasarathy S, et al. Anxiolytic-like effects of mitragynine in the open-field and elevated plus-maze tests in rats. *J Physiol Sci.* 2014;64(3):161–169.

136. Matsumoto K, Mizowaki M, Takayama H, et al. Suppressive effect of mitragynine on the 5-methoxy-N,N-dimethyltryptamine-induced head-twitch response in mice. *Pharmacol Biochem Behav.* 1997;57(1–2):319–323.

137. Kowalski R. Studies of selected plant raw materials as alternative sources of triterpenes of oleanolic and ursolic acid types. *J Agric Food Chem.* 2007;55(3):656–662.

138. Fajemiroye JO, Galdino PM, Florentino IF, et al. Plurality of anxiety and depression alteration mechanism by oleanolic acid. *J Psychopharmacol.* 2014;28(10):923–934.

139. Yi LT, Li J, Liu Q, et al. Antidepressant-like effect of oleanolic acid in mice exposed to the repeated forced swimming test. *J Psychopharmacol.* 2013;27(5):459–468.

140. Yi LT, Li J, Liu BB, et al. BDNF-ERK-CREB signaling mediates the role of miR-132 in the regulation of the effects of oleanolic acid in male mice. *J Psychiatry Neurosci.* 2014;39(5):348–359.

141. Elix JA, Wardlaw JH. Lusitanic acid, peristictic acid and verrucigeric acid. Three new β-orcinol depsidones from the lichens Relicina sydneyensis and Xanthoparmelia verrucigera. *Aust J Chem.* 2000;53(9):815–818.

142. Ramchandani D, Ganeshpurkar A, Bansal D, et al. Protective effect of Curculigo Orchioides extract on cyclophosphamide-induced neurotoxicity in murine model. *Toxicol Int.* 2014;21(3):232–235.

143. Ge JF, Gao WC, Cheng WM, et al. Orcinol glucoside produces antidepressant effects by blocking the behavioural and neuronal deficits caused by chronic stress. *Eur Neuropsychopharmacol.* 2014;24(1):172–180.

144. Jiang WL, Chen XG, Zhu HB, et al. Paeoniflorin inhibits systemic inflammation and improves survival in experimental sepsis. *Basic Clin Pharmacol Toxicol.* 2009;105(1):64–71.

145. Qiu FM, Zhong XM, Mao QQ, et al. The antidepressant-like effects of paeoniflorin in mouse models. *Exp Ther Med.* 2013;5(4):1113–1116.

146. Cui G-Z. Effect of paeoniflorin on reserpine-induced depression model in mice. *Chin J Exp Tradit Formulae.* 2012;2012:22.

147. Li J, Huang S, Huang W, et al. Paeoniflorin ameliorates interferon-alpha-induced neuroinflammation and depressive-like behaviors in mice. *Oncotarget.* 2017;8(5): 8264–8282.

148. Qiu FM, Zhong XM, Mao QQ, et al. Antidepressant-like effects of paeoniflorin on the behavioural, biochemical, and neurochemical patterns of rats exposed to chronic unpredictable stress. *Neurosci Lett.* 2013;541:209–213.

149. Li W, Wang J, Zhu Y, et al. The antidepressant-like effects of paeoniflorin in rats model exposed to chronic restraint stress. *Planta Med.* 2015;81(11):PV2.

150. Reierson GW, Guo S, Mastronardi C, et al. CGMP signaling, phosphodiesterases and major depressive disorder. *Curr Neuropharmacol.* 2011;9(4):715–727.

151. Huang H, Zhao J, Jiang L, et al. Paeoniflorin improves menopause depression in ovariectomized rats under chronic unpredictable mild stress. *Int J Clin Exp Med.* 2015;8(4):5103–5111.

152. Mao QQ, Zhong XM, Qiu FM, et al. Protective effects of paeoniflorin against corticosteroneinduced neurotoxicity in PC12 cells. *Phytother Res.* 2012;26(7):969–973.

153. Mao QQ, Zhong XM, Li ZY, et al. Paeoniflorin protects against NMDA-induced neurotoxicity in PC12 cells via Ca2+ antagonism. *Phytother Res.* 2011;25(5): 681–685.

154. Zhu W, Ma S, Qu R, et al. Antidepressant-like effect of paeonol. *Pharm Biol.* 2006;44(3):229–235.

155. Zhu XL, Chen JJ, Han F, et al. Novel antidepressant effects of paeonol alleviate neuronal injury with concomitant alterations in BDNF, Rac1 and RhoA levels in chronic unpredictable mild stress rats. *Psychopharmacology.* 2018;235(7):2177–2191.

156. Tao W, Wang H, Su Q, et al. Paeonol attenuates lipopolysaccharide-induced depressive-like behavior in mice. *Psychiatry Res.* 2016;238:116–121.

157. Wang YM, Zhao LB, Lin SL, et al. Determination of berberine and palmatine in cortex Phellodendron and Chinese patent medicines by HPLC. *Acta Pharm Sin.* 1989;24(4):275–279.

158. Lee SS, Kim YH, Lee MK. Inhibition of monoamine oxidase by palmatine. *Arch Pharm Res.* 1999;22(5):529–531.

159. Dhingra D, Kumar V. Memory-enhancing activity of palmatine in mice using elevated plus maze and morris water maze. *Adv Pharmacol Sci.* 2012;2012:Article ID 357368.

160. Dhingra D, Bhankher A. Behavioral and biochemical evidences for antidepressant-like activity of palmatine in mice subjected to chronic unpredictable mild stress. *Pharmacol Rep.* 2014;66(1):1–9.

161. Shen Y, Guan S, Ge H, et al. Effects of palmatine on rats with comorbidity of diabetic neuropathic pain and depression. *Brain Res Bull.* 2018;139:56–66.

162. Luo P, Wong YF, Ge L, et al. Anti-inflammatory and analgesic effect of plumbagin through inhibition of nuclear factor-κB activation. *J Pharmacol Exp Ther.* 2010;335(3):735–742.

163. Pai SA, Munshi RP, Panchal FH, et al. Plumbagin reduces obesity and nonalcoholic fatty liver disease induced by fructose in rats through regulation of lipid metabolism, inflammation and oxidative stress. *Biomed Pharmacother.* 2019;111:686–694.

164. Dhingra D, Bansal S. Antidepressant-like activity of plumbagin in unstressed and stressed mice. *Pharmacol Rep.* 2015;67(5):1024–1032.

165. Gonçalves AE, Bürger C, Amoah SKS, et al. The antidepressant-like effect of Hedyosmum brasiliense and its sesquiterpene lactone, podoandin in mice: evidence for the involvement of adrenergic, dopaminergic and serotonergic systems. *Eur J Pharmacol.* 2012;674(2–3):307–314.

166. Ghantous A, Gali-Muhtasib H, Vuorela H, et al. What made sesquiterpene lactones reach cancer clinical trials? *Drug Discov Today.* 2010;15(15–16):668–678.

167. Amoah SKS, Dalla Vecchia MTD, Pedrini B, et al. Inhibitory effect of sesquiterpene lactones and the sesquiterpene alcohol aromadendrane-4β,10α-diol on memory impairment in a mouse model of Alzheimer. *Eur J Pharmacol.* 2015;769:195–202.

168. Amorim MHR, Gil da Costa RMG, Lopes C, et al. Sesquiterpene lactones: adverse health effects and toxicity mechanisms. *Crit Rev Toxicol.* 2013;43(7):559–579.

169. Dhingra D, Valecha R. Evidence for involvement of the monoaminergic system in antidepressant-like activity of an ethanol extract of Boerhaavia diffusa and its isolated constituent, punarnavine, in mice. *Pharm Biol.* 2014;52(6):767–774.

170. Dhingra D, Valecha R. Punarnavine, an alkaloid isolated from ethanolic extract of Boerhaavia diffusa Linn. reverses depression-like behaviour in mice subjected to chronic unpredictable mild stress. *Indian J Exp Biol.* 2014;52(8):799–807.

171. Smoliga JM, Baur JA, Hausenblas HA. Resveratrol and health—a comprehensive review of human clinical trials. *Mol Nutr Food Res.* 2011;55(8):1129–1141.

172. Wang Z, Gu J, Wang X, et al. Antidepressant-like activity of resveratrol treatment in the forced swim test and tail suspension test in mice: the HPA axis, BDNF expression and phosphorylation of ERK. *Pharmacol Biochem Behav.* 2013;112:104–110.

173. Ge L, Liu L, Liu H, et al. Resveratrol abrogates lipopolysaccharide-induced depressive-like behavior, neuroinflammatory response, and CREB/BDNF signaling in mice. *Eur J Pharmacol.* 2015;768:49–57.

174. Yu Y, Wang R, Chen C, et al. Antidepressant-like effect of trans-resveratrol in chronic stress model: behavioral and neurochemical evidences. *J Psychiatr Res.* 2013;47(3):315–322.

175. Hurley LL, Akinfiresoye L, Kalejaiye O, et al. Antidepressant effects of resveratrol in an animal model of depression. *Behav Brain Res.* 2014;268:1–7.

176. Evans HM, Howe PR, Wong RH. Clinical evaluation of effects of chronic resveratrol supplementation on cerebrovascular function, cognition, mood, physical function and general well-being in postmenopausal women—rationale and study design. *Nutrients.* 2016;8(3):150.

177. Davinelli S, Scapagnini G, Marzatico F, et al. Influence of equol and resveratrol supplementation on health-related quality of life in menopausal women: a randomized, placebo-controlled study. *Maturitas.* 2017;96:77–83.

178. Xu Y, Zhang C, Wu F-Y, et al. Piperine potentiates the effects of trans-resveratrol on stress-induced depressive-like behavior: involvement of monoaminergic system and cAMP-dependent pathway. *Metab Brain Dis.* 2016;31(4):837–848.

179. Melo CTV, de Carvalho AMR, Moura BA, et al. Evidence for the involvement of the serotonergic, noradrenergic, and dopaminergic systems in the antidepressant-like action of riparin III obtained from Aniba riparia (Nees) Mez (Lauraceae) in mice. *Fundam Clin Pharmacol.* 2013;27(1):104–112.

180. De Sousa FC, Oliveira IC, Silva MI, et al. Involvement of monoaminergic system in the antidepressant-like effect of riparin I from Aniba riparia (Nees) Mez (Lauraceae) in mice. *Fundam Clin Pharmacol.* 2014;28(1):95–103.

181. de Carvalho AMR, Rocha NFM, Vasconcelos LF, et al. Evaluation of the anti-inflammatory activity of riparin II (O-methil-N-2-hidroxi-benzoyl tyramine) in animal models. *Chem Biol Interact.* 2013;205(3):165–172.

182. Bandoniene D, Murkovic MM, Venskutonis PR. Determination of rosmarinic acid in sage and borage leaves by high-performance liquid chromatography with different detection methods. *J Chromatogr Sci.* 2005;43(7):372–376.

183. Boonyarikpunchai W, Sukrong S, Towiwat P. Antinociceptive and anti-inflammatory effects of rosmarinic acid isolated from Thunbergia laurifolia Lindl. *Pharmacol Biochem Behav.* 2014;124:67–73.

184. Kondo S, El Omri A, Han J, et al. Antidepressant-like effects of rosmarinic acid through mitogen-activated protein kinase phosphatase-1 and brain-derived neurotrophic factor modulation. *J Funct Food.* 2015;14:758–766.

185. Lin SH, Chou ML, Chen WC, et al. A medicinal herb, *Melissa officinalis* L. ameliorates depressive-like behavior of rats in the forced swimming test via regulating the serotonergic neurotransmitter. *J Ethnopharmacol.* 2015;175:266–272.

186. Jin X, Liu P, Yang F, et al. Rosmarinic acid ameliorates depressive-like behaviors in a rat model of CUS and up-regulates BDNF levels in the hippocampus and hippocampal-derived astrocytes. *Neurochem Res.* 2013;38(9):1828–1837.

187. Pereira P, Tysca D, Oliveira P, et al. Neurobehavioral and genotoxic aspects of rosmarinic acid. *Pharmacol Res.* 2005;52(3):199–203.

188. Hosseinzadeh H, Karimi G, Niapoor M. Antidepressant effect of Crocus sativus L. stigma extracts and their constituents, crocin and safranal, in mice. *Acta Hortic.* 2004;650:435–445.

189. Hosseinzadeh H, Noraei NB. Anxiolytic and hypnotic effect of Crocus sativus aqueous extract and its constituents, crocin and safranal, in mice. *Phytother Res.* 2009;23(6):768–774.

190. Hosseinzadeh H, Sadeghnia HR. Safranal, a constituent of Crocus sativus (saffron), attenuated cerebral ischemia induced oxidative damage in rat hippocampus. *J Pharm Sci.* 2005;8(3):394–399.

191. Mao Y, Li Y, Yao N. Simultaneous determination of salidroside and tyrosol in extracts of Rhodiola L. by microwave assisted extraction and high-performance liquid chromatography. *J Pharm Biomed Anal.* 2007;45(3):510–515.

192. Zhu L, Wei T, Gao J, et al. Salidroside attenuates lipopolysaccharide (LPS) induced serum cytokines and depressive-like behavior in mice. *Neurosci Lett.* 2015;606:1–6.

193. Yang SJ, Yu HY, Kang DY, et al. Antidepressant-like effects of salidroside on olfactory bulbectomy-induced pro-inflammatory cytokine production and hyperactivity of HPA axis in rats. *Pharmacol Biochem Behav.* 2014;124:451–457.

194. Power FB, Salway AH. Chemical examination of sarsaparilla root. *J Chem Soc Trans.* 1914;105:201–219.

195. Wang Y, Feng F, Wang Z. Determination of selected elements in aqueous extractions of a traditional Chinese medicine formula by ICP-MS and FAAS: evaluation of formula rationality. *Anal Lett.* 2010;43(6):983–992.

196. Nedeljkovic M, Tian L, Ji P, et al. Effects of acupuncture and Chinese herbal medicine (Zhi Mu 14) on hot flushes and quality of life in postmenopausal women: results of a four-arm randomized controlled pilot trial. *Menopause.* 2014;21(1):15–24.

197. Schnyer EN, Flaws B. *Curing Depression Naturally with Chinese Medicine.* Boulder, CO: Blue Poppy Press; 1998.

198. Ren LX, Luo YF, Li X, Zuo DY, Wu YL. Antidepressant-like effects of sarsasapogenin from Anemarrhena asphodeloides BUNGE (Liliaceae). *Biol Pharm Bull.* 2006;29(11):2304–2306.

199. Feng B, Zhao XY, Song YZ, et al. Sarsasapogenin reverses depressive-like behaviors and nicotinic acetylcholine receptors induced by olfactory bulbectomy. *Neurosci Lett.* 2017;639:173–178.

200. Ren L, Luo Y, Song S, et al. Experimental study of antidepressant effects of total timosaponin. *Tradit Chin Drug Res Clin Pharmacol.* 1993;(1).

201. Lu M-Z, Zhang Z-Q, Yi J, et al. Study on the effect and mechanisms of timosaponin B-II on antidepressant. *J Pharm Pract.* 2010:2010–2004.

202. Jiang W, Guo J, Xue R, et al. Anti-depressive activities and biotransformation of timosaponin B-III and its derivatives. *Nat Prod Res.* 2014;28(18):1446–1453.

203. Guo F, Zhang B, Fu Z, et al. The rapid antidepressant and anxiolytic-like effects of YY-21 involve enhancement of excitatory synaptic transmission via activation of mTOR signaling in the mPFC. *Eur Neuropsychopharmacol.* 2016;26(7):1087–1098.

204. Zhang Q, Guo F, Fu ZW, et al. Timosaponin derivative YY-23 acts as a non-competitive NMDA receptor antagonist and exerts a rapid antidepressant-like effect in mice. *Acta Pharmacol Sin.* 2016;37(2):166–176.

205. Cheng L, Pan GF, Sun XB, et al. Evaluation of anxiolytic-like effect of aqueous extract of asparagus stem in mice. *Evid Based Complement Alternat Med.* 2013;2013:Article ID 587260.

206. Capra JC, Cunha MP, Machado DG, et al. Antidepressant-like effect of scopoletin, a coumarin isolated from Polygala sabulosa (Polygalaceae) in mice: evidence for the involvement of monoaminergic systems. *Eur J Pharmacol.* 2010;643(2–3):232–238.

207. Zhao Y, Liu F, Lou HX. Studies on the chemical constituents of Solanum nigrum. *Zhong Yao Cai.* 2010;33(4):555–556.

208. Ma J, Jones SH, Hecht SM. A coumarin from Mallotus resinosus that mediates DNA cleavage. *J Nat Prod.* 2004;67(9):1614–1616.

209. Ouzir M, El Bairi K, Amzazi S. Toxicological properties of fenugreek (Trigonella foenum graecum). *Food Chem Toxicol.* 2016;96:145–154.
210. Verma A, Dewangan P, Kesharwani D, et al. Hypoglycemic and hypolipidemic activity of scopoletin (coumarin derivative) in streptozotocin induced diabetic rats. *Int J Pharm Sci Rev Res.* 2013;22(1):79–83.
211. Basu M, Mayana K, Xavier S, et al. Effect of scopoletin on monoamine oxidases and brain amines. *Neurochem Int.* 2016;93:113–117.
212. Wu S, Gao Q, Zhao P, et al. Sulforaphane produces antidepressant- and anxiolytic-like effects in adult mice. *Behav Brain Res.* 2016;301:55–62.
213. Zhang JC, Yao W, Dong C, et al. Prophylactic effects of sulforaphane on depression-like behavior and dendritic changes in mice after inflammation. *J Nutr Biochem.* 2017;39:134–144.
214. Martín-de-Saavedra MD, Budni J, Cunha MP, et al. Nrf2 participates in depressive disorders through an anti-inflammatory mechanism. *Psychoneuroendocrinology.* 2013;38(10):2010–2022.
215. Gao J, Xiong B, Zhang B, et al. Sulforaphane alleviates lipopolysaccharide-induced spatial learning and memory dysfunction in mice: the role of BDNF-mTOR signaling pathway. *Neuroscience.* 2018;388:357–366.
216. Pańczyszyn-Trzewik P, Sowa-Kućma M, Stachowicz K, et al. Antidepressant-like effect of (R,S)-sulforaphane (an activator of Nrf2) in the olfactory bulbectomy model in mice. *Eur Neuropsychopharmacol.* 2019;29:S128–S129.
217. Liu T, Liu X, Li W. Tetrandrine, a Chinese plant-derived alkaloid, is a potential candidate for cancer chemotherapy. *Oncotarget.* 2016;7(26):40800–40815.
218. Gao S, Cui YL, Yu CQ, et al. Tetrandrine exerts antidepressant-like effects in animal models: role of brain-derived neurotrophic factor. *Behav Brain Res.* 2013;238:79–85.
219. Sugimoto Y, Nishimura K, Itoh A, et al. Serotonergic mechanisms are involved in antidepressant-like effects of bisbenzylisoquinolines liensinine and its analogs isolated from the embryo of Nelumbo nucifera Gaertner seeds in mice. *J Pharm Pharmacol.* 2015;67(12):1716–1722.
220. Semwal DK, Semwal RB, Vermaak I, et al. From arrow poison to herbal medicine—the ethnobotanical, phytochemical and pharmacological significance of Cissampelos (Menispermaceae). *J Ethnopharmacol.* 2014;155(2):1011–1028.
221. Yin C, Gou L, Liu Y, et al. Antidepressant-like effects of L-theanine in the forced swim and tail suspension tests in mice. *Phytother Res.* 2011;25(11):1636–1639.
222. Wakabayashi C, Numakawa T, Ninomiya M, et al. Behavioral and molecular evidence for psychotropic effects in l-theanine. *Psychopharmacology.* 2012;219(4):1099–1109.
223. Hidese S, Ota M, Wakabayashi C, et al. Effects of chronic l-theanine administration in patients with major depressive disorder: an open-label study. *Acta Neuropsychiatr.* 2017;29(2):72–79.
224. Ferraz ABF, Schripsema J, Pohlmann AR, et al. Uliginosin B from Hypericum myrianthum. *Biochem Syst Ecol.* 2002;30(10):989–991.
225. Stolz ED, Hasse DR, von Poser GL, et al. Uliginosin B, a natural phloroglucinol derivative, presents a multimediated antinociceptive effect in mice. *J Pharm Pharmacol.* 2014;66(12):1774–1785.
226. Stein AC, Viana AF, Müller LG, et al. Uliginosin B, a phloroglucinol derivative from Hypericum polyanthemum: a promising new molecular pattern for the development of antidepressant drugs. *Behav Brain Res.* 2012;228(1):66–73.
227. Jäger S, Trojan H, Kopp T, et al. Pentacyclic triterpene distribution in various plants—rich sources for a new group of multi-potent plant extracts. *Molecules.* 2009;14(6):2016–2031.

228. Machado DG, Neis VB, Balen GO, et al. Antidepressant-like effect of ursolic acid isolated from Rosmarinus officinalis L. in mice: evidence for the involvement of the dopaminergic system. *Pharmacol Biochem Behav.* 2012;103(2):204–211.
229. Colla AR, Oliveira A, Pazini FL, et al. Serotonergic and noradrenergic systems are implicated in the antidepressant-like effect of ursolic acid in mice. *Pharmacol Biochem Behav.* 2014;124:108–116.
230. Colla AR, Rosa JM, Cunha MP, et al. Anxiolytic-like effects of ursolic acid in mice. *Eur J Pharmacol.* 2015;758:171–176.
231. Shoeb A, Chowta M, Pallempati G, et al. Evaluation of antidepressant activity of vanillin in mice. *Indian J Pharmacol.* 2013;45(2):141–144.
232. Abo-Youssef AM. Possible antidepressant effects of vanillin against experimentally induced chronic mild stress in rats. *Benisuef Univ J Basic Appl Sci.* 2016;5(2):187–192.
233. Xu J, Xu H, Liu Y, et al. Vanillin-induced amelioration of depression-like behaviors in rats by modulating monoamine neurotransmitters in the brain. *Psychiatry Res.* 2015;225(3):509–514.

11 Choosing Herbal Treatments

Having to choose one treatment from among many possibilities is a common task in medicine. In the pharmacological treatment of MDD, there is a variety of antidepressants from which to choose. Many are quite similar in their mechanisms of actions and, after figuring in considerations of side effects and tolerability, almost all have a roughly equal likelihood of benefiting the patient.[1]

In many cases, prescribers simply choose the antidepressant medication they have become most familiar with, and they vary from this choice only under specific clinical demands. For example, while a clinician's usual choice to treat depression may be fluoxetine, if the patient is losing weight and having difficulty sleeping, they may instead choose mirtazapine. On the other hand, if the patient suffers pain, they may instead select duloxetine, venlafaxine, or milnacipran. If the initial choice is ineffective or intolerable, it is reasonable to switch to another class of antidepressant. Algorithms, such as STAR*D, have been developed to facilitate such changes in treatment.[2] The fact that nature provides so many medicinal herbs to choose from does not alter the basic paradigm by which an herb or combination of herbs is selected as treatment for a patient. The paradigm continues to be the choosing of a treatment out of regard for efficacy, special clinical demands arising from psychiatric or medical comorbidities, and safety.

The major difference between standard antidepressant medications and herbal treatments is that all antidepressants approved by the United States Food and Drug Administration (FDA) have been shown to be efficacious in the treatment of MDD. As has been alluded to in earlier chapters, the efficacy of antidepressants may be less than hoped for. Nonetheless, no provider would be taken to task for choosing any one of them as an initial treatment. Thus, establishing efficacy is the primary concern in choosing an herbal remedy.

Hundreds of different herbs have been used throughout human history in folk medicines, Traditional Chinese Medicine (TCM), Ayurvedic Medicine, and Western herbalism for the treatment of symptoms consistent with the diagnosis of MDD. One of the most promising developments in neuroscience is the use of modern techniques to evaluate the efficacy of the traditional treatments and discover the neurochemical mechanisms by which they may act. It has been shown that many herbs act against oxidative stress, inflammation, insulin resistance, and other components of metabolic syndrome. These properties help prevent the pathologies that lead to systemic decompensation, neural dysfunction, dendritic atrophy, and development of MDD. There is also evidence that flavonoids, terpenes, and other constituent phytochemicals of herbs act directly upon neurons and microglia. Some may act upon the trkB receptors that mediate the effects of BDNF,[3] or at the TLR4 receptors that modulate inflammatory responses.[4] These phytochemicals also interact with various levels of the MAPK system,[5] that in turn may mediate some effects of flavonoids on the PI3K/akt/mTOR pathway,[6] GSK-3β,[7] BDNF,[8] NFkB[9] Nrf2,[10] the NLRP3 inflammasome[11] and other intracellular pathways that have become associated with antidepressant effects.

Many herbs have produced antidepressant-like effects in animals, and in many of these preclinical studies, those behavioral changes are shown to occur concomitantly with neurochemical changes similar to those of ketamine and standard antidepressant treatments. Unfortunately, in relying upon animal data and likely physiological mechanisms, one runs the risk of quickly outpacing the limited collection of human data. Studies and case reports in the literature have shown many herbs to improve mood states in human subjects. However, studies have often been marginal in quality. In only a handful of studies have herbal treatments been subjected to the usual scientific standards of randomized, placebocontrolled, double-blinded evaluations of antidepressant effects in patients diagnosed with MDD by standard instruments.

11.1   EFFICACY OF HERBAL TREATMENTS OF MDD

The two herbs whose anti-depressant defects are best studied and have the strongest scientific support are *Hypericum perforatum* (St John's wort) and *Crocus sativus* (saffron). There are also well-designed studies showing antidepressant effects of curcumin, the primary active component of *Curcuma longa* (turmeric); species of the genus *Lavandula* (lavender); *Matricaria recutita* (chamomile); *Humulus lupulus* (hops); and *Rhodiola rosea*.

In a Cochrane review of *Hypericum perforatum*, it was concluded that "hypericum extracts tested in the included trials are superior to placebo in patients with MDD; are similarly effective as standard antidepressants; and have fewer side effects than standard antidepressants."[12] However, in studies that included severely depressed patients, *Hypericum perforatum* has not been so effective.[13] The likelihood of interactions with standard medications must also be considered. Several reviews have also been supportive of the antidepressant effects of *Crocus sativus*.[14–16] As with *Hypericum perforatum*, several studies have found it equally effective as standard antidepressants – e.g., imipramine and fluoxetine – but with fewer side effects.

There has also been compelling evidence for antidepressant effects of curcumin. At least two randomized, double-blind, placebo-controlled studies have shown curcumin to be effective in the treatment of patients with MDD.[17] Although effects of *Lavandula* have been inconsistent, in at least one randomized, placebo-controlled study, the herb produced significant antidepressant effects in human subjects.[18]

Matricaria recutita, perhaps best known as an herb used for the making of calming tea, has had antidepressant effects in placebo-controlled human studies. In one case, the subjects suffered comorbid anxiety,[19] whereas in the other, subjects suffered a constellation of symptoms as part of the syndrome of Post-partum Depression.[20] A single small but randomized, placebo-controlled, double-blind study showed the mildly sedating herb, *Humulus lupulus*, to also significantly improve anxiety and mild depression per the Depression Anxiety Stress Scale-21.[21] In a comparison with sertraline, *Rhodiola rosea* did not exhibit significant antidepressant effect, though the antidepressant did.[22] However, the herb was effective in patients with mild to moderate degrees of MDD.[23] Perhaps consistent with its effects in milder forms of depression, *Rhodiola rosea* was in two studies shown to reverse the effects of so-called "burnout."[24,25]

In some cases, clinical evidence of antidepressant effects of herbs has been suggestive but the data were less compelling than those above. Experimental designs were weak, or nonstandard methods for measuring mood states had been utilized. Mood states were sometimes merely suggestive of MDD, and were reported in vague terms such as improvement in mood, quality of life, or social functioning. *Eleutherococcus senticosus* (Siberian Ginseng) was found to improve sleep, well-being, appetite, stamina, cognitive function, and mood in "neurotic" patients.[26] Neuroticism is often used to describe dysthymia rather than MDD. *Bacopa monnieri* (water hyssop) was tested in elderly adults in a randomized, double-blind, placebo-controlled clinical trial. Although those subjects were not formally diagnosed with MDD, *Bacopa monnieri* improved depression scores, per the Center for Epidemiologic Studies Depression scale.[27] In a double-blind, placebo-controlled cross-over study of healthy adults, an extract of *Bacopa monnieri* improved cognitive performance as well as producing "positive mood effects" by assessment with the Bond-Lader visual analogue scales.[28] In a study of *Ocimum sanctum* (holy basil), none of the subjects had been diagnosed as suffering MDD. However, the depression index scores in the Brief Psychiatric Rating Scale dropped significantly, by 30%, over the two-month period of treatment.[29] In a randomized, double-blind study, *Rosmarinus officinalis* (Rosemary) improved anxiety and, to a minor degree, mood per the Hospital Anxiety and Depression Scale.[30] In an open, randomized study, extract of *Withania somnifera* (Ashwagandha) was compared against 40 mg per day dosing of fluoxetine in patients diagnosed with MDD by DSM-IV criteria. After 42 days of treatment, patients in both groups experienced similar relief from symptoms.[31] Finally, a preparation of high cocoa liquor/polyphenol rich chocolate, prepared from *Theobroma cacao*, administered over eight weeks also relieved symptoms in patients diagnosed with Chronic Fatigue Syndrome by the British Centres for Disease Control and Prevention criteria. None of the subjects were suffering comorbid psychiatric illnesses as defined by the DSM-IV. Nonetheless, by the end of the study, the chocolate significantly lowered not only fatigue, but also anxiety and depression scores on the Hospital Anxiety and Depression Scale.[32]

Although the above reports of efficacy in human subjects may be weak, they can be bolstered by longstanding, traditional use of those herbs to treat depression. In a study of the use of antidepressant herbs by experienced herbalists – all members of the American Herbalist Guild – a number of herbs were held by consensus to be effective in the treatment of depressed mood.[33] *Eleutherococcus senticosus*, *Bacopa monnieri*, *Ocimum sanctum*, *Rosmarinus officinalis*, and *Withania somnifera* noted in paragraphs immediately above were included in that list. Moreover, preclinical studies of those specific herbs have shown strong indications of antidepressant effects, as well as possible mechanisms of action in the brain. Those data strengthen the argument for their use.

Certainly, preventing and treating MDD are two different things. Nonetheless, studies have also shown that daily use of the *Coffea arabica*-based beverage, coffee,[34] is associated with reduced risk of MDD. A randomized, double-blind, placebo-controlled experiment showed that consumption of green tea extract also improved scores both the Montgomery-Asberg Depression Rating Scale and 17-item Hamilton Rating Scale for Depression.[35] However, none of the subjects had been diagnosed with MDD at any time during the five-week study. In some individuals, caffeine can cause anxiety. This is particularly the case in individuals with pre-existing anxiety disorders.[36] Nonetheless, for those that can tolerate it, it seems reasonable to add several cups a day of coffee, tea, the above-noted cocoa, or possibly the stimulant drink, *Ilex paraguariensis* (Yerba mate), to the treatment regimen.

Finally, epidemiological studies have shown that consumption of plant foods rich in flavonoids – many of which are found in medicinal herbs – is associated with reduced risk of MDD.[37,38] Thus, the daily intake of flavonoids from herbal medicines or food is beneficial for individuals suffering or at risk for MDD.

11.2   HERBS FOR WHICH THERE IS LESS THAN COMPELLING EVIDENCE OF EFFICACY

It is most prudent to recommend herbs for which there are studies establishing efficacy in human patients. Unfortunately, there have been few incentives to scientifically study herbs. As is sometimes noted, lack of evidence is not evidence of lack. In any case, aside from those herbs noted above for which there is varying degree of evidence of efficacy in humans, there are also many traditionally used herbs for which there is no experimental data in humans. The herbs discussed in this text have all met the criteria of both being recognized as traditional herbal treatments of depression-like conditions, and having at least two animal studies supporting the likelihood of them exerting antidepressant effects. They also exert anti-oxidant and anti-inflammatory effects, and in some cases are shown to act upon intracellular pathways that affect MDD. Many of the herbs discussed in this text that meet those criteria, but do not have human clinical data, are on the above-noted list of herbs recognized by experienced members of the prestigious American Herbalist Guild as being useful for the treatment of MDD. Those herbs include *Angelicae sinensis* (angelica root), *Avena sativa* (oats), *Centella asiatica* (gotu kola), *Echium amoenum* (European borage), *Ginkgo biloba* (ginkgo), *Melissa officinalis* (lemon balm), *Mimosa pudica* (mimosa), *Paeonia lactiflora* (Chinese peony), *Panax ginseng* (Asian ginseng), *Schisandra chinensis* (schizandra), *Scutellaria lateriflora* (skullcap), and *Vitex agnus-castus* (chaste tree). These criteria provide at least some scientific and practical bases for use when there are no contraindications in terms of safety or interactions with medications.

For completeness sake, the above-mentioned list of herbs recommended by members of the American Herbalist Guild as treatments of MDD also includes herbs for which there can be found no scientific evidence – not even animal data – of antidepressant efficacy. These herbs, none of which were discussed in this text, include: Apple fruit (*Malus domestica*), artichoke (*Cynara scolymus*), calendula (*Calendula officinalis*), cat's claw (*Uncaria tomentosa*), Chaga mushroom (*Inonotus Obliquus*), Chuan Xin Lian (*Andrographis herba*), cleavers (*Galium aparine*), cranberry (*Vaccinium macrocarpon*), damiana (*Turnera diffusa*), echinacea root (*Echinacea angustifolia*), elephant head (*Pedicularis* spp.), evening primrose

(*Oenothera biennis*), Oregon grape (*Mahonia aquifolium*), Indian pipe (*Monotropa uniflora*), Japanese knotweed (*Fallopia japonica*), mugwort (*Artemisia vulgaris*), peach leaf (*Amygdalus persica*), pine pollen (*Tsuga canadensis*), prairie mimosa (*Desmanthus illinoensis*), prickly ash bark (*Zanthoxylum americanum*), red clover flower (*Trifolium pratense*), Solomon's seal (*Polygonatum odoratum*), teasel root (*Dipsacus japonica*), wild lettuce (*Lactuca virosa*), willow bark (*Salix alba*), wood betony (*Stachys officinalis*), yarrow (*Achillea millefolium*), and yellow dock (*Rumex crispus*). However, mere "expert opinion" is the lowest level of evidence-based medicine.[39] Given the wide range of therapeutic options that are supported by experimental evidence, one should avoid relying upon expert opinion that is otherwise unsubstantiated.

11.3 COMBINATIONS OF HERBS

While polypharmacy is discouraged in mainstream Western medicine, the combining of herbs is not only standard treatment in TCM, but also common in traditional Western herbalism. David Hoffman, an authoritative member of the American Herbalist Guild, goes to some length in his book, *Medical Herbalism*, to describe the process of selecting herbs to combine for the treatments of various ailments.[40] The German E Commission also approves a variety of herbal combinations for a wide range of physical and emotional complaints.[41]

Published studies, mainly out of China, have found various combinations of herbs to significantly improve symptoms of MDD in human patients. The classical herbal combination *Yueju* has antidepressant effects similar to those of venlafaxine in depressed patients, but with fewer side effects.[42] The traditional Japanese herbal medicine *kamishoyosan* contains a similar mix of *Atractylodis lanceae*, *Zingiberis Rhizoma*, *Menthae Herba*, *Glycyrrhizae Radix*, *Moutan Cortex*, *Bupleuri Radix*, *Poria cocos*, *Paeoniae Radix*, *Angelicae Radix*, and *Gardenia jasminoides*. It relieved anxiety and depression in menopausal women in a manner similar to that of paroxetine.[43] *Xiao yao san* improved depression in a majority of patients.[44] In a meta-analysis of randomized controlled trials, the combination *Chai hu shu gan* was significantly better than standard antidepressants in improving depressive symptoms.[45] In a meta-analysis of ten Chinese clinical studies, *Gan mai da zao* compared well with standard antidepressants, but with fewer side-effects.[46] In other reports of varying quality, the combinations *Chai hu gui zhi gan jiang tang*,[47] *Kai xin san*[48] and *Wu ling*[49] significantly improved MDD in human subjects.

Shu gan jie yu is a traditional Chinese treatment that is the simple combination of *Hypericum perforatum* and *Eleutherococcus senticoccus*. A meta-analysis of studies showed it to be more effective than paroxetine, citalopram, sertraline, or fluoxetine in the treatment of mild-to-moderate depression.[50]

Whereas curcumin and *Crocus sativa* are each effective as antidepressants on their own, a randomized, double-blind, placebo-controlled study showed the combination of the two herbs to be particularly effective for treatment of MDD.[51] It may be noted that the bioavailability of curcumin and other active phytochemicals in herbs can be substantially increased by adding piperine, a phytochemical from *Piper nigrum* (black pepper).[52]

In a Western study of treatment of insomnia with the combination of *Melissa offi-cinalis* and *Nepeta menthoides*, there was significant improvement of both insomnia and scores on the Beck Depression index. However, none of the subjects were for-mally diagnosed as suffering MDD.[53] Four weeks of supplementation with a combi-nation of *Magnolia officinalis* and *Phellodendron amurense* extracts reduced stress, tension, depression, anger, fatigue, and confusion. Again, however, none of the sub-jects were diagnosed with MDD.[54]

It is not clear if the herbs noted above act additively or synergistically in their effects on MDD, thus making it necessary, or at least advantageous, to use them in combination. Certainly, in TCM, the doctrine of *Jun-Chen-Zuo-Shi* requires a practitioner to prescribe herbs in harmonious combinations. However, in the effects of the herbal combination *Yueju* on mice, the extract of just one of its herbal compo-nents, *Gardenia jasminoides*, replicated the antidepressant effects and rapid upregu-lation of hippocampal BDNF produced by the entire combination.[55] Moreover, it was subsequently found that geniposide, a terpenoid glycoside extracted from *Gardenia jasminoides*, was able on its own to increase BDNF in mouse brain.[56]

As a final caveat, while there is a dearth of information about interactions between herbs and medications, there is even less reliable information about possible interac-tions herbs may have with other herbs. Thus, one must remain circumspect. The use of traditional rather than impromptu combinations may be prudent.

11.4   ADDRESSING COMORBIDITIES

11.4.1   ANXIETY AND INSOMNIA

The existence of certain symptoms or comorbidities may also be considered in choos-ing an herb for monotherapy or for inclusion in a combination. Anxiety is a common comorbidity of MDD. In one study of patients with MDD, it was found that 29% had a history of panic attacks, 62% had experienced at least moderate psychic anxiety, and 72% complained of persistent worry.[57] In a study of patients with chronic MDD, 24% of the patients had at least one lifetime comorbid anxiety disorder.[58] The pres-ence of comorbid anxiety affects treatment response. Such patients respond more poorly to antidepressant treatment. They tend to have a slower response to medica-tion and more often fail to achieve remission of symptoms.[59] Insomnia is another serious and common comorbidity of MDD. Up to 75% of patients suffering MDD also suffer sleep disturbance.[60] Indeed, it has been noted that sleep disturbance is often the reason that depressed patients first seek help. It is also one of the few proven risk factors for suicide.[61]

A number of herbs have also been found to exert both anxiolytic and soporific effects. In "*A Delphi Study on Herbs Used to Address Depression and Anxiety According to Master Herbalists*," Einerson lists 17 herbs for which master herbalists showed moderate to strong agreement regarding their use in the treatment of anxi-ety. In descending order of strength of this agreement, those herbs are *Lavandula* spp. (lavender), *Leonurus cardiaca* (motherwort), *Passiflora* spp. (passion-flower), *Scutellaria lateriflora* (skullcap), *Matricaria recutita* (chamomile), *Avena sativa* (oats), *Ganoderma lucidum* (Reishi), *Schisandra chinensis* (schizandra),

Cymbopogon citratus (lemongrass), *Melissa officinalis* (lemonbalm), *Escholzchia californica* (California poppy), *Humulus lupulus* (hops), *Piper methysticum* (kava, see caveat, Chapter 7), *Paeonia lactiflora* (Chinese peony), *Tilia platyphyllos* (linden), *Asparagus racemosus* (Shatavari), and *Piscidia erythrina* (Jamaica dogwood). Of note, ten of these herbs are discussed in Chapter 7 as also having preclinical and, in some cases, clinical evidence of antidepressant effects.

The impressions of the Master Herbalists are largely born out in scientific, clinical studies. Thus, it does appear that dosing of herbs can adjusted to provide anxiolytic relief in the daytime and soporific aid at bedtime. In patients suffering comorbid anxiety and depression, *Lavandula* significantly reduced anxiety per the Hamilton Anxiety Rating Scale. Treatment improved clinical outcomes, daily living skills, and health-related quality of life.[62] *Lavandula angustifolia* essential oil, taken orally or by inhalation, has been found improve sleep latency and duration in such varied groups as benzodiazepine-withdrawing patients, hospitalized elderly patients, middle-aged women with insomnia, and patients with subsyndromal anxiety.[63] The basis of the anxiolytic and soporific effects of *Lavandula* is likely its ability to enhance GABAergic activity in the brain.[64]

In Iran and other parts of the Middle East, *Passiflora incarnata* has long been used as a folk remedy for anxiety. In a double-blind randomized trial, extract of *Passiflora incarnata* was as effective as oxazepam in the treatment of patients diagnosed with generalized anxiety disorder using DSM-IV criteria. Although oxazepam showed a more rapid onset of action, it caused more impairment of job performance than did the herb.[65] Extract of the herb also alleviated pre-operative anxiety in surgical patients without affecting psychomotor motor performance.[66] Along with anxiolytic effects, *Passiflora incarnata* also relieves insomnia. In a placebo-controlled study, tea made from the herb improved self-reported sleep quality, and this finding was verified by polysomnography.[67]

In at least two randomized, double-blinded studies, extract of *Matricaria recutita* was found to significantly reduce symptoms in patients diagnosed with Generalized Anxiety Disorder.[68,69] It was also found to have significant antidepressant effects in patients with comorbid anxiety and MDD.[70] Tea from *Matricaria recutita* is widely used as an aid to relaxation and sleep, but only a few studies have scientifically evaluated its soporific effects. Unfortunately, these studies were inconsistent in the conclusions. Whereas studies suggested improved sleep in menopausal women[71] and men without specific sleep complaints,[72] a third found no differences from placebo[73] Nonetheless, it is suspected that the flavonoid apigenin, found in *Matricaria recutita*, can produce sedative effects through modulation of GABA receptors.[74]

Humulus lupulus was found to significantly reduce depression, anxiety, and stress scores on the Depression Anxiety Stress Scale-21 in young adults that had complained of mild distress in those three symptom categories.[21] *Melissa officinalis* was found to significantly reduce anxiety and complaints of heart palpitations by the General Health Questionnaire.[75] However, aside from that study, there are no formal clinical studies of effects of the herb on anxiety or insomnia.

The *Scutellaria* species has long been used by herbalists to treat anxiety, and there is substantial preclinical evidence of anxiolytic-like effects.[76] It should be noted that at least one phytochemical constituent of *Scutellaria*, wogonin, has been found to be an agonist active at benzodiazepine receptors.[77] Unfortunately, there are no formal studies of anxiolytic effects of *Scutellaria* in humans. The herb is also quite frequently prescribed for insomnia, particularly *Scutellaria baicalensis*, by Chinese herbalists. However, there are no experimental data in human subjects to support this practice. There are also no scientific studies of effects of *Avena sativa, Schisandra chinensis*, or *Paeonia lactiflora* on anxiety or insomnia in human subjects. Despite favor for *Tilia species* among herbalists – and inclusion of *Tilia cordata* flowers as an ingredient in the well-known Sleepytime tea – there are no scientific studies of effects of any *Tilia* species on anxiety or insomnia in human subjects. It is here worth noting that while *Valeriana officinalis* is an herb with a strong reputation as an anxiolytic and soporific agent, the results of scientific studies of its effects have been disappointing. There is no clear evidence that the herb is more effective than placebo for the treatment of either anxiety or insomnia. Thus, *Lavendula, Passiflora incarnata, Matricaria recutita, Humulus lupulus*, and possibly *Melissa officinalis* are the herbs with the strongest scientific evidence for adding one or more to a treatment regimen to improve symptoms of anxiety and insomnia.

As a final note, combinations of herbs are commonly prescribed by experienced herbalists for the treatment of anxiety and insomnia. For example, in a case report, the naturopath, herbalist, and author Jerome Sarris N.D. describes a combination concocted for a young woman with chronic anxiety and insomnia that included *Withania somnifera, passiflora incarnata, Scutellaria lateriflora, Matricaria recutita*, and *Piper mythysticum*.[78] The German Commission E lists several different combinations of *Valeriana officinalis, Humulus lupulus, Passiflora incarnata*, or *Melissa officinalis* for treatment of "insomnia and unrest."[79] Unfortunately, it is not made clear why three or four ostensibly sedating/anxiolytic herbs are combined rather than simply giving a larger dose of a single such herb.

11.4.2 OBSESSIVE-COMPULSIVE DISORDER

There is significant overlap between Obsessive-Compulsive Disorder (OCD) and MDD. One study found a 22% incidence of obsessive-compulsive symptoms in depressed patients.[80] The condition is notoriously difficult to treat.[81] Thus, comorbidity of OCD and MDD may guide choice of herbal treatments.

At least two randomized, placebo-controlled studies have found *Echium amoenum* to benefit patients suffering OCD. In the first case, patients showed significant improvement after six weeks of treatment as measured by the Yale-Brown Obsessive Compulsive.[82] These results were replicated by others in a subsequent and very similar study.[83] Silymarin from *Silybum marianum* (milk thistle) may also be useful in the treatment of OCD in human subjects. In a double-blind, randomized, placebo-controlled study, it was found to be as effective as fluoxetine.[84]

In an initial, open-label study of monotherapy with *Hypericum perforatum*, patients suffering OCD appeared to experience significant improvement of symptoms.[85] However, these results were not replicated in a subsequent randomized double-blind, placebo-controlled study.[86]

11.4.3 PREMENSTRUAL AND PERIMENOPAUSAL SYMPTOMS

Mood disorders and, at times, MDD are associated with hormonal transitions in women. Several herbs for which there is preclinical or clinical evidence of antidepressant effects, have also been used to treat such hormone-related disorders. Thus, the presence of premenstrual or perimenopausal mood symptoms can prompt their use.

Foeniculum vulgare (fennel) has been reported to be useful in women with perimenopausal mood symptoms. In one study, eight weeks of treatment with fennel oil brought significant improvements in self-reported general quality of life, including psychological and sexual aspects.[87] In a similar study, menopausal women chronically treated with *Foeniculum vulgare* showed improvements in anxiety and depression scores, but these were not statistically significant.[88]

Many studies have shown *Cimicifuga racemosa* (black cohosh) to relieve common discomforts of menopause, including anxiety and uncomfortable mood states. In one such study, extract of *Cimicifuga racemosa* improved the scores of menopausal women in a questionnaire for self-evaluation of depression and in the Hamilton Anxiety Scale. However, as has been the general case with such studies, none of the subjects were formally diagnosed with MDD.[89] *Vitex agnus-castus* (chaste tree) was similarly found to reduce both physical and "psychological" symptoms of menopausal women in a selfcare survey that cannot be considered a formal evaluation of MDD.[90]

Far more work has been done in evaluating the use of *Vitex agnus-castus* to treat Premenstrual Dysphoric Disorder (PMDD). Studies have uniformly found that chronic administration of extract of *Vitex agnus-castus* berries gives significant improvement of emotional and neurovegetative symptoms of PMDD. When comparisons were made, the treatment compared favorably with other treatments, such as fluoxetine and pyridoxine.[91–93] In a randomized trial in young women suffering PMDD, chronic treatment with *Melissa officinalis* (lemon balm) significantly reduced psychosomatic symptoms, anxiety, insomnia, and social function disorder per the General Health Standard Questionnaire.[94]

Foeniculum vulgare, which has been reported to improve mood states in perimenopausal women, has also been found to ameliorate symptoms of PMDD. However, results have been inconsistent. In one study, twice daily treatment with the herb for a month provided significant improvement in mood.[95] In another study, *Foeniculum vulgare* significantly improved symptoms of premenstrual syndrome that included mood components, but the results did not include which specific components of the syndrome improved.[96] In yet another study, modest improvements in mood and anxiety were noted, but they did not reach significance.[97]

Also reported is a trial of extract of *Borago officinalis* (borage) in women for the treatment of Premenstrual Syndrome.[98] One of the outcomes in that study was an improvement in emotional wellbeing. However, the extract of *Borago officinalis* seeds was used because of its high content of γlinolenic acid. It is possible that mere replenishment of an essential fatty acid was responsible for this effect.

11.4.4 DEMENTIA

Several herbs have been found to improve cognitive function in the elderly, and may be of use when dementia or mere cognitive decline are comorbid with MDD.

In patients suffering from chronic cerebrovascular insufficiency, *Ginkgo biloba* improved motor activity, speech comprehension and production, and mood.[99] In a similar study of elderly patients with mild cognitive impairment, *Ginkgo biloba* increased motivation and interest in activities of daily living.[100]

Hericium erinaceus has been found to improve function in elderly diagnosed with mild cognitive impairment.[101] Unfortunately, data concerning effects on mood and MDD are scant and weak, consisting largely of a case report. Interestingly, a study suggested that the acetylcholinesterase inhibitor, galantamine, in addition to an antidepressant, improved both mood and cognitive early in treatment. However, the treatment was poorly tolerated – the dropout rate in the galantamine group was 63% – and differences from placebo were lost.[102] Although currently a prescribed medication, galantamine was first isolated from bulbs of *Galanthus nivalis*, the common snowdrop.[103]

Huperzine, extracted from *Huperzia serrata*, has been touted as an anti-Alzhemier's drug. However, addition of huperzine to an antidepressant in the treatment of elderly patients suffering both MDD and cognitive impairment has led to mixed results. In one case,[104] addition of huperzine appeared to enhance the antidepressant effect, whereas in another,[105] quality of life was improved in most of those that received combined treatment. In a third study, cognitive improvements were noted, but there were no additional benefits for mood.[106] It is important to state that in a large Phase II trial, huperzine A dosed at 200 µg twice daily was found not to be effective in the treatment of mild to moderate Alzheimer's disease. Although small differences from placebo were noted, these were not significant.[107]

11.4.5 DIABETES AND METABOLIC SYNDROME

In a recent review it was noted that the herbs most frequently cited in the literature as being useful in the treatment of diabetes type II were *Momordica charantia* (bitter melon), *Trigonella foenum graecum* (fenugreek), *Gymnema sylvestre* (gurmar), *Coccinia indica* (ivy gourd), *Opuntia* spp. (nopal), *Panax ginseng* (ginseng), *Artemisia dracunculus* (Russian tarragon), *Cinnamomum cassia* (cinnamon), *Plantago ovata* (psyllium), and *Allium sativum* (garlic).[108] The mechanisms of action for these herbs were described as including regulation of insulin signaling pathways, translocation of GLUT-4 receptor and/or activation the PPARγ. All of those mechanisms have been associated with anti-depressant effects.[109–111]

Interestingly, most of these herbs, including *Momordica charantia*,[112] *Coccinia indica*,[113] *Opuntia dilleniid*,[114] *Artemisia dracunculus*[115] *Trigonella foenum graecum*,[116] and *Allium sativum*[117] have shown antidepressant-like effects in animal models. However, from among them only *Panax ginseng* and *Cinnamomum cassia* have shown antidepressant effects in humans by enhancing the effects of antidepressants in treatment-resistant patients.

Panax ginseng has been demonstrated to have hypoglycemic effects in human patients. A meta-analysis of 16 randomized controlled trials in people with and without diabetes concluded that the herb significantly improved fasting blood glucose in people both with and without diabetes.[118] *Cinnamomum cassia* has also been shown to be a useful add-on therapy in managing diabetes type II, though further trials are needed to fully establish its efficacy and safety.[119] Thus, in cases where diabetes type II, Metabolic Syndrome, or so-called pre-diabetes is comorbid with depression, either *Panax ginseng* or *Cinnamomum cassia* would be reasonable additions to an herbal regimen.

11.4.6 Fatigue, Lack of Resiliency, and General Malaise

Often, patients complain of a general sense of fatigue, emotional exhaustion, lack of resiliency, and general malaise. These are a common, almost defining set of symptoms in MDD. This constellation also typifies the condition often referred to as burnout, which can precede or be comorbid with MDD.[120]

The term "adaptogen" was coined by the Russian researcher, Dr Israel Brekhman, and used to describe a class of herbs that had the general effect of increasing resiliency under a variety of conditions of stress. In his initial work, the 1960 paper "A New Medicinal Plant of the Family of *Araliaceae* – The Spiny *Eleutherococcus*,"[121] Brekhman focused on the adaptogenic properties of *Eleutherococcus senticosus*. In subsequent years, scientists, primarily in Russia and Eastern Europe, have studied "adaptogenic" effects of *Eleutherococcus senticosus*, *Rhodiola rosea*, *Schisandra chinensis*, *Scutellaria baicalensis*, and others. During this time, the conceptualization of adaptogen has expanded to become a substance that can enhance resistance to stress, and increase concentration, performance, and endurance during fatigue. The adaptogen is to be distinguished from a mere stimulant, in that there is no habituation to the adaptogen, nor are there withdrawal effects. Indeed, it appears that this therapeutic entity largely serves as a buffer in the stress response system that restores the ability to respond to new physiological challenges that can be lost during chronic stress.[122] Interestingly, most adaptogen plant species, such as *Panax ginseng* and *Scutellaria*, contain pentacyclic triterpenoids that structurally resemble glucocorticoids.[123] Thus, it is possible that adaptogens may act directly to reduce glucocorticoid-resistance that prolongs the stress response and exacerbates MDD.[124] Such action would reduce the allostatic load that is increased in MDD and other psychiatric illnesses.[125]

The notion of there being adaptogens has never gained a firm foothold in Western medical science, perhaps due to a common, *a priori* dismissal of the effectiveness of herbal treatment. Nonetheless, many well-managed scientific investigations have provided support for there being such a class of herbs. For example, a review of randomized, placebo-controlled clinical studies of *Rhodiola rosea* found the general enhancing effects that an adaptogen would be expected to exhibit. Improvements were noted in physical performance, mental performance, and emotional states.[126] *Schisandra* and *Eleutherococcus*, both of which have been shown to have antidepressant effects in humans, have also been found to improve physical and cognitive performance while reducing feelings of fatigue.[127]

In older, primarily herb-based medical traditions, the use of adaptogens has been well accepted. In Traditional Chinese Medicine, there are tonic herbs that have long been used to improve low body resistance and weak constitution, or to add support when the body is finding it difficult to fight severe diseases.[128] These various herbs have been categorized as those that strengthen *Chi*, blood, *Yin*, or *Yang* in the body.

Among these tonic herbs are *Panax ginseng, Rhodiola crenulata, Astragalus membranaceus, Glycyrrhiza uralensis, Eleutherococcus senticosus, Atractylodes macrocephala, Angelica sinensis, Epimedium brevicornu, Psoralea corylifolia, Stachys geobombycis, Asparagus cochinchinensis,* and *Polygonatum odoratum.* Many of those herbs are found on various lists of adaptogenic herbs as well as in TCM combinations used for the treatment of depression-like syndromes. In Ayurveda, the ancient traditional medicine of India, adaptogenic herbs are referred to as *rasayana,* and have included *Tinospora cordifolia, Asparagus racemosus, Emblica officinalis, Withania somnifera, Piper longum, Terminalia chebula,* and others.[129]

In traditional Western herbalism, there has never been a specific category of herbs defined as "antidepressants." Rather, there have been herbs known as "nervines" that act upon the nervous system as tonics, relaxants, or stimulants. Dr Marisa Marciano, a naturopath, herbalist, and author, has noted similarities between nervines and adaptogens. "Nervine tonics are perhaps the most important contribution herbal medicine can make in the area of stress and anxiety, as they will strengthen and feed the nervous system in cases of nervous debility and exhaustion. Adaptogens may also be considered in this group due to their ability to aid the whole body and mind to cope with demands made upon it."[130] David Hoffmann, also a Western herbalist and member of the American Herbalist Guild, provided a list of adaptogens in his book, *Medical Herbalism.*[131] This list includes: *Acanthopanax sessiliflorum, Albizzia julibrissin, Aralia alata, Aralia manchuria, Aralia schmidtii, Cicer arietinum, Codonoposis pilosula, Echinopanax elatus, Eleutherococcus senticosus, Eucommia ulmoides, Ganoderma lucidum, Hoppea dicotoma, Leuzea carthamoides, Ocimum sanctum, Panax ginseng, Panax quinquefolius, Rhodiola rosea, Schisandra chinensis, Tinospora cordifolia, Trichopus zeylanicus,* and *Withania somnifera.* This list overlaps with other such lists, both modern and traditional. Many of these herbs have not been evaluated for antidepressant or other effects in human subjects. However, as may be recognized, several have been found to have significant antidepressant effects in clinical studies, whereas others are ingredients of Chinese herbal antidepressant combinations.

11.5 AUGMENTATION OF STANDARD ANTIDEPRESSANT TREATMENT WITH HERBS

When considering herbal treatments, a difficult question is whether herbs are suitable for the treatment of severe MDD. Of course, considering the fact that the majority of patients do not achieve remission of MDD with standard antidepressant treatment, it is not unreasonable to ask if such antidepressants are themselves suitable for treatment of severe depression. In any case, having to choose between herbs and standard antidepressants for the treatment of MDD may be unnecessary, as the combination of the two types of treatment can be both safe and efficacious. Indeed, the augmentation of standard antidepressants with herbal treatments, particularly those chosen to address unremitting comorbidities, may bring treatment-resistant or only partially responsive patients into full remission.

Some herbs and herbal combinations that have been shown to exert antidepressant effects in human patients as monotherapy, have also been reported to enhance the effects of standard antidepressant treatment in patients who were poorly responsive or treatment resistant. Curcumin, which has significant antidepressant effects as monotherapy, also enhances the efficacy of standard antidepressant treatment. Supplementation with curcumin significantly improved symptoms of depression in patients already receiving antidepressants. The addition of curcumin also reduced serum levels of IL-1β and TNFα, increased plasma BDNF levels, and decreased salivary cortisol concentrations.[132]

Results of studies of the antidepressant effects of species of *Lavandula* as monotherapy are mixed. However, the addition of tincture of *Lavandula angustifolia* to treatment with imipramine showed a significantly greater effect than treatment with imipramine alone.[133] In a similar study, the herb significantly enhanced the antidepressant effects of citalopram.[134]

Addition of the Chinese herbal treatment *Yueju* to fluoxetine resulted in a more rapid antidepressant response in human patients with MDD.[135] The herbal combination, *Chai hu shu gan*, also significantly enhanced the antidepressant effects of fluoxetine, paroxetine, and doxepin.[45] Similar enhancement of the effects of antidepressants was seen with *Gan mai da zao*.[46] It was noted that the latter data concerning *Gan mai da zao* were randomized and controlled, but still deemed to be of low quality by the standards of the Cochrane reviews.

In at least four other cases, herbs that have not been evaluated for efficacy as monotherapy for treatment of MDD enhanced the effects of standard antidepressants. For example, in individuals that had not responded to fluoxetine alone, the addition of essential oil of *Cinnamomum cassia* over eight weeks significantly improved response to antidepressant treatment per Beck Depression scale scores.[136] In two studies, *Panax ginseng* also significantly improved the residual symptoms of depression in women whose antidepressants provided only partial remission.[137,138] In elderly patients with MDD and comorbid cerebral insufficiency, the addition of *Ginkgo biloba* extract improved response to standard antidepressant treatment.[139]

Huperzine A, a sesquiterpene alkaloid, is believed to be the active component of the medicinal herb, *Huperzia serrata*. Like *Ginkgo biloba*, it has been used to enhance cognitive function in the elderly. Data concerning the ability of huperzine to augment antidepressants have been mixed. The addition of huperzine to venlafaxine significantly improved scores on the Hamilton Depression Scale.[140] In another study, the addition of huperzine A to fluoxetine "improved quality of Life," as measured by the World Health Organization Quality of Life assessment. However, it did not alter scores in the Hamilton Depression test.[105] In the third study, the addition of huperzine A to fluoxetine improved cognitive function, but made no difference in Hamilton Depression scores.[106]

11.6   SAFETY

The final, but perhaps most important consideration, is safety. A significant difference between herbal treatments and standard, prescribed antidepressant medications is that the FDA has insured a reasonable degree of safety of medications that it has approved for use in patients. None of the herbal treatments discussed in this text are approved by the FDA as treatments for MDD. This is not due to lack of efficacy, but rather to their classification as, "dietary supplements." The FDA does not evaluate the efficacy of dietary supplements. They simply demand that manufacturers and distributors of such products follow good manufacturing practices to ensure that supplements are processed consistently and meet quality standards. These regulations are intended to keep the wrong ingredients and contaminants out of supplements, as well as making sure that the right ingredients are included in appropriate amounts. If the FDA finds reason to declare a product to be unsafe, it can take action against the manufacturer or distributor or both. However, whereas pharmaceutical manufacturers must provide the FDA with proof of efficacy for a prescription drug to enter the market, the FDA must establish proof of toxicity for a supplement to be removed from public sale. Because of this relatively *laissez faire* approach of the FDA, the provider takes considerable responsibility upon him or herself in prescribing herbs as treatment.

The herbs discussed in Chapter 7 have all been found to be reasonably safe with prudent use. There are a few notable exceptions, such as *Piper methysticum* and *Salvia divinorum*. Although *Piper methysticum* has earned a reputation as being useful for treatment of mood disorders, it can in rare cases cause severe liver injury. *Salvia divinorum* is dangerous for a different reason, that being its propensity to cause psychotic reactions. It must be used, if at all, with extreme caution and under experienced guidance. Generally, the toxicities of the herbs discussed in this text are low, with available LD50 data uniformly showing lethal doses to be far beyond typical human therapeutic ranges. There have been a number of studies published in regard to safety of specific herbs in pregnancy, and basic information is provided for the specific herbs discussed. An excellent source of information for herbs that includes information on safety during pregnancy and lactation is the *Botanical Safety Handbook*.[141] It includes an appendix that is a long list of herbs prescribed for various purposes for which pregnancy is considered a contraindication. That source was frequently cited in this text, and it should be reviewed if safety concerns arise.

Basic information on interactions with drugs are given in the discussions of each herb. In many cases, no known interactions with drugs have been found. However, there are notable exceptions. Aside from the *Botanical Safety Handbook*, there are many well-researched reviews of herb–drug interaction in the literature. In one such report, it was generally noted that the largest percentage of interactions were between herbal treatments – of various kinds, not only for treatment of mood – and medications affecting the central nervous system or cardiovascular system.[142] Nearly half of those interactions were due to altered pharmacokinetics, and about a quarter of interactions were deemed "major interactions." In specifics of that report, *Hypericum perforatum* and *Ginkgo biloba* were among the herbs most frequently reported to interact with standard medications. Warfarin, insulin, aspirin, digoxin, and ticlopidine were medications that had the greatest number of reported interactions with herbal treatments. Another such review similarly states that most herb–drug interactions are pharmacokinetic in nature, and that *Hypericum perforatum* and warfarin are the most frequently reported participants in such interactions.[143] In this regard, it must be again noted that the addition of *Hypericum perforatum* to standard antidepressant treatment may increase the risk of serotonin syndrome. Thus, while circumspection must always be maintained, certain useful herbs have more than average propensities for drug interaction, mostly due to effects on specific cytochrome P450 enzymes. Those initial considerations of potential interactions between herbs and medications may guide one away from the use of a particular herb. Finally, the concentrations of both active and potentially toxic components of herbs may vary with locale, growing conditions, and times of year. Thus, while foraging for herbs or producing one's own extracts from raw product may be an admirable pursuit, it is wisest to depend upon standardized extracts from well-known, dependable, commercial sources.

REFERENCES

1. Cipriani A, Furukawa TA, Salanti G, et al. Comparative efficacy and acceptability of 12 new-generation antidepressants: a multiple-treatments meta-analysis. *Lancet.* 2009;373(9665):746–758.
2. Sinyor M, Schaffer A, Levitt A. The sequenced treatment alternatives to relieve depression (STAR*D) trial: a review. *Can J Psychiatry.* 2010;55(3):126–135.
3. Jang SW, Liu X, Yepes M, et al. A selective TrkB agonist with potent neurotrophic activities by 7,8dihydroxyflavone. *Proc Natl Acad Sci USA.* 2010;107(6):2687–2692.
4. Liu D, Cao G, Han L, et al. Flavonoids from Radix Tetrastigmae inhibit TLR4/MD-2 mediated JNK and NF-κB pathway with anti-inflammatory properties. *Cytokine.* 2016;84:29–36.
5. Spencer JPE. The interactions of flavonoids within neuronal signaling pathways. *Genes Nutr.* 2007;2(3):257–273.
6. Wang R, Sun Y, Huang H, et al. Rutin, a natural flavonoid protects PC12 cells against sodium nitroprusside-induced neurotoxicity through activating PI3K/Akt/mTOR and ERK1/2 pathway. *Neurochem Res.* 2015;40(9):1945–1953.
7. Gong EJ, Park HR, Kim ME, et al. Morin attenuates tau hyperphosphorylation by inhibiting GSK-3β. *Neurobiol Dis.* 2011;44(2):223–230.
8. German-Ponciano LJ, Rosas-Sánchez GU, Rivadeneyra-Domınguez E, et al. Advances in the preclinical study of some flavonoids as potential antidepressant agents. *Scientifica (Cairo).* 2018;2018:2963565.
9. Zhao B. Natural antioxidants protect neurons in Alzheimer's disease and Parkinson's disease. *Neurochem Res.* 2009;34(4):630–638.
10. Xi YD, Yu HL, Ding J, et al. Flavonoids protect cerebrovascular endothelial cells through Nrf2 and PI3K from β-amyloid peptide-induced oxidative damage. *Curr Neurovasc Res.* 2012;9(1):32–41.
11. Ahn H, Lee GS. Isorhamnetin and hyperoside derived from water dropwort inhibits inflammasome activation. *Phytomedicine.* 2017;24:77–86.
12. Linde K, Berner MM, Kriston L. St John's wort for major depression. *Cochrane Database Syst Rev.* 2008(4):Article ID CD000448.
13. Shelton RC, Keller MB, Gelenberg A, et al. Effectiveness of St John's wort in major depression: a randomized controlled trial. *JAMA.* 2001;285(15):1978–1986.
14. Hausenblas HA, Saha D, Dubyak PJ, et al. Saffron (*Crocus sativus* L.) and major depressive disorder: a meta-analysis of randomized clinical trials. *J Integr Med.* 2013;11(6):377–383.
15. Lopresti AL, Drummond PD. Saffron (*Crocus sativus*) for depression: a systematic review of clinical studies and examination of underlying antidepressant mechanisms of action. *Hum Psychopharmacol.* 2014;29(6):517–527.
16. Kamalipour M, Jamshidi AH, Akhondzadeh S. Antidepressant effect of *Crocus sativus*: an evidence-based review. *J Med Plant.* 2010;9(6):35–38.
17. Lopresti AL, Maes M, Maker GL, et al. Curcumin for the treatment of major depression: a randomized, double-blind, placebo-controlled study. *J Affect Dis.* 2014;167:368–375.

18. Kasper S, Volz HP, Dienel A, et al. Efficacy of Silexan in mixed anxiety-depression—a randomized, placebo-controlled trial. *Eur Neuropsychopharmacol.* 2016;26(2):331–340.

19. Amsterdam JD, Shults J, Soeller I, et al. Chamomile (*Matricaria recutita*) may provide antidepressant activity in anxious, depressed humans: an exploratory study. *Altern Ther Health Med.* 2012;18(5):44–49.

20. Chang SM, Chen CH. Effects of an intervention with drinking chamomile tea on sleep quality and depression in sleep disturbed postnatal women: a randomized controlled trial. *J Adv Nurs.* 2016;72(2):306–315.

21. Kyrou I, Christou A, Panagiotakos D, et al. Effects of a hops (*Humulus lupulus* L.) dry extract supplement on self-reported depression, anxiety and stress levels in apparently healthy young adults: a randomized, placebo-controlled, double-blind, crossover pilot study. *Hormones.* 2017;16(2):171–180.

22. Mao JJ, Xie SX, Zee J, et al. *Rhodiola rosea* versus sertraline for major depressive disorder: a randomized placebo-controlled trial. *Phytomedicine.* 2015;22(3):394–399.

23. Darbinyan V, Aslanyan G, Amroyan E, et al. Clinical trial of *Rhodiola rosea* L. extract SHR-5 in the treatment of mild to moderate depression. *Nord J Psychiatry.* 2007;61(5):343–348.

24. Olsson EMG, von Schéele B, Panossian AG. A randomized, double-blind, placebo-controlled, parallel-group study of the standardized extract SHR-5 of the roots of *Rhodiola rosea* in the treatment of subjects with stress-related fatigue. *Planta Med.* 2009;75(2):105–112.

25. Kasper S, Dienel A. Multicenter, open-label, exploratory clinical trial with *Rhodiola rosea* extract in patients suffering from burnout symptoms. *Neuropsychiatr Dis Treat.* 2017;13:889–898.

26. Panossian AG. Adaptogens in mental and behavioral disorders. *Psychiatr Clin North Am.* 2013;36(1):49–64.

27. Calabrese C, Gregory WL, Leo M, et al. Effects of a standardized *Bacopa monnieri* extract on cognitive performance, anxiety, and depression in the elderly: a randomized, double-blind, placebo-controlled trial. *J Altern Complement Med.* 2008;14(6):707–713.

28. Benson S, Downey LA, Stough C, et al. An acute, double-blind, placebo-controlled cross-over study of 320 mg and 640 mg doses of *Bacopa monnieri* (CDRI 08) on multitasking stress reactivity and mood. *Phytother Res.* 2014;28(4):551–559.

29. Bhattacharyya D, Sur TK, Jana U, et al. Controlled programmed trial of *Ocimum sanctum* leaf on generalized anxiety disorders. *Nepal Med Coll J.* 2008;10(3):176–179.

30. Nematolahia P, Mehrabani M, Karami-Mohajer S, et al. Effects of *Rosmarinus officinalis* L. on memory performance, anxiety, depression, and sleep quality in university students: a randomized clinical trial. *Complement Ther Clin Pract.* 2018;30:24–28.

31. Fulzele A, Hudda N. Selective ayurvedic therapy for the management of major depressive disorder: a randomised control trial. *Anc Sci Life.* 2012;32(Suppl 1):S41.

32. Sathyapalan T, Beckett S, Rigby AS, et al. High cocoa polyphenol rich chocolate may reduce the burden of the symptoms in chronic fatigue syndrome. *Nutr J.* 2010;9:55.

33. Einerson LS. *A Delphi Study on Herbs Used to Address Depression and Anxiety According to Master Herbalists.* Lubbock, TX: Texas Tech University; 2017.

34. Tenore GC, Daglia M, Orlando V, et al. Coffee and depression: a short review of literature. *Curr Pharm Des.* 2015;21(34):5034–5040.

35. Zhang Q, Yang H, Wang J, et al. Effect of green tea on reward learning in healthy individuals: a randomized, double-blind, placebo-controlled pilot study. *Nutr J.* 2013;12:84.

36. Bruce M, Scott N, Shine P, et al. Anxiogenic effects of caffeine in patients with anxiety disorders. *Arch Gen Psychiatry.* 1992;49(11):867–869.

37. Chang SC, Cassidy A, Willett WC, et al. Dietary flavonoid intake and risk of incident depression in midlife and older women. *Am J Clin Nutr.* 2016;104(3):704–714.

38. Samieri C, Sun Q, Townsend MK, et al. Dietary flavonoid intake at midlife and healthy aging in women. *Am J Clin Nutr.* 2014;100(6):1489–1497.
39. Burns PB, Rohrich RJ, Chung KC. The levels of evidence and their role in evidence-based medicine. *Plast Reconstr Surg.* 2011;128(1):305–310.
40. Hoffmann D. *Medical Herbalism: The Science and Practice of Herbal Medicine.* Rochester, VT: Healing Arts Press; 2003. p. 236–257.
41. Blumenthal M, Busse WR, Goldberg A, et al. *The Complete German Commission E Monographs.* Austin, TX: American Botanical Council; 1998.
42. Zhouke G. The effect of Yu-Le decoction to major depression of stagnated heat type of heart-liver and level change of 5-HT and NE in the blood plasma. *Chin Arch Tradit Chin Med.* 2009;8.
43. Yasui T, Yamada M, Uemura H, et al. Changes in circulating cytokine levels in midlife women with psychological symptoms with selective serotonin reuptake inhibitor and Japanese traditional medicine. *Maturitas.* 2009;62(2):146–152.
44. Zhang LD, Zhang YL, Xu SH, et al. Traditional Chinese medicine typing of affective disorders and treatment. *Am J Chin Med.* 1994;22(3–4):321–327.
45. Wang Y, Fan R, Huang X. Meta-analysis of the clinical effectiveness of traditional Chinese medicine formula *Chaihu-Shugan-San* in depression. *J Ethnopharmacol.* 2012;141(2):571–577.
46. Yeung WF, Chung KF, Ng KY, et al. A meta-analysis of the efficacy and safety of traditional Chinese medicine formula Ganmai Dazao decoction for depression. *J Ethnopharmacol.* 2014;153(2):309–317.
47. Ushiroyama T, Ikeda A, Sakuma K, et al. Chai-Hu-Gui-Zhi-Gan-Jiang-Tang regulates plasma interleukin-6 and soluble interleukin-6 receptor concentrations and improves depressed mood in climacteric women with insomnia. *Am J Chin Med.* 2005;33(5):703–711.
48. Bao ZX, Zhao GP, Sun W, et al. Clinical curative effects of *kaixin* powder on depression with mild or moderate degree. *Chin Arch Tradit Chin Med.* 2011;28(5):987–988.
49. Peng L, Zhang X, Kang DY, et al. Effectiveness and safety of *Wuling* capsule for post stroke depression: a systematic review. *Complement Ther Med.* 2014;22(3):549–566.
50. Huang L, Chen LJ, Liu LL, et al. A systematic review of therapeutic efficacy and safety of *shugan jieyu* capsules in the treatment of mild to moderate depression. *China Pharm.* 2013;24(32):3043–3046.
51. Lopresti AL, Drummond PD. Efficacy of curcumin, and a saffron/curcumin combination for the treatment of major depression: a randomized, double-blind, placebo-controlled study. *J Affect Disord.* 2017;207(1):188–196.
52. Shoba G, Joy D, Joseph T, Majeed M, Rajendran R, Srinivas PS. Influence of piperine on the pharmacokinetics of curcumin in animals and human volunteers. *Planta Med.* 1998;64(4):353–356.
53. Ranjbar M, Firoozabadi A, Salehi A, et al. Effects of herbal combination (*Melissa officinalis* L. and *Nepeta menthoides* Boiss. & Buhse) on insomnia severity, anxiety and depression in insomniacs: randomized placebo-controlled trial. *Integr Med Res.* 2018;7(4):328–332.
54. Talbott SM, Talbott JA, Pugh M. Effect of *Magnolia officinalis* and *Phellodendron amurense* (Relora®) on cortisol and psychological mood state in moderately stressed subjects. *J Int Soc Sports Nutr.* 2013;10(1):37.
55. Ren L, Tao W, Zhang H, et al. Two standardized fractions of *Gardenia jasminoides* Ellis with rapid antidepressant effects are differentially associated with BDNF up-regulation in the hippocampus. *J Ethnopharmacol.* 2016;187(1):66–73.
56. Wang J, Duan P, Cui Y, et al. Geniposide alleviates depression-like behavior via enhancing BDNF expression in hippocampus of streptozotocin-evoked mice. *Metab Brain Dis.* 2016;31(5):1113–1122.

57. Fawcett J, Kravitz HM. Anxiety syndromes and their relationship to depressive illness. *J Clin Psychiatry.* 1983;44(8):8–11.

58. Keller MB, Boland RJ. Implication of failing to achieve successful long-term mainte-nance treatment of recurrent unipolar depression. *Biol Psychiatry.* 1998;44(5):348–360.

59. Kornstein SG, Schneider RK. Clinical features of treatment-resistant depression. *J Clin Psychiatry.* 2001;62(Suppl 16):18–25.

60. Stewart R, Besset A, Bebbington P, et al. Insomnia comorbidity and impact and hyp-notic use by age group in a national survey population aged 16 to 74 years. *Sleep.* 2006;29(11):1391–1397.

61. Nutt D, Wilson S, Paterson L. Sleep disorders as core symptoms of depression. *Dialogues Clin Neurosci.* 2008;10(3):329–336.

62. Kasper S, Volz HP, Dienel A, et al. Efficacy of Silexan in mixed anxiety-depression—a randomized, placebo-controlled trial. *Eur Neuropsychopharmacol.* 2016;26(2):331–340.

63. Greenberg MJ, Slyer JT. Effectiveness of Silexan oral lavender essential oil compared to inhaled lavender essential oil aromatherapy on sleep in adults: a systematic review protocol. *JBI Database Syst Rev Implement Rep.* 2017;15(4):961–970.

64. Huang L, Abuhamdah S, Howes MJ, et al. Pharmacological profile of essential oils derived from *Lavandula angustifolia* and *Melissa officinalis* with anti-agitation prop-erties: focus on ligand-gated channels. *J Pharm Pharmacol.* 2008;60(11):1515–1522.

65. Akhondzadeh S, Naghavi HR, Vazirian M, et al. Passionflower in the treatment of generalized anxiety: a pilot double-blind randomized controlled trial with oxazepam. *J Clin Pharm Ther.* 2001;26(5):363–367.

66. Movafegh A, Alizadeh R, Hajimohamadi F, et al. Preoperative oral *Passiflora incar-nata* reduces anxiety in ambulatory surgery patients: a double-blind, placebo-con-trolled study. *Anesth Analg.* 2008;106(6):1728–1732.

67. Ngan A, Conduit R. A double-blind, placebo-controlled investigation of the effects of *Passiflora incarnata* (passionflower) herbal tea on subjective sleep quality. *Phytother Res.* 2011;25(8):1153–1159.

68. Amsterdam JD, Li Y, Soeller I, et al. A randomized, double-blind, placebo-controlled trial of oral *Matricaria recutita* (chamomile) extract therapy of generalized anxiety disorder. *J Clin Psychopharmacol.* 2009;29(4):378–382.

69. Mao JJ, Xie SX, Keefe JR, et al. Long-term chamomile (*Matricaria chamomilla* L.) treatment for generalized anxiety disorder: a randomized clinical trial. *Phytomedicine.* 2016;23(14):1735–1742.

70. Amsterdam JD, Shults J, Soeller I, et al. Chamomile (*Matricaria recutita*) may have antidepressant activity in anxious depressed humans: an exploratory study. *Altern Ther Health Med.* 2012;18(5):44–49.

71. Kupfersztain C, Rotem C, Fagot R, et al. The immediate effect of natural plant extract, *Angelica sinensis* and (Climex) for the treatment of hot flushes during menopause. A preliminary report. *Clin Exp Obstet Gynecol.* 2003;30(4):203–206.

72. Kakuta H, Yano-Kakuta E, Moriya K. Psychological and physiological effects in humans of eating chamomile jelly. In: *ISHS Acta Horticulture 749/International Symposium on Chamomile Research, Development and Production;* 2007.

73. Zick SM, Wright BD, Sen A, et al. Preliminary examination of the efficacy and safety of a standardized chamomile extract for chronic primary insomnia: a randomized pla-cebo-controlled pilot study. *BMC Complement Altern Med.* 2011;11:78.

74. Viola H, Wasowski C, Levi de Stein M, et al. Apigenin, a component of *Matricaria recutita* flowers, is a central benzodiazepine receptors-ligand with anxiolytic effects. *Planta Med.* 1995;61(3):213–216.

75. Alijaniha F, Naseri M, Afsharypuor S, et al. Heart palpitation relief with *Melissa offi-cinalis* leaf extract: double blind, randomized, placebo-controlled trial of efficacy and safety. *J Ethnopharmacol.* 2015;164(22):378–384.

76. Brock C, Whitehouse J, Ihab Tewfik I, et al. American skullcap (*Scutellaria lateriflora*): an ancient remedy for today's anxiety? *Br J Wellbeing.* 2010;1(4):25–30.

77. Hui KM, Huen MSY, Wang HY, et al. Anxiolytic effect of wogonin, a benzodiazepine receptor ligand isolated from *Scutellaria baicalensis* Georgi. *Biochem Pharmacol.* 2002;64(9):1415–1424.

78. Sarris J, Wardle J. *Clinical Naturopathy: In Practice.* Chatswood, UK: Elsevier; 2017. p. 67.

79. Blumenthal M, Busse WR, Goldberg A, et al. *The Complete German Commission E Monograph: Therapeutic Guide to Herbal Medicines.* Austin, TX: American Botanical Council; 1998. p. 134.

80. Kendell RE, DiScipio WJ. Obsessional symptoms and obsessional personality traits in patients with depressive illness. *Br J Psychiatry.* 1980;136:1–25.

81. Schruers K, Koning K, Luermans J, et al. Obsessive-compulsive disorder: a critical review of therapeutic perspectives. *Acta Psychiatr Scand.* 2005;111(4):261–271.

82. Saiiah-bargard M, Boostani H, Saiiah M, et al. Efficacy of aqueous extract of Echium amoenum L. in the treatment of mild to moderate obsessive: compulsive disorder. *J Med Plants.* 2005;3(15):43–50.

83. Sayyah M, Boostani H, Pakseresht S, et al. Efficacy of aqueous extract of *Echium amoenum* in treatment of obsessive-compulsive disorder. *Prog Neuropsychopharmacol Biol Psychiatry.* 2009;33(8):1513–1516.

84. Sayyah M, Boostani H, Pakseresht S, et al. Comparison of *Silybum marianum* (L.) Gaertn. with fluoxetine in the treatment of obsessive–compulsive disorder. *Prog Neuropsychopharmacol Biol Psychiatry.* 2010;34(2):362–365.

85. Taylor LH, Kobak KA. An open-label trial of St. John's wort (*Hypericum perforatum*) in obsessivecompulsive disorder. *J Clin Psychiatry.* 2000;61(8):575–578.

86. Kobak KA, Taylor LV, Bystritsky A, et al. St John's wort versus placebo in obsessive-compulsive disorder: results from a double-blind study. *Int Clin Psychopharmacol.* 2005;20(6):299–304.

87. Kian RF, Bekhradi R, Rahimi R, et al. Evaluating the effect of fennel soft capsules on the quality of life and its different aspects in menopausal women: a randomized clinical trial. *Nurs Pract Today.* 2017;4(2):87–95.

88. Ghazanfarpour M, Mohammadzadeh F, Shokrollahi P, et al. Effect of *Foeniculum vulgare* (fennel) on symptoms of depression and anxiety in postmenopausal women: a double-blind randomised controlled trial. *J Obstet Gynaecol.* 2018;38(1): 121–126.

89. Warnecke G. Influence of a phytopharmaceutical on climacteric complaints. *Meizinisch Welt.* 1985;36:871–874.

90. Lucks BC, Sørensen J, Veal L. *Vitex agnus-castus* essential oil and menopausal balance: a self-care survey. *Complement Ther Nurs Midwifery.* 2002;8(3):148–154.

91. Propping D, Burger HG, Teede HJ. Vitex agnus castus. The treatment of gynaecological syndromes. *Therapeutikon.* 1991;5:581–585.

92. Lauritzen CH, Reuter HD, Repges R, et al. Treatment of premenstrual tension syndrome with Vitex agnus castus: controlled, double-blind study versus pyridoxine. *Phytomedicine.* 1997;4:183–189.

93. Loch EG, Selle H, Boblitz N. Treatment of premenstrual syndrome with a phytopharmaceutical formulation containing *Vitex agnus castus. J Women's Health Gend Based Med.* 2000;9(3):315–320.

94. Heydari N, Dehghani M, Emamghoreishi M, et al. Effect of *Melissa officinalis* capsule on the mental health of female adolescents with premenstrual syndrome: a clinical trial study. *Int J Adolesc Med Health.* 2018;15:2191–0278.

95. Omidali F. The effect of pilates exercise and consuming fennel on pre-menstrual syndrome symptoms in non-athletic girls. *CMJA.* 2015;5(2):1203–1213.

96. Pazoki H, Bolouri G, Farokhi F, et al. Comparing the effects of aerobic exercise and *Foeniculum vulgare* on pre-menstrual syndrome. *Middle East Fertil Soc J.* 2016;21(1):61–64.

97. Delaram M, Kheiri S, Hodjati MR. Comparing the effects of *Echinophora-platyloba*, fennel and placebo on pre-menstrual syndrome. *J Reprod Infertil.* 2011;12(3): 221–226.

98. Gama CRB, Lasmar R, Gama GF, et al. Premenstrual syndrome: clinical assessment of treatment outcomes following Borago officinalis extract therapy. *RBM.* 2014;71:211–217.

99. Eckmann F, Schlag H. Kontrollierte Doppelbind-Studie zum Wirksamkeitsnachweis von tebonin forte bei Patienten mit zerebrovaskulärer Insuffizienz. *Fortschr Med.* 1982;100(31–32):1474–1478.

100. Wesnes K, Simmons D, Rook M, et al. A double-blind placebo-controlled trial of Tanakan in the treatment of idiopathic cognitive impairment in elderly. *Hum Psychopharmacol.* 1987;2(3):159–169.

101. Mori K, Inatomi S, Ouchi K, Azumi Y, et al. Improving effects of the mushroom *Yamabushitake (Hericium erinaceus)* on mild cognitive impairment: a double-blind placebo-controlled clinical trial. *Phytother Res.* 2009;23(3):367–372.

102. Holtzheimer PE, Meeks TW, Kelley ME, et al. A double blind, placebo-controlled pilot study of galantamine augmentation of antidepressant treatment in older adults with major depression. *Int J Geriatr Psychiatry.* 2008;23(6):625–631.

103. Tewari D, Stankiewicz AM, Mocan A, et al. Ethnopharmacological approaches for dementia therapy and significance of natural products and herbal drugs. *Front Aging Neurosci.* 2018;10:3.

104. Liu SZ, Wang PJ, Yin AJ, et al. Effects of huperzine A combined with venlafaxine for patients with depression. *Zhongguo Shi Yong Yi Yao.* 2010;5(11):151–152.

105. Gao YF, Li J, Meng H. Effects of huperzine on cognition, function and life quality of patients with depression. *Chongqing Yi Xue.* 2007;36(6):483–485.

106. Yang ZB, Deng XM, Zhang GX, et al. The study of huperzine combined with fluoxetine on cognition function of patients with depression. *Lin Chuang Jing Shen Yi Xue Zhi.* 2010;20(6):418–419.

107. Rafii MS, Walsh S, Little JT, et al. A phase II trial of huperzine A in mild to moderate Alzheimer disease. *Neurology.* 2011;76(16):1389–1394.

108. Ota A, Ulrih NP. An overview of herbal products and secondary metabolites used for management of type two diabetes. *Front Pharmacol.* 2017;8:Article ID 436.

109. Mendelson SD. *Metabolic Syndrome and Psychiatric Illness: Interactions, Pathophysiology, Assessment and Treatment.* London, UK: Academic Press; 2008.

110. Rasgon NL, McEwen BS. Insulin resistance-a missing link no more. *Mol Psychiatry.* 2016;21(12):1648–1652.

111. Zhao Q, Wu X, Yan S, et al. The antidepressant-like effects of pioglitazone in a chronic mild stress mouse model are associated with PPARγ-mediated alteration of microglial activation phenotypes. *J Neuroinflammation.* 2016;13(1):259–276.

112. Ishola IO, Akinyede AA, Sholarin AM. Antidepressant and anxiolytic properties of the methanolic extract of *Momordica charantia* Linn (*Cucurbitaceae*) and its mechanism of action. *Drug Res (Stuttg).* 2014;64(7):368–376.

113. Randhawa K, Kumar D, Jamwal A, Kumar S. Screening of antidepressant activity and estimation of quercetin from *Coccinia indica* using TLC densitometry. *Pharm Biol.* 2015;53(12):1867–1874.

114. Ismail MO, Dar A, Faizi S, et al. Antidepressant-like actions of *Opuntia dillenii* butanol fraction in rodents. *Pak J Pharmacol.* 2010;27(2):9–14.

115. Khosravi H, Rahnema M, Asle RM. Anxiolytic and antidepressant effects of tarragon (*Artemisia dracunculus*) hydroalcoholic extract in male rats exposed to chronic restraint stress. *Nova Biol Reperta.* 2017;4(1):1–8.

116. Pawar VS, Hugar S, Gawade B, et al. Evaluation of antidepressant-like activity of *Trigonella foenum graecum* Linn. Seeds in mice. *Pharmacologyonline*. 2008;1:455–465.

117. Singh A, Singh A. Antidepressant activity of aqueous extract of *Allium sativum* Linn. in albino rats. *Int J Sci Res*. 2018;7(4):36–38.

118. Shishtar E, Sievenpiper JL, Djedovic V, et al. The effect of ginseng (the genus panax) on glycemic control: a systematic review and meta-analysis of randomized controlled clinical trials. *PLoS One*. 2014;9(9):e107391.

119. Medagama AB. The glycaemic outcomes of cinnamon, a review of the experimental evidence and clinical trials. *Nutr J*. 2015;14:108.

120. Freudenberger HJ. Staff burnout. *J Soc Issues*. 1974;30(1):159–165.

121. Brekhman II. A new medicinal plant of the family *Araliceae* the spiny *Eleutherococcus*. *Iz Sibir Otdel Akad Nauk USSR*. 1960;9:113–120.

122. Panossian A, Wikman G. Evidence-based efficacy of adaptogens in fatigue, and molecular mechanisms related to their stress-protective activity. *Curr Clin Pharmacol*. 2009;4(3):198–219.

123. Pawar VS, Shivakumar H. A current status of adaptogens: natural remedy to stress. *Asian Pac J Trop Dis*. 2012;2(Suppl 1):S480–S490.

124. Jarcho MR, Slavich GM, Tylova-Stein H, et al. Dysregulated diurnal cortisol pattern is associated with glucocorticoid resistance in women with major depressive disorder. *Biol Psychol*. 2013;93(1):150–158.

125. McEwen BS. Allostasis and allostatic load: implications for neuropsychopharmacology. *Neuropsychopharmacology*. 2000;22(2):108–124.

126. Hung SK, Perry R, Ernst E. The effectiveness and efficacy of *Rhodiola rosea* L.: a systematic review of randomized clinical trials. *Phytomedicine*. 2011;18(4):235–244.

127. Panossian A, Wikman G. Effects of adaptogens on the central nervous system and the molecular mechanisms associated with their stress—protective activity. *Pharmaceuticals*. 2010;3(1):188–224.

128. Liao LY, He YF, Li L, et al. A preliminary review of studies on adaptogens: comparison of their bioactivity in TCM with that of ginseng-like herbs used worldwide. *Chin Med*. 2018;13:57.

129. Rege NN, Thatte UM, Dahanukar SA. Adaptogenic properties of six rasayana herbs used in Ayurvedic medicine. *Phytother Res*. 1999;13(4):275–291.

130. Marciano M, Vizniak NA. *Botanical Medicine*. Toronto, ON: Prohealth; 2016. p. 52.

131. Hoffmann D. *Medical Herbalism: The Science and Practice of Herbal Medicine*. Rochester, VT: Healing Arts Press; 2003. p. 483.

132. Yu JJ, Pei LB, Zhang Y, et al. Chronic supplementation of curcumin enhances the efficacy of antidepressants in major depressive disorder: a randomized, double-blind, placebo-controlled pilot study. *J Clin Psychopharmacol*. 2015;35(4):406–410.

133. Akhondzadeh S, Kashani L, Fotouhi A, et al. Comparison of *Lavandula angustifolia* Mill. tincture and imipramine in the treatment of mild to moderate depression: a double-blind, randomized trial. *Prog Neuropsychopharmacol Biol Psychiatry*. 2003;27(1):123–127.

134. Nikfarjam M, Parvin N, Assarzadegan N, et al. The effects of *Lavandula angustifolia* mill infusion on depression in patients using citalopram: a comparison study. *Iran Red Crescent Med J*. 2013;15(8):734–739.

135. Wu R, Zhu D, Xia Y, et al. A role of *Yueju* in fast-onset antidepressant action on major depressive disorder and serum BDNF expression: a randomly double-blind, fluoxetine-adjunct, placebo-controlled, pilot clinical study. *Neuropsychiatr Dis Treat*. 2015;11:2013–2021.

136. Ghaderi H, Nikan R, Rafieian-Kopaei M, et al. The effect of *Cinnamon zeylanicum* essential oil on treatment of patients with unipolar nonpsychotic major depression disorder treated with fluoxetine. *Pharmacophore*. 2017;8(3):24–31.

137. Jeong HG, Ko YH, Oh SY, et al. Effect of Korean Red ginseng as an adjuvant treatment for women with residual symptoms of major depression. *Asia Pac Psychiatry.* 2015;7(3):330–336.

138. Li L, Li L, Yang L. Clinical observation on ginseng Tiaopi powder and venlafaxine in treating 30 cases of depression. *J Fujian Univ Tradit Chin Med.* 2008;2.

139. Schubert H, Halama P. Depressive episode primarily unresponsive to therapy in elderly patients: efficacy of *Ginkgo biloba* extract (EGb 761) in combination with antidepressants. *Geriatr Forsch.* 1993;1:45–53.

140. Liu SZ, Wang PJ, Yin AJ, et al. Effects of huperzine A combined with venlafaxine for patients with depression. *Zhongguo Shi Yong Yi Yao.* 2010;5(11):151–152.

141. Gardner Z, McGuffin M, editors. *Botanical Safety Handbook.* 3rd ed. 2013. Boca Raton, FL: CRC Press; 2013.

142. Tsai HH, Lin HW, Simon Pickard A, et al. Evaluation of documented drug interactions and contraindications associated with herbs and dietary supplements: a systematic literature review. *Int J Clin Pract.* 2012;66(11):1056–1078.

143. Brazier NC, Levine MAH. Drug-herb interaction among commonly used conventional medicines: a compendium for health care professionals. *Am J Ther.* 2003;10(3):163–169.

Index

2,4-Bis(2-hydroxy-5-prop-2-enylphenyl)- 6-prop-2-enylphenol (macranthol), 363
3,6'-Disinapoyl sucrose, 357
3-n-Butylphthalide, 354
4-Hydroxyisoleucine, 362
5-HIAA, *see* 5-Hydroxyindoleacetic acid
5-Hydroxyindoleacetic acid (5-HIAA), 9
7,8,Dihydroxyflavone, 333–334

Alkaloids, 35–36
Alles, Gordon, 5
Allium sativum (garlic)
    antidiabetic/anti-metabolic syndrome effects, 51
    anti-inflammatory effects, 50–51
    antioxidant effects, 50
    dosage of, 52
    drug interactions of, 53
    human antidepressant effects, 52
    overview of, 49–50
    preclinical antidepressant-like effects, 51–52
    safety in pregnancy, 53
    toxicity of, 53
Amentoflavone, 331
American Journal of Psychiatry, 9
Amyrins, 352–353
Angelica sinensis
    antidiabetic/anti-metabolic syndrome effects, 56
    anti-inflammatory effects, 56
    antioxidant effects, 55
    dosage of, 57
    drug interactions of, 57
    human antidepressant effects, 57
    overview of, 55
    preclinical antidepressant-like effects, 56–57
    safety in pregnancy, 57
    toxicity of, 57
Antidepressants, 7–10
Anxiety, 392–394
Apigenin, 332
Apium graveolens (celery)
    antidiabetic/anti-metabolic syndrome effects, 60
    anti-inflammatory effects, 59–60
    antioxidant effects, 59
    dosage of, 61
    drug interactions of, 61–62
    human antidepressant effects, 61
    overview of, 59

    preclinical antidepressant-like effects, 60–61
    safety in pregnancy, 61
    toxicity of, 61
Ashwagandha, *see Withania somnifera*
Assessment, antidepressant effects
    antidiabetic/anti-metabolic syndrome effects, 41
    anti-inflammatory effects, 40–41
    antioxidant effects, 39–40
    preclinical antidepressant-like effects
        forces swim test, 42–44
        sucrose consumption test, 44–45
        tail suspension test, 44
        test conditions, 45–46
Astilbin, 332
Astragalus membranaceus
    antidiabetic/anti-metabolic syndrome effects, 64
    anti-inflammatory effects, 64
    antioxidant effects, 63–64
    dosage of, 65
    drug interactions of, 66
    human antidepressant effects, 65
    overview of, 63
    preclinical antidepressant-like effects, 64–65
    safety in pregnancy, 65
    toxicity of, 65
Atractylodes macrocephala
    antidiabetic/anti-metabolic syndrome effects, 68
    anti-inflammatory effects, 68
    antioxidant effects, 67
    dosage of, 69
    drug interactions of, 69
    human antidepressant effects, 68–69
    overview of, 67
    preclinical antidepressant-like effects, 68
    safety in pregnancy, 69
    toxicity of, 69
Augmentation of standard antidepressant herbal treatment, 398–399
Avena sativa (common oat)
    antidiabetic/anti-metabolic syndrome effects, 71
    anti-inflammatory effects, 70–71
    antioxidant effects, 70
    dosage of, 71
    drug interactions of, 72
    human antidepressant effects, 71

overview of, 70
preclinical antidepressant-like effects, 71
safety in pregnancy, 72
toxicity of, 72
Axelrod, Julius, 5

Bacon, Francis, 3
Bacopa monnieri
    antidiabetic/anti-metabolic syndrome
        effects, 74
    anti-inflammatory effects, 74
    antioxidant effects, 73–74
    dosage of, 75
    drug interactions of, 75
    human antidepressant effects, 75
    overview of, 73
    preclinical antidepressant-like effects, 74
    safety in pregnancy, 75
    toxicity of, 75
Bacopasides, 353
Baicalein, 332
Baicalin, 332
Ban xia hou pu, 322
BDNF, *see* Brain-derived neurotrophic factor
Berberine, 353–354
β-Carotene, 355
β-Caryophyllene, 356
Black cohosh, *see Cimicifuga racemosa*
Black pepper, *see Piper nigrum*
Borago officinalis (European Borage)
    antidiabetic/anti-metabolic syndrome
        effects, 78
    anti-inflammatory effects, 77
    antioxidant effects, 77
    dosage of, 78
    drug interactions of, 79
    human antidepressant effects, 78
    overview of, 77
    preclinical antidepressant-like effects, 78
    safety in pregnancy, 79
    toxicity of, 78–79
Brahmi, *see Bacopa monnieri*
Brain-derived neurotrophic factor (BDNF),
        14, 21
British Journal of General Practice, 9
Bupleurum chinense
    antidiabetic/anti-metabolic syndrome
        effects, 81
    anti-inflammatory effects, 80–81
    antioxidant effects, 80
    dosage of, 82
    drug interactions of, 82
    human antidepressant effects, 81
    overview of, 80
    preclinical antidepressant-like effects, 81
    safety in pregnancy, 82
    toxicity of, 82

Cade, John, 4, 16
Caffeic acid, 354–355
Cajal, Ramon y, 4
Camellia sinensis (tea)
    antidiabetic/anti-metabolic syndrome
        effects, 84
    anti-inflammatory effects, 84
    antioxidant effects, 84
    dosage of, 85
    drug interactions of, 86
    human antidepressant effects, 85
    overview of, 83–84
    preclinical antidepressant-like
        effects, 84–85
    safety in pregnancy, 86
    toxicity of, 85–86
Cannabis
    antidiabetic/anti-metabolic syndrome
        effects, 88–89
    anti-inflammatory effects, 88
    antioxidant effects, 88
    dosage of, 91
    drug interactions of, 92
    human antidepressant effects, 90–91
    overview of, 87–88
    preclinical antidepressant-like
        effects, 89–90
    safety in pregnancy, 92
    toxicity of, 91–92
Carbohydrates, 31–32
Carlsson, Arvid, 5
Carvacrol, 355
Cecropia
    antidiabetic/anti-metabolic syndrome
        effects, 95–96
    anti-inflammatory effects, 95
    antioxidant effects, 95
    dosage of, 97
    drug interactions of, 97
    human antidepressant effects, 96
    overview of, 95
    preclinical antidepressant-like effects, 96
    safety in pregnancy, 97
    toxicity of, 97
Celery, *see Apium graveolens*
Celsus, Aulus Cornelius, 2
Centella asiatica (Gotu Kola)
    antidiabetic/anti-metabolic syndrome
        effects, 99
    anti-inflammatory effects, 99
    antioxidant effects, 99
    dosage of, 100
    drug interactions of, 101
    human antidepressant effects, 100
    overview of, 98–99
    preclinical antidepressant-like effects, 100
    safety in pregnancy, 101

toxicity of, 101
Chai hu jia long gu mu li, 322
Chai hu shu gan, 320
Chamomile, *see Matricaria recutita*
Chaste tree, *see Vitex agnus-castus*
Cheese effect, 7
Chlorogenic acid, 356
Chocolate, *see Theobroma cacao*
Chronic stress, 23–24
Chrysactinia mexicana
  antidiabetic/anti-metabolic syndrome
    effects, 103
  anti-inflammatory effects, 103
  antioxidant effects, 103
  dosage of, 104
  drug interactions of, 104
  human antidepressant effects, 103
  overview of, 102
  preclinical antidepressant-like effects, 103
  safety in pregnancy, 104
  toxicity of, 104
Chrysin, 332–333
Cimicifuga racemosa (black cohosh)
  antidiabetic/anti-metabolic syndrome
    effects, 106–107
  anti-inflammatory effects, 106
  antioxidant effects, 105–106
  dosage of, 108
  drug interactions of, 108
  human antidepressant effects, 107–108
  overview of, 105
  preclinical antidepressant-like effects, 107
  safety in pregnancy, 108
  toxicity of, 108
Cinnamomum zeylanicum (Cinnamon)
  antidiabetic/anti-metabolic syndrome
    effects, 111–112
  anti-inflammatory effects, 111
  antioxidant effects, 110–111
  dosage of, 113
  drug interactions of, 113
  human antidepressant effects, 112–113
  overview of, 110
  preclinical antidepressant-like effects, 112
  safety in pregnancy, 113
  toxicity of, 113
Cinnamon, *see Cinnamomum zeylanicum*
Club drug, 13
Cocoa, *see Theobroma cacao*
Coffea arabica (coffee)
  antidiabetic/anti-metabolic syndrome
    effects, 116
  anti-inflammatory effects, 116
  antioxidant effects, 115–116
  dosage of, 117–118
  drug interactions of, 118
  human antidepressant effects, 117

overview of, 115
  preclinical antidepressant-like effects,
    116–117
  safety in pregnancy, 118
  toxicity of, 118
Coffee, *see Coffea arabica*
Common oat, *see Avena sativa*
Coriander, *see Coriandrum sativum*
Coriandrum sativum (coriander)
  antidiabetic/anti-metabolic syndrome
    effects, 121
  anti-inflammatory effects, 121
  antioxidant effects, 120–121
  dosage of, 122
  drug interactions of, 122
  human antidepressant effects, 122
  overview of, 120
  preclinical antidepressant-like effects,
    121–122
  safety in pregnancy, 122
  toxicity of, 122
Corydalis yanhusuo
  antidiabetic/anti-metabolic syndrome
    effects, 124
  anti-inflammatory effects, 124
  antioxidant effects, 124
  dosage of, 125
  drug interactions of, 126
  human antidepressant effects, 125
  overview of, 123–124
  preclinical antidepressant-like effects, 125
  safety in pregnancy, 125
  toxicity of, 125
Crocin, 356
Crocus sativa (saffron)
  antidiabetic/anti-metabolic syndrome
    effects, 128
  anti-inflammatory effects, 127–128
  antioxidant effects, 127
  dosage of, 129
  drug interactions of, 130
  human antidepressant effects, 129
  overview of, 127
  preclinical antidepressant-like effects,
    128–129
  safety in pregnancy, 130
  toxicity of, 129
Culpeper, Nicholas, 3
Curcuma longa (turmeric)
  antidiabetic/anti-metabolic syndrome
    effects, 133
  anti-inflammatory effects, 132
  antioxidant effects, 132
  dosage of, 133
  drug interactions of, 134
  human antidepressant effects, 133
  overview of, 132

preclinical antidepressant-like effects, 133
safety in pregnancy, 134
toxicity of, 133
Curcumin, 356–357
Cyperus rotundus
   antidiabetic/anti-metabolic syndrome
      effects, 136
   anti-inflammatory effects, 136
   antioxidant effects, 135–136
   dosage of, 137
   drug interactions of, 137
   human antidepressant effects, 137
   overview of, 135
   preclinical antidepressant-like effects,
      136–137
   safety in pregnancy, 137
   toxicity of, 137

Dale, Henry Sir, 5
Daylily, *see Hemerocallis citrina*
Dementia, 395–396
Diabetes, 396–397

Echium amoenum
   antidiabetic/anti-metabolic syndrome
      effects, 139
   anti-inflammatory effects, 139
   antioxidant effects, 139
   dosage of, 140
   drug interactions of, 140
   human antidepressant effects, 140
   overview of, 139
   preclinical antidepressant-like effects,
      139–140
   safety in pregnancy, 140
   toxicity of, 140
eEF2, *see* Eukaryotic elongation factor 2
Elavil, 8
Eleutherococcus senticoccus (Siberian Ginseng)
   antidiabetic/anti-metabolic syndrome
      effects, 142–143
   anti-inflammatory effects, 142
   antioxidant effects, 142
   dosage of, 143
   drug interactions of, 144
   human antidepressant effects, 143
   overview of, 141–142
   preclinical antidepressant-like effects, 143
   safety in pregnancy, 144
   toxicity of, 144
Ellagic acid, 357–358
The English Physician (Culpeper), 3
Epimedium brevicornum (horny goat weed)
   antidiabetic/anti-metabolic syndrome
      effects, 146–147
   anti-inflammatory effects, 146
   antioxidant effects, 145–146

dosage of, 148
drug interactions of, 148
human antidepressant effects, 147
overview of, 145
preclinical antidepressant-like effects, 147
safety in pregnancy, 148
toxicity of, 148
Erspamer, Vittorio, 5
Eugenol, 358
Eukaryotic elongation factor 2 (eEF2), 14–15
European Borage, *see Borago officinalis*

Fall through the K-hole, 13
Fatigue, 397–398
Fennel, *see Foeniculum vulgare*
Fenugreek, *see Trigonella foenum-graecum*
Ferulic acid, 358–359
Fisetin, 334
Flavonoids
   antidepressant action mechanism, 340–343
   definition of, 34
   overview of, 331
   in phenolics, 34
   with preclinical antidepressant-like
      effects, 331–343
   synthetic, 340
Fluoxetine, 8
Foeniculum vulgare (fennel)
   antidiabetic/anti-metabolic syndrome
      effects, 150
   anti-inflammatory effects, 150
   antioxidant effects, 150
   dosage of, 151
   drug interactions of, 152
   human antidepressant effects, 151
   overview of, 149–150
   preclinical antidepressant-like effects, 151
   safety in pregnancy, 152
   toxicity of, 151
Forces swim test, 42–44
Freud, Sigmund, 4

Galileo, 3
Gallic acid, 359
γ-Glutamylethylamide, 372
Gan mai da zao, 320–321
Garlic, *see Allium sativum*
Gastrodin, 360
General malaise, 397–398
Genipin, 360–361
Ginkgo biloba
   antidiabetic/anti-metabolic syndrome
      effects, 154
   anti-inflammatory effects, 154
   antioxidant effects, 154
   dosage of, 156
   drug interactions of, 156

human antidepressant effects, 155–156
overview of, 153–154
preclinical antidepressant-like effects,
    154–155
safety in pregnancy, 156
toxicity of, 156
Ginseng, *see Panax ginseng*
Ginsenoside Rg1, 361
Glycogen synthase kinase 3beta (GSK-3β),
    14–15
Glycyrrhiza (licorice)
antidiabetic/anti-metabolic syndrome
    effects, 159
anti-inflammatory effects, 158–159
antioxidant effects, 158
dosage of, 159
drug interactions of, 160
human antidepressant effects, 159
overview of, 158
preclinical antidepressant-like effects, 159
safety in pregnancy, 160
toxicity of, 159–160
Glycyrrhizin, 361
Gotu Kola, *see Centella asiatica*
Griesinger, Wilhelm, 4
GSK-3β, *see* Glycogen synthase kinase 3beta
Gui pi, 321

Hedyosmum brasiliense
antidiabetic/anti-metabolic syndrome
    effects, 162
anti-inflammatory effects, 162
antioxidant effects, 162
dosage of, 162
drug interactions of, 163
human antidepressant effects, 162
overview of, 161–162
preclinical antidepressant-like effects, 162
safety in pregnancy, 163
toxicity of, 163
Hemerocallis citrina (daylily)
antidiabetic/anti-metabolic syndrome
    effects, 164
anti-inflammatory effects, 164
antioxidant effects, 164
dosage of, 165
drug interactions of, 166
human antidepressant effects, 165
overview of, 163–164
preclinical antidepressant-like effects,
    164–165
safety in pregnancy, 165
toxicity of, 165
Heptomethoxyflavone, 334
Herbal treatments
augmentation of standard antidepressant
    treatment, 398–399

combinations of, 391–392
comorbidities
anxiety and insomnia, 392–394
dementia, 395–396
diabetes, 396–397
fatigue, 397–398
general malaise, 397–398
lack of resiliency, 397–398
metabolic syndrome, 396–397
obsessive-compulsive disorder, 394
perimenopausal symptom, 395
premenstrual symptom, 395
compelling evidence of efficacy, 390–391
efficacy of, 388–390
safety of, 399–401
Herbs with antidepressant effects
Allium sativum (garlic), 49–53
Angelica sinensis, 55–57
Apium graveolens (celery), 59–62
Astragalus membranaceus, 63–66
Atractylodes macrocephala, 67–69
Avena sativa (common oat), 70–72
Bacopa monnieri, 73–75
Borage officinalis (European Borage), 77–79
Bupleurum chinense, 80–82
Camellia sinensis (tea), 83–86
Cannabis, 87–92
Cecropia, 95–97
Centella asiatica (Gotu Kola), 98–101
Chrysactinia mexicana, 102–104
Cimicifuga racemosa (Black Cohosh),
    105–108
Cinnamomum zeylanicum (Cinnamon),
    110–113
Coffea arabica (coffee), 115–118
Coriandrum sativum (coriander), 120–122
Corydalis yanhusuo, 123–126
Crocus sativa (saffron), 127–130
Curcuma longa (turmeric), 132–134
Cyperus rotundus, 135–137
Echium amoenum, 139–140
Eleutherococcus senticoccus (Siberian
    Ginseng), 141–144
Epimedium brevicornum (Horny Goat
    Weed), 145–148
Foeniculum vulgare (Fennel), 149–152
Ginkgo biloba, 153–156
Glycyrrhiza (Licorice), 158–160
Hedyosmum brasiliense, 161–163
Hemerocallis citrina (daylily), 163–166
Hericium erinaceus (Lion's Mane), 167–170
Hibiscus rosa-sinensis (hibiscus), 171–173
Humulus lupulus (hops), 174–176
Huperzia serrata, 178–180
Hypericum perforatum (St John's wort),
    181–185
Ilex paraguariensis (yerba mate), 187–189

Lavandula (lavender), 190–193
Ligusticum chuanxiong, 195–197
Magnolia officinalis, 198–201
Matricaria recutita (chamomile), 203–206
Melissa officinalis (lemon balm), 208–210
Mimosa pudica, 211–214
Ocimum basilicum (sweet basil), 215–217
Origanum vulgare (oregano), 219–221
Paeonia lactiflora (peony), 222–225
Panax ginseng (ginseng), 226–229
Passifloraceae incarnata (passionflower), 231–233
Piper methysticum (kava), 234–236
Piper nigrum (black pepper), 238–240
Polygala tenuifolia, 240–245
Poria cocos, 247–249
Psoralea corylifolia, 250–253
Rhodiola rosea, 255–258
Rosmarinus officinalis (rosemary), 260–262
Salvia divinorum, 264–267
Sceletium tortuosum, 269–271
Schisandra chinensis, 273–275
Scutellaria lateriflora (skullcap), 276–278
Silybum marianum (milk thistle), 280–283
Theobroma cacao (chocolate), 284–287
Tilia (linden), 287–291
Trigonella foenum-graecum (fenugreek), 292–296
Valeriana officinalis (valerian), 298–301
Verbena officinalis (vervain), 301–305
Vitex agnus-castus (chaste tree), 305–309
Withania somnifera (Ashwagandha), 311–313
Hericium erinaceus (Lion's Mane)
  antidiabetic/anti-metabolic syndrome effects, 168
  anti-inflammatory effects, 167–168
  antioxidant effects, 167
  dosage of, 169
  drug interactions of, 170
  human antidepressant effects, 169
  overview of, 167
  preclinical antidepressant-like effects, 168
  safety in pregnancy, 170
  toxicity of, 169–170
Hesperidin, 334–335
Hesperitin, 334–335
Hibiscus, *see Hibiscus rosa-sinensis*
Hibiscus rosa-sinensis (hibiscus)
  antidiabetic/anti-metabolic syndrome effects, 172
  anti-inflammatory effects, 172
  antioxidant effects, 171–172
  dosage of, 173
  drug interactions of, 173
  human antidepressant effects, 172–173
  overview of, 171

  preclinical antidepressant-like effects, 172
  safety in pregnancy, 173
  toxicity of, 173
Hillarp, Nils-Ake, 5
Hippocrates, 2
Hops, *see Humulus lupulus*
Horny goat weed, *see Epimedium brevicornum*
Humulus lupulus (hops)
  antidiabetic/anti-metabolic syndrome effects, 175
  anti-inflammatory effects, 175
  antioxidant effects, 174–175
  dosage of, 176
  drug interactions of, 176
  human antidepressant effects, 176
  overview of, 174
  preclinical antidepressant-like effects, 175–176
  safety in pregnancy, 176
  toxicity of, 176
Huperzia serrata
  antidiabetic/anti-metabolic syndrome effects, 178
  anti-inflammatory effects, 178
  antioxidant effects, 178
  dosage of, 180
  drug interactions of, 180
  human antidepressant effects, 179–180
  overview of, 178
  preclinical antidepressant-like effects, 179
  safety in pregnancy, 180
  toxicity of, 180
Hyperfoliatin, 362
Hypericum perforatum (St John's wort)
  antidiabetic/anti-metabolic syndrome effects, 182–183
  anti-inflammatory effects, 182
  antioxidant effects, 182
  dosage of, 185
  drug interactions of, 185
  human antidepressant effects, 184
  overview of, 181–182
  preclinical antidepressant-like effects, 183–184
  safety in pregnancy, 185
  toxicity of, 185
Hyperoside, 335

Icariin, 335
Ilex paraguariensis (yerba mate)
  antidiabetic/anti-metabolic syndrome effects, 188
  anti-inflammatory effects, 188
  antioxidant effects, 187–188
  dosage of, 189
  drug interactions of, 189
  human antidepressant effects, 189

overview of, 187
preclinical antidepressant-like effects,
188–189
safety in pregnancy, 189
toxicity of, 189
Imipramine, 7
Inflammation, in MDD, 21–23
Insomnia, 392–394
Insulin resistance, in MDD, 24–25
The Interpretation of Dreams (Freud), 4
Isoliquirtin, 336
Isosakurentin-5-O-rutinoside, 335

Jun-Chen-Zuo-Shi approach, 318

Kaempferol, 335
Kai xin san, 323
Kava, *see Piper methysticum*
Ketamine, 13–16
Kraepelin, Emil, 4

Lamotrigine, 16
Lange, Carl, 4
Lange, Fritz, 4
Largactil, 7
Lavandula (lavender)
antidiabetic/anti-metabolic syndrome
effects, 191–192
anti-inflammatory effects, 191
antioxidant effects, 191
dosage of, 193
drug interactions of, 193
human antidepressant effects, 192–193
overview of, 190–191
preclinical antidepressant-like effects, 192
safety in pregnancy, 193
toxicity of, 193
Lavender, *see Lavandula*
Lectures on Clinical Psychiatry (Kraepelin), 4
Lemon balm, *see Melissa officinalis*
Licorice, *see Glycyrrhiza*
Ligusticum chuanxiong
antidiabetic/anti-metabolic syndrome
effects, 196
anti-inflammatory effects, 196
antioxidant effects, 195
dosage of, 196
drug interactions of, 197
human antidepressant effects, 196
overview of, 195
preclinical antidepressant-like effects, 196
safety in pregnancy, 197
toxicity of, 197
Linalool, 362
Linden, *see Tilia*
Lion's Mane, *see Hericium erinaceus*
Lipids, 32

Liquiritin, 336
Lister, Joseph, 3
Loewi, Otto, 5
L-theanine, 372
Luteolin, 336

Macranthol, 363
Magnolia officinalis
antidiabetic/anti-metabolic syndrome
effects, 199
anti-inflammatory effects, 198–199
antioxidant effects, 98
dosage of, 200
drug interactions of, 201
human antidepressant effects, 200
overview of, 198
preclinical antidepressant-like effects,
199–200
safety in pregnancy, 201
toxicity of, 200–201
Magnolia-vein, *see Schisandra chinensis*
Major depressive disorder (MDD)
herbal treatments (*see* Herbal treatments)
nature and causes of
chronic stress, 23–24
inflammation, 21–23
insulin resistance, 24–25
metabolic syndrome, 25–26
nitrosative damage, 19–21
oxidative damage, 19–21
severity of, 1
traditional Chinese Medicine herbs,
325–326
Western medical treatment history, 1–6
Mammalian target of rapamycin (mTOR),
14–15, 23
MAOIs, *see* Monoamine oxidase inhibitors
Marsilid, 7
Matricaria recutita (chamomile)
antidiabetic/anti-metabolic syndrome
effects, 204
anti-inflammatory effects, 203–204
antioxidant effects, 203
dosage of, 205
drug interactions of, 206
human antidepressant effects, 205
overview of, 203
preclinical antidepressant-like effects,
204–205
safety in pregnancy, 206
toxicity of, 206
MDD, *see* Major depressive disorder
Medical "theory of everything," 326–327
Melancholia, 2, 4
Melissa officinalis (lemon balm)
antidiabetic/anti-metabolic syndrome
effects, 209

anti-inflammatory effects, 208–209
antioxidant effects, 208
dosage of, 210
drug interactions of, 210
human antidepressant effects, 209–210
overview of, 208
preclinical antidepressant-like effects, 209
safety in pregnancy, 210
toxicity of, 210
Metabolic syndrome
    herbal treatments, 396–397
    in MDD, 25–26
Methyl jasmonate, 363
Meyer, Adolf, 5
Milk thistle, *see Silybum marianum*
Mimosa pudica
    antidiabetic/anti-metabolic syndrome
        effects, 212
    anti-inflammatory effects, 212
    antioxidant effects, 212
    dosage of, 213
    drug interactions of, 214
    human antidepressant effects, 213
    overview of, 211–212
    preclinical antidepressant-like effects,
        212–213
    safety in pregnancy, 213
    toxicity of, 213
Miquelianin, 336
Mitragynine, 363–364
Monoamine oxidase inhibitors (MAOIs), 7
mTOR, *see* Mammalian target of rapamycin
Myricetin, 337

Naringenin, 337
Naringin, 337
Nitrosative damage, 19–21
N-methyl-D-aspartate (NMDA) receptor, 14
Nobiletin, 337–338
Non-flavonoid phenols, 34–35

Obsessive-compulsive disorder (OCD), 394
OCD, *see* Obsessive-compulsive disorder
Ocimum basilicum (sweet basil)
    antidiabetic/anti-metabolic syndrome
        effects, 215–216
    anti-inflammatory effects, 215
    antioxidant effects, 215
    dosage of, 217
    drug interactions of, 217
    human antidepressant effects, 217
    overview of, 215
    preclinical antidepressant-like effects, 216
    safety in pregnancy, 217
    toxicity of, 217
Oleanolic Acid, 364
On Medicinal Substances (Celsus), 2

Orcinol, 364
Oregano, *see Origanum vulgare*
Orientin, 338
Origanum vulgare (oregano)
    antidiabetic/anti-metabolic syndrome
        effects, 220
    anti-inflammatory effects, 219
    antioxidant effects, 219
    dosage of, 221
    drug interactions of, 221
    human antidepressant effects, 220–221
    overview of, 219
    preclinical antidepressant-like effects, 220
    safety in pregnancy, 221
    toxicity of, 221
Oxidative damage, 19–21

Paeonia lactiflora (peony)
    antidiabetic/anti-metabolic syndrome
        effects, 223
    anti-inflammatory effects, 223
    antioxidant effects, 223
    dosage of, 224
    drug interactions of, 225
    human antidepressant effects, 224
    overview of, 222–223
    preclinical antidepressant-like effects,
        223–224
    safety in pregnancy, 225
    toxicity of, 224
Paeoniflorin, 365
Paeonol, 365–366
Page, Irvine, 5
Palmatine, 366
Panax ginseng (ginseng)
    antidiabetic/anti-metabolic syndrome
        effects, 227
    anti-inflammatory effects, 227
    antioxidant effects, 226–227
    dosage of, 229
    drug interactions of, 229
    human antidepressant effects, 228
    overview of, 226
    preclinical antidepressant-like effects,
        227–228
    toxicity of, 229
Passifloraceae incarnata (passionflower)
    antidiabetic/anti-metabolic syndrome
        effects, 231–232
    anti-inflammatory effects, 231
    antioxidant effects, 231
    dosage of, 232
    drug interactions of, 233
    human antidepressant effects, 232
    overview of, 231
    preclinical antidepressant-like effects, 232
    safety in pregnancy, 233

toxicity of, 232–233
Passionflower, *see Passifloraceae incarnata*
Pavlov, Ivan, 3
PCP, *see* Phencyclidine
Peony, *see Paeonia lactiflora*
Perimenopausal symptoms, 395
Phencyclidine (PCP), 13
Phenolics
  definition of, 33
  flavonoids, 34
  non-flavonoid phenols, 34–35
Phytochemicals
  alkaloids, 35–36
  carbohydrates, 31–32
  classification of, 31
  lipids, 32
  phenolics
    definition of, 33
    flavonoids, 34
    non-flavonoid phenols, 34–35
  terpenes, 32–33
Piper methysticum (kava)
  antidiabetic/anti-metabolic syndrome
    effects, 235
  anti-inflammatory effects, 235
  antioxidant effects, 235
  human antidepressant effects, 236
  overview of, 234–235
  preclinical antidepressant-like effects,
    235–236
  safety, 236–237
  toxicity of, 236–237
Piper nigrum (black pepper)
  antidiabetic/anti-metabolic syndrome
    effects, 239
  anti-inflammatory effects, 239
  antioxidant effects, 238–239
  dosage of, 240
  drug interactions of, 240–241
  human antidepressant effects, 240
  overview of, 238
  preclinical antidepressant-like effects, 239–240
  safety in pregnancy, 240
  toxicity of, 240
Plumbagin, 366–367
Podoandin, 367
Polygala tenuifolia
  antidiabetic/anti-matabolic syndrome
    effects, 243
  anti-inflammatory effects, 243
  antioxidant effects, 242–243
  dosage of, 245
  drug interactions of, 245–246
  human antidepressant effects, 245
  overview of, 242
  preclinical antidepressant-like effects,
    243–245

safety in pregnancy, 245
toxicity of, 245
Poria cocos
  antidiabetic/anti-metabolic syndrome
    effects, 248
  anti-inflammatory effects, 248
  antioxidant effects, 248
  dosage of, 249
  drug interactions of, 249
  human antidepressant effects, 248
  overview of, 247–248
  preclinical antidepressant-like effects, 248
  safety in pregnancy, 249
  toxicity of, 249
Preclinical antidepressant-like effects
  flavonoids with, 331–343
  forces swim test, 42–44
  sucrose consumption test, 44–45
  tail suspension test, 44
  test conditions, 45–46
Premenstrual symptoms, 395
Project for a Scientific Psychology (Freud), 4
Prozac, 8
Psoralea corylifolia
  antidiabetic/anti-metabolic syndrome
    effects, 251–252
  anti-inflammatory effects, 251
  antioxidant effects, 250–251
  dosage of, 253
  drug interactions of, 253
  human antidepressant effects, 253
  overview of, 250
  preclinical antidepressant-like effects,
    252–253
  safety in pregnancy, 253
  toxicity of, 253
Punarnavine, 367

Quercetin, 338–339

Reactive oxidative species (ROS), 20
Resiliency, lack of, 397–398
Resveratrol, 367–368
Rhodiola rosea
  antidiabetic/anti-metabolic syndrome
    effects, 256–257
  anti-inflammatory effects, 256
  antioxidant effects, 256
  dosage of, 258
  drug interactions of, 258
  human antidepressant effects, 257–258
  overview of, 255
  preclinical antidepressant-like effects, 257
  safety in pregnancy, 258
  toxicity of, 258
Riparin, 368
Robitzek, Edward, 7

ROS, *see* Reactive oxidative species
Rosemary, *see Rosmarinus officinalis*
Rosmarinic acid, 369
Rosmarinus officinalis (rosemary)
    antidiabetic/anti-metabolic syndrome
        effects, 261
    anti-inflammatory effects, 260–261
    antioxidant effects, 260
    dosage of, 262
    drug interactions of, 262–263
    human antidepressant effects, 262
    overview of, 260
    preclinical antidepressant-like effects,
        261–262
    safety in pregnancy, 262
    toxicity of, 262
Rush, Benjamin, 3, 5

Safety of herbal treatments, 399–401
Saffron, *see Crocus sativa*
Safranal, 369
Salidroside, 369–370
Salvia divinorum
    antidiabetic/anti-metabolic syndrome
        effects, 265
    anti-inflammatory effects, 265
    antioxidant effects, 265
    dosage of, 267
    drug interactions of, 267
    human antidepressant effects, 266–267
    overview of, 264–265
    preclinical antidepressant-like effects,
        265–266
    safety in pregnancy, 267
    toxicity of, 267
Sarsasapogenin, 370–371
Sceletium tortuosum
    antidiabetic/anti-metabolic syndrome
        effects, 270
    anti-inflammatory effects, 269–270
    antioxidant effects, 269
    dosage of, 271
    drug interactions of, 271
    human antidepressant effects, 270–271
    overview of, 269
    preclinical antidepressant-like effects, 270
    safety in pregnancy, 271
    toxicity of, 271
Schisandra, *see Schisandra chinensis*
Schisandra chinensis (magnolia-vine/
        schisandra)
    antidiabetic/anti-metabolic syndrome
        effects, 273–274
    anti-inflammatory effects, 273
    antioxidant effects, 273
    dosage of, 274

    drug interactions of, 275
    human antidepressant effects, 274
    overview of, 273
    preclinical antidepressant-like effects, 274
    safety in pregnancy, 274–275
    toxicity of, 274
Scopoletin, 371
Scutellaria lateriflora (skullcap)
    antidiabetic/anti-metabolic syndrome
        effects, 277
    anti-inflammatory effects, 276
    antioxidant effects, 276
    dosage of, 278
    drug interactions of, 278
    human antidepressant effects, 277–278
    overview of, 276
    preclinical antidepressant-like effects, 277
    safety in pregnancy, 278
    toxicity of, 278
Selikoff, Irving, 7
Sherrington, Charles Sir, 4
Shi wei wen dan tang, 321–322
Shu gan jie yu, 323–324
Siberian Ginseng, *see Eleutherococcus
        senticoccus*
Silybum marianum (milk thistle)
    antidiabetic/anti-metabolic syndrome
        effects, 281
    anti-inflammatory effects, 280–281
    antioxidant effects, 280
    dosage of, 282
    drug interactions of, 283
    human antidepressant effects, 282
    overview of, 280
    preclinical antidepressant-like
        effects, 281–282
    safety in pregnancy, 283
    toxicity of, 283
Si ni san, 324
Skullcap, *see Scutellaria lateriflora*
St John's wort, *see Hypericum perforatum*
Sucrose consumption test, 44–45
Sulphoraphane, 371
Sweet basil, *see Ocimum basilicum*
Synthetic flavonoids, 340

Tail suspension test, 44
Tang shen kang, 323
TCAs, *see* Tricyclic antidepressants
Tea, *see Camellia sinensis*
Terpenes, 32–33
Tetrandrine, 371–372
Theobroma cacao (cocoa/chocolate)
    antidiabetic/anti-metabolic syndrome
        effects, 285–286
    anti-inflammatory effects, 285

antioxidant effects, 285
dosage of, 287
drug interactions of, 287
human antidepressant effects, 286–287
overview of, 284–285
preclinical antidepressant-like effects, 286
safety in pregnancy, 287
toxicity of, 287
Thorazine, 7
Tianeptine, 16
Tiao qi, 322
Tilia (linden)
    antidiabetic/anti-metabolic syndrome
        effects, 290
    anti-inflammatory effects, 289–290
    antioxidant effects, 289
    dosage of, 290–291
    drug interactions of, 291
    human antidepressant effects, 290
    overview of, 289
    preclinical antidepressant-like effects, 290
    safety in pregnancy, 291
    toxicity of, 291
Tofranil, 7
Traditional Chinese Medicine (TCM) herbs
    Ban xia hou pu, 322
    Chai hu jia long gu mu li, 322
    Chai hu shu gan, 320
    Gan mai da zao, 320–321
    Gui pi, 321
    Kai xin san, 323
    in MDD treatment, 325–326
    medical "theory of everything," 326–327
    overview of, 317–318
    Shi wei wen dan tang, 321–322
    Shu gan jie yu, 323–324
    Si ni san, 324
    Tang shen kang, 323
    Tiao qi, 322
    Wu ling, 324–325
    Xiao yao san, 319–320
    Yi pi, 323
    Yueju, 318–319
Tricyclic antidepressants (TCAs), 8
Trigonella foenum-graecum (fenugreek)
    antidiabetic/anti-metabolic syndrome
        effects, 294
    anti-inflammatory effects, 293–294
    antioxidant effects, 293
    dosage of, 295
    drug interactions of, 296
    human antidepressant effects, 295
    overview of, 292–293
    preclinical antidepressant-like effects,
        294–295
    safety in pregnancy, 296

toxicity of, 295–296
Turmeric, *see Curcuma longa*
Twarog, Betty, 5

Uliginosin B, 372
Ursolic acid, 372

Valerian, *see Valeriana officinalis*
Valeriana officinalis (valerian)
    antidiabetic/anti-metabolic syndrome
        effects, 299
    anti-inflammatory effects, 299
    antioxidant effects, 298–299
    dosage of, 300
    drug interactions of, 301
    human antidepressant effects, 300
    overview of, 298
    preclinical antidepressant-like effects,
        299–300
    safety in pregnancy, 301
    toxicity of, 301
Vanillin, 373
Verbena officinalis (vervain)
    antidiabetic/anti-metabolic syndrome
        effects, 304
    anti-inflammatory effects, 303–304
    antioxidant effects, 303
    dosage of, 305
    drug interactions of, 305
    human antidepressant effects, 305
    overview of, 302–303
    preclinical antidepressant-like effects, 304
    safety in pregnancy, 305
    toxicity of, 305
Vervain, *see Verbena officinalis*
Virchov, Rudolf, 3
Vitex agnus-castus (chaste tree)
    antidiabetic/anti-metabolic syndrome
        effects, 307
    anti-inflammatory effects, 307
    antioxidant effects, 307
    dosage of, 309
    drug interactions of, 309
    human antidepressant effects, 308–309
    overview of, 307
    preclinical antidepressant-like effects, 308
    safety in pregnancy, 309
    toxicity of, 309
Vitexin, 339

Water hyssop, *see Bacopa monnieri*
Willis, Thomas, 3
Withania somnifera (Ashwagandha)
    antidiabetic/anti-metabolic syndrome
        effects, 312
    anti-inflammatory effects, 311–312

antioxidant effects, 311
dosage of, 313
drug interactions of, 313
human antidepressant effects, 313
overview of, 311
preclinical antidepressant-like effects,
    312–313
safety in pregnancy, 313
toxicity of, 313

Wogonin, 339–340
Wogonoside, 339–340
Wu ling, 324–325

Xiao yao san, 319–320

Yerba mate, *see Ilex paraguariensis*
Yi pi, 323
Yueju, 318–319

Printed in the United States
by Baker & Taylor Publisher Services